A BIRD-FINDING GUIDE TO MEXICO

A Bird-Finding Guide to Mexico

STEVE N. G. HOWELL

Comstock Publishing Associates
A Division of
CORNELL UNIVERSITY PRESS
Ithaca, New York

First published 1999 by Cornell University Press
First printing, Cornell Paperbacks 1999
Printed in the United States of America

Library of Congress Cataloging-in-Publication Data
Howell, Steve N. G.
A bird finding guide to Mexico / Steve N.G. Howell.
p. cm.
Includes bibliographical references (p.)
and index.
ISBN 0-8014-8581-9 (pbk. : alk. paper)
1. Bird watching—Mexico—Guidebooks.
2. Birds—Mexico. 3. Mexico—Guidebooks.
I. Title.
QL686.H65 1998
598'.07'23472—dc21 98-28783

Cornell University Press strives to use environmentally responsible suppliers and materials to the fullest extent possible in the publishing of its books. Such materials include vegetable-based, low-VOC inks and acid-free papers that are also recycled, totally chlorine free, or partly composed of nonwood fibers.

Paperback printing
10 9 8 7 6 5 4 3 2 1

ISBN 0-8014-8581-9

TO RICHARD AND ELDA,
for their hospitality and remarkable
knowledge of Mexico

CONTENTS

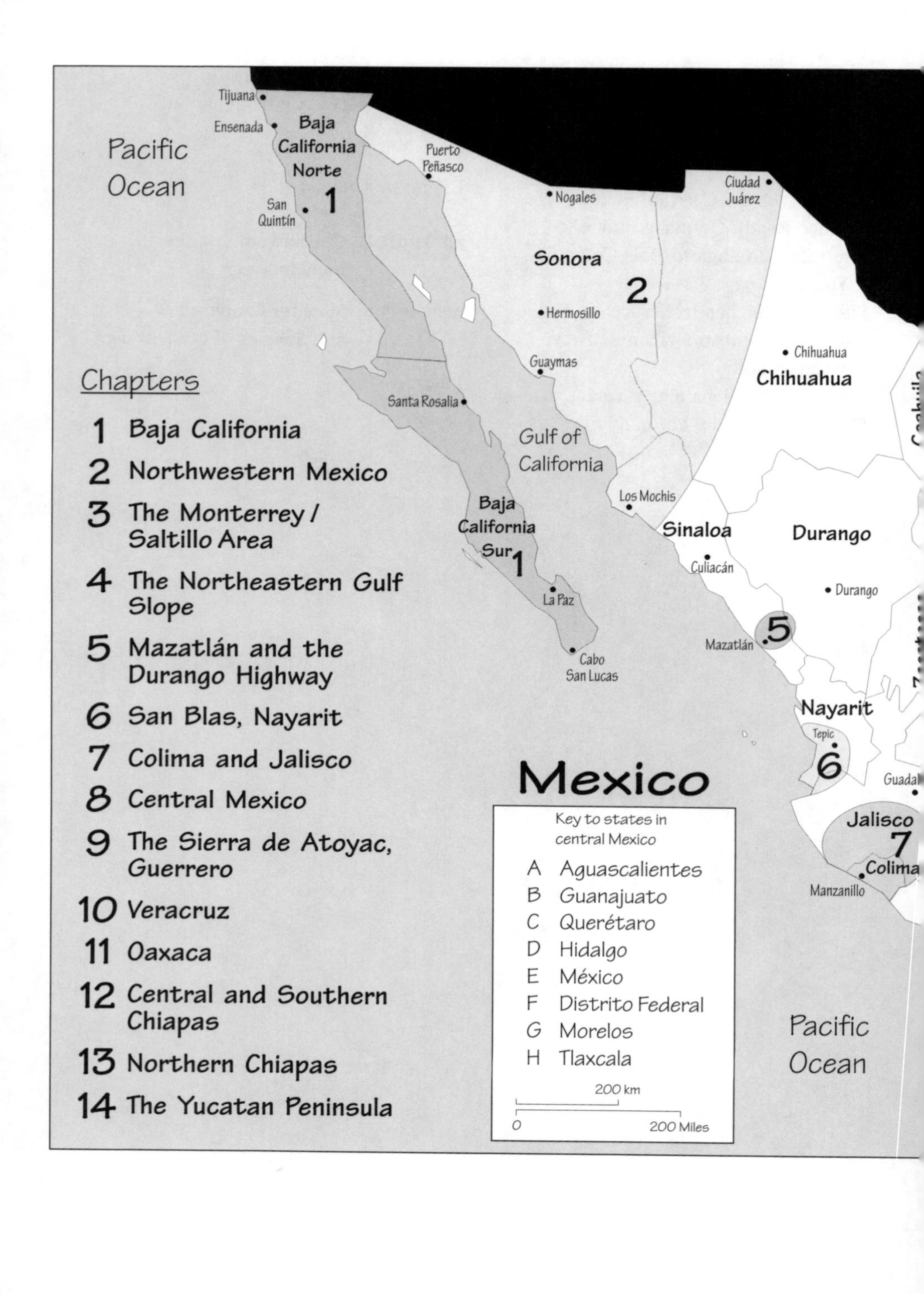

Tijuana
Ensenada
Baja California Norte
1
San Quintín
Pacific Ocean
Puerto Peñasco
Nogales
Ciudad Juárez
Sonora
2
Hermosillo
Guaymas
Chihuahua
Chihuahua
Santa Rosalia
Gulf of California
Los Mochis
Baja California Sur
1
La Paz
Cabo San Lucas
Sinaloa
Culiacán
Durango
Durango
Mazatlán
5
Nayarit
Tepic
6
Guadal
Jalisco
7
Colima
Manzanillo
Pacific Ocean
Mexico
Chapters
1 Baja California
2 Northwestern Mexico
3 The Monterrey / Saltillo Area
4 The Northeastern Gulf Slope
5 Mazatlán and the Durango Highway
6 San Blas, Nayarit
7 Colima and Jalisco
8 Central Mexico
9 The Sierra de Atoyac, Guerrero
10 Veracruz
11 Oaxaca
12 Central and Southern Chiapas
13 Northern Chiapas
14 The Yucatan Peninsula
Key to states in central Mexico
A Aguascalientes
B Guanajuato
C Querétaro
D Hidalgo
E México
F Distrito Federal
G Morelos
H Tlaxcala
200 km
0
200 Miles

UNITED STATES
Coahuila
Nuevo Laredo
Nuevo León
Saltillo
Monterrey
3
4
Matamoros
Gulf of Mexico
Cuidad Victoria
Tamaulipas
San Luis Potosí
San Luis Potosí
Tampico
Guanajuato
14
Mérida
Yucatán
Cancún
Isla de Cozumel
B
C
D
Pachuca
Bahía de Campeche
Campeche
Quintana Roo
Morelia
8
Michoacán
Mexico City
E
F
Toluca
G
Cuernavaca
H
Puebla
Puebla
10
Veracruz
Veracruz
Chetumal
Campeche
Coatzacoalcos
Villahermosa
Guerrero
9
Acapulco
Oaxaca
Oaxaca
11
13
Tuxtla Gutiérrez
Chiapas
12
BELIZE
GUATEMALA
HONDURAS
Golfo de Tehuantepec
EL SALVADOR

MAPS

ABBREVIATIONS

Fol.-gleaner	Foliage-gleaner
ft	feet
km	kilometer(s)
km/h	kilometers per hour
m	meter(s)
mi	mile(s)
N.	Northern
N.-Thrush	Nightingale-Thrush
Rough-w.	Rough-winged
sp.	species
ssp.	subspecies
Yellow-hd.	Yellow-headed

SPANISH TERMS

aguada	pond
autopista	motorway (usually toll)
barranca	large canyon
bienvenido	welcome
cabañas	cabins
cambios	exchange booths
cascadas	falls
caseta	small building, booth
cenote	limestone sinkhole
centro	center/downtown
chivizcoyo	Bearded Wood-Partridge
colectivo	collective (communal) taxi
colectivo combis	VW minibuses
cumbre	summit
cuota	toll
Don	term of respect for (usually oder) men
Doña	term of respect for (usually older) women
ejido	common land/ cooperative village
estero	estuary
finca	plantation
glorieta(s)	traffic circle(s)
grutas	caves
lancha(s)	open fishing launch(es)
libre	free
malecon	waterfront street or boulevard
mercado	market
microondas	radio towers
milpa(s)	agricultural plot(s)
mirador	overlook
norte(s)	norther(s)
palapa(s)	thatched open hut(s)
Pemex	gas station
periferico	bypass
propina	tip
pueblo	village, town
puente	bridge
puerto	port
rancho(s)	ranch(es)
río	river
topes	speed bumps
zócalo	town square, plaza
Zona Hotelera	hotel zone

PREFACE

Mexico is a large and diverse country with a rich avifauna and is many birders' first taste of the Neotropics. New and exciting birds may include tinamous, sungrebes, parrots, toucans, motmots, woodcreepers, antbirds, cotingas, and manakins. Families well represented in Mexico include parrots (22 species, 7 endemic), hummingbirds (61 species, 16 endemic), woodpeckers (24 species, 6 endemic), jays and crows (23 species, 8 endemic), wrens (32 species, 12 endemic), mockingbirds and thrashers (18 species, 5 endemic), vireos (24 species, 4 endemic), and sparrows and seedeaters (75 species, 16 endemic).

Of 1069 species (1059 AOU species; see AOU 1995, 1997) recorded in Mexico through March 1998, around 425 do not occur in the U.S.A. or Canada except, in the case of a handful (such as Aztec Thrush and Crimson-collared Grosbeak), as occasional strays to southern Arizona or Texas. About 115 species nest only in Mexico (i.e., are endemic breeders), while a further 25 are shared endemics, nesting elsewhere only in adjacent Guatemala and/or Belize. Most of the endemic breeders are resident, although four species, all seabirds, are regular post-breeding visitors to California: Black-vented Shearwater, Least Storm-Petrel, Yellow-footed Gull, and Craveri's Murrelet.

So if you want to see Long-tailed Wood-Partridge, Maroon-fronted Parrot, Eared Poorwill, Short-crested Coquette, Pileated Flycatcher, Tufted Jay, Giant Wren, Ocellated Thrasher, Slaty Vireo, Red Warbler, Rosita's Bunting, Sierra Madre Sparrow, and many others, Mexico's the place to go.

This book aims to provide the visiting or resident birder with a number of options for birding trips that, put together, allow a chance to see over 950 species, including virtually all of the endemics (except a few on offshore islands) and regional specialities. It is not intended as a guide to every single birding spot where one might, or can, see a certain species. Clearly, this would be impractical for a country and avifauna of this size. And remember: like your bird identification books, this is only a guide, and good birding areas can be found anywhere you go.

ACKNOWLEDGMENTS

On 30 November 1981, I took a bus from Brownsville, Texas, across the border to Mexico and the start of a four-month birding trip to unfamiliar lands. I spoke basically no Spanish, and my travel plans changed almost daily, as did my knowledge of the birds and the country. Even now, after seventeen years of travel throughout almost all corners of Mexico, there is still much to learn—and always a reason to go back! Over the years, I and Sophie Webb (my coauthor and illustrator of *A Guide to the Birds of Mexico and Northern Central America*, published in 1995 by Oxford University Press) got to visit some great birding areas, many of them passed down through the birding grapevine, many of them newly discovered. This guide is the product of our travels and those of the birding community that has been increasingly active in Mexico in recent years: I thank all who have birded in Mexico and shared their information, and I hope this book will help others explore and appreciate the wonders of birding "south of the border." I also express a special thanks to Will Russell, of WINGS Inc., for facilitating my travel opportunities throughout Mexico in recent years.

Unlike some bird-finding guides for Latin America, all of the sites covered herein have been visited (most of them in recent years) by the author; this approach is preferred over trying to gather and edit a mixture of information, some of it vague at best, from numerous contributors. At the same time, a project like this is the product of many persons, and in particular I thank the following for their review, comment, and/or help in field-checking sites:

John C. Arvin (The Northeastern Gulf Slope)

Parker Backstrom (Colima and Jalisco, The Sierra de Atoyac, Guerrero)

Robert A. Behrstock (The Monterrey/Saltillo Area, Oaxaca)

Bob Berman and Cindy Lippincott (Northwestern Mexico, Central Mexico)

Chris Corben (The Yucatan Peninsula)

Richard A. Erickson (Baja California)

Steve Ganley (Northwestern Mexico)

Hector Gomez de Silva G. (Central Mexico, Veracruz)

Ing. Ricardo Guerra de la Garza and members of the Club de Observadores de Aves del Noreste (The Monterrey/Saltillo Area, The Northeastern Gulf Slope)

Robert A. Hamilton (Baja California, Northwestern Mexico)

Loren Hays (Mazatlán and the Durango Highway)

Lisa Hug (The Yucatan Peninsula)

Joe Keenan (The Yucatan Peninsula)

Barbara Mackinnon (The Yucatan Peninsula)

John P. and Jeanette Martin (Northern Chiapas, The Yucatan Peninsula)

Bernie Master (Baja California, The Monterrey/Saltillo Area, The Northeastern Gulf Slope, Veracruz, Central Mexico)

Peter Metropulos (Colima and Jalisco, Oaxaca)

Borja Mila (The Monterrey/Saltillo Area)

Jim Paton (Northwestern Mexico)

Ernesto Ruelas (Veracruz)

Andres M. Sada (The Monterrey/Saltillo Area)

Eric Salzman (Oaxaca, Central and Southern Chiapas)

Mike San Miguel (Northwestern Mexico)

Rick Taylor (Northwestern Mexico)

Francis Toldi (Colima and Jalisco, Oaxaca)

Guy Tudor (Introduction)

Richard G. Wilson (Introduction, Central Mexico, The Sierra de Atoyac, Guerrero, Veracruz, Oaxaca, Central and Southern Chiapas)

Thomas E. Wurster (Baja California, Colima and Jalisco).

I especially thank Cindy Lippincott and Bob Berman for their help and interest in this project, particularly Cindy for crafting the maps, and Sophie Webb for contributing the beautiful black-and-white wash drawings that help relieve the text.

INTRODUCTION

This guide separates Mexico into fourteen birding regions, each comprising a chapter. The first four regions (Baja California; Northwestern Mexico; the Monterrey/Saltillo Area; the Northeastern Gulf Slope) can be visited easily by driving from the U.S.A. The following ten regions, which also can be visited by driving from the U.S.A., are reached more often by flying in and renting a car or traveling by public transport. These regions are: Mazatlán and the Durango Highway; San Blas, Nayarit; Colima and Jalisco; Central Mexico; the Sierra de Atoyac, Guerrero; Oaxaca; Veracruz; Central and Southern Chiapas; Northern Chiapas; and the Yucatan Peninsula.

Each chapter opens with a general introduction to the region covered, including a list of some of the more interesting species possible, an overview of general access (e.g., which airports serve the region), an idea of how long you could spend there (many destinations translate into a one-week or two-week birding trip), a note when an area can be combined easily with another region, and, in some cases, suggested itineraries.

Choosing which sites to include and which to omit was difficult. In a country the size of Mexico, and with such a large avifauna, it simply is not practical to try and cover all possible sites in a single book. I have tried to give a good geographic spread and include areas that are popular destinations for birders and, to some extent, for non-birding tourists. Unlike in some bird-finding guides for Latin America, all of the sites covered have been visited by the author; this approach is preferred over trying to gather and edit a mixture of information, of varying accuracy, from numerous contributors. At the same time, much of the information has been checked recently by other persons familiar with various sites (see the Acknowledgments).

Important Note: The information used to write the site accounts was gathered over a period of seventeen years, and inevitably some sites have been researched more thoroughly than others. The dates listed at the start of each site indicate the seasonality (helpful when interpreting bird lists) and degree of coverage as well as noting when a site has been visited most recently. A notation such as "checked February 1997" indicates how recently directions and the site have been checked by other birders. An effort has been made to have all sites visited and/or reviewed from 1996 through early 1998. Inevitably, some sites have not achieved this level of field checking. Rather than wait until these sites could be checked (by which time information on other sites would be out of date!), I decided to include all of the information, noting the level of checking that has been possible. Birders visiting any site, in particular those whose directions include the occasional use of (?) after distances and the like, are encouraged to update and correct the information provided.

At the same time, even for sites checked recently, things can (and do) change rapidly: new toll highways may be constructed, forests may be cleared, areas may be closed off to public access. Remember, common sense is a prerequisite when using the information in this guide.

BIRDING ETHICS

The American Birding Association (ABA) promotes a code of birding ethics that, one hopes, should be innate to most if not all of us. One of the first principles of this code is "In any conflict of interest between birds and birders, the welfare of the birds and their environment comes first." Arguably, if this were adhered to strictly, no birders would use tape playback to see birds. As this seems an unlikely scenario (unless perhaps the ABA allows birders only a "half bird" for birds seen with the use of tape playback!) it is worth repeating the words of Harold Holt in *A Birder's Guide to Colorado:* "Birders gradually are coming to realize that attracting birds with bird-sound recordings is a little like catching trout in a bathtub." Beyond simply this aesthetic loss, there are more serious ethical issues involved. An increasing body of evidence is accumulating to show that use of tape playback can be detrimental to birds: as Buckingham (1997) wrote, "Playback always stops the bird from doing whatever it was doing—and exposes it, and its nestlings to increased risk." Baptista and Gaunt (1997) note that "investigators need to know the species they study well enough to be aware of any consequences that might accrue to the animal when playback is used," and they report a case where playback apparently increased the aggression of the birds to the point that males pecked their nestlings to death! Ireland (1997) also echoes the need to know the social status of both the bird you have on tape and the bird to which you are playing tape. Clearly, birders visiting an area briefly are unlikely to know the consequences their use of tapes may have on the birds targeted. However, as Buckingham (1997) notes: "Ethics are not absolute. If you care for the birds as much as for yourself you'll be very sparing with playback. If you regard birds largely as objects then you'll be relatively unconcerned about the harm that may, or may not, ensue."

Tapes and other forms of recordings are valuable tools for birders: listening to them before you go, or in your hotel room or car during a trip, can help you to learn and identify songs and calls, which can enhance greatly your birding holiday.

HOW TO USE THIS BOOK

Site information is conveyed by means of up to four sections, discussed below. Maps at the beginning of each chapter locate all of the birding sites in a region, and additional maps show greater detail for some sites. These maps should be used in conjunction with information in the Birding Sites sections.

Introduction/Key Species

This section gives background information on the site, such as habitats, the best times to visit, and a list of the main bird species of interest (noting seasonal status if relevant).

Access

This section covers how to get to the site in relation to nearby cities and/or other birding sites, and what types of accommodation can be found at or near the site. Access may be combined with Birding Sites when relevant.

Note that the distances and elevations given are only as good as the equipment used in taking the readings. Mileages should be viewed as having a 1 to (rarely) 10 percent potential level of "error," and elevations are generally to within ± 50 to 100 m. As an example, on a recent tour with two vehicles I checked odometer readings: one vehicle read 12.6 km to a site, while the other read 14.0 km! Directions should be interpreted with this in mind—if you've driven the requisite 28.5 km to an apparent left turn and there's no sign of it, try going a little farther before worrying about whether or not you're lost. Likewise, driving times should be taken only as a rough idea—everyone has their own driving speed and road conditions are prone to change.

In general, for sites in the first four chapters, miles (mi) are used to designate distances (with kilometers in parentheses) because most birders visiting these areas will be driving down from the U.S.A., and their cars will read in miles. For sites in the remaining chapters, usually birded courtesy of rental vehicles, distances are given in kilometers (km).

The names and availability of potential accommodations—not to mention their levels of maintenance, comfort, and price—are prone to change frequently. For example, while scouting one chapter recently I found a recommended hotel of two years earlier closed down (and abandoned!) and a new one built in a nearby town, while at another site a formerly rundown hotel had been upgraded and was now quite reasonable. For such reasons, while the types and levels of accommodation at or near a site are indicated, mention of specific accommodation may be avoided unless it relates to a small town unlikely to be covered by general tourist guidebooks (see Additional Literature, p. 17). And, for those hotels that are mentioned, be aware of the caveat above. Restaurants are even more prone than hotels to an ephemeral existence and are thus not covered herein; hotel personnel and/or your own nose are often the best guides. Numerous general guidebooks cover these subjects (see Additional Literature, p. 17) and should be consulted when planning a birding trip.

Birding Sites

This section describes the areas best for certain species at the site and may give other tips on finding birds of interest. In the case of sites with relatively few species, or for sites which have considerable overlap with another site treated in more depth, the more interesting and typical birds may be listed under Birding Sites, rather than as a separate bird list. Often this section is combined with Access.

Bird Lists

For any site, the bird lists are based on few trips relative to most North American or European bird-finding guides, which usually deal with well-known areas. In most cases, however, the lists should include all or almost all species of note and 90 percent or more of the species you can expect to find on a given visit (at the same season). Much of my birding in Mexico has been between November and March, so visiting many sites during the height of spring and/or autumn migration probably will add species to site lists.

The birds of Mexico are far from well known in terms of just about everything, including their distribution and abundance. Thus, the bird list for any site would be variably incomplete even if I collated lists from other observers. Rather than walk the thin ice of attempting to vet sightings from numerous observers, the bird lists include only species I have recorded at a site. However, when I am aware of reliable sightings of additional species of interest, these are mentioned in the Birding Sites section, with details of where (and when) to expect them. For each site, the dates of my visits are noted to give an indication of thoroughness and seasonality of coverage; I have visited most sites at least twice.

The abundance (and seasonal status) of a given species can be determined broadly from the species accounts in Howell and Webb (1995); rare (usually migrant) and vagrant species are listed for completeness but are marked as such; these species should not be expected on an average day's birding. Pages 75–77 ("Status and Distribution") in Howell and Webb (1995) discuss factors that contribute to the (apparent or real) abundance of species and, consequently, how likely you are to find them.

I have not listed what might be considered the "commoner" or more "characteristic" birds of a region or site, beyond the species of primary interest. This would be nice, but because many of the sites are home to at least 200–300 species, most of which occur there regularly, such lists would be unwieldy and would largely duplicate the full site lists. Note, though, that Appendix A lists by chapter all species included and can be checked to gain an idea of which species can be found in sites covered by any given chapter.

Taxonomy and Nomenclature

Almost all guidebooks use different names for some species of birds. This largely reflects two related issues: first, opinions change, and second, book publishers cannot realistically keep up with this near-constant change of opinions. Thus, although the American Ornithologists' Union (AOU) Checklist and its supplements are often taken as the standard for taxonomy and nomenclature of North American birds, no field guide in print reflects the latest opinions of the AOU Committee on Classification and Nomenclature. Furthermore, while the AOU Checklist has the potential to reflect committee benefits such as stability and relative conservatism, it is also prone to reflect other inevitable committee characteristics, such as inertia and political machinations. As an example, seven name changes and splits accepted by Howell and Webb (1995) were adopted recently by the AOU (1997), highlighting the futility of trying to play a game when the goalposts keep moving!

For such reasons, I have chosen to follow the taxonomy used in Howell and Webb (1995), as modified in Howell (1996), which incorporated changes based on the Fortieth Supplement (1995) to the AOU Checklist. This has the advantage of making the bird lists in this bird-finding guide comparable with a widely accessible field guide and field checklist. In a few cases, my taxonomy differs from that of the AOU (1983–1995). Such differences are explained by Howell (1996), and a summary of the main differences between Howell and Webb (1995) and the AOU (1983–1995) was provided by Eric Salzman in the March 1997 (vol. 9, no. 3) issue of *Winging It*, the newsletter of the American Birding Association.

Boldfaced bird names in site accounts and bird lists indicate species endemic (as breeders) to Mexico and northern Central America, i.e., the region covered by the Howell and Webb (1995) guide.

Judging from the Forty-first Supplement (1997) to the AOU Checklist, the forthcoming seventh edition of the AOU *Checklist of North American Birds* promises to introduce some fairly radical changes in the sequence of families, as well as the usual suite of inconsistent and at times unsupported changes and re-changes to species-level taxonomy. Rather than adopt these changes here, I have retained a sequence that will be familiar to most birders and ornithologists, because stability is one merit of taxonomic lists that committees often tend to overlook. English names used in this guide that differ from those of the AOU (1983–1995) and that might be confusing

Names used in this guide	AOU (1983–1995) names
American Flamingo (split from Greater Flamingo)	Greater Flamingo
White-fronted Goose	Greater White-fronted Goose
White-breasted Hawk* (split from Sharp-shinned Hawk)	Sharp-shinned Hawk
Yucatan Bobwhite	Black-throated Bobwhite
Grey-headed Dove* (split from Grey-fronted Dove)	Grey-fronted Dove
Aztec Parakeet (split from Olive-throated Parakeet)	Olive-throated Parakeet
Yucatan Parrot	Yellow-lored Parrot
Mountain Pygmy-Owl*	Northern Pygmy-Owl
Cape Pygmy-Owl (both split from Northern Pygmy-Owl)	Northern Pygmy-Owl
Colima Pygmy-Owl†	Least Pygmy-Owl
Tamaulipas Pygmy-Owl†	Least Pygmy-Owl
Central American Pygmy-Owl† (all split from Least Pygmy-Owl)†	Least Pygmy-Owl
Mexican Whip-poor-will (split from Whip-poor-will)	Whip-poor-will
Mexican Hermit (split from Long-tailed Hermit)	Long-tailed Hermit
Salvin's Emerald (split from Canivet's Emerald)	Canivet's Emerald
Doubleday's Hummingbird (split from Broad-billed Hummingbird)	Broad-billed Hummingbird
Cinnamon-sided Hummingbird (split from Green-fronted Hummingbird)	Green-fronted Hummingbird
Sparkling-tailed Woodstar	Sparkling-tailed Hummingbird
Eared Quetzal	Eared Trogon
Pygmy Kingfisher	American Pygmy Kingfisher
Yucatan Woodpecker	Red-vented Woodpecker
Arizona Woodpecker (split from Strickland's Woodpecker)	Strickland's Woodpecker
Bronze-winged Woodpecker (split from Golden-olive Woodpecker)	Golden-olive Woodpecker
Scaly-throated Foliage-gleaner	Spectacled Foliage-gleaner
Mexican Antthrush (split from Black-faced Antthrush)	Black-faced Antthrush
Common Tufted Flycatcher	Tufted Flycatcher
Western Pewee	Western Wood-Pewee
Eastern Pewee	Eastern Wood-Pewee
Western Flycatcher ‡	Pacific-Slope Flycatcher
Western Flycatcher ‡	Cordilleran Flycatcher
Thrushlike Mourner	Thrushlike Manakin (Schiffornis†)
Tamaulipas Crow†	Mexican Crow

Ridgway's Rough-winged Swallow*	Northern Rough-winged Swallow
(split from N. Rough-winged Swallow)	
Grey-breasted Jay	Mexican Jay
(six species of jays are endemic to Mexico; this is not one of them)	
Northern Raven	Common Raven
Sumichrast's Wren†	Slender-billed Wren
Nava's Wren†	Slender-billed Wren
(both split from Slender-billed Wren)†	
Cozumel Wren	House Wren
(split from House Wren)	
Black Thrush	Black Robin
Mountain Thrush	Mountain Robin
Clay-colored Thrush	Clay-colored Robin
White-throated Thrush	White-throated Robin
Rufous-backed Thrush	Rufous-backed Robin
Grey Silky	Gray Silky-flycatcher
Painted Whitestart	Painted Redstart
Slate-throated Whitestart	Slate-throated Redstart
Chestnut-capped Warbler	Rufous-capped Warbler
(split from Rufous-capped Warbler)	
Cabanis' Tanager	Azure-rumped Tanager
Golden-hooded Tanager†	Golden-masked Tanager
Rosita's Bunting	Rose-bellied Bunting
Slate-blue Seedeater	Blue Seedeater
(split from Blue Seedeater)	
Sumichrast's Sparrow	Cinnamon-tailed Sparrow
Baird's Junco	Yellow-eyed Junco
(split from Yellow-eyed Junco)	
Abeille's Oriole	Black-backed Oriole
(twelve of the sixteen oriole species in Mexico have black backs)	

* indicates species also split by Sibley and Monroe (1990)
† indicates changes adopted by AOU (1997)
‡ (Because these two types intergrade extensively in Washington and British Columbia, migrants cannot be identified safely, even by voice, as birds giving "Pacific-Slope" call notes can give "Cordilleran" songs! [Cannings and Hunn, unpubl. data]; Mexican breeding birds can be identified to "species" by range; cf. Howell and Webb 1995.)

to users are as follow. (Minor spelling differences such as Common Black Hawk vs. Common Black-Hawk, Grey Vireo vs. Gray Vireo, etc., are not listed here.)

To convey more information in the bird lists, the following distinctive forms are referred to by English names that differ from the species' overall name. *This does not mean that these forms are recognized as full species*, but being aware of their existence can increase the interest of a trip—particularly if some of these later get split as species! In particular, Cozumel Bananaquit, Godman's Euphonia, and Cinnamon-rumped Seedeater are perhaps best treated as full species, while AOU (1997) split Gilded Flicker from Northern Flicker, split the Plain Titmouse into the California ("Oak") and Grey ("Juniper") titmice, and split the Solitary Vireo into three species:

English names used in this guide	"Parent species"
Mexican Duck	Mallard
Tuxtla Quail-Dove	Purplish-backed Quail-Dove
Red-shafted Flicker	Northern Flicker
Guatemalan Flicker	Northern Flicker
Gilded Flicker	Northern Flicker
California Titmouse	Plain Titmouse (western form)
Grey Titmouse	Plain Titmouse (interior form)
Black-crested Titmouse	Tufted Titmouse
White-browed Wren	Carolina Wren
San Lucas Robin	American Robin
Northern House Wren	House Wren
Brown-throated Wren	House Wren
Southern House Wren	House Wren
Blue-headed Vireo	Solitary Vireo
Cassin's Vireo	Solitary Vireo
Plumbeous Vireo	Solitary Vireo
Mangrove Warbler	Yellow Warbler
Golden Warbler	Yellow Warbler
Myrtle Warbler	Yellow-rumped Warbler
Audubon's Warbler	Yellow-rumped Warbler
Cozumel Bananaquit	Bananaquit
Godman's Euphonia	Scrub Euphonia
Plain-breasted Brushfinch	Chestnut-capped Brushfinch
Cinnamon-rumped Seedeater	White-collared Seedeater
Timberline Sparrow	Brewer's Sparrow
Large-billed Sparrow	Savannah Sparrow
Sooty Fox Sparrow	Fox Sparrow
Slaty Fox Sparrow	Fox Sparrow
Sierran Fox Sparrow	Fox Sparrow
Oregon Junco	Dark-eyed Junco
Pink-sided Junco	Dark-eyed Junco
Grey-headed Junco	Dark-eyed Junco
Mexican Junco	Yellow-eyed Junco
Chiapas Junco	Yellow-eyed Junco
Guatemalan Junco	Yellow-eyed Junco
Bicolored Blackbird	Red-winged Blackbird
Ochre Oriole	Orchard Oriole
Dickey's Oriole	Audubon's Oriole

Blue-headed Vireo, Plumbeous Vireo, and Cassin's Vireo (Mexican breeding populations were, however, ignored in the analyses). See Howell (1996) and Howell and Webb (1995) for details of these forms and their distributions.

THINGS TO KNOW BEFORE YOU GO

Many organized birding tours go to Mexico and allow you to maximize your time and minimize your effort. Birding tours are great, but they're not for everyone. For me, at least, much of the enjoyment is finding and identifying a bird for myself—or so I try to tell myself when I've spent two long, hot, sweaty, nonbirding days under the car with a mechanic in a small village six hours by dirt road from East Comatosepec!

So if you're prepared to venture on a birding trip to Mexico, on your own or with a group of friends (and I do recommend such a trip), I'll assume you are not the type to worry about being in unfamiliar areas. Nonetheless, a few important points bear repeating for foreigners before they take a birding trip to Mexico. Yes, foreigners, because that's what you will be. Remember: it will be Mexico, *not* the U.S.A., Canada, or Europe. You are a visitor in another country, and it is only common sense and courtesy to respect the views of the people there.

Entry Requirements

All foreigners entering Mexico must stop at the border (or airport) to show proof of citizenship. While U.S. and Canadian citizens can use a certified copy of their birth certificate for proof of citizenship, a passport is the recommended way to go, especially if you plan to venture off the beaten track, i.e., away from the direct line between your hotel room and the beach! Citizens of some countries may need visas; check before you go.

Unless you're just crossing at border towns, mainly in Baja California (Tijuana, Tecate, Mexicali), you need a tourist card to visit the "interior" of Mexico. These are free and are provided by airlines or at the border crossings. These single-entry cards are valid for up to 180 days, although usually they get stamped for just 90 or even 30 days. If you're planning a long trip, check how many days you have been granted and, if necessary, get it changed (filling out a new card may be easiest). You should carry your tourist card with you at all times; failure to do so could, in theory, result in a fine.

If you plan to travel south of northern Baja (or other border towns, where you also can be just waved through if driving), make sure you stop and fill out the relevant paperwork then and there rather than think, "Hey, that was easy, great!": when you reach another point—for example, trying to take a ferry across from Baja to the mainland—you may encounter problems.

Driving into Mexico

If you plan to drive down in your own vehicle, the first piece of advice would be: *learn some Spanish.* This will help greatly in dealing with whatever comes your way on your travels. Beyond that, regulations (and the attendant paperwork) about entering with your own vehicle are prone to change, and you should check the latest situation in advance of your trip. For driving in northern Baja and to Puerto Peñasco, Sonora (Site 2.1), you don't (as of 1997) need car paperwork, but you should get insurance.

Membership of the American Automobile Association (AAA) is recommended as a good way to find out the latest information, get help with the various forms needed, and get the mandatory Mexican auto insurance at reasonable rates. Auto insurance also can be obtained easily at various places (gas stations, general stores) near most U.S. border crossings. AAA also publishes a map of Mexico and of Baja California and an annual travel book on Mexico (full of information), while the Automobile Club of Southern California, a regional branch of AAA, also puts out a useful driving guide for the Baja California peninsula. If you're driving beyond border areas where car permits aren't needed, make sure you hand your paperwork in to the Mexican authorities when you leave the country, and get your passport stamped, or you may have problems later.

Note: See the following sections on Driving in Mexico (p. 10) and Birding from the Highway (p. 11) and the warnings about driving restrictions in Mexico City (Chapter 8, p. 154).

Maps

In general, any large, self-respecting bookstore near where you live is likely to sell, or be able to obtain, a map or maps that cover Mexico, and some have regional maps (mainly for Baja and the Yucatan). The American Automobile Association also puts out a Mexico map (approximately 1:4,000,000), updated periodically; this can suffice on a driving trip, although it is a little too small to show many birding sites clearly; AAA also puts out a fairly good Baja map (1:800,000). In Mexico, many larger city supermarkets as well as stores in major airports (especially Mexico City) sell road atlases for the entire country, such as the one published by Guia Roji (1:1,000,000); these are updated periodically and are reasonably accurate.

Note, however, that many maps of Mexico look very attractive but aren't necessarily too accurate, perhaps because most people know where they're going and which roads do and don't exist. A recent series of state-by-state road maps (1:400,000) put out by the SCT (Secretaria de Comunicaciones y Transportes) appears to be reasonably accurate and should suffice for almost all birding purposes. However, these maps are not widely available and can be downright problematic to obtain!

If you're lost, or think you are, showing a map to most locals can be ineffectual, as they may have little or no grasp of the concept and may simply cock their head and look puzzled.

Car Rental

To rent a car in Mexico you'll need a credit card (except perhaps with some local companies, which will charge hefty deposits), although you often can pay in cash when you return the car. Most companies have you sign two blank credit card vouchers. One is a security deposit—be sure to get it back at the end and tear it up. You'll also need a driver's license—in most places a U.S., Canadian, or even British license is acceptable. Other nationalities should check in advance to see if an international driver's license (which, ironically, may be met with blank stares by Mexican rental companies!) is recommended.

In much of Mexico car rental is expensive, especially relative to U.S. prices. Also, automatic and four-wheel-drive vehicles are rarely if ever available at most sites; the former are not good anyway on many dirt or mountain roads. To counterbalance this, the past ten years or so have seen an exponential improvement in many roads, and numerous freeway-like toll (*cuota*) highways help considerably in cutting down on long drives—well worth the relatively high tolls. But, of course, many of the good birding sites are still on bad dirt roads!

Rental vehicles can be booked and prepaid in advance (e.g. from the U.S.A. or Europe) via the major international/U.S. car rental companies or through travel agents, but make sure you get written confirmation and a reservation number. In doing this, the deals you get can be better, or worse, than what you'd get by just turning up at the airport desk. Be aware that there's a fair chance when you turn up they won't have your reservation or the type of vehicle you requested. Most airports have a suite of rental companies, however, and you can almost invariably find a car somewhere, often with a local company for cheaper than your original reservation. If you're traveling over Christmas or Easter, though, finding a vehicle may be a challenge.

Note that prices quoted in the U.S.A. usually don't include a fixed daily insurance rate (optional, but best for peace of mind) or the 10–15 percent government tax, which all help add significantly to what you might have thought was a reasonable deal. Ask in advance to avoid any unpleasant surprises.

Besides checking that the odometer works (not infrequently it doesn't; and note the caveat about directions under Access, in the How to Use This Book section, p. 2), it's a good idea to check the vehicle fairly carefully, even if you've booked with a well-known international company. Someone with the rental company will go over the car with you, making sure that all dents, broken lights, missing wheels, and so on are noted on a piece of paper before you drive off. In particular, things to look for are cracked or pocked windows, the presence of a jack and the thing to remove the wheel nuts, and windows and doors that don't close and/or lock properly—insist on getting them fixed (assuming you want them to work) and/or change vehicles. Remember, you will be charged at the end for anything that is perceived as missing or damaged, so a thorough check before you drive off is worth an extra five minutes.

Driving and Roads

Driving in Mexico requires the same considerations as anywhere else: be alert and respect other road users. Some people enjoy driving in Mexico more than in the U.S.A., as one is more in touch with the environment (such as potholed roads!), and common sense, not a plethora of road signs, dictates how one drives. Speed limits (in km/h) vary with the roads you are on (and often seem absurdly low), and it is best to use the speed of other traffic as a measure of how fast you may drive.

Learning when to actually stop at stop (*Alto*) signs is a little like learning when to use the subjunctive mode in Spanish: there are a few rules, but mainly it's a matter of intuition and observing what those around you are doing. In general, do stop at *Alto* signs in border areas such as Tijuana and Ensenada, but do not stop (slow down, though, and prepare to yield) at *Alto* signs at (mostly disused-looking) railway crossings, unless of course there's a train coming! *Alto* often seems to be interpreted as "Yield" or "Give Way" rather than a full stop, which could get you rear-ended at a railway crossing. You may see *Alto* or *Alto Total* signs for *Inspeccion Fitosanitaria*, *Inspeccion Zoosanitaria*, *Ganado*, and so on, and if it looks like most cars aren't stopping, just go by—these are for commercial and agricultural vehicles.

One common form of highway signaling in Mexico (and most of Latin America) is for vehicles ahead of you to use their directional flashers to indicate that you can pass safely (or what they consider to be safely). However, using the left (outside) flasher means *siga* (please pass), while the right (inside) flasher means *alto* (don't pass), which may be the opposite of what you're used to or what may seem logical. Of course, the left flasher also means "I'm turning left," so if you think you can pass and the car or truck in front pulls out into your path, watch out! The illogic of this ambiguous signaling system seems not to bother the Mexicans, who presumably pick up on subliminal clues to which gringos are insensitive. If in doubt, hand signals aren't a bad idea at obscure left turns. Mexicans also seem to have better night vision than gringos, so you may be "flashed" frequently if you put your lights on early in the evening.

Contrary to gringo folklore, Mexican police are unlikely to stop you for something you didn't do, except perhaps in some notorious areas of Mexico City, such as the route to the Puebla highway from the airport. Thus, it may not have been well signed, but you probably did go the wrong way down a one-way street! If you are pulled over, be polite, apologetic, and friendly. Do not act anxiously or you may open yourself to being ripped off.

Frequently one can get away without a fine, but if the policeman insists, make him or her take you to the police station and do things legally, unless you feel you know the ropes and want to negotiate a settlement—but don't "overtip." Remember, pesos are pesos, and people earn and buy in pesos, not in U.S. dollars. This also goes for many parts of the world where well-meaning tourists have helped unbalance a local economy. Also see the notes about the "one day without a car" policy in Mexico City (Chapter 8, p. 155).

Highway checkpoints can be common (increasingly so in southern Mexico in recent years) but erratic in occurrence. At state line, military, and police checkpoints you should stop, unless they wave you on—beware of what may look like an advance guard waving you on when he is in fact waving you to stop up ahead. If it looks like you tried to run a checkpoint, be prepared for a serious vehicle search! Checkpoints can be a pain, but overall they serve a useful purpose, so be patient and polite and they'll be over all the more quickly.

You have a choice of gasoline (petrol) in Mexico—you either use Pemex (the nationalized Petroleo Mexicano) or run out of gas. The word for gas station is *gasolinera,* but most people will understand if you ask *¿Donde esta la Pemex?* ("Where is the Pemex?"). For short trips from the U.S.A. you may be able to get by with a full tank before you leave, as Mexican gasoline is not as rich as some U.S. brands. Gas stations in Mexico (with exceptions) sell only gasoline (although this is changing). They often don't have mechanics or even air pumps, let alone convenience stores. To get air, go to a *vulcanizadora* (literally, a vulcanizer, i.e., a tire-fixing place), of which there tends to be no shortage. They will check/fill your tires, and a tip of 2–4 pesos (25–50 cents) is customary, depending on how much they do. Usually there is someone (often a gang of kids) at a gas station to clean your car windows (even if you don't need it!). If your windows are clean, a firm *¡No!* or *¡No le pago!* ("I'm not paying you") in advance may be needed; shaking your head doesn't necessarily work. A tip of about a peso (or 2 pesos if you ask someone at a gas station) is customary, although some kids will complain no matter what you give them.

There are three types of gas, plus diesel: regular leaded (*Nova*, at the blue pumps) and 87 octane unleaded (*Magna Sin*, at the green pumps); a recent addition in many areas is *Premium*, a higher grade (93 octane) of unleaded fuel. For many years, unleaded (*sin plomo*) gas could be hard to find, but now *leaded* gas is being phased out rapidly and may soon disappear from all but the most remote, rural areas. Be aware that gas stations run out of gas more often than you'd like; if you're making long drives through sparsely populated areas (such as central Baja California) or spending time in a remote area, it doesn't hurt to fill up when you have the chance, or to carry a spare gas can.

There's a good chance that if you drive in Mexico long enough you'll get a flat tire or something will go wrong with your car. Don't worry, particularly if it's a Volkswagen. There are mechanics (*mecanicos*) all over Mexico, in almost every town (usually on the outskirts), and many people you wave down on the highway have an impressive command of the workings of an engine. This reflects the value people put on their vehicles, and on making them last. If your car can be fixed, they'll fix it, usually on the spot rather than make an appointment for next Thursday, and the cost should be quite reasonable. However, if you have a fancy, automatic, all-electric model, it might prove harder (or impossible) to fix.

Birding from the Highway

In many areas, the best birding spots involve being on roads. Again, common sense is really all you need, but a few points might be helpful. First, know where your emergency flashers are: find them on a rental car right away. Then, if you're not on a seriously busy, two-lane highway and do see something you "have" to stop for, put on the flashers immediately and pull over (preferably briefly) to take a look. Cars break down frequently in Mexico, so stopping by the roadside is quite common, and drivers tend to be accommodating—they've been there before.

On most toll roads there is a shoulder that can be used to pull over and stop for birding, and often these roads seem to go through good marshes. Although stopping might be technically illegal, there's perhaps as much chance of seeing a Harpy Eagle as there is of a policeman stopping and asking you to move on. If that happens (the policeman, not the eagle!), be polite and apologetic (*Lo siento* means "I'm sorry"), maybe make some confused noises about car problems, and they may let you stay and watch birds ("crazy gringo").

If you have to park anywhere, pull off safely from the road, perhaps allowing an extra foot or two if there's room. If you get out, birding from in front of your car generally is safer than birding from the back, and also attracts less attention from passing cars in your lane. If you walk far out of sight from your car, common sense dictates that you leave little more than an empty Coke bottle in view.

It seems to be a truism that the sloth manifest by some waiters and hotel personnel is compensated for when they drive. Cargo trucks and buses may also travel faster than you would on narrow winding roads, so be particularly alert when birding from such roads. However, contrary to popular belief, local drivers are not out to kill all and any pedestrians, although driver awareness of pedestrians may not be what you are used to at home. Drivers often honk their horns at pedestrians (and not just foreigners), which may be considered rude in some cultures and can take a while to get used to. It is simply a way of saying "don't step into the road now," or even just "hello." Wave at drivers (or ignore them if you're a single female), and almost invariably they will wave back. Indeed, if you wave first, you may get a wave and avoid being honked at.

If you drive at night, be aware that many bicycles (and some cars) do not have lights, that livestock can be found wandering or lying in the warm roads, and that drunk drivers may be at large on Friday nights, holidays, and weekends. Thus, nocturnal driving is advised against by many guidebooks, but with care it is rarely eventful.

Buses and Camping/Dossing

This book is written mostly with the driving birder in mind, and I apologize to birders who travel on a lower budget, using public transport and camping out (especially since I spent many years using the latter modus operandi). However, Mexico is served by an impressive volume of public buses, featuring many companies, so you can reach just about anywhere in the country, including the birding sites described herein. Of course, figuring out exact distances will be problematic, and it may be more difficult to get to some places in the early morning unless you sleep out near or at the site.

Most travel guides (see Additional Literature, p. 17) cover transport and other forms of accommodation so I won't dwell upon these here. In general, first-class buses (the newer models often have a toilet) tend to go from point A to point B, maybe with a few major stops in between, but they *do not stop* if you try to wave them down in the middle of nowhere. If you're making long journeys you may be subjected to loud-volume, low-grade U.S. movies which, sadly, are an increasingly common symbol of "first-class" public transport in Latin America. Some would prefer a drunk passing out beside them, or a pig squealing in the aisle. Second- (and lower-) class buses go just about everywhere, but more slowly and noisily, and, unlike the first-class buses, stop (usually) if you flag them down on the highway. Once at or near your destination you'll be relying on second-class buses, *colectivos* (collective taxis, with fixed rates and routes), and hitching rides.

There are few official campgrounds in Mexico away from Baja California and the Northwest, and these tend to cater more to trailers and camper vans than to tent campers, anyway. Such campgrounds are rarely near good birding spots, but, when relevant, they may be mentioned.

If you're not too fussy, however, you can camp out just about anywhere in Mexico, such as beside a quiet dirt road or in a field. If you're obviously on somebody's land it is wise to ask permission, assuming they're around, and most rural families could care less if you camped out in their yard. If you're traveling alone and feel a little uneasy about this you can usually find somewhere in the "bush" away from people and, usually, away from that ubiquitous human commensal, the nocturnal barking dog. (Barking dogs and all-night crowing chickens can be a problem, though, even at nice hotels.) It is also possible to find a bed or floor or hammock in just about any village. Basically, if you're used to this style of traveling and birding, what the British term "dossing," then you'll figure it out in Mexico with little problem.

Time Zones

Most of Mexico is on Central Standard Time (Greenwich Mean Time [GMT] minus six hours); Northern Baja California is on Pacific Standard Time (GMT minus eight), while the states of Baja California Sur, Sonora, Sinaloa, and Nayarit are on Mountain Standard Time (GMT minus seven) and the state of Quintana Roo switched recently to Eastern Standard Time (GMT minus five). In 1996 Mexico instigated summer time, using the same dates as those in the U.S.A.

Money

The unit of currency in Mexico is the peso, divided into 100 centavos. Over the past fifteen years the peso has seen some major devaluations; at press time there were about 8 pesos to the U.S. dollar. While you can change money at banks, it is usually easier and quicker (except perhaps at airports) to change at exchange booths (*cambios*), which can be found widely throughout the country, even in some not-so-touristy towns. When in doubt though, change money when you can rather than wait until you're in some out-of-the-way place. U.S. dollars (and dollar traveler checks) are the best way to carry funds—trying to change other currencies can be anywhere from a frustrating to a futile exercise. Many hotels and nicer restaurants, especially in tourist areas, also take credit cards.

Language

Mexico is a country where English is not spoken widely (except in major resorts), so if you plan to travel away from tourist centers, at least a rudimentary command of Spanish will be helpful. But even if you speak next to no Spanish you'll survive—it might just be more challenging sometimes! And remember, just as English spoken in the U.S. varies from Texas to Maine, so Spanish in Mexico varies regionally, and some dialects may be harder to understand than others. In some parts of the south, particularly Chiapas and Yucatán, many people speak little Spanish, so unless your Maya is in working order, meaningful conversations may be problematic. If you did get permission to enter someone's land, and they're still there when you leave (e.g., they live there), it is good to go and thank them and be enthusiastic about their place and the birds there.

A few very basic Spanish phrases/words that may be of use when birding are as follow:

los binoculares (para larga vista) (Binoculars [to see long distance])

un pájaro/un ave and *los pájaros/las aves* (bird/birds). *Pájaros* is more casual, *aves* more formal; one can be used to support the other if the person you are talking to doesn't seem to understand.

¿Disculpame, sabe a donde va este camino/sendero? ("Excuse me, do you know where this path/trail goes?")

Estoy/Estamos mirando a los pájaros/las aves ("I/We are looking at birds"). Showing a bird guide can help but may get you into a longer conversation. This phrase can be used with the following request:

Buenos dias/Buenas tardes. ¿Puedo/Podemos buscar pájaros/aves en su campo/en este lugar/aqui? ("Good morning/afternoon. Can I/Can we look (search) for birds in your field/in this place/here?")

No, solamente mirando ("No, only looking"). I.e., you don't want to catch, or photograph, the birds; this can be accompanied by waving your binoculars and saying *para larga vista*, because many people, not just Mexicans, mistake binoculars for a camera.

(muchas) gracias, pero acabo de comer/tomar ("[Many] Thanks but I've just eaten/had a drink"). A polite way to decline an invitation.

Muchas gracias ("Many thanks.")

Este es un lugar interesante/bueno para pájaros/aves ("This is an interesting/good place for birds"). Even if it isn't particularly, this a polite way to thank someone.

People and the Land

Mexicans, like most people, are generally friendly, honest, and helpful, and you'd have to go out of your way to upset them. However, as with any part of the world, there is a chance of being short-changed, especially in or around major resorts, and one should bear this in mind.

Mexico is a macho country, and there's nothing you can do about it, so if that's not to your taste, just grin and bear it. If you are a lone female, Mexican men (especially drivers) will shout and whistle, which can be annoying, but they are extremely unlikely to try anything. Nonetheless, it is only common sense for a lone female birder to dress inconspicuously (no shorts away from tourist areas) and not to court unwanted attentions.

In rural areas, many *campesinos* (country people) carry impressive machetes or rusty-looking rifles slung over their back. These are carried in much the same way you might carry a pocketknife or binoculars and should not be viewed as threatening—although one's first impression on meeting a band of machete-wielding people on a remote trail might be otherwise! The group of people will pass, usually smiling, and bid you *buenos dias* (good morning) or *buenas tardes* (good afternoon), and you should reply in kind. In fact, rural people may be scared of you (!), so getting in the first *buenos dias* can break the ice, although many women and children may be too shy or frightened to answer.

Birding is foremost a hobby and requires leisure time (although often it doesn't seem too leisurely!), so in poorer countries one does not find many "birders" as we might know them. Thus, you can attract attention as a novelty, and people may ask you what you are doing. They often ask where you're from, how far away that is (they may think in terms of hours or days walking), how much your binoculars cost, and so on. These are questions of polite curiosity, not designed to see how much they could get for your binoculars! At the same time, I'd underplay the cost if you own Leicas or Zeiss, and say "not much" (*no caro*) or "around $50" (*mas or menos cincuenta dolares*) rather than $1,000! Local people may find it strange that you simply watch birds, not kill them or eat them, but in talking to you they'll have good stories to share with families and friends: "Hey, I met this crazy gringo . . . "

Also, don't underestimate the observational skills and local knowledge of country people. They may not have binoculars or bird guides, but they often know the birds, have keener eyes, and can help with finding some species. For example, my best views ever of Military Macaw are thanks to an old man who insisted that I follow him to some fruiting trees that were "good for birds at that time of year"! However, in an effort to please, some people may just say yes to many of your leading questions, so think before you phrase a question. And because Mexico is a macho country, there is a tendency to make up an answer rather than admit "I don't know," as you may soon discover if you walk around an unfamiliar town asking for directions to an obscure hotel!

Property in Mexico is generally not viewed in the same way as it is in much of North America, and surprise (at being asked) is often the reaction obtained when one asks permission to enter or cross someone's land. Indeed, you may often find yourself declining invitations of hospitality (e.g., coming back to the ranch for food or a drink) when you just want to be off birding. But it is always best to err on the side of courtesy and, of course, to respect private property signs (*Propiedad Privada*, *Prohibido el Paso*, etc.) when you see them.

The word *mañana* might come to mind when you think of Mexico. In much of the country, people's lifestyles are a lot less fast paced (although they do work hard) than the stressful Western lifestyles some birders may be used to. While this can be frustrating when you want to get something done quickly, you're unlikely to change a national trait in one trip, so it's best to go with the flow and relax: you're on holiday. Remember: things are done differently, not necessarily better, where you come from.

Climate and Timing Your Trip

In most parts of Mexico there are two seasons: dry and wet. Over most of the country the dry season is from November/December to April/May. For a more thorough overview of regional variation in climate, you can peruse the chapter on climate and habitat in Howell and Webb (1995).

Many birders visit Mexico between December and April, which corresponds to the dry season, colder weather in temperate latitudes, and the widespread presence of wintering migrant birds from North America. This is perhaps the best period to visit most areas (especially February/April, when more resident birds are starting to sing and nest) and can be taken as assumed in the site guides for each area, unless specific notes on seasonal timing are given. But birding can be good almost anywhere at any season, and even in the wet season the rain is rarely prolonged enough to be a major inconvenience, except for making some dirt roads impassable.

In terms of clothing, think about where you are going. Not everywhere in Mexico is hot and sunny, although rarely even in the mountains do you need more than a light jacket and gloves for the first few hours of the morning.

Food and Water

Approaches to eating and drinking in Mexico run the gamut from total paranoia to the "live by the sword, die by the sword" school of thought; neither is recommended. And while just thinking about Mexico can bring to mind images of the notorious *turista* or "Aztec two-step," sanitary conditions throughout the country have improved greatly in recent years, and you would be fatefully unlucky to contract any serious illness—assuming, of course, that you take reasonable precautions.

It's just a fact of life that some people have more sensitive stomachs than others, and even a simple change in eating habits, such as between Britain and the U.S.A., can cause minor stomach upsets. Bottled, purified water is widely available in Mexico at stores and some gas stations and is cheap and worth the price of an easy conscience. Ice in restaurants is made from purified water (ironically, people often shun ice while pouring their drink into a glass washed in local tap water!), as is ice cream at ice cream shops (street venders may be another story, though).

No matter what you do, even if you follow the adage of eating only cooked foods and avoiding things that have not been peeled, you still may experience a morning of internal discomfort. This shouldn't put you off traveling—some great birds have been seen by people answering the (unwanted) call of nature!

Other Health

No vaccinations are officially required to visit Mexico per se, but if you're coming from an area with yellow fever, you should have a vaccination certificate. Having said this, shots for typhoid, cholera, hepatitis, polio, and tetanus are always worth having, although vaccinations should not be treated as a substitute for careful eating and drinking habits. The chance of contracting malaria is minimal, although dengue fever is more common; use of long sleeves and pants plus insect repellent are the best protection, in conjunction with anti-malarial prophylaxis if you're going to be in areas where this disease occurs (it's present, but uncommon to rare, in most of the tropical lowlands). Before your trip, check with your doctor about precautions and medication.

One thing that a few birders (still a small minority) seem to get, and something with which most Western doctors seem unfamiliar, is bot-fly larvae (from eggs transmitted by mosquitoes). Not knowing what you have can be most of the problem and can be scary. Symptoms are a bite that doesn't seem to heal up and remains as a small, hard, red or purplish spot, followed by a major swelling (which soon goes down) and then pain (like a red-hot poker or a razor blade being turned inside you!). While it should be nothing to worry about (in terms of serious injury or disease), it can be painful.

Many bot-fly remedies have been proposed: cut it out with a penknife (ouch!), hook it out with a needle (less painful but more skillful, often done by locals), surgical removal (not so easy in remote areas), tape a piece of animal fat over the hole and let the worm transfer itself to that, or kill it inside you and pop it out. Ways to kill the larva are with tobacco resin, nail varnish, or typing correction fluid (e.g., White-out, Tippex). Apply the White-out or whatever to the spot, let it dry, and wait a few hours. Then try to pop the larva out. If it's dead, it should come without too much effort and might be an explosive experience. Make sure you get the whole worm out; don't leave something to fester. Once the larva is out, clean the wound and let it heal. I've had about four bot-fly larvae, and after the fright of the first (mainly fear of the unknown), they've all been a straightforward matter.

Other Wildlife

As a rule, unless you visit fairly remote and/or protected areas, you won't see much in the way of native mammals in Mexico, save for squirrels and bats. And if you see one snake, let alone a poisonous one, on a two-week birding trip without making a special effort, you're lucky (honestly!). I see far more snakes and mammals in California. Lizards, some of them spectacular (e.g., the big iguanas), can be common but are all basically harmless.

I encounter scorpions and poisonous spiders about as often as snakes, so again, this shouldn't be a major concern. Anywhere in the tropical lowlands, even in fancy hotels, it is common sense (but I usually forget) to check under and in one's bed before tucking in for the night.

The insect realm, however, is where you are likely to strike it rich. Mosquitoes, chiggers, biting gnats, and biting or simply annoying flies can be locally numerous, particularly in the lowlands, and especially in the wet season—another reason to visit during the dry season. While ticks can be annoying, they are potentially more dangerous in New England or California, where Lyme disease is prevalent. The usual rules apply: long sleeves and pants, insect repellent when needed. Also see under Other Health, just above.

If you're interested in butterflies, they can more than compensate for any annoying insects. Although, like the latter, the greatest numbers and diversity of butterflies occur in the wet season, numerous spectacular species can be seen year-round in many areas. To identify most butterflies requires a combination of books (see Additional Literature, below), but you'll still see some you can't identify, particularly if you're warped enough to look at skippers.

Additional Literature

Many books have been published on virtually all aspects of Mexico, and here I include only a few, almost miscellaneous titles that may be of interest or use to the visiting birder. Sometimes natural history books can be difficult to obtain, and recommended stores with extensive inventories include:

Buteo Books, 3130 Laurel Road, Shipman, VA 22971, U.S.A. Phone: 800-722-2460; fax 804-263-4842; e-mail, buteo@comet.net

Los Angeles Audubon Society, 7377 Santa Monica Boulevard, Los Angeles, CA 90046, U.S.A. Phone: 213-876-0202; fax 213-876-7609; e-mail, laas@IX.netcom.com

Natural History Bookstore (NHBS), 2–3 Wills Road, Totnes, Devon TQ9 5XN, U.K. Phone 01803-865913; fax 01803–865280; e-mail, nhbs@nhbs.co.uk

Subbuteo Books, Pistyll Farm, Nercwys, Nr. Mold, Flintshire, CH7 4EW, U.K. Phone: 01352-756551; fax 01352-756004; e-mail, sales@subbooks.demon.co.uk

ABA Sales, P.O. Box 6599, Colorado Springs, CO 80934, U.S.A. Phone: 800-634-7736; fax 800-590-2473; e-mail, abasales@abasales.com

General Travel Books. These can be found in any large, self-respecting bookstore. The following is a selection of the wide choice of travel books available, most of which are updated regularly, some on an annual basis.

AAA Mexico TravelBook. Published by the American Automobile Association and available from their offices.

Fodor's Guide series. Fodor Travel Publications Inc.

Frommer's Frugal Travel Guide series. Published by Macmillan Travel.

Lonely Planet Travel Survival Kit. Lonely Planet Publications.

The People's Guide to Mexico. John Muir Publications.

Spanish Language. Many phrase books and other publications are designed to help the beginner understand and be understood in Spanish-speaking countries. Once you've reached a certain level, though, where do you go next? An excellent book for those with some command of Spanish is J. J. Keenan, *Breaking out of Beginner's Spanish* (Austin: University of Texas Press, 1994).

Bird Books. Several guides cover the birds of Mexico. Note, however, that none illustrates all of the North American migrants, so you will probably also want to take along a North American field guide, such as *The National Geographic Society Field Guide to the Birds of North America*. Guides to the birds of Mexico include:

Howell, S. N. G., and S. Webb. 1995. *A Guide to the Birds of Mexico and Northern Central America*. New York: Oxford University Press. This is by far the most up-to-date and complete guide, featuring range maps and full descriptions of all species. It is, however, somewhat weighty. Recommended!

Peterson, R. T., and E. L. Chalif. 1973. *A Field Guide to Mexican Birds*. Boston: Houghton Mifflin Co. Out of date and lightweight; don't expect to be able to identify too many immature raptors, female hummingbirds, or flycatchers with this guide.

Howell, S. N. G. 1996. *A Checklist of the Birds of Mexico*. Golden Gate Audubon Society, 2530 San Pablo Avenue, Berkeley, CA, U.S.A. Phone: 510-843-2222; fax: 510-843-5351. An attractive and up-to-date list of all species, with scientific names and explanation of differences between Howell and Webb taxonomy and AOU. Eight columns for checking off lists. Recommended.

Other Natural History. While there are no guides that cover fully the mammals, reptiles, amphibians, butterflies, and/or plants of Mexico, I have found that the following list (by no means exhaustive) can help greatly in figuring out some of what you see, at least in certain regions.

Alvarez del Toro, M. 1982. *Los Reptiles de Chiapas*. Tuxtla Gutierrez, Chiapas: Instituto de Historia Natural. In Spanish and possibly hard to obtain, but useful for Chiapas and adjacent regions.

Berry, F., and W. J. Kress. 1991. *Heliconia: An Identification Guide*. Washington: Smithsonian Institution Press. A color-photo guide to 200 forms of these spectacular tropical plants.

Brown, J. W., H. G. Real, and D. K. Faulkner. 1992. *Butterflies of Baja California*. Beverly Hills, Calif.: Lepidoptera Research Foundation. Useful, but lack of cross-referencing between the plates and text reflects inexcusable ineptitude.

Campbell, J. A., and W. W. Lamar. 1989. *The Venomous Reptiles of Latin America*. Ithaca, N.Y.: Cornell University Press. Interesting and useful, even if you hope you don't need it! Lots of color photos.

De la Maza R., R. 1987. *Mariposas Mexicanas*. Mexico, D.F.: Fondo de la Cultura Economica. A beautiful coffee-table book, not easy to obtain but very useful for identifying Mexican butterflies, especially if used in conjunction with DeVries 1987.

DeVries, P. J. 1987. *The Butterflies of Costa Rica and Their Natural History: Papilionidae, Pieridae, Nymphalidae*. Princeton, N.J.: Princeton University Press. Useful in southern Mexico for these three families, which include most of the conspicuous species you're likely to encounter. See also De la Maza 1987.

Garel, T., and S. Matola. 1995. *A Field Guide to the Snakes of Belize*. Belize: Belize Zoo and Tropical Education Center. A color-photo guide also of use in adjacent Mexico.

Leopold, A. S. 1959. *Wildlife of Mexico*. Berkeley: University of California Press. Out of print but useful for what the author terms "the game and fur-bearing mammals," basically anything larger than and including rabbits and squirrels.

Mason, C. T., and P. B. Mason. 1987. *A Handbook of Mexican Roadside Flora*. Tucson: University of Arizona Press. A useful guide for identifying over 200 of the more conspicuous roadside plants in Mexico.

Meyer, J. R., and C. F. Foster. 1996. *A Guide to the Frogs and Toads of Belize*. Malabar, Fla.: Krieger. A color-photo guide also of use in adjacent Mexico.

Reid, F. A. 1997. *A Field Guide to the Mammals of Central America and Southeast Mexico*. New York: Oxford University Press. This excellent new guide covers all Central American species north to the Isthmus of Tehuantepec.

Roberts, N. C. 1989. *Baja California Plant Field Guide*. La Jolla, Calif.: Natural History Publishing Co. Covers 550 of the commoner and more distinctive plants of Baja.

Stebbins, R. C. 1985. *Peterson Series Field Guide to Western Reptiles and Amphibians*. Boston: Houghton Mifflin. Covers Baja California and is also useful in mainland border areas.

CHAPTER 1 BAJA CALIFORNIA

Xantus' Hummingbird

Mexico's rugged, 800-mile-long Baja California peninsula (often referred to simply as "Baja") is frequently contrasted to "mainland Mexico," reflecting how it stands apart from the rest of the country in terms of both its physical and human geography. While much of the peninsula is desert and coastline, there are also some high mountains in the north, with extensive conifer forests, and the tropical Cape District at the southern tip is a minor center of endemism.

From a birder's point of view, Baja has few species not found in adjacent (Alta) California, U.S.A.; consequently, it tends not to be a priority destination. However, the southern tip of the peninsula does have five endemics, although an overnight hiking and camping trip is usually needed to see Cape Pygmy-Owl and Baird's

Junco in the Sierra de la Laguna. The other three endemics (Xantus' Hummingbird, Grey Thrasher, Belding's Yellowthroat) can be seen easily during a day in the Cape, while the thrasher can also be seen within a day's drive of the U.S. border.

Other than these endemics, the main reason to visit Baja is to see species not found, or found less easily, in the rest of Mexico. Breeding birds include Black-vented Shearwater, Brandt's and Pelagic cormorants, Black Oystercatcher, Western Gull, Cassin's Auklet, California and Mountain quail, Anna's Hummingbird, Williamson's Sapsucker, Nuttall's Woodpecker, Pinyon Jay, American Crow, Mountain Chickadee, California Gnatcatcher, Wrentit, California Thrasher, Grey Vireo, California Towhee, Sage and Fox sparrows, Dark-eyed Junco, Cassin's Finch, and Lawrence's Goldfinch. Winter migrants include loons, Horned Grebe, Tundra Swan, Brant, Canada Goose, Eurasian Wigeon, scoters, Mountain Plover, gulls, alcids, Red-naped and Red-breasted sapsuckers, White-throated and Golden-crowned sparrows, and Purple Finch.

For the avid Mexico lister, Baja has proven to be the premier hunting ground for vagrants, particularly in autumn and winter: species found in recent years include Yellow-billed and Arctic loons, Eurasian Dotterel, Bar-tailed Godwit, Marbled Murrelet, Dusky and Arctic warblers, Yellow and White wagtails, Olive-backed and Red-throated pipits, Harris' Sparrow, Rusty Blackbird, and Common Grackle.

The two main ways to reach Baja are by driving from the U.S.A. (good for the northern third to half of the peninsula) or by flying into Los Cabos at the southern tip, where cars can be rented readily from a number of companies at the airport. If you have two weeks or longer, driving the length of the peninsula and back makes for an interesting trip; this drive can be done in less time but would be so hurried as to not be worth it.

As a consequence of logistical access, this chapter is broken into three parts. First come areas that can be reached in a day or as trips of up to a week by driving in from adjacent California (south to the Sierra San Pedro San Mártir and Bahía San Quintín). Then, for those driving the length of the peninsula, a brief site-by-site summary is given for the long stretch of mid peninsula between San Quintín and La Paz. Last, at the southern tip, come Los Cabos (Cabo San Lucas and San José del Cabo), a resort destination often visited by air.

Another transport option when birding in the southern part of the peninsula is to take a ferry to/from the Mexican mainland. Ferries run between Santa Rosalia, Baja, and Guaymas, Sonora, and from La Paz, Baja, to both Topolobampo and Mazatlán, Sinaloa (see Chapter 5). If you plan to make a ferry crossing with a vehicle, book at least a month ahead during holiday periods (a week before at other times usually works) when you can reconfirm the latest routes and schedules. Be aware that ferry information (schedules in particular, and even the operating companies!) is prone to change. Up-to-date information may be obtained through the Automobile Club of Southern California, which also puts out a useful driving guide to the peninsula entitled *Baja California*. Foot passengers on the ferries can usually just buy their ticket(s) a day in advance.

Many sites in this chapter (particularly northern Baja) are best reached with your own vehicle, especially the sierras (Sites 1.4, 1.7), to which you could hitch a ride, although traffic can be sparse except on (and even during) weekends. Many parts of the large estuary-bay complexes (Sites 1.3, 1.8) also are difficult to bird without your own transport. However, most sites in mid peninsula and at Los Cabos can be reached easily by public transport.

Suggested Itineraries (short trips to the northwest)

Crossing at Tijuana before noon, bird along the northwest coast, south to Ensenada (Sites 1.1, 1.2); arrive at Ensenada in time (best before 7 P.M.) to arrange a pelagic boat trip for the next morning (Site 1.2). There are numerous hotels and restaurants in and around Ensenada. After a morning boat trip, you can drive to Laguna Hanson (Site 1.4) for an overnight trip (camping, so take all your supplies) or south to the Sierra San Pedro Mártir (Site 1.7), where at least one night should be spent (camping is easy, but take all your supplies), or to Bahía San Quintín (several hotels and campsites). If you are not doing a boat trip, you can bird the Estero Punta Banda area (Site 1.3) in the morning before heading east or south. From any of the end points (Sites 1.4, 1.7, 1.8), it's an easy day's drive (after birding in the morning) back to the United States; remember to allow time for getting back to the highway from the Sierra San Pedro Mártir.

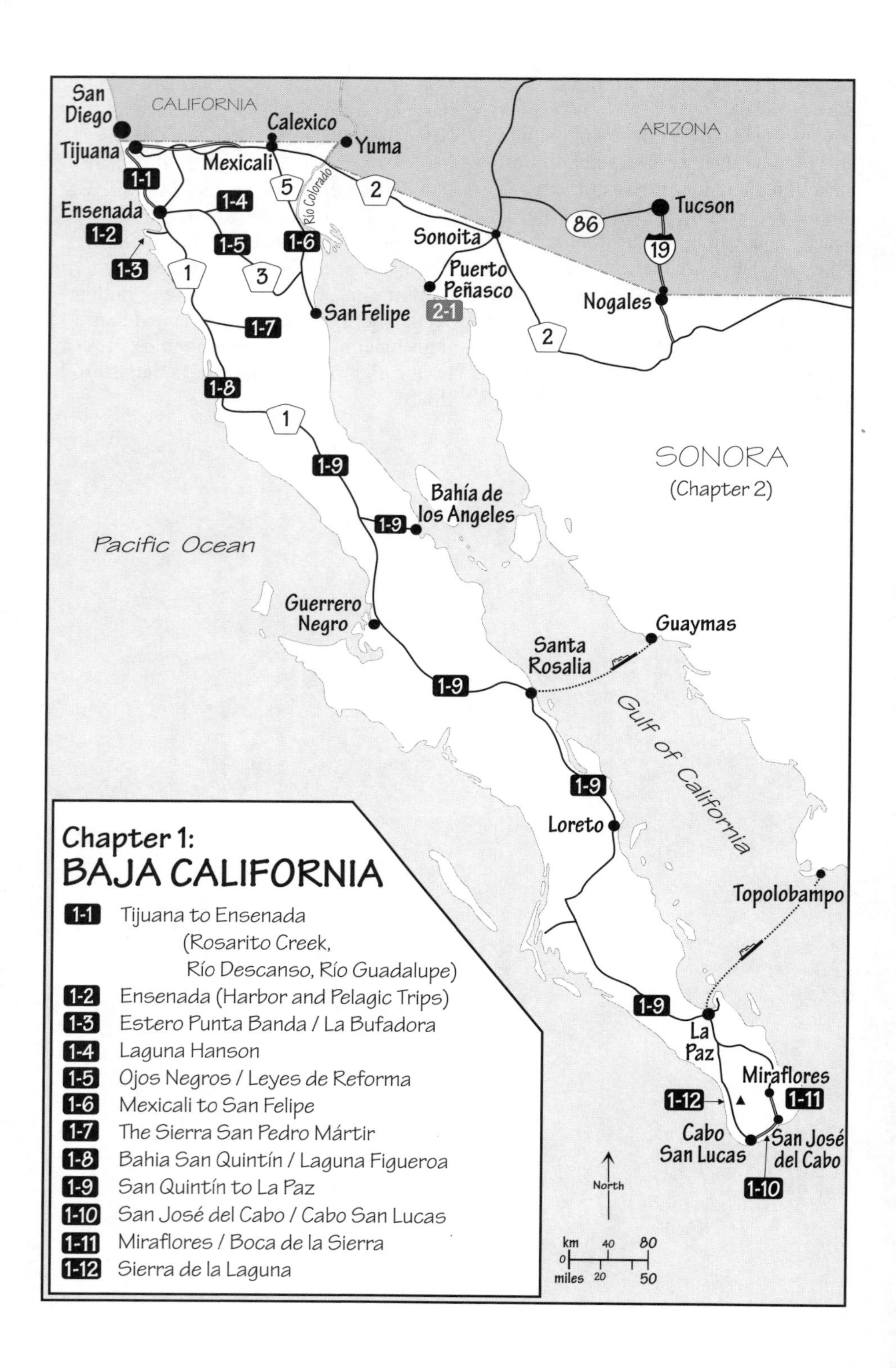

San Diego
CALIFORNIA
Calexico
Yuma
ARIZONA
Tijuana
Mexicali
1-1
5
Río Colorado
2
Ensenada
1-4
Tucson
86
1-2
1-5
1-6
Sonoita
19
1-3
1
3
Puerto Peñasco
Nogales
San Felipe
2-1
1-7
2
1-8
1
SONORA
(Chapter 2)
1-9
Bahía de los Angeles
1-9
Pacific Ocean
Guerrero Negro
Guaymas
Santa Rosalia
1-9
Gulf of California
1-9
Loreto
Topolobampo
1-9
La Paz
Miraflores
1-12
1-11
Cabo San Lucas
San José del Cabo
1-10
North
km 0 40 80
miles 20 50
Chapter 1:
BAJA CALIFORNIA
1-1 Tijuana to Ensenada (Rosarito Creek, Río Descanso, Río Guadalupe)
1-2 Ensenada (Harbor and Pelagic Trips)
1-3 Estero Punta Banda / La Bufadora
1-4 Laguna Hanson
1-5 Ojos Negros / Leyes de Reforma
1-6 Mexicali to San Felipe
1-7 The Sierra San Pedro Mártir
1-8 Bahia San Quintín / Laguna Figueroa
1-9 San Quintín to La Paz
1-10 San José del Cabo / Cabo San Lucas
1-11 Miraflores / Boca de la Sierra
1-12 Sierra de la Laguna

NORTHERN BAJA SITE LIST

Site 1.1 Tijuana to Ensenada (Rosarito Creek, Río Descanso, Río Guadalupe)
Site 1.2 Ensenada (Harbor and Pelagic Trips)
Site 1.3 Estero Punta Banda/La Bufadora
Site 1.4 Laguna Hanson
Site 1.5 Ojos Negros/Leyes de Reforma
Site 1.6 Mexicali to San Felipe
Site 1.7 The Sierra San Pedro Mártir
Site 1.8 Bahía San Quintín/Laguna Figueroa

MID PENINSULA SITE LIST

Site 1.9 San Quintín to La Paz (Cataviña, Bahía de Los Angeles, Guerrero Negro, San Ignacio, Santa Rosalia, Mulege, Loreto, La Paz to Los Cabos)

LOS CABOS SITE LIST

Site 1.10 San José del Cabo/Cabo San Lucas
Site 1.11 Miraflores/Boca de la Sierra
Site 1.12 Sierra de la Laguna

NORTHERN BAJA

Site 1.1: Tijuana to Ensenada (Rosarito Creek, Río Descanso, Río Guadalupe)

(May 1989, June, October 1991, January 1993, 1994, January, September, October 1995, January, October 1996, January, May, August, September 1997, February 1998)

Introduction/Key Species

Most birders heading down the peninsula enter Baja via San Diego at the San Ysidro/Tijuana border crossing. From here there is a choice of roads to Ensenada, about 70 mi to the south: the faster *cuota* (toll) highway (about $5.00 total from Tijuana to Ensenada) or the old *libre* (free) highway. The coast alternates between high rocky cliffs and sandy beaches with numerous creeks and river mouths. Seemingly unplanned "development" has largely ruined this stretch of coast from an aesthetic viewpoint, but there are a few areas worth birding.

Stops at coastal overlooks and creeks with riparian vegetation can produce loons (winter), **Black-vented Shearwater**, Brandt's and Pelagic cormorants, scoters (winter), gulls, Anna's Hummingbird, Nuttall's Woodpecker, Red-naped and Red-breasted sapsuckers (winter), American Crow, Wrentit, California Thrasher, California Towhee, Fox, White-throated, and Golden-crowned sparrows (winter), and Purple Finch (winter).

Access/Birding Sites/Bird Lists

The *cuota* highway runs along the coast for more of its length than does the *libre*, and there are several exits where you can cross between the two highways to access birding sites. There are many ways to get through Tijuana, and many ways to get lost. The easiest approach is to follow signs for "Rosarito" and "Ensenada scenic road" (Route 1D).

From the San Ysidro crossing, immediately after you enter Mexico stay straight on the main road, up over an overpass and across a concrete-banked canal, staying in the right-hand center lane, as indicated by the signs; at 0.6 mi from the border crossing you fork right, then bend down and around to the left, back under the highway, and fork left (signed "Rosarito" and "Ensenada scenic road") to put you on a main road that heads west, right beside the border fence. At 4.5 mi, watch to bear right (signed Rosarito, etc.), versus straight on Route 2 to Tecate, at 5.8 mi pass the right exit to Playas de Tijuana, and at 6.5 mi you reach the first toll booth, where you should reset to zero.

If you wish to avoid paying toll charges (and miss some good rocky coast overlooks), continue straight (rather than bearing right) after the San Ysidro border crossing and transit Tijuana via a maze of junctions, following signs for Ensenada and Rosarito (good luck!).

Along the *cuota*, basically any coastal overlook, river mouth, or riparian wash with vegetation can be productive. In the first few miles there are several good rocky coast overlooks where you can stop. For example, at 3.0 mi take the exit signed to La Joya and pull over almost immediately on the shoulder to view the ocean and rocks (you can U-turn easily to get back on the *cuota*). Birds to look

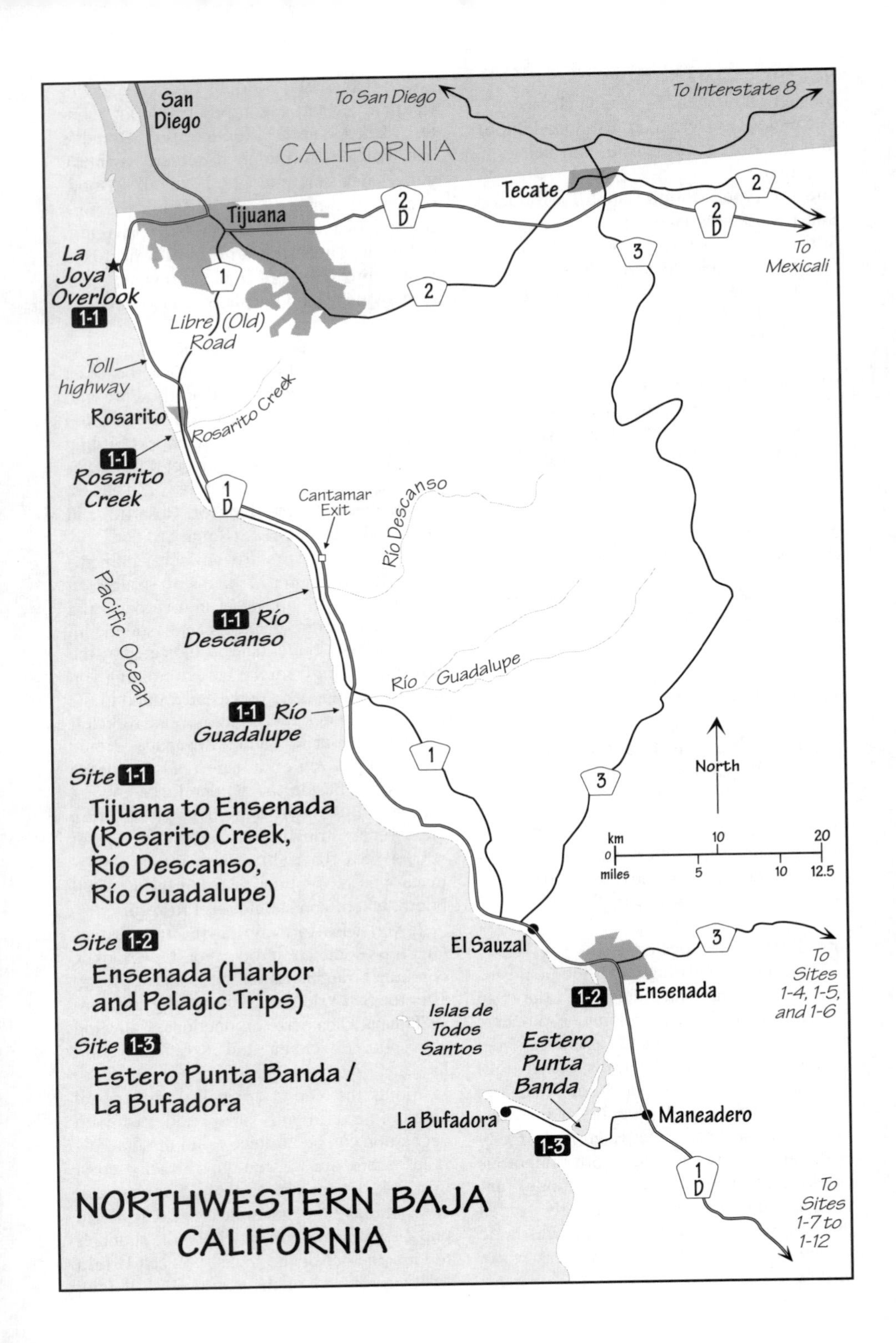

San Diego
To San Diego
To Interstate 8
CALIFORNIA
Tecate
Tijuana
2 D
2
3
1
To Mexicali
La Joya Overlook
1-1
Libre (Old) Road
Toll highway
Rosarito
Rosarito Creek
1-1 Rosarito Creek
1 D
Cantamar Exit
Río Descanso
Pacific Ocean
1-1 Río Descanso
Río Guadalupe
1-1 Río Guadalupe
North
km
0
10
20
miles
5
10
12.5
Site 1-1
Tijuana to Ensenada (Rosarito Creek, Río Descanso, Río Guadalupe)
Site 1-2
Ensenada (Harbor and Pelagic Trips)
Site 1-3
Estero Punta Banda / La Bufadora
El Sauzal
Ensenada
1-2
To Sites 1-4, 1-5, and 1-6
Islas de Todos Santos
Estero Punta Banda
La Bufadora
Maneadero
1-3
To Sites 1-7 to 1-12
NORTHWESTERN BAJA CALIFORNIA

for here include loons (winter), Western and Clark's grebes (mainly winter), cormorants, scoters (including Black in winter), Black and American oystercatchers (beware of intergrades), and Surfbird (winter).

At 11.5 mi you start to see signs for exits to Rosarito. At 14.5 mi take the southernmost (third) exit (the second such signed Rosarito Centro) and continue 0.2 mi to the right until you hit the main street (the *libre* road, in fact) through town. Turn left here and continue 0.2 mi to a small creek (known among birders as **Rosarito Creek**) on the south side of town, with the *cuota* highway bridge very close on your left. There is plenty of room to park off the road on the right, just before the creek. From here you can cross the creek (a trickle) and walk via a broad track along the south side for about 350 m to the beach. This short riparian stretch is often good for migrants in spring and autumn. Other birds to look for here include Red-shouldered Hawk, Spotted Dove, Anna's Hummingbird, Purple Finch (irregular in winter), and, on the rocks off the beach, cormorants and Wandering Tattler (migrant). The beach-dammed river mouth usually has some gulls and shorebirds, and, depending on season, loons, grebes, **Black-vented Shearwaters** (mainly August to January), and scoters can be seen off the beach (as is true of almost any stretch of the coast between Tijuana and Ensenada). You can rejoin the southbound *cuota* 0.3 mi south from the Rosarito Creek bridge (follow the signs for Ensenada Scenic Road).

Another spot worth checking, although continued habitat destruction has made this site less attractive in recent years, is the **Río Descanso** drainage. From Rosarito you can stay on the *libre* road for 12.9 mi (not much longer, time-wise, than the *cuota*) until you come into the Descanso Valley, as described below. If you get back on the *cuota* at Rosarito, take the exit signed for Cantamar at 11.0 mi south of the Rosarito toll booth and head south on the *libre* road; in about a mile you come over a rise and look down over the Descanso Valley, with the *cuota* road paralleling you to the left.

There are two ways to bird the riparian areas here: from the north side or the south side. **North side access:** as you come into the valley there is a large sign for a plant nursery (Vivero La Central) on your left; pull off to the left here on the broad gravel verge and curve back under the *cuota* on a dirt road past a few houses, a school, and the nursery, curving back right (south) and then sharp left at 0.4 mi, after which you should stay in the right-hand lane; in another 0.2 mi there is a dirt road to the right, which, in another 0.2 mi, hits the creek at some willows. You can park here and explore on foot up and down the creek, and also cross to the south side via an earth dam. **South side access:** stay on the *libre* highway past the nursery. You will soon cross the Río Descanso (a creek, really), at the small Puente La Posta; just after this bridge you can turn left or, if traffic is heavy, pull over to the right and wait for a good opportunity to cross to the left, where a dirt road runs up along the creek for a few miles and ends, effectively, at a potentially locked gate, beyond which are some gravel works. Stopping and birding at good-looking sites and/or when you see things is the best way to bird this road, although "development" has destroyed much of the habitat in the last few years.

Species to look for at Río Descanso include Red-shouldered Hawk, California Quail, Anna's Hummingbird, Nuttall's Woodpecker, Red-naped and Red-breasted sapsuckers (winter), American Crow (autumn), California Thrasher, Bell's Vireo (summer), Wrentit (adjacent chaparral), Sooty Fox [Fox] (rare) and Golden-crowned sparrows (winter), Tricolored Blackbird (irregular), Purple Finch (irregular in winter), and a variety of migrant flycatchers, vireos, and warblers in season.

A third spot easy to check from the *cuota* highway is the overlook of the beach-dammed **Río Guadalupe** mouth near La Misión, shortly north of which the *libre* road heads inland (American Crows can be found along the *libre* road between La Misión and Ensenada). To get back on the *cuota* from Descanso, go back north to the Cantamar junction and take the southbound (Ensenada) lane. At 9.8 mi the *cuota* highway runs right behind the beach, with a large parking lot on the right. You can park here and walk across the *cuota* (with care, because traffic is often fast) to the inland side, overlooking the river. From here it's about 26 mi to Ensenada.

Birds at the Río Guadalupe mouth can include Western and Clark's grebes (mainly winter), Reddish Egret (migrant), Canada Goose (winter), Mallard (breeds), a good variety of wintering ducks, Snowy Plover (nests on the beach), and gulls, which have included Glaucous.

Site 1.2: Ensenada (Harbor and Pelagic Trips)

(December 1982, 1983, May 1989, June, October 1991, January 1993, 1994, January, September, October 1995, January, October 1996, January, May, August, September 1997, February 1998)

Introduction/Key Species

Although undergoing some rapid new construction development (started in 1996), the Ensenada Harbor remains a good site (mainly during autumn through winter) for loons, grebes (including Horned), sea ducks, shorebirds, gulls (including Mew), Black Skimmer, and even alcids, and it's always worth a check if you're passing through town.

In addition, it is easy to arrange short pelagic trips out to and beyond the Islas de Todos Santos; possibilities include several shearwaters, storm-petrels, Sabine's Gull, and alcids.

Access/Birding Sites

Coming via the *cuota* highway from Tijuana you reach the northern breakwater of Ensenada Harbor about 5.3 mi south of the El Sauzal/Tecate (Route 3) overpass junction, and 63.0 mi from the Tijuana toll booth. The highway runs along the rocky coast for a short stretch before the breakwater, and there is a large *parador* (parking overlook) on the right: stopping here to scan can be productive (loons, grebes, scoters, gulls). A telescope is useful if not essential, as scoter flocks often move up and down along the outer edge of the breakwater, which is off limits to birders, as is the entry to the port at the head of the breakwater.

To view inside the harbor, continue 0.5 mi along the highway beside the walled-off port facilities to your right and fork right into downtown Ensenada on Boulevard Lázaro Cárdenas (i.e., the first main junction, with a Pemex station across the street). Mark zero as you get on to Cárdenas Boulevard. (For arranging pelagic trips, turn right again in one block, just past the traffic lights, and onto a minor street through fish stalls a short distance to the sport fishing piers.)

At 0.7 mi on Lázaro Cárdenas, after crossing a small creek at 0.3 mi, you pass the Corona Hotel on the right; take the first right (Calle Las Rocas) one block past this hotel, or the second right (Calle Diamante) two blocks past it, and in about 100 m you come to a fenced-off view of the harbor at a small creek inflow. From here you can go right or left on the dirt road along the edge of the harbor. The creek inflows and mudflats have been good for shorebirds, gulls, terns, and, usually, a flock of Black Skimmers. The tidal flats by the northern creek inflow and the small jetty can be good for gulls (including Mew), and the rocky breakwaters and wrecks should be checked for Black Oystercatcher. Scan the harbor for loons, grebes, and ducks.

Ensenada Harbor Bird List

(waterbirds only)

65 sp. R: rare; V: vagrant (not to be expected).

Red-throated Loon
Common Loon
Pied-billed Grebe
Horned Grebe
Eared Grebe
Western Grebe
Clark's Grebe
Brown Pelican
Double-crested Cormorant
Great Blue Heron
Great Egret
Snowy Egret
Reddish Egret
Green Heron
Black-crowned Night-Heron
Yellow-crowned Night-Heron (R)
Lesser Scaup
Surf Scoter
White-winged Scoter
Bufflehead
Red-breasted Merganser
American Coot
Black-bellied Plover
Snowy Plover
Semipalmated Plover
Killdeer
Black Oystercatcher
Black-necked Stilt
American Avocet
Greater Yellowlegs
Willet
Spotted Sandpiper
Whimbrel
Long-billed Curlew
Marbled Godwit
Ruddy Turnstone
Black Turnstone
Red Knot
Sanderling

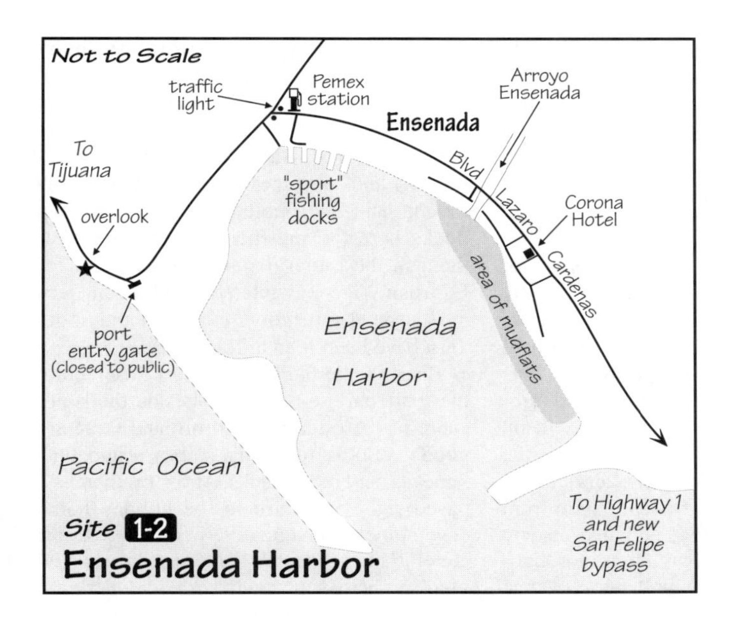

Semipalmated Sandpiper (R)
Western Sandpiper
Least Sandpiper
Pectoral Sandpiper
Dunlin
Short-billed Dowitcher
Long-billed Dowitcher
Red Phalarope
Parasitic Jaeger
Bonaparte's Gull
Heermann's Gull
Mew Gull
Ring-billed Gull
California Gull
Thayer's Gull
Herring Gull
Western Gull
Glaucous-winged Gull
Caspian Tern
Royal Tern
Elegant Tern
Common Tern
Forster's Tern
Black Skimmer
Marbled Murrelet (V)
Ancient Murrelet (R)

Pelagic trips can be arranged at the sport-fishing piers (see directions above), at Gordo's or one of the adjacent companies. A good trip is to leave around 7 A.M. for a five- to six-hour trip, first out to the Islas de Todos Santos and then offshore from there for thirty minutes to an hour before heading back to Ensenada. The cost of such trips in 1997 was $25 U.S./hour, which works out quite reasonably for three or four persons. It is also possible to get on tuna-fishing boats (prices negotiable for birders) and go farther offshore (September 1995: Leach's Storm-Petrel, Long-tailed Jaeger, Arctic Tern), but you are pretty much at the mercy of the fishermen, and it can be frustrating.

Birds seen on five short trips (October 1991, September, October 1995, January 1996, September 1997) include Northern Fulmar, Pink-footed, Sooty, and **Black-vented shearwaters**, Wilson's (rare), Ashy (uncommon), Black, and **Least storm-petrels**, Red-necked and Red phalaropes, Pomarine and Parasitic jaegers, Sabine's Gull, Xantus' and **Craveri's murrelets**, Cassin's and Rhinoceros auklets, plus loons, grebes, cormorants, scoters, gulls, and terns, as can be seen from shore.

Site 1.3: Estero Punta Banda/ La Bufadora

(December 1983, May 1989, June, October 1991, January 1993, 1994, January, September, October 1995, January, October 1996, September 1997, February 1998)

Introduction/Key Species

The Estero Punta Banda, a large estuary just south of Ensenada, is an excellent site in autumn through spring for loons, grebes (including Horned), ducks, raptors, shorebirds, gulls, terns, and Black Skimmer, as well as

Clapper Rails and Large-billed [Savannah] Sparrows (winter). Continuing out the Punta Banda road to its end at La Bufadora provides access to some rocky coast for Brandt's and Pelagic cormorants and American and Black oystercatchers.

Access/Birding Sites

There are numerous viewing points for this large estuary, which is fifteen to thirty minutes south of Ensenada. Head south from Ensenada on Route 1, and 4.0 mi south of the overpass of the San Felipe bypass, note the traffic light and paved right-hand turn signed to Estero Beach. Continue south on Route 1 and you soon drop down to the Maneadero Plain, a wide agricultural valley backing the Estero Punta Banda; 1.0 mi from the Estero Beach turning, and just as you drop down to the plain, there is a minor highway bridge (Puente Chapultepec) over a small dry wash, with a dirt road off to the right at the north side of the bridge. You can turn here and in 1.2 mi come to the edge of the salt marsh along the northeast corner of the *estero* (estuary), across from its mouth. At high tide this site often has good roosts of shorebirds and terns (the latter mainly in autumn) and can also be good for Tricolored Heron, Reddish Egret, geese (winter), and raptors (winter).

Back at the highway, continue south on Route 1 and through the town of Maneadero to a major right-hand junction (signed to La Bufadora), with traffic lights, at 4.0 mi south of the Estero Beach turn. Turn right here and mark zero at the turn. The road runs through large crop fields bordered by weedy, seasonally wet ditches. Flocks of shorebirds and blackbirds can be worth checking anywhere you find them, and White-faced Ibis occur locally. Shortly after the first sharp left-hand bend, there is a bridge over a (usually) dry sandy wash at 2.7 mi. A dirt road runs along the south side of this wash and heads out to the salt marshes at the southeast corner of the Estero Punta Banda in about a mile. The lines of trees along the dirt road are worth checking for landbirds, especially in autumn (Dusky Warbler has been found here!). The marshes and mud flats at the end can be good at any state of tide but are better at high tide or on the rising tide, when Clapper Rail and Large-billed [Savannah] Sparrow (usually far outnumbered by "regular" and Belding's Savannah Sparrows) are pushed out of the marsh. The fields bordering the marsh are good for other sparrows (including Grasshopper) and in winter for owls (Barn, Burrowing, and Short-eared), while any wigeon flocks here (or anywhere you can get near them in the estuary) should be checked for Eurasian Wigeon (winter). Shorebird and tern roosts can also be good (e.g., Ruff and Sooty Tern have been seen at this spot).

Back on the highway out to La Bufadora, the road starts to run right beside the south shore of the estuary at 5.0 mi, and there are good overlooks for loons, grebes, scaup, buffleheads, and other ducks. At 6.5 mi there is a paved right-hand turn to the holiday home development (imaginatively signed "Resort Hotel Sandy Beach") that now marks the sand spit across the southwest side of the estuary. Turning right here allows good views of the estuary on your right (Horned Grebe in winter), and the ponds on the left in 0.5 mi are worth checking. The first and shallower pond often has shorebirds (especially at high tide), while in winter the second and deeper pond often has loons, grebes, and ducks.

If you don't turn down to these ponds but stay on the main road, in 1.0 mi you go through the village and trailer-park settlement of Punta Banda (the trees around the park on your south and the adjacent neighborhoods should be checked in autumn for migrants; more than twenty species of warblers have been recorded), and then the road winds up through chaparral-covered slopes (California Quail, White-throated Swift, California Gnatcatcher, Wrentit, California Thrasher, Rufous-crowned Sparrow) and out to the small fishing and tourist settlement of La Bufadora, at 13.0 mi. The road ends here at a couple of parking lots and (seasonally) an army of blanket and trinket stalls, whence you can look into a rocky cove. The rocks often have Pelagic and Brandt's cormorants, both American and Black oystercatchers (and intergrades), Wandering Tattler (winter), Black Turnstone (winter), and Surfbird (winter). Looking offshore you may see streams of **Black-vented Shearwaters** passing by and, with luck (and a telescope), perhaps Cassin's and Rhinoceros auklets.

Estero Punta Banda Bird List
(including Punta Banda village): 202 sp.
R: rare; V: vagrant (not to be expected). †See pp. 5–7 for notes on English names.

Red-throated Loon
Pacific Loon
Common Loon
Pied-billed Grebe
Horned Grebe
Eared Grebe
Western Grebe
Clark's Grebe
Brown Pelican
Double-crested Cormorant
Pelagic Cormorant (R)
Great Blue Heron
Great Egret
Snowy Egret
Tricolored Heron
Reddish Egret
Cattle Egret
Green Heron
Black-crowned Night-Heron
White-faced Ibis
White-fronted Goose†
Snow Goose
Brant
Canada Goose
Green-winged Teal
Northern Pintail
Blue-winged Teal
Cinnamon Teal
Northern Shoveler
Gadwall
Eurasian Wigeon (R)
American Wigeon
Redhead
Greater Scaup (R)
Lesser Scaup
Surf Scoter
Common Goldeneye (R)
Bufflehead
Red-breasted Merganser
Ruddy Duck
Turkey Vulture
Osprey
White-tailed Kite
Northern Harrier
Sharp-shinned Hawk
Cooper's Hawk
Red-shouldered Hawk
Broad-winged Hawk (V)
Red-tailed Hawk
American Kestrel
Merlin
Peregrine Falcon
California Quail
Clapper Rail
Sora
Common Moorhen
American Coot
Black-bellied Plover
Snowy Plover
Semipalmated Plover
Killdeer
Black-necked Stilt
American Avocet
Greater Yellowlegs
Lesser Yellowlegs
Willet
Spotted Sandpiper
Whimbrel
Long-billed Curlew
Marbled Godwit
Ruddy Turnstone
Black Turnstone
Red Knot
Sanderling
Semipalmated Sandpiper (R)
Western Sandpiper
Least Sandpiper
Pectoral Sandpiper
Dunlin
Ruff (V)
Short-billed Dowitcher
Long-billed Dowitcher
Common Snipe
Wilson's Phalarope
Red-necked Phalarope
Red Phalarope
Parasitic Jaeger
Bonaparte's Gull
Heermann's Gull
Mew Gull
Ring-billed Gull
California Gull
Herring Gull
Thayer's Gull (R)
Western Gull
Glaucous-winged Gull
Caspian Tern
Royal Tern
Elegant Tern
Common Tern
Forster's Tern
Least Tern
Black Tern (R)
Black Skimmer
White-winged Dove
Mourning Dove
Common Ground-Dove
Greater Roadrunner
Barn Owl
Burrowing Owl
Short-eared Owl
Anna's Hummingbird
Costa's Hummingbird
Rufous Hummingbird
Belted Kingfisher
Yellow-bellied Sapsucker (V)
Red-naped Sapsucker
Red-shafted Flicker†
Western Pewee†
Willow Flycatcher
Western Flycatcher†
Black Phoebe
Say's Phoebe
Vermilion Flycatcher
Ash-throated Flycatcher
Cassin's Kingbird
Western Kingbird
Horned Lark
Tree Swallow
N. Rough-winged Swallow
Bank Swallow
Cliff Swallow
Barn Swallow
Western Scrub Jay
Northern Raven†
Bushtit
Bewick's Wren
Northern House Wren†
Marsh Wren
Golden-crowned Kinglet (R)
Ruby-crowned Kinglet
Blue-grey Gnatcatcher
Swainson's Thrush
Hermit Thrush
Northern Mockingbird
California Thrasher
Red-throated Pipit (V)
American Pipit
Loggerhead Shrike
European Starling
Bell's Vireo
Cassin's Vireo†
Warbling Vireo
Yellow-green Vireo (V)
Tennessee Warbler (V)
Orange-crowned Warbler
Nashville Warbler
Virginia's Warbler (V)
Lucy's Warbler (V)
Northern Parula (V)
Yellow Warbler
Audubon's Warbler†
Myrtle Warbler†
Black-throated Grey Warbler
Townsend's Warbler
Hermit Warbler
Blackburnian Warbler (V)
Prairie Warbler (V)
American Redstart

Northern Waterthrush
MacGillivray's Warbler
Common Yellowthroat
Wilson's Warbler
Yellow-breasted Chat
Scarlet Tanager (V)
Western Tanager
Blue Grosbeak
Lazuli Bunting
Indigo Bunting (V)
Green-tailed Towhee
Spotted Towhee
California Towhee
Vesper Sparrow
Lark Sparrow
Savannah Sparrow
Large-billed Sparrow†
Grasshopper Sparrow
Song Sparrow
Lincoln's Sparrow
White-throated Sparrow (R)
Golden-crowned Sparrow
White-crowned Sparrow
Oregon Junco†
Red-winged Blackbird
Tricolored Blackbird
Western Meadowlark
Brewer's Blackbird
Brown-headed Cowbird
Orchard Oriole (V)
Hooded Oriole
Bullock's Oriole
House Finch
Lesser Goldfinch
House Sparrow

Site 1.4: Laguna Hanson

(May 1989, June, October 1991, January 1994, January, October 1996, May, August 1997)

Introduction/Key Species

This large shallow lake lies amid a spectacular setting of conifer forests and boulder hillsides at the heart of the Parque Nacional Constitución de 1857. While the lake (which can be dry for months or even years) often has vast numbers of ducks in winter, all of the waterfowl species can be found more easily elsewhere. The lake does attract wintering Bald Eagles, however, and the surrounding conifer forests and chaparral en route have Mountain Quail, Mountain Chickadee, California [Plain] Titmouse, Pygmy Nuthatch, Wrentit, Western Bluebird, California Thrasher, Grey Vireo (summer), Black-chinned (summer), Sage, and Fox (winter) sparrows, Oregon [Dark-eyed] Junco, and Lawrence's Goldfinch.

Access

Laguna Hanson is about 55 mi (1.5 to 2 hours) east of Ensenada, whence it is reached via Route 3 (the highway to San Felipe). Head east from Ensenada on Route 3 until a signed dirt road to the left (north) at about Km Post 55.5 (i.e., 10.0 mi east of the paved left-turn into the town of Ojos Negros, and some 35 mi from Ensenada). From this turn it is about 21 mi to the lake, which also can be reached from Ojos Negros (see below) or via dirt roads off Route 2 to the north (about 40 mi), although high-clearance and/or four-wheel-drive vehicles may be needed for this latter access route. The route to Laguna Hanson from Ojos Negros involves 5.7 more mi on dirt roads but lacks one steeper grade found on the shorter (and quicker) route described below (see Birding Sites). For access via Ojos Negros, turn left off Route 3 on the paved road into Ojos Negros and turn right at the "downtown" crossroads at 1.2 mi, where the pavement ends, and mark zero. This starts as a graded dirt road through farmland: stay straight on the "main road" (some of the farm roads off to the left are at least as wide), crossing a cattle guard at 3.8 mi and bearing right and then left at 4.2 mi, after which the road leaves the valley and climbs into arid scrub and rocky hillsides; at 6.7 mi stay right at a fork, and at 9.7 mi stay straight/left (here the road described below joins in from the right); after this, follow the directions given below (see Birding Sites).

You can camp in the park (e.g., around the lake), for which a small fee may be payable (if there is a park warden on duty); you will need to take all of your supplies and be aware that it can be cold at night (down to below freezing, even through May!). On weekends and holidays there are sometimes numbers of campers and picnickers, but this doesn't affect the birding too much, except perhaps on summer weekends.

Birding Sites

Mark zero as you turn onto the Laguna Hanson dirt (sand) road off Route 3 (also signed for Rancho Gongora). The first 4 mi go through fairly open scrub, rocky hills, and chaparral, which have Greater Roadrunner, Phainopepla (mainly summer), Black-throated Sparrow, and Scott's Oriole (summer). After the cattle guard at 0.7 mi stay straight past a few minor tracks and stay straight at the right turn signed "Rancho Gongora 10 km" at 3.0 mi; after fording a small stream at 3.3 mi you come to a major

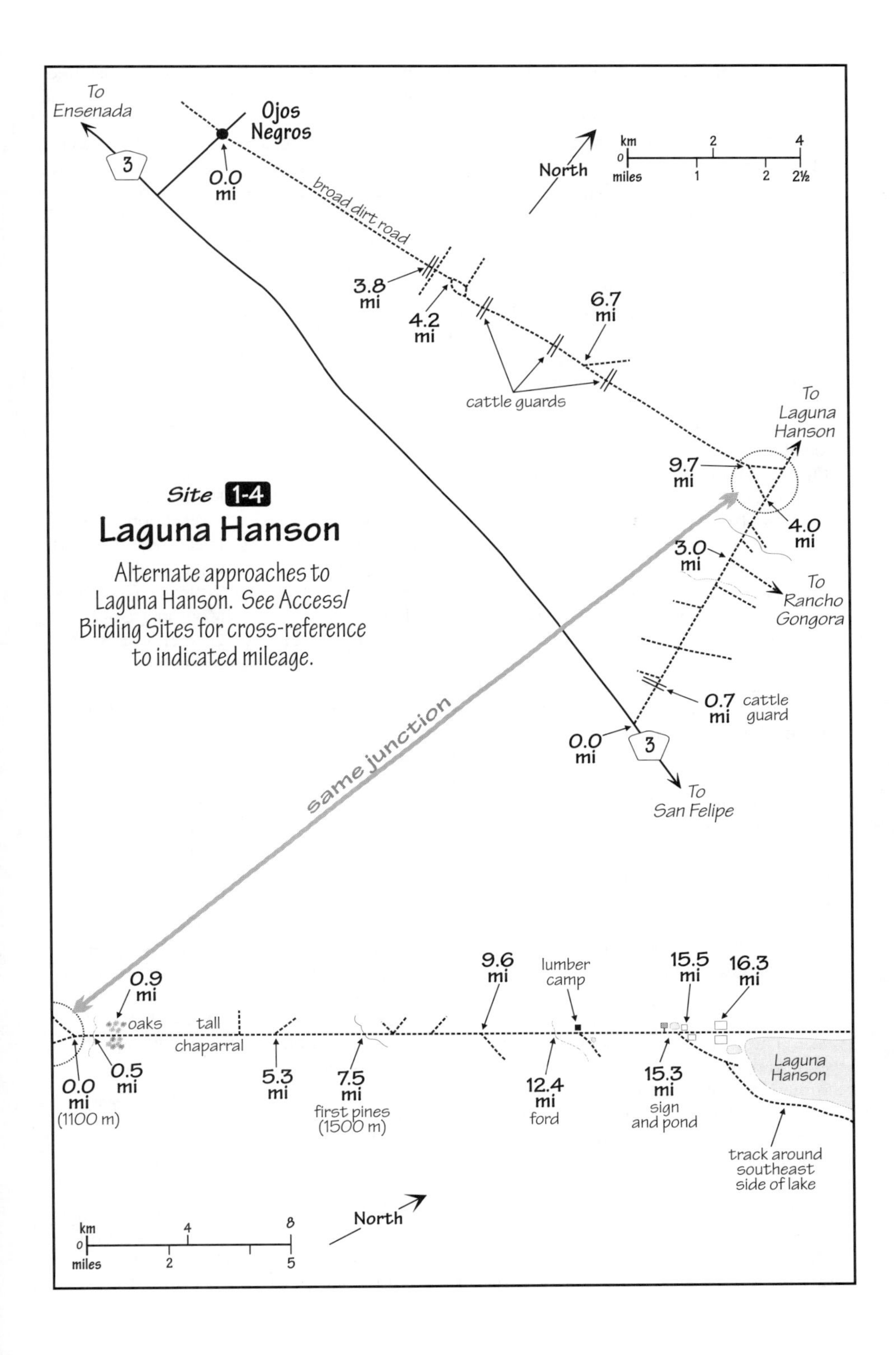

To Ensenada
3
Ojos Negros
0.0 mi
broad dirt road
North
km
0
2
4
miles
1
2
2½
3.8 mi
4.2 mi
6.7 mi
cattle guards
To Laguna Hanson
9.7 mi
4.0 mi
Site 1-4
Laguna Hanson
Alternate approaches to Laguna Hanson. See Access/ Birding Sites for cross-reference to indicated mileage.
3.0 mi
To Rancho Gongora
0.7 mi
cattle guard
0.0 mi
3
To San Felipe
same junction
0.9 mi
oaks
tall chaparral
9.6 mi
lumber camp
15.5 mi
16.3 mi
0.0 mi
(1100 m)
0.5 mi
5.3 mi
7.5 mi
first pines
(1500 m)
12.4 mi
ford
15.3 mi
sign and pond
Laguna Hanson
track around southeast side of lake
North
km
0
4
8
miles
2
5

fork at 4.0 mi: bear right, and stay straight in another 0.2 mi, where you should reset to zero (this left-hand Y junction goes back to Ojos Negros via 10 mi of dirt roads; see above, under Access).

Shortly after the Ojos Negros junction you cross a seasonal stream (0.5 mi) and go through a small oak grove (0.9 mi, 1200 m; California [Plain] Titmouse, Hutton's Vireo), after which the road runs through several miles of tall, dense chaparral (1300–1400 m; Wrentit, California Thrasher, Phainopepla [summer], Grey Vireo [summer], Black-chinned [summer] and Sage sparrows). After staying right at a fork at 5.3 mi you get to the first scattered pines and a semi-permanent small stream at 7.5 mi (1500 m; Mountain Quail, Mountain Chickadee, California [Plain] Titmouse, Fox and Golden-crowned sparrows [winter]), Lawrence's Goldfinch).

Stay on the main dirt road, which gets into pine woods (Pinyon Jay), passing a few tracks to the left off to ranches; keep left at a more major fork at 9.6 mi, and at 12.4 mi you ford a stream that can be fairly deep—if in doubt get out and check; this can be good for birds drinking, including Lawrence's Goldfinch, and Pinyon Jays often are in this area. Shortly after the ford stay straight/bear left (keeping on the main road) through a small lumber camp. At 15.3 mi you pass a sign for the Parque Nacional and go through a meadow (the seasonal ponds to the left here can have a good variety of ducks in winter) and come to some buildings at 15.5 mi (1600 m; Lawrence's Goldfinch is often along this last mile, at least in summer), beyond which is Laguna Hanson.

Depending on the angle of the sun, from these first buildings you can either stay left on the main dirt road, which skirts a small lake, passes a derelict-looking hostel and some park buildings at 16.3 mi, and then runs through open pine woods along the west side of Laguna Hanson; or you can bear right on a more minor-looking dirt road, which goes around the east side of Laguna Hanson and joins up with the main road at the north end of the lake. If you haven't yet run into them, flocks of Pinyon Jays usually can be found in the woods around the lake, and other species to look for include Williamson's Sapsucker (winter), Clark's Nutcracker (irregular), Red-breasted Nuthatch (irregular in winter), and Red Crossbill.

Bald Eagles can be seen anywhere around the lake (the warden usually knows if eagles are around), and there are numerous good vantage points for checking the waterfowl. If the lake is dry you can crisscross the lake bed via numerous tracks and check for pipits and anything else that might be there!

Laguna Hanson Bird List

(from Route 3): 133 sp. R: rare; V: vagrant (not to be expected). †See pp. 5–7 for notes on English names.

Pied-billed Grebe
Eared Grebe
Western Grebe
Great Blue Heron
Great Egret
White-fronted Goose†
Green-winged Teal
Mallard
Northern Pintail
Blue-winged Teal
Cinnamon Teal
Northern Shoveler
Gadwall
American Wigeon
Canvasback
Redhead
Lesser Scaup
Ring-necked Duck
Bufflehead
Ruddy Duck
Turkey Vulture
Bald Eagle
Sharp-shinned Hawk
Red-shouldered Hawk
Red-tailed Hawk
American Kestrel
Merlin
California Quail
Mountain Quail
American Coot
Killdeer
Black-necked Stilt
Greater Yellowlegs
Spotted Sandpiper
Pectoral Sandpiper
Dunlin
Long-billed Dowitcher
Red-necked Phalarope (R)
Bonaparte's Gull (R)
California Gull
White-winged Dove
Mourning Dove
Greater Roadrunner
Barn Owl
Western Screech-Owl
Great Horned Owl
Lesser Nighthawk
Common Poorwill
Anna's Hummingbird
Costa's Hummingbird
Acorn Woodpecker
Nuttall's Woodpecker
Hairy Woodpecker
Red-shafted Flicker†
Western Pewee†
Olive-sided Flycatcher
Hammond's Flycatcher
Western Flycatcher†
Ash-throated Flycatcher
Black Phoebe
Say's Phoebe
Western Kingbird
Horned Lark
Purple Martin

Tree Swallow
Violet-green Swallow
N. Rough-winged Swallow
Cliff Swallow
Barn Swallow
Western Scrub Jay
Pinyon Jay
Clark's Nutcracker
Northern Raven†
Mountain Chickadee
California Titmouse†
Bushtit
Red-breasted Nuthatch
White-breasted Nuthatch
Pygmy Nuthatch
Rock Wren
Canyon Wren
Bewick's Wren
Northern House Wren†
Ruby-crowned Kinglet
Western Bluebird
Swainson's Thrush
American Robin
Wrentit
Northern Mockingbird
California Thrasher
American Pipit
Phainopepla
Loggerhead Shrike
European Starling
Grey Vireo
Cassin's Vireo†
Hutton's Vireo
Warbling Vireo
Orange-crowned Warbler
Yellow Warbler
Audubon's Warbler†
Myrtle Warbler†
Black-throated Grey Warbler
Townsend's Warbler
Hermit Warbler
Palm Warbler (V)
Wilson's Warbler
Western Tanager
Black-headed Grosbeak
Green-tailed Towhee
Spotted Towhee
California Towhee
Chipping Sparrow
Black-chinned Sparrow
Lark Sparrow
Black-throated Sparrow
Sage Sparrow
Slaty Fox Sparrow†
Sierran Fox Sparrow†
Song Sparrow
Lincoln's Sparrow
Golden-crowned Sparrow
White-crowned Sparrow
Oregon Junco†
Western Meadowlark
Brewer's Blackbird
Brown-headed Cowbird
Bullock's Oriole
Scott's Oriole
Cassin's Finch
House Finch
Red Crossbill
Pine Siskin
Lesser Goldfinch
Lawrence's Goldfinch

Site 1.5: Ojos Negros/Leyes de Reforma

(December 1983, May 1989, June, October 1991, January 1993, 1994, January, October 1995, January, October 1996, May 1997, February 1998)

Introduction/Key Species

The Ojos Negros agricultural valley and some lakes to the south along Route 3 can provide good winter birding, especially for raptors and waterfowl. Depending on water levels (the lakes have been increasingly dry in recent years), a variety of ducks and blackbirds also nests locally in the area.

Birds of interest include Tundra Swan (rare in winter), Canada Goose (winter), Common Merganser (winter), Ferruginous Hawk (winter), Golden Eagle, Merlin (winter), Prairie Falcon (winter), Virginia Rail, Mountain Plover (winter), Mountain Bluebird (winter), Chestnut-collared Longspur (winter), and Tricolored and Yellow-headed (irregular in summer) blackbirds.

Access/Birding Sites

These areas are 45 minutes to 1.5 hours east of Ensenada along Route 3 (the San Felipe highway). After winding up through the oak and chaparral on the climb out of Ensenada, the highway descends into the large, level valley surrounding the small farming town of **Ojos Negros** about 24 mi (forty-five minutes to an hour) from Ensenada. The paved left turn into Ojos Negros is well signed just past Km Post 39. Turn left here and it is 1.2 mi to "downtown" Ojos Negros and a main crossroads. From here you can turn right on a graded dirt road that runs through open fields for several miles, and you can also take graded roads branching off to the left (e.g., at 0.3, 2.6, and 3.0 mi) through more of the same farmland. Exploring along these farm roads in winter should produce Ferruginous Hawk, Merlin, Prairie Falcon, and blackbird flocks including Tricoloreds. The reed-fringed ditches and scattered farm ponds can be good for ducks (Eurasian Wigeon and Hooded Merganser have been seen!) and rails (Virginia, Sora).

Continuing south on Route 3 toward San Felipe, marking zero at the paved road into Ojos Negros, you pass the left turn to Laguna Hanson at 10.0 mi, go through a belt of pinyon-juniper scrub, and then come into more farmland as you approach **Ejido Heroes de la Independencia**, the center of which is at 32.4 mi (Km Post 91). On the west side of the village, a half mile before the church, is a dirt road off to the right (south), signed to Campamento SCT. Turn right here, then left on a major dirt road at 0.25 mi, and take the first right at 0.3 mi (passing the school on your left). Stay straight on the main dirt road, which bends right (seemingly through someone's yard!) and comes up to the bank of a small reservoir at 0.8 mi. This has been mostly dry in the last few years, but one good, wet winter could easily change things. You can walk left (toward the dam and a reed-choked secondary pond that can have nesting Red-winged, Tricolored, and Yellow-headed blackbirds) or right along the banked-up south side of the reservoir (which has held some good birds, including Common Merganser and Heermann's Gull). During migration the willows at the creek inflow off to the right can be good for flycatchers, warblers, and so on, and California [Plain] Titmouse occurs here also.

Continuing on Route 3, 2.5–3 mi south past Heroes there is a large shallow lake to the east of the highway. Much of the lake's east end can be viewed easily from the highway at 35.8 mi (**Km Post 97**), and/or you can park in a pulloff on the left (east) of the highway at 35.0 mi, get through the fence, and walk about 300 m to the dam and view the west end of the lake from there. This has been a site in winter for Tundra Swan, geese (best in late afternoon), a good variety of ducks (many of which also oversummer), and also for shorebirds during migration.

The last site along this stretch is a shallow, ephemeral lake north of the highway at **Ejido Leyes de Reforma** at 38.7 mi (Km Post 102). To view this lake, turn left (east) off Route 3 at 38.7 mi on a broad, graded dirt road that runs diagonally through the village, past a small, dilapidated church on the left, and at 0.6 mi you reach a fence adjacent to the north end of the lake. You can park here and scan, and also get through the fence and walk. Often this lake is dry; consequently, it is a good site in winter for Mountain Plover, Mountain Bluebird, and longspurs (Chestnut-collared and Lapland have been seen with the pipit and lark flocks).

Ojos Negros/Leyes de Reforma Bird List

141 sp. R: rare (not to be expected). †See pp. 5–7 for notes on English names.

Pied-billed Grebe
Eared Grebe
Western Grebe (R)
Double-crested Cormorant
Great Blue Heron
Great Egret
Snowy Egret
Green Heron
Black-crowned Night-Heron
Tundra Swan
White-fronted Goose†
Snow Goose
Ross' Goose (R)
Canada Goose
Green-winged Teal
Mallard
Northern Pintail
Blue-winged Teal
Cinnamon Teal
Northern Shoveler
Gadwall
Eurasian Wigeon (R)
American Wigeon
Canvasback
Redhead
Lesser Scaup
Ring-necked Duck
Bufflehead
Common Merganser (R?)
Ruddy Duck
Turkey Vulture
White-tailed Kite
Northern Harrier
Sharp-shinned Hawk
Cooper's Hawk
Red-tailed Hawk
Ferruginous Hawk
Golden Eagle
American Kestrel
Merlin
Peregrine Falcon
Prairie Falcon
California Quail
Virginia Rail
Sora
American Coot
Killdeer
Snowy Plover
Mountain Plover
Black-necked Stilt
American Avocet
Greater Yellowlegs
Solitary Sandpiper
Spotted Sandpiper
Western Sandpiper
Least Sandpiper
Baird's Sandpiper
Pectoral Sandpiper
Long-billed Curlew
Dunlin
Long-billed Dowitcher
Red Phalarope (R)
Heermann's Gull (R?)
Forster's Tern
Mourning Dove
Greater Roadrunner
Barn Owl
Great Horned Owl
Burrowing Owl
Lesser Nighthawk
Common Poorwill
Anna's Hummingbird
Red-naped Sapsucker
Red-shafted Flicker†
Western Pewee†
Hammond's Flycatcher
Grey Flycatcher
Western Flycatcher†
Vermilion Flycatcher
Black Phoebe
Say's Phoebe
Cassin's Kingbird
Horned Lark
Tree Swallow
Violet-green Swallow
N. Rough-winged Swallow
Bank Swallow
Cliff Swallow
Barn Swallow
Western Scrub Jay
Northern Raven†
California Titmouse†
Bewick's Wren
Northern House Wren†

Marsh Wren
Ruby-crowned Kinglet
Blue-grey Gnatcatcher
Western Bluebird
Mountain Bluebird
Hermit Thrush
Northern Mockingbird
Sage Thrasher
American Pipit
Cedar Waxwing
Phainopepla
Loggerhead Shrike
European Starling
Warbling Vireo
Orange-crowned Warbler
Yellow Warbler
Audubon's Warbler†
Townsend's Warbler
Hermit Warbler
MacGillivray's Warbler
Wilson's Warbler
Common Yellowthroat
Western Tanager
Black-headed Grosbeak
Lazuli Bunting
California Towhee
Chipping Sparrow
Brewer's Sparrow
Vesper Sparrow
Lark Sparrow
Black-throated Sparrow
Savannah Sparrow
Song Sparrow
Lincoln's Sparrow
White-crowned Sparrow
Oregon Junco†
Chestnut-collared Longspur
Red-winged Blackbird
Tricolored Blackbird
Yellow-headed Blackbird
Western Meadowlark
Brewer's Blackbird
Great-tailed Grackle
Brown-headed Cowbird
House Finch
Lesser Goldfinch
Lawrence's Goldfinch

Site 1.6: Mexicali to San Felipe

(December 1983, May 1989, June 1991, January 1994, September 1995, May 1997, February 1998)

Introduction/Key Species

Route 5 runs south along the western edge of the Mexicali Valley and Río Colorado delta to San Felipe, a small fishing port on the Gulf of California some 120 mi (2.5–3 hours) south of the Calexico/Mexicali U.S.A./Mexico border crossing.

Unless you're keen on your Mexico and/or Baja lists, there isn't much reason to visit this part of Baja, which is notoriously hot at any time of year. Most of the land between the Río Colorado and Río Hardy (usually referred to as the Mexicali Valley) has long since been converted to a vast area of intensive agriculture crossed by a maze of roads. Species of potential interest here include Common Pheasant (locally common), Gambel's Quail, and Abert's Towhee. Birds around San Felipe include **Least Storm-Petrel** (summer), **Yellow-footed Gull**, Gull-billed Tern (migrant), Le Conte's Thrasher, and Black-tailed and California gnatcatchers.

Access/Birding Sites

Mark zero as you cross into Mexico at Calexico/Mexicali and bear right immediately in response to signs for San Felipe and San Luis R.C.; stay on this six-lane divided highway past a major traffic-light junction at 4.0 mi (right to Tijuana). At 5.0 mi you come to a *glorieta* (traffic circle), where you bear off right to San Felipe and continue south out of Mexicali. San Felipe can also be reached via Route 3, which cuts across the peninsula from Ensenada, about 150 mi (3.5–4 hours) distant.

Any brushy thickets along Route 5 are likely to have Abert's Towhee and possibly Gambel's Quail. About 29 mi (by Km Post 39) south of Mexicali there is a paved road left (east) signed to Colonia V. Carranza. In 1.0 mi this road crosses the **Río Hardy**, where you can pull off and look up and down the river (Gull-billed Tern in spring/summer); the next 5 mi or so, through and beyond Ejido Durango, give a taste of what most of the valley is like, and it is possible to drive along reed-fringed ditches like the one to the right (south) at 5.1 mi after the Río Hardy crossing (breeding birds here include Least Bittern, Burrowing Owl, Marsh Wren, and Yellow-headed Blackbird); Abert's Towhee is common in this area.

Back on Route 5, continuing south from the Ejido Durango junction, you leave the agricultural areas behind in 5–6 mi and get into desert with tamarisk and mesquite along the Río Hardy off to your left (east). One spot worth checking is **Campo Mosqueda** (a recreational camp area), the left-hand turn to which is 9 mi south of the Durango junction, near Km Post 53. (Beware: there is a sign that says "Campo Mosqueda 3 km" only about 1 km before you turn off from Route 5). Turn left onto the dirt road signed for Campo Mosqueda and follow the road to the left, past other camps off to the right, then bear left into the gate to Campo Mosqueda at 1.3

mi from Route 5. You can park by the buildings and walk around the trees and oxbow lake. Resident species are few but include Cactus Wren and Abert's Towhee; Great Blue Herons nest here, and Hooded Mergansers have occurred in winter. The area can be good for migrants in autumn through spring.

South from here the highway runs through barren desert, passes the junction with Route 3 from Ensenada at 92.5 mi, and, in another 30 mi, enters San Felipe. This small fishing port on the Gulf of California is a popular weekend and holiday spot for Californians and Mexicans alike, offering beaches, sun, and fishing. As you come into town, stay straight at both *glorietas* and at 124.2 mi you'll hit the waterfront, where you should turn left on the one-way street; you can park here and scan the beach and offshore waters for loons and, especially when fishing boats are coming in, boobies and Magnificent Frigatebirds; also, mainly in windy conditions during spring and summer, **Least Storm-Petrels** can be seen close inshore off the beaches. Other species of interest that can be seen here include Laughing, Heermann's, and **Yellow-footed gulls** and Gull-billed and Elegant terns. In winter, the beach should be checked for gulls (which have included Western and Thayer's).

The road south from town to the airport is worth birding. From the waterfront, head back out 0.6 mi to the first *glorieta* and turn left (south); alternatively, turn right at the second *glorieta* when coming into town; reset to zero as you make the turn. At 2.5 mi you can turn left 250 m to the harbor (signed to Capitania de Puerto) and check for loons, grebes, gulls, and other waterbirds. If you stay south on the main road, there are several good views of the beaches and offshore waters, and the small river mouth at 3.5 mi can have gatherings of gulls. At 6.2 mi the road to the airport goes straight on while the road south along the Gulf goes left. The desert around this junction has Cactus Wren and Le Conte's Thrasher; the small dry wash right (west) of the road about 200 m south of the junction has both California and Gambel's quail and Black-tailed and California gnatcatchers!

Site 1.7: The Sierra San Pedro Mártir

(December 1983, May 1989, June, October 1991, January 1993)

Introduction/Key Species

Much of this area comprises a national park centered around the Picacho del Diablo, a peak rising to 3095 m in the Sierra San Pedro Mártir, and the presence of an astronomical observatory in the park helps keep the approach road in passable condition for most of the year. Habitats on the drive up from Route 1 start in cactus desert and farmland, grade through chaparral and oaks, and culminate in pine-fir forests with meadows and aspen groves around the observatory. This is a spectacular site, and on clear days you can look down to both the Pacific Ocean and the Gulf of California. For a Mexican birder, this is a "must," since the Sierra is home to species not found easily anywhere else in Mexico.

Birds of interest here include Mountain Quail, Northern Saw-whet Owl, Red-naped (winter) and Williamson's sapsuckers, Olive-sided Flycatcher (summer), Purple Martin (summer), Pinyon Jay, Mountain Chickadee, Pygmy Nuthatch, Townsend's Solitaire (winter), **Grey** and California **thrashers**, Bell's and Grey vireos (summer), Black-chinned (summer), Sage, and Sierran Fox [Fox] sparrows, Dark-eyed Junco, Cassin's Finch, Lawrence's Goldfinch, and Red Crossbill.

Access

To get to the Sierra San Pedro Mártir, drive south from Ensenada on Route 1 for about 75 mi (1.7–2.2 hours) to the town of Colonet. It's another 9.0 mi south on Route 1 from the Pemex station (where you should fill up with gas) on the north side of Colonet to the Puente San Telmo and the turnoff (east) on a graded dirt road up to the park. At the junction there should be a sign for Observatorio and/or Meling Guest Ranch as well as for the town of San Telmo, about 5.8 mi along the road. The road is unpaved and often teeth-gratingly washboarded but, unless there have been recent storms and rain, it is likely to be passable to the observatory in a regular car.

A guest ranch (Meling Ranch) along a wooded, permanent stream in the foothills outside the park makes a very comfortable (but expensive) base from which to explore the sierra, or you can camp. Take all of your supplies if you plan to camp, and be aware that nights can be cold (to below freezing), even in mid summer! Snow is regular on the higher peaks in winter.

Birding Sites

Highway to Meling Ranch: Mark the turn-off from Route 1 as zero. At about 2.6 mi there is a willow-edged stream beside the road to the right (Bell's Vireo [summer], Com-mon Yellowthroat, Song Sparrow) and hills with cacti on the left. **Grey Thrasher** and California Gnatcatcher (paler than the U.S. race) are fairly easy to see here—look for them especially as you get into the first hills, the thrashers in areas with cacti. Sage Thrasher occurs here in winter, and California Thrasher is resident. Also watch for Tricolored Blackbirds in the fields between Route 1 and San Telmo.

At 18.9 mi the road climbs out of agricultural fields and into hills covered in desert scrub and, higher up, chaparral. A stop anywhere between 25 and 29 mi is likely to produce Wrentit, Phainopepla (summer), Lazuli Bunting (summer), and Black-chinned (summer) and Sage sparrows. At 31.4 mi a sign points down the right-hand fork to Meling Ranch, a short distance from the "main road" and set amid cottonwoods along a stream. Continuing on the main road you cross the Meling Ranch stream at 32.6 mi. This riparian grove and surrounding desert (700 m) offers good birding at all seasons: Red-shouldered Hawk, Western Screech-Owl, Long-eared Owl (has bred, no recent records), Lesser Nighthawk (summer), Lewis' Woodpecker (irregular in winter, no recent reports), Red-naped Sapsucker (winter), Nuttall's Woodpecker, Bell's Vireo (summer), Yellow Warbler (summer), Yellow-breasted Chat (summer), and Fox and Golden-crowned sparrows (winter). Rarities found at Meling Ranch include Yellow-bellied Sapsucker, Eastern Phoebe, and Blackpoll Warbler.

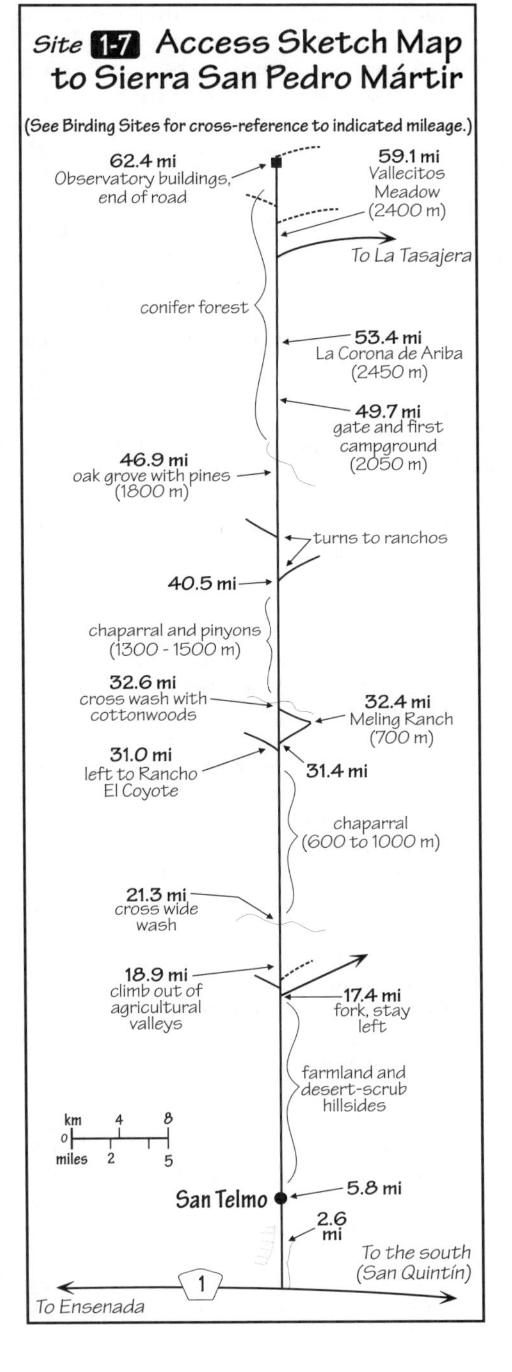

Meling Ranch to the Observatory. The road climbs into pinyons at 35.4 mi (900 m), and Grey Vireos (summer) become common. The fork (stay left/straight) at 40.5 mi (1250 m) is a good point to stop in summer for California Thrasher, Grey Vireo, Black-chinned Sparrow, and Scott's Oriole; it is also about where Mountain Quail start (California Quail occur all the way up into the park). On a stretch of remnant, potholed pavement at 46.9 mi (1800 m) you pass through a grove of oaks by a stream (California [Plain] Titmouse, Warbling Vireo [summer], Western Tanager [summer], Black-headed Grosbeak [summer], Lawrence's Goldfinch). The park entry gate and first campground are at 49.7 mi (2050 m), whence it is 12.7 mi up to the observatory buildings. Birds around the first campground include Mountain Quail, Purple Martin (summer), Violet-green Swallow, Olive-sided Flycatcher (summer), Western Pewee (summer), Mountain Chickadee, Pygmy Nuthatch, Western Bluebird, and Dark-eyed Junco (the resident breeding race *townsendi* looks closer to Pink-sided than to typical Oregon).

From the entry gate the road climbs steadily to La Corona de Arriba at 53.4 mi (2450 m) and then winds up and down through conifer forests to Vallecitos Meadow at 59.1 mi (2400 m). This stretch is good for Northern Saw-whet Owl (a recent discovery; Erickson et al. 1994), Williamson's Sapsucker, Townsend's Solitaire (winter), Sierran Fox [Fox] Sparrow (recently found nesting here; Erickson and Wurster 1998). From Vallecitos Meadow the road climbs to some buildings at 62.4 mi, where you can park and walk (1.2 mi) to the observatory (2830 m); this walk is the best stretch for Cassin's Finch. From the observatory you can look southeast down to a grove of aspens (take care finding a path), the only place in Mexico where Dusky Flycatcher, Hermit Thrush, and Green-tailed Towhee are known to nest. Around Vallecitos Meadow look for roaming flocks of Pinyon Jays (if they haven't already found you). Unfortunately, the high-elevation meadows are overgrazed by cattle, so flowers for Calliope Hummingbirds (summer) are hard to find.

Sierra San Pedro Mártir Bird List

(Route 1 to observatory): 147 sp. R: rare; V: vagrant (not to be expected). †See p. 5–6 for notes on English names.

Pied-billed Grebe
Great Blue Heron
Great Egret
Snowy Egret
Black-crowned Night-Heron
Green-winged Teal
Mallard
Cinnamon Teal
Turkey Vulture
White-tailed Kite
Northern Harrier
Sharp-shinned Hawk
Cooper's Hawk
Red-shouldered Hawk
Red-tailed Hawk
Golden Eagle
American Kestrel
Prairie Falcon
California Quail
Mountain Quail
Sora
American Coot
Killdeer
Black-necked Stilt
Greater Yellowlegs
Long-billed Dowitcher
Band-tailed Pigeon
White-winged Dove
Mourning Dove
Common Ground-Dove
Greater Roadrunner
Barn Owl
Great Horned Owl
Burrowing Owl
Western Screech-Owl
Lesser Nighthawk
Common Poorwill
White-throated Swift
Black-chinned Hummingbird
Anna's Hummingbird
Costa's Hummingbird
Calliope Hummingbird (R?)
Rufous Hummingbird
Belted Kingfisher
Acorn Woodpecker
Ladder-backed Woodpecker
Nuttall's Woodpecker
Hairy Woodpecker
Red-naped Sapsucker
Williamson's Sapsucker
Red-shafted Flicker†
Western Pewee†
Olive-sided Flycatcher
Dusky Flycatcher
Western Flycatcher†
Vermilion Flycatcher
Ash-throated Flycatcher
Black Phoebe
Say's Phoebe
Cassin's Kingbird
Western Kingbird
Horned Lark
Violet-green Swallow
N. Rough-winged Swallow
Cliff Swallow
Barn Swallow
Western Scrub Jay
Pinyon Jay
Northern Raven†
Mountain Chickadee
California Titmouse†
Bushtit
White-breasted Nuthatch
Pygmy Nuthatch
Cactus Wren
Rock Wren
Canyon Wren

Bewick's Wren
Northern House Wren†
Ruby-crowned Kinglet
Blue-grey Gnatcatcher
California Gnatcatcher
Western Bluebird
Townsend's Solitaire
Swainson's Thrush
Hermit Thrush
American Robin
Wrentit
Northern Mockingbird
Grey Thrasher
California Thrasher
American Pipit
Cedar Waxwing
Phainopepla
Loggerhead Shrike
European Starling
Bell's Vireo
Grey Vireo
Cassin's Vireo†
Warbling Vireo
Tennessee Warbler (V)
Orange-crowned Warbler
Yellow Warbler
Audubon's Warbler†
Black-throated Grey Warbler
Townsend's Warbler
Hermit Warbler
Blackpoll Warbler (V)
Wilson's Warbler
Common Yellowthroat
Yellow-breasted Chat
Western Tanager
Black-headed Grosbeak
Blue Grosbeak
Lazuli Bunting
Green-tailed Towhee
Spotted Towhee
California Towhee
Rufous-crowned Sparrow
Chipping Sparrow
Black-chinned Sparrow
Lark Sparrow
Black-throated Sparrow
Sage Sparrow
Sierran Fox Sparrow†
Slaty Fox Sparrow†
Song Sparrow
Lincoln's Sparrow
Golden-crowned Sparrow
White-crowned Sparrow
Oregon Junco†
Pink-sided Junco†
Western Meadowlark
Red-winged Blackbird
Tricolored Blackbird
Brewer's Blackbird
Great-tailed Grackle
Brown-headed Cowbird
Hooded Oriole
Bullock's Oriole
Scott's Oriole
Cassin's Finch
Purple Finch
House Finch
Red Crossbill
Pine Siskin
Lesser Goldfinch
Lawrence's Goldfinch
House Sparrow

Site 1.8: Bahía San Quintín/Laguna Figueroa

(December 1983, May 1989, June 1991, January 1994, January, October 1996, August, September 1997, February 1998)

Introduction/Key Species

Bahía San Quintín, a largely unspoiled and "wild" bay/estuary complex, is famous as a major wintering site for Black Brant and also hosts a good variety of other waterfowl, loons, grebes, gulls, and so forth, mainly in winter. Note that hunting is allowed here, so waterfowl are generally wary, and that most of the same species can be seen at Estero Punta Banda (Site 1.3). Nonetheless, if you're in the area it can be worth visiting. Black Rails have been found recently in the salt marshes here, and the only specimen of Nelson's Sharp-tailed Sparrow for Mexico is from San Quintín.

Laguna Figueroa, a short way north of San Quintín, is a shallow saline lagoon that often has a good variety of shorebirds (in winter including up to or more than 100 Snowy Plovers, Stilt Sandpipers), and the surrounding plains have a wintering flock of Mountain Plovers and large pipit and Horned Lark flocks, among which Red-throated Pipit and Lapland Longspur have been found.

Access/Birding Sites

The town of Valle de San Quintín is about 120 mi (2.5–3 hours) south of Ensenada, along Route 1. There are a few motels along the highway and others around the bay, and it is also possible to camp at trailer parks or out in the desert or along the shore.

There are two main ways to view the bay: from the north, whence a graded but washboarded gravel road runs around the peninsula, which projects into the bay, and ends up on the outer (Pacific) coast; and from the south, via a hard-sand road that runs out the spit across the south side of the bay to its mouth.

The North Side. Coming from the north, continue south through San Quintín on Route 1 for about 2 mi to a large military camp on your right (west) at the north side of the settlement of Lázaro Cárdenas. Immediately south of the military camp and just where a stretch of divided highway starts (1.7 mi south of Puente San Quintín and 0.3 mi north of the only traffic light in Lázaro Cárdenas) is a graded dirt road off to the right (west). Turn here, and in 3.7 mi you cross a creek and go through a small village, where

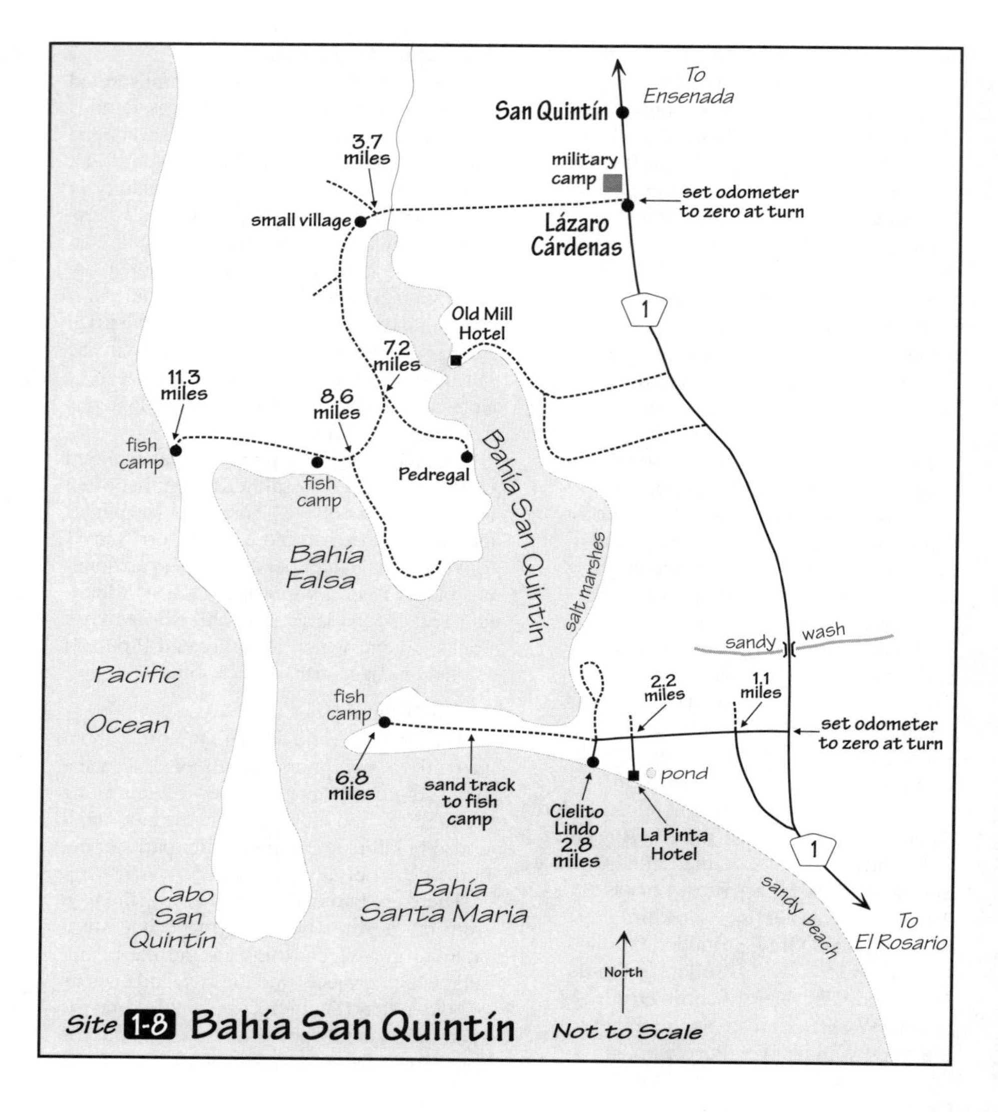

Site 1-8 Bahía San Quintín

you fork left and go alongside the northernmost arm of the bay on your left. Stay left/straight at a minor fork at 5.0 mi, and at 7.2 mi, with a rocky volcanic hill straight ahead, you have a choice: you can stay right on the main road (see below) or fork left onto a minor sand road, bearing left again at a minor junction to reach salt marshes along the bay at an area known as Pedregal. This whole northernmost arm can be good for ducks (including wigeon, which should be checked for Eurasian) and Brant, and the salt marshes have Clapper Rail.

If you stay right at the Pedregal junction, another sand road forks off left at 8.6 mi, giving good views into the bay. Stay right on the main gravel road; you go along the north edge of the northwest arm of the bay and, to all intents and purposes for most vehicles and birders, end up at 11.3 mi in a small fishing village on the outer coast, where long Pacific breakers roll in on the rocky coast. There is also a nice view offshore to Isla San Martín.

The South Side. From San Quintín, continue south on Route 1. At 1.1 mi south of the river bridge on the south side of the San Quintín/Lázaro Cárdenas strip of development, and near Km Post 11, there is a paved (if potholed) road to the right, which should be signed to the Hotel La Pinta. Turn right here, go through a crossroads at 1.1 mi, and in another 1.1 mi a paved left-hand turn goes 0.9 mi to the Hotel La Pinta. The beach off this hotel can have large numbers of Red-throated Loons and scoters (including White-winged) in winter, and the saline ponds on the left just before the hotel are worth a look for ducks and shorebirds; in summer, nesting birds here include Snowy Plover, Black-necked Stilt, American Avocet, and Forster's Tern. The lines of trees bordering the road near the La Pinta junction can be good in autumn and winter for migrant warblers.

Staying straight at the La Pinta junction, the road turns to dirt, and in 0.6 mi there is a left turn to the Cielito Lindo Motel and trailer park; stay straight, following the main, hard sand road (ignore minor tracks off to the right unless you want to explore the salt marsh, where there is a chance of Black Rail), and 4.0 mi after the Cielito Lindo turn you come to a fishing camp near the end of the spit across the south side of the bay. This is a gathering point for gulls, terns, and Black Skimmers, while the lagoon outflow here is often good for loons, grebes, and sea ducks.

Laguna Figueroa. Coming from the north (Ensenada, etc.), about 2 mi south of Colonia Vicente Guerrero you crest a rise on Route 1, with a view down over the San Quintín plain; off to the right you may see Laguna Figueroa, just behind the beach. On the left (east) look for Km Post 175 and a sign for a right turn to Ejido Zarahemla. Turn right here on a long, straight dirt road that runs toward the coast. Stay straight, passing right-hand turns into the *ejido*, or cooperative farm (where the flowering eucalyptus trees can be good in late winter and spring for hummingbirds), and in 2.5 mi you hit the north end of Laguna Figueroa. In the bushy salt flats between the *ejido* and lagoon watch in winter for Mountain Plovers and pipit and lark flocks. The main lagoon, off to your left (south), is often mostly dry, but the smaller lagoon on the right usually has some water and birds. The main road bends left between the two lagoons and ends shortly at the beach dunes. You can also take minor tracks that wend around through the flats to the north and south, but beware that the ground here can be very soft, despite a hard-caked top layer, and it is easy to get your vehicle stuck (even with four-wheel drive).

Coming north from San Quintín, the Laguna Figueroa turn is on the left, 8.4 mi north of the Restaurant Misión Santa Isabel, which is on the east side of the highway near the north edge of San Quintín. The turn is just as you start to climb out of the plain and from this direction is also signed to Ejido Zarahemla.

Bahía San Quintín/Laguna Figueroa Bird List

161 sp. R: rare; V: vagrant (not to be expected). †See pp. 5–7 for notes on English names.

Red-throated Loon
Arctic Loon (V)
Pacific Loon
Common Loon
Pied-billed Grebe
Horned Grebe
Eared Grebe
Western Grebe
Clark's Grebe
Sooty Shearwater
Black-vented Shearwater
Brown Pelican
Double-crested Cormorant
Brandt's Cormorant
Magnificent Frigatebird (R)
Great Blue Heron
Great Egret
Snowy Egret
Little Blue Heron
Tricolored Heron
Reddish Egret
Cattle Egret
Black-crowned Night-Heron
Brant
Green-winged Teal
Mallard
Northern Pintail
Cinnamon Teal
Northern Shoveler
Gadwall
American Wigeon
Redhead
Canvasback
Lesser Scaup
Black Scoter (R)
Surf Scoter
White-winged Scoter
Common Goldeneye
Bufflehead
Red-breasted Merganser
Ruddy Duck
Turkey Vulture
Osprey
White-tailed Kite
Northern Harrier
Sharp-shinned Hawk
Cooper's Hawk
Red-shouldered Hawk
Swainson's Hawk (V)
Red-tailed Hawk
Ferruginous Hawk

Golden Eagle
American Kestrel
Peregrine Falcon
Prairie Falcon
California Quail
Clapper Rail
Common Moorhen
American Coot
Black-bellied Plover
Snowy Plover
Semipalmated Plover
Mountain Plover
Killdeer
Black-necked Stilt
American Avocet
Greater Yellowlegs
Lesser Yellowlegs
Willet
Spotted Sandpiper
Whimbrel
Long-billed Curlew
Marbled Godwit
Ruddy Turnstone
Black Turnstone
Red Knot
Sanderling
Western Sandpiper
Least Sandpiper
Dunlin
Stilt Sandpiper (R?)
Short-billed Dowitcher
Long-billed Dowitcher
Common Snipe
Wilson's Phalarope
Red-necked Phalarope
Parasitic Jaeger
Bonaparte's Gull
Heermann's Gull
Mew Gull (R)
Ring-billed Gull
California Gull
Herring Gull
Thayer's Gull (R)
Western Gull
Glaucous-winged Gull
Caspian Tern
Royal Tern
Elegant Tern
Forster's Tern
Black Skimmer
Mourning Dove
Common Ground-Dove
Barn Owl
Burrowing Owl
Lesser Nighthawk
Anna's Hummingbird
Costa's Hummingbird
Allen's Hummingbird
Belted Kingfisher
Willow Flycatcher
Western Flycatcher†
Black Phoebe
Say's Phoebe
Vermilion Flycatcher
Cassin's Kingbird
Horned Lark
Tree Swallow
Cliff Swallow
Barn Swallow
Northern Raven†
Cactus Wren
Rock Wren
Bewick's Wren
Northern House Wren†
Marsh Wren
Ruby-crowned Kinglet
California Gnatcatcher
Northern Mockingbird
Sage Thrasher
California Thrasher
Yellow Wagtail (V)
American Pipit
Loggerhead Shrike
European Starling
Bell's Vireo
Orange-crowned Warbler
Yellow Warbler
Myrtle Warbler†
Audubon's Warbler†
Townsend's Warbler
Blackpoll Warbler
Common Yellowthroat
Blue Grosbeak
California Towhee
Sage Sparrow
Vesper Sparrow
Savannah Sparrow
Large-billed Sparrow†
Song Sparrow
Lincoln's Sparrow
White-crowned Sparrow
Bobolink (V)
Red-winged Blackbird
Yellow-headed Blackbird
Brewer's Blackbird
Western Meadowlark
Brown-headed Cowbird
Bullock's Oriole
House Finch
Lesser Goldfinch
House Sparrow

MID PENINSULA

Site 1.9: San Quintín to La Paz

After passing through El Rosario, about 36 mi (35–50 minutes) south of San Quintín, Route 1 cuts inland through the desert, leaving behind the northwest corner of Baja with its distinctive avifauna. From El Rosario it is a long, hot, but often spectacular drive of about 700 mi (16–20 hours) to La Paz, capital of the state of Baja California Sur, and another 100 mi (2.5–3 hours) to Cabo San Lucas at the southern tip of the peninsula. In terms of birds there is no real reason to make this drive except to get to the Cape District, although there always are birds along the way.

The first spot that can be worth a stop is the small oasis of **Cataviña** (December 1983, May 1989, June, October 1991, October 1995, 1996), which is about 110 mi (2.5–3 hours) south of San Quintín, near Km Post 175 (which measures from zero at San Quintín). There is one fairly expensive hotel (La Pinta), a trailer park and camping area at the nearby Rancho Santa Ynez (see below), and a Pemex station. At 0.3 mi south of the hotel and Pemex the highway fords a permanent wash, where you could camp (truck traffic can be noisy, though). The wash acts as a magnet for migrants and provides nesting habitat for Bell's Vireo (summer) and Song Sparrow. If you're passing through during migration periods it's always worth a stop to check for migrants and vagrants, for which Cataviña/Santa Ynez are famous: among the 100+ migrant species recorded here are Yellow-bellied, Red-naped, and Red-breasted sapsuckers, Clark's Nutcracker, Varied Thrush, Olive-backed and Red-throated pipits, Philadelphia Vireo, Dusky Warbler, Scarlet Tanager, and Harris' Sparrow!

A good approach to birding Cataviña is to walk upstream (left, or east) from the highway about 100 m, to below the tall palms and dense vegetation by a small *rancho* (ranch) and spring, and to walk downstream about 500–700 m, to where the taller vegetation peters out; from the downstream walk you can cut back up to the right (north) along the small drainage to the back of the La Pinta hotel. There is no single, easy-to-follow trail along the wash, and you might have to get your feet wet at some points.

At 0.4 mi south of the stream crossing a paved road forks left off Route 1 and goes to the trailer campsite at **Rancho Santa Ynez** in 0.6 mi. The trees around the ranch can be good for migrants; ask permission (usually given freely) to look around the yard before you wander too far. The water drips here are worth watching if there are lots of migrants around. **Grey Thrasher** occurs at both Cataviña and Santa Ynez.

Continuing 65 mi (1.5 hours) south of Cataviña, you reach the junction to the Gulf of California shores at **Bahía de Los Angeles** (June 1991), 42 mi (an hour) away by paved highway. In summer it is possible to arrange boat trips from Bahía de Los Angeles out to **Isla Rasa** (ask at the waterfront and, as always, agree on a price beforehand), with its spectacular nesting colonies of tens of thousands of Heermann's Gulls and Elegant and Royal terns. The boat trip (in an open skiff) to Isla Rasa takes a little under two hours, and, once there, you can take a tour with Mexican biologists who spend the nesting season on the island. Allow a full half day (6–7 hours) for the trip. Species to look out for en route to/from the island include Pink-footed, Sooty, and **Black-vented shearwaters, Least Storm-Petrel**, and **Craveri's Murrelet**, plus the usual Blue-footed and Brown boobies. The Isla Rasa trip is well worth doing and is also a good bet for seeing whales.

From the Bahía de Los Angeles junction it is another 80 mi (1.5–2 hours), through increasingly bleak desert, to **Guerrero Negro** (December 1982, 1983, June 1986, June, October 1991, October 1995, 1996), gateway to the state of Baja California Sur (note the change from Pacific to Mountain time and make sure you eat any fresh fruit before the border check station, or you'll be asked to give it up). Guerrero Negro is best known as a base for whale-watching trips (from shore or by boat) in the nearby Laguna Ojo de Liebre (Scammon's Lagoon), and Ospreys nest commonly atop poles in and around town. There are several hotels and restaurants in town, or you could camp out in the nearby desert, which is generally bleak and cool under a sheet of coastal cloud, reflecting the cold offshore Pacific waters not far away.

Coming from the north, you cross the state line at an imposing if somewhat bizarre monument, 2.0 mi after which there is a well-signed, paved right turn into Guerrero Negro. Mark zero at this turn. In 3.0 mi, after a mile or so of busy main street, the road hits a green park just past an entrance to the saltworks on your left. This park is maintained by watering and often has good numbers of migrants in spring and autumn. Species recorded here include Common Ground-Dove, Anna's and Costa's hummingbirds, numerous migrant warblers (American Redstart is often quite common in autumn), and Summer Tanager.

The road along the near (north) edge of the park continues (as well-graded dirt road) west through salt flats, ponds, and some excellent salt marsh and ends in about 6.5 mi at a disused wharf overlooking a large tidal lagoon, the Estero de San José. Birds along the drive and visible from the dock include loons, Little Blue and Tricolored herons, Reddish Egret, Yellow-crowned Night Heron, Brant (winter), Common Goldeneye (winter), Red-breasted Merganser (winter), Clapper Rail, Snowy and Wilson's plovers, and Least Tern (summer).

Another area near town that's worth birding is along Route 1 between 1.0 and 7.0 mi south of the Guerrero Negro turnoff. In this stretch, low hills with sparse bushes relieve the monotony of vast salt flats; birds around these roadside hills include Cactus Wren, California Gnatcatcher, Le Conte's Thrasher, Loggerhead Shrike, and Sage Sparrow.

As you head south on Route 1 from Guerrero Negro, the desert becomes more vegetated after Vizcaino, and Harris' Hawks and

Crested Caracaras become more frequent. About 88 mi (1.7–2.2 hours) south of Guerrero Negro you pass **San Ignacio** (December 1982, 1983, June 1991), a large oasis of lush date palms that provide an abrupt contrast to the surrounding desert. If you turn right (south) toward San Ignacio, which is 1.6 mi off the highway, in 0.4 mi you cross a causeway between two parts of a dammed, reed-fringed, freshwater lagoon where birds include Least Bittern and **Belding's Yellowthroat** (and migrant Commons in winter). Note that this is the dull, northern race (*goldmani*) of Belding's, which looks somewhat intermediate between the brighter southern nominate race and some races of Common Yellowthroat. Other species possible here and in the date "forest" include Western and Clark's grebes (winter), Lesser Nighthawk, Red-naped Sapsucker (winter), Purple Martin (summer), and Hooded and Scott's orioles.

Forty-five miles (1–1.5 hours) beyond the San Ignacio turnoff, Route 1 drops down to the shores of the Gulf of California at **Santa Rosalia** (where the ferry to Guaymas, Sonora, docks). Blue-footed Boobies can often be seen off the beach here, **Yellow-footed Gulls** are common, and in summer there is a chance of seeing **Craveri's Murrelets** feeding inshore.

About 38 mi (0.7–1 hour) south of Santa Rosalia is the small oasis town of **Mulege** (December 1982, June 1991), which sits along the north side of a palm-fringed river about 2 mi upstream from the Gulf. There are several hotels, trailer parks, and restaurants in town and along the nearby beach.

One area for birding is along the south side of the river en route to the old **Misión Santa Rosalia de Mulege**, which provides a good overlook of the area. To get to the mission requires negotiating Mulege's one-way system, or you could just walk the half mile from the *zócalo* (plaza). To reach the mission, turn off Route 1 on the paved road into Mulege, bear right on the one-way system, cross Calle Zaragoza (which leads to the river, if you're walking), and at the next right cut sharply back to the *zócalo*, at the far side of which you turn left on Zaragoza (now two-way), follow it under the Route 1 highway bridge over a small bridge across the river, and turn sharp right on the dirt road that runs along the river and up to the mission in about a quarter mile. You can park at the mission, which is a good place from which to watch frigatebirds coming in to drink, to check the river for Least Grebes, and to scan for Zone-tailed Hawks (in winter, at least). You can follow paths upstream and downstream from below the mission, where birds include Gila Woodpecker, Gilded Flicker, Vermilion Flycatcher, Phainopepla, **Belding's Yellowthroat** (still of the race *goldmani*), and Song Sparrow.

Some 84 mi (1.7–2.2 hours) south of Mulege, and backed by the peaks of the Sierra de la Giganta, is **Loreto** (December 1982, June 1991), a resort town on the shores of the Gulf of California. From the south end of the waterfront road in Loreto you can walk along the beach a short distance to a river mouth which, depending on tide and season, can have a variety of waterbirds, including Reddish Egret, Snowy and Wilson's plovers, **Yellow-footed Gull,** and Elegant Tern.

The dirt road inland (west) to the **Misión San Javier**, off Route 1 just south of Loreto, can be impassable, but it is usually possible to drive at least a few miles to the first broad desert wash, where wintering birds include Grey Flycatcher and Grey Vireo. Other birds here include Gila and Ladder-backed woodpeckers, Western Scrub Jay, Verdin, California Gnatcatcher, and Northern Cardinal.

From Loreto, Route 1 parallels the coast for a ways before climbing inland through the rugged face of the spectacular Sierra de la Giganta and then dropping through cactus desert to the Magdalena Plain. In the 1960s this plain was opened up to intensive agriculture, centered on the town of Ciudad Constitución, 89 mi (2–2.5 hours) from Loreto. With the habitat change have come avian colonizers such as White-tailed Kite and Western Meadowlark. South of Ciudad Constitución you get back into desert after a while, and it is 130 miles (2.5–3 hours) from Ciudad Constitución to La Paz, which is the state capital of Baja California Sur and offers a wide range of accommodations and restaurants.

With the completion of a paved road between Todos Santos and Cabo San Lucas, there is now a choice of routes from **La Paz to Los Cabos**: Route 1 (the old highway, longer and more winding) through the eastern foothills of the Cape District mountains (140 mi to Cabo San Lucas; 3–4 hours), and Route 19, via Todos Santos (100 miles to Cabo San Lucas; 1.7–2.2 hours). Birding areas at the tip are covered under Los Cabos (Site 1.10).

If you're driving from La Paz to Los Cabos, the older route offers some good birding—for example, the thorn scrub (green in late summer through autumn) around and between the villages of El Triunfo and San Antonio, 32–37 mi (45 minutes to 1.2 hours) south of La Paz. Birds in this stretch (December 1982, 1983, February 1998) include **Xantus'** and Costa's **hummingbirds**, Gilded Flicker, Gila and Ladder-backed woodpeckers, Grey (winter), Western (winter), and Ash-throated flycatchers, Western Scrub Jay, Verdin, Cactus Wren, **Grey Thrasher**, Grey Vireo (winter), Orange-crowned, MacGillivray's, and Wilson's warblers (all winter), Western Tanager (winter), Northern Cardinal, Pyrrhuloxia, Black-headed Grosbeak (winter), Varied Bunting, Green-tailed (winter) and California towhees, and Black-throated, Clay-colored (winter), Black-chinned (winter), and Lark (winter) sparrows.

If you're driving via Todos Santos, stops anywhere in the desert can be productive. One spot that gets you off the highway is the signed dirt road toward Presa Santa Ynez: this road leaves Route 19 to the east (inland) near Km Post 42, about 5 mi north of Todos Santos and about 45 mi (45 minutes to an hour) south of La Paz. Taking this dirt road over the first low rise cuts down on traffic noise, and you can park at the fork in the road here and walk in either direction and along tracks through the desert. Birds (June 1991) include California Quail, **Xantus' Hummingbird**, Gilded Flicker, Gila and Ladder-backed woodpeckers, Ash-throated Flycatcher, Purple Martin (summer), Violet-green Swallow, Western Scrub Jay, Verdin, Cactus Wren, Blue-grey and California gnatcatchers, **Grey Thrasher**, Phainopepla, Northern Cardinal, Pyrrhuloxia, California Towhee, and Hooded and Scott's orioles.

LOS CABOS

Site 1.10: San José del Cabo/ Cabo San Lucas

(December 1982, 1983, February 1988, June 1991, April 1992, October 1993, September 1995, February 1998)

Introduction/Key Species

Los Cabos comprise Cabo San Lucas and San José del Cabo, and the name also refers to the international airport serving these popular resort destinations. The Cape is dominated by the Sierra Victoria (see Sites 1.11, 1.12), the vegetation on the lower slopes of which is a mixture of Baja desert and of thorn forest suggesting western Mexico. For those familiar with western Mexico, however, the Cape's thorn forest will seem strikingly depauperate in bird diversity. Unlike the Mediterranean climate of northwestern Baja, the Cape's rainy season is late summer and autumn (July to October), when the vegetation's dead gray appearance is exchanged for a verdant cloak and thousands of migrant passerines pass through and stage here, taking advantage of the ephemeral food bounty.

Birds of interest in the Cape include Black and **Least storm-petrels**, Brown Booby, Zone-tailed Hawk, Crested Caracara, **Xantus'** and Costa's **hummingbirds**, Gilded Flicker, Grey Flycatcher (winter), California Gnatcatcher (a very pale race, like Black-tailed), **Grey Thrasher**, Bell's Vireo (winter), **Belding's Yellowthroat** (the bright nominate race), Varied and Lazuli (mainly autumn) buntings, Clay-colored and White-crowned sparrows (winter), and Hooded and Scott's orioles.

Access

If you haven't driven down the peninsula (see Mid Peninsula, Site 1.9), the main gateway to Los Cabos is via the airport of the same name, which is about 10 km (6 mi) north of San José del Cabo off Route 1 to La Paz (and at about Km Post 42, i.e., 42 km [25 mi] from Cabo San Lucas). Cars can be rented readily (if not cheaply) at the airport (reservations may be advisable during holiday periods). The area offers a wide range of

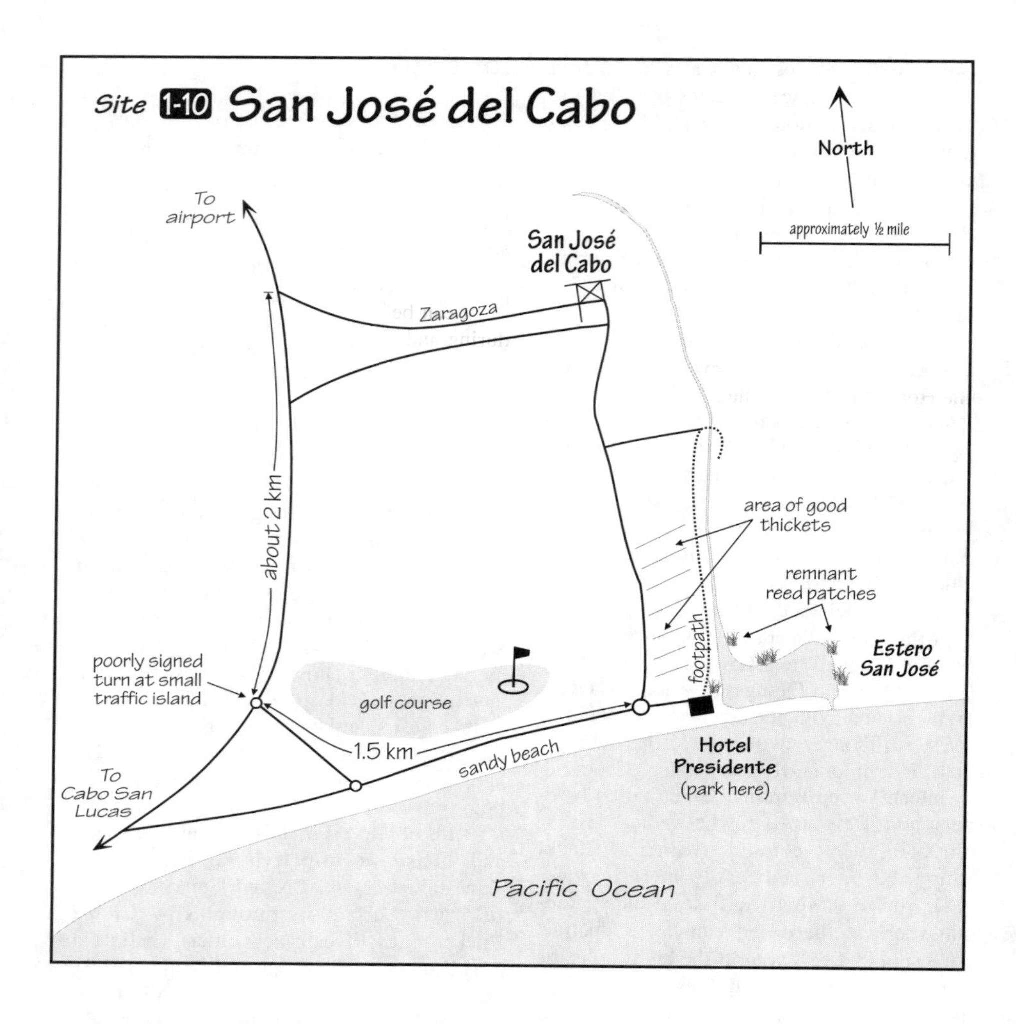

hotels (although low-budget ones are not so easy to come by) and campgrounds (mostly for trailers), and you can also camp out on beaches or in the surrounding desert.

Birding Sites

One site worth birding is a broad desert wash near the airport. From the airport turning, head north on Route 1 for about 1.5 km (0.9 mi), to where you drop into the wash. At the near (south) side watch for a sand road to the left, opposite a small shrine and Tecate sign on the right. One can bird along this dirt road for several kilometers, although most species occur within the first 1–2 km (0.6–1.2 mi). **Grey Thrasher** is quite common in the desert here, and other species include Greater Roadrunner, **Xantus'** (January-February, at least) and Costa's **hummingbirds**, Gilded Flicker, California Gnatcatcher, Cactus Wren, Northern Cardinal, and Pyrrhuloxia.

One of the best spots for birding in Los Cabos is (or has been; see below) the **Estero San José** in San José del Cabo. This is a mostly beach-dammed estuary on the southeast corner of town, adjacent to the Hotel Presidente. Coming from the airport, head south toward Los Cabos on Route 1 and continue basically past San José del Cabo till you get to the expansive tourist developments and a golf course on the southwest side of town, en route to Cabo San Lucas. Watch for a left-hand turn poorly signed to the *Zona Hotel-*

era, or hotel zone (about 2 km south of the junction off Route 1 into downtown San José via Calle Zaragoza), and turn here, via a traffic circle in the median strip, onto what shortly becomes a palm-lined boulevard behind the beach. In about 1.5 km (0.9 mi) you come to a small *glorieta*, whence you could go left and back to downtown San José; continue straight for 200 m and you come to the Hotel Presidente, where you can park.

A lagoon and slow-moving, shallow river (the Estero San José) are immediately east of the Hotel Presidente. The reeds that formerly bordered the near side of the lagoon (and supported a good population of **Belding's Yellowthroats**) have all but been cleared (in October 1995 the hotel tried to blame this on the recent hurricane!). The birds were common and easy to see here, and a pair was found in the remnant patch of reeds near the hotel in February 1998. If you have no luck here, it is possible to cross the river a few hundred meters upstream and look for the yellowthroats in the reed beds along the east side. Ironically, the reed clearing was part of opening up the *estero* for public visibility as an "ecological area"! Birders might wish to register complaints with the hotel management, and you could certainly avoid staying at the hotel.

There also used to be reed beds (with **Belding's Yellowthroat**) along the lagoon shore behind the beach east of the hotel, but in 1998 this area was denuded also. This beach side of the lagoon, and the seasonal river mouth a few hundred meters east of the hotel, can be good for shorebirds, gulls, and terns. The whole *estero* is also often good for grebes, Neotropic Cormorant, herons, and ducks. The scrub and thickets bordering the west side of the *estero* between the hotel and town have Common and Ruddy (recent colonist) ground-doves, Phainopepla, Northern Cardinal, Pyrrhuloxia, Varied Bunting, Hooded Oriole, and in season a good variety of migrants, including Bell's Vireo, warblers, and sparrows. **Xantus' Hummingbird** (December to February at least) and **Grey Thrasher** occur here but are easier to see elsewhere.

The desert around San José, where it hasn't been built upon, and along the highway to Cabo San Lucas has the typical selection of Cape desert birds; Crested Caracaras are fairly common, and you should also watch for Zone-tailed Hawk (in winter, at least).

Birding around **Cabo San Lucas**, 30 km (15–30 minutes) west of San José, can be quite good, but the area (formerly a sleepy fishing village!) is suffering even greater exponential building construction and consequent habitat destruction. The harbor at Cabo itself can be good in winter for gulls, while during and after storms (mainly autumn and early winter) storm-petrels can be seen close inshore off the beaches. The gardens in town, especially of the houses and condominiums on the slopes to the southwest, can be good for wintering migrants and vagrants, and **Xantus' Hummingbirds** (December to February, at least) are attracted to flowers in town. **Grey Thrashers** occur in the desert in and around town. The rocky peninsula and arch southeast of town, beyond the old ferry terminal, often have roosting Brown Boobies (best seen from a boat) and Wandering Tattler (winter).

For those interested in pelagics, boats can be rented to explore offshore, or you could try to get on a sport-fishing boat to the rich grounds off the Cape. Birds possible closer inshore include Pink-footed, Wedge-tailed, Sooty, and **Black-vented shearwaters**, Black and **Least storm-petrels**, boobies (including Masked), phalaropes, jaegers, and Sabine's Gull. Possibilities farther offshore include Cook's Petrel (April, at least), **Townsend's Shearwater**, and Leach's and Galapagos storm-petrels.

See also La Paz to Los Cabos, at the end of the Mid Peninsula section (Site 1.9).

San José del Cabo/Cabo San Lucas Bird List

171 sp. R: rare; V: vagrant (not to be expected). †See pp. 5–7 for notes on English names. Does not include offshore pelagic species.

Pacific Loon
Pied-billed Grebe
Eared Grebe
Western Grebe
Clark's Grebe
Northern Fulmar (R)
Black Storm-Petrel
Least Storm-Petrel
Brown Booby
Brown Pelican
Double-crested Cormorant
Neotropic Cormorant
Brandt's Cormorant
Magnificent Frigatebird

Least Bittern
Great Blue Heron
Great Egret
Snowy Egret
Reddish Egret
Cattle Egret
Green Heron
Black-crowned Night-Heron
White-faced Ibis
Black-bellied Whistling-Duck (V)
Green-winged Teal
Northern Pintail
Blue-winged Teal
Cinnamon Teal
Northern Shoveler
Gadwall
American Wigeon
Canvasback
Redhead
Ring-necked Duck
Lesser Scaup
Ruddy Duck
Red-breasted Merganser
Turkey Vulture
Osprey
Sharp-shinned Hawk
Cooper's Hawk
Zone-tailed Hawk
Red-tailed Hawk
Crested Caracara
American Kestrel
Peregrine Falcon
California Quail
Sora
Common Moorhen
American Coot
Snowy Plover
Semipalmated Plover
Killdeer
Black-necked Stilt
Greater Yellowlegs
Lesser Yellowlegs
Solitary Sandpiper
Willet
Wandering Tattler
Spotted Sandpiper
Marbled Godwit
Red Knot
Sanderling
Western Sandpiper
Least Sandpiper
Pectoral Sandpiper
Dunlin
Short-billed Dowitcher
Long-billed Dowitcher
Common Snipe
Red Phalarope
Laughing Gull
Bonaparte's Gull
Heermann's Gull
Ring-billed Gull
California Gull
Herring Gull
Yellow-footed Gull
Western Gull
Glaucous-winged Gull
Caspian Tern
Elegant Tern
Common Tern
Forster's Tern
Least Tern
White-winged Dove
Mourning Dove
Common Ground-Dove
Ruddy Ground-Dove
Greater Roadrunner
Burrowing Owl
Lesser Nighthawk
Xantus' Hummingbird
Ruby-throated Hummingbird (V)
Black-chinned Hummingbird (R)
Costa's Hummingbird
Belted Kingfisher
Gila Woodpecker
Ladder-backed Woodpecker
Gilded Flicker†
Grey Flycatcher
Western Flycatcher†
Black Phoebe
Vermilion Flycatcher
Ash-throated Flycatcher
Tropical Kingbird (R)
Cassin's Kingbird
Thick-billed Kingbird (V)
Tree Swallow
Violet-green Swallow
N. Rough-winged Swallow
Bank Swallow
Cliff Swallow
Barn Swallow
Western Scrub Jay
Northern Raven†
Verdin
Cactus Wren
Rock Wren
Northern House Wren†
Marsh Wren
Blue-grey Gnatcatcher
California Gnatcatcher
Northern Mockingbird
Grey Thrasher
American Pipit
Cedar Waxwing
Phainopepla
Loggerhead Shrike
European Starling
Bell's Vireo
Plumbeous Vireo†
Tennessee Warbler (V)
Orange-crowned Warbler
Nashville Warbler
Yellow Warbler
Audubon's Warbler†
Palm Warbler (V)
Black-and-white Warbler
MacGillivray's Warbler
Common Yellowthroat
Belding's Yellowthroat
Wilson's Warbler
Yellow-breasted Chat
Summer Tanager
Western Tanager
Northern Cardinal
Pyrrhuloxia
Black-headed Grosbeak
Blue Grosbeak
Lazuli Bunting
Varied Bunting
Dickcissel (R)
Green-tailed Towhee
California Towhee
Black-throated Sparrow
Clay-colored Sparrow
Lark Sparrow
Lark Bunting
Savannah Sparrow
Lincoln's Sparrow
White-crowned Sparrow
Brewer's Blackbird
Brown-headed Cowbird
Hooded Oriole
Bullock's Oriole
Scott's Oriole
House Finch
Pine Siskin (R)
Lesser Goldfinch
House Sparrow

Site 1.11: Miraflores/Boca de la Sierra

(October 1993, September 1995, January 1998)

Introduction/Key Species

While **Xantus' Hummingbird** can be found at least seasonally in the lowlands around Los Cabos, generally it is not common there, and its center of abundance lies in the foothills. One site for this attractive hummer is in and around the village of Boca de la Sierra. **Grey Thrasher** can also be found along the road between Miraflores and Boca de la Sierra, and **Cape Pygmy-Owl** has been reported in Miraflores.

Access/Birding Sites

From San José del Cabo head north on Route 1, past the airport, and mark zero at the airport junction. About 28 km (17 mi; 30–45 minutes from San José), after winding for a while through the first low hills and across a broad wash or two, you come to a paved junction with a Pemex station on the left. This turn is just after Km Post 70 and should be signed to Miraflores. Turn left here, and in 2 km (1.2 mi) the paved road turns to graded dirt in the small town of Miraflores, where **Cape Pygmy-Owl** has been reported recently on the southwest edge of town. To reach Boca de la Sierra, stay straight where the pavement ends, turn right at 2.5 km (1.5 mi), and fork left at 2.9 km as you leave town. Stay on the main road, and after about 4 km (2.4 mi) of desert (lush and full of migrants in September-October) you come to Boca de la Sierra, famous for its freshwater spring. Park at the beginning of the village and walk along the main street, watching and listening for Xantus' Hummingbirds at the flowers. The slopes above town have oaks higher up, and flocks of Band-tailed Pigeons wander down to these at times.

Site 1.12: Sierra de la Laguna

(June 1991)

Introduction/Key Species

The Cape of southern Baja is dominated by the Sierra Victoria, which, unlike the sierras of northern Baja, offers no easy (driving) access to the high mountain forests. It is possible, however, to hike into the northern mountains, known as the Sierra de la Laguna, for which it would be wise to use a guide. For birding you should plan to camp at least two nights.

Species in the Sierra de la Laguna and on the hike up include Band-tailed Pigeon, **Cape Pygmy-Owl**, Mexican Whip-poor-will, **Xantus' Hummingbird**, Acorn Woodpecker, Western Flycatcher, California [Plain] Titmouse, White-breasted Nuthatch, **San Lucas** [American] **Robin**, and **Baird's Junco**.

Access

There are two routes of access: from near Todos Santos in the west (a steeper but shorter route) and from the east off Route 1, a gentler gradient but a longer overall climb (of which I have no first-hand experience). While it may be possible to follow the directions below (no guarantees!), I recommend finding a guide for the trail and bringing a pack animal to help carry your gear and supplies. I've heard that some companies offer tours to the sierra, and for this it may be best to make enquiries in Los Cabos and/or La Paz. If you ask around in Todos Santos, you may be able to find a guide for the steep trail up to the western face.

In 1991 we employed Miguel Dominguez, a resident of Todos Santos, as a guide, and I would recommend him if he still takes people up to La Laguna. He brought a mule that carried our packs, and we camped in the La Laguna meadow and hiked down on our own, having taken care to note the route of the trail on the ascent. There was running water at La Laguna, which should be purified, and we took a tent and all our food.

If your guide suggests leaving Todos Santos very early in the morning, it's worth it—the hike is strenuous and steadily uphill, with little to no shade. Consequently, the earlier you start (preferably before dawn), the sooner you're up in the pine-oak, with some shade from the baking sun. We crossed only one stream with water, so take plenty of water for the ascent. We started the hike from Rancho La Burrera, a small ranch in the foothills about an hour's drive south of Todos Santos, and the hike to La Laguna took 4.5 hours, with brief stops for birding.

To reach Rancho La Burrera (465 m elevation), where you could camp out overnight (Western Screech-Owl, Elf Owl, Lesser Nighthawk), head south on Route 1 out of Todos Santos toward Cabo San Lucas and, just south of Km Post 54 (about 2 km [1.2 mi] south of Todos Santos), at a rise in the highway, look for a gated track off to the east (inland). Turn left onto this hard sand track, bear right at 3.8 km (2.2 mi) and left at 6.6 km (4.0 mi), go through another gate at 9.2 km (5.5 mi) and another at 12.3 km (7.4 mi), and you come to a fairly major fork at 13.5 km (8.1 mi). Go left and at 15.1 km (9.1 mi) stay right at a fork (the left fork may be signed to San Martin); at 21 km (12.6 mi) you come to another gate, this one possibly locked. Rancho La Burrera is 2 km (1.2 mi) beyond this gate; bear right and stay on the main track to a couple of ranch buildings and some trees. Obviously, you should ask permission for access to the ranch. The trail up to La Laguna starts as a broad, driveable track opposite the Rancho La Burrera buildings, which were alive with birds (including **Xantus' Hummingbirds**) during our visit.

Birding Sites

From Rancho La Burrera (where we left our car) you can bird all the way up to La Laguna, starting in brushy desert and climbing into thorn forest and then oaks and pine-oak-madrone forest. The following description of the trail gives some idea of the route and birds along the hike.

Leaving Rancho La Burrera (California Quail, Common Ground-Dove, Ash-throated Flycatcher, **Grey Thrasher**, Hooded Oriole) at 6:10 A.M. (earlier would have been better), we came to a fork at 6:30 and took a trail to the left, signed "Agua 50 m"; basically this was the trail all the way up to La Laguna, with no major junctions, although the trail split and rejoined sometimes. At 7:25 we crossed a stream (the last chance to refill water bottles) with some oaks but still mostly in thorn forest (850 m; **Xantus' Hummingbird**, Ladder-backed Woodpecker, Western Pewee, Blue-grey Gnatcatcher, Cassin's Vireo, Varied Bunting). At 8:35 we took a 10-minute break in a cleared area amid brushy thickets (1350 m; Spotted Towhee); the first **Baird's Junco** was at 1470 m, and at 9:45 the trail thankfully leveled out somewhat in pine-oak-madrone forest (1675 m; Bushtit, **San Lucas Robin**, **Baird's Junco**), ran along a crest at 1800 m (**Cape Pygmy-Owl**), and then dropped to a small clearing (1675 m) with a stream that runs through to the adjacent and much larger La Laguna meadow. The woods surrounding La Laguna meadow had all of the main species, most of which were common and conspicuous during our visit.

Sierra de la Laguna Bird List

(ascent and forest around La Laguna): 48 sp.
†See pp. 5–7 for notes on English names.

Turkey Vulture
Red-tailed Hawk
American Kestrel
California Quail
Band-tailed Pigeon
White-winged Dove
Mourning Dove
Common Ground-Dove
Greater Roadrunner
Great Horned Owl
Cape Pygmy-Owl†
Elf Owl
Lesser Nighthawk
Mexican Whip-poor-will†
White-throated Swift
Xantus' Hummingbird
Acorn Woodpecker
Gila Woodpecker
Ladder-backed Woodpecker
Gilded Flicker†
Western Pewee†
Western Flycatcher†
Black Phoebe
Ash-throated Flycatcher
Violet-green Swallow
Western Scrub Jay
Northern Raven†
California Titmouse†
Bushtit
White-breasted Nuthatch
Canyon Wren
Blue-grey Gnatcatcher
San Lucas Robin†
Grey Thrasher
Loggerhead Shrike
Cassin's Vireo†
Hutton's Vireo
Warbling Vireo
Northern Cardinal
Varied Bunting
Spotted Towhee
California Towhee
Rufous-crowned Sparrow
Baird's Junco†
Hooded Oriole
Scott's Oriole
House Finch
Lesser Goldfinch

CHAPTER 2 NORTHWESTERN MEXICO

Eared Quetzal

This chapter covers areas that can be reached easily from the adjacent southwestern U.S.A.—that is, within a day of driving from the border. These sites are divided into those on the Sonora Coastal Slope (reached most easily from the Nogales, Arizona/Sonora, border crossing) and those in the interior and Sierra Madre. New toll highways beginning near the Nogales crossing make sites in Sonora much easier and quicker to reach than they were even ten years ago. There is also a good highway from the El Paso, Texas/Ciudad Juárez, Chihuahua, crossing to Ciudad Chihuahua (Chihuahua City). Because of a relatively high volume of North American tourist traffic, numerous motels, trailer parks, and campsites can be found in this part of Mexico, particularly the Sonora Coastal Slope.

Much of this region is desert, similar to the southwestern U.S.A., but the Gulf of California adds an ocean element (boobies, frigatebirds, gulls), and the foothills of central Sonora mark the northern limit of tropical thorn forest, with a variety of species not found, or barely found, in the U.S.A. (such as Elegant Quail, Sinaloa Wren, Black-capped Gnatcatcher, Yellow Grosbeak, and Five-striped Sparrow). The Alamos area, in southeastern Sonora, has long been recognized as the northern tip of the Pacific Slope iceberg of the neotropical avifauna, with species such as Laughing Falcon, Rufous-bellied Chachalaca, Mexican Parrotlet, White-fronted and Lilac-crowned parrots, Squirrel Cuckoo, Ivory-billed Woodcreeper, Masked Tityra, Black-throated Magpie-Jay, and Yellow-winged Cacique.

In general, all of these tropical species are much more easily seen farther south in western Mexico, and Alamos, like the rest of Sonora, is of most interest for introductory birding trips to western Mexico, or for combined birding and non-birding holidays.

The forests of the northern Sierra Madre Occidental, which separates the tropical coastal slopes from the temperate interior plateau, have an avifauna similar to that of mountain ranges in the southwestern U.S.A., with all of the "Southeast Arizona specialities" (White-eared Hummingbird, Mexican Chickadee, Red-faced and Olive warblers, Mexican [Yellow-eyed] Junco, etc.) mixed with an increasingly different element as one moves south (Mountain Trogon, Eared Quetzal, Slate-throated Whitestart, etc.). From the point of view of a Mexico list, there are a few interesting high-elevation specialities in the Sierra, such as Northern Goshawk, Townsend's Solitaire, and Evening Grosbeak.

The deserts, grasslands, and lakes of the Mexican Plateau, too, are similar to areas in the adjacent southwestern U.S.A., and from a birding perspective they are of most interest for their numbers of wintering waterfowl, cranes, and sparrows.

The sites in this chapter enable a visiting birder to sample the variety of habitats in northwestern Mexico. Obviously, many more good sites exist, and are being discovered, as an increasing body of birders continues to explore the avifauna of Sonora. One recommended publication that covers much of the recent period of explorations, as well as the historical perspective, is the recently published *The Birds of Sonora*, by Steve Russell and Gale Monson. This work will help you find birds and interpret your findings in a wider perspective. Another publication of interest, now long out of print, is Peter Alden's pioneering work *Finding the Birds in Western Mexico* (1969), which covers numerous sites in the states of Sonora, Sinaloa, and Nayarit. Obviously this book is somewhat outdated, but if you want suggestions for other birding sites (and additional bird lists for sites covered in this guide), it may be worth trying to get hold of a copy.

Suggested Itineraries

Although sites in this chapter could be combined into any itinerary, most sites comprise their own trip. That is, Puerto Peñasco (Site 2.1), Agua Prieta (Site 2.5), and Janos (Site 2.6) can be done as easy day trips from adjacent Arizona; Bahía Kino (Site 2.2), the Sahuaripa Road (Site 2.3), and Cascadas de Cusárare are easy overnight or weekend trips; and Alamos can be done as a long weekend trip from Arizona.

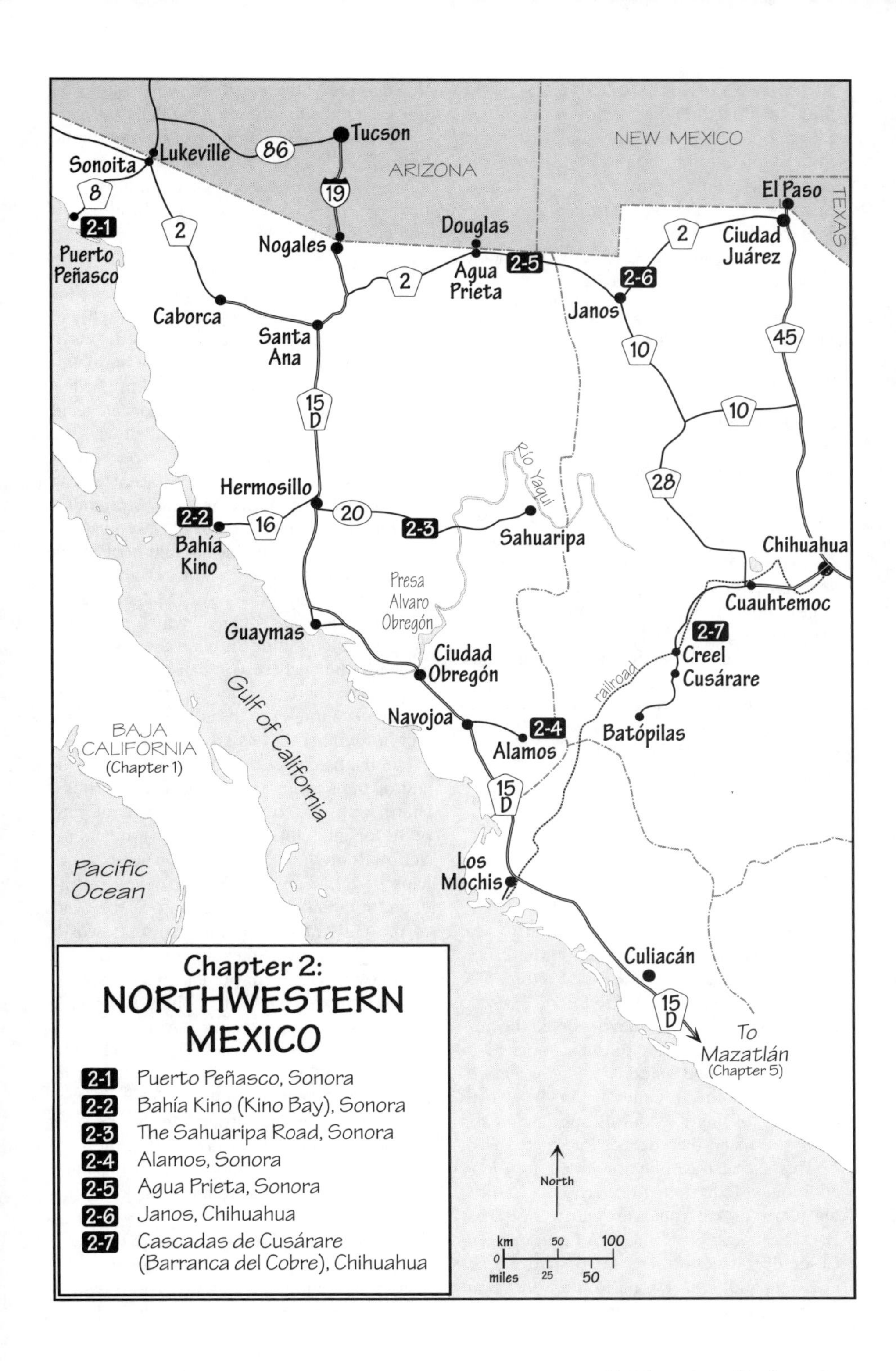
Tucson
NEW MEXICO
ARIZONA
Sonoita
Lukeville
86
8
19
El Paso
TEXAS
2-1
Puerto Peñasco
2
Nogales
Douglas
2
Ciudad Juárez
2
Agua Prieta
2-5
2-6
Caborca
Janos
Santa Ana
10
45
15 D
10
Río Yaqui
28
Hermosillo
2-2
16
20
2-3
Sahuaripa
Bahía Kino
Chihuahua
Presa Alvaro Obregón
Cuauhtemoc
Guaymas
2-7
Ciudad Obregón
Creel
Cusárare
railroad
Gulf of California
Navojoa
2-4
Batópilas
BAJA CALIFORNIA
(Chapter 1)
Alamos
15 D
Pacific Ocean
Los Mochis
Culiacán
15 D
To Mazatlán
(Chapter 5)
Chapter 2:
NORTHWESTERN MEXICO
2-1 Puerto Peñasco, Sonora
2-2 Bahía Kino (Kino Bay), Sonora
2-3 The Sahuaripa Road, Sonora
2-4 Alamos, Sonora
2-5 Agua Prieta, Sonora
2-6 Janos, Chihuahua
2-7 Cascadas de Cusárare (Barranca del Cobre), Chihuahua
North
km 0 50 100
miles 25 50

NORTHWESTERN MEXICO SITE LIST

Site 2.1 Puerto Peñasco, Sonora
Site 2.2 Bahía Kino (Kino Bay), Sonora
Site 2.3 The Sahuaripa Road, Sonora
Site 2.4 Alamos, Sonora
Site 2.5 Agua Prieta, Sonora
Site 2.6 Janos, Chihuahua
Site 2.7 Cascadas de Cusárare (Barranca del Cobre), Chihuahua

Site 2.1: Puerto Peñasco, Sonora

(June 1991, January 1994, May, June 1997)

Introduction/Key Species

Puerto Peñasco is a small fishing and resort town at the head of the Gulf of California. If you are planning to go farther south along the Gulf Coast, there may be no reason to visit this site, but it is a good spot for an easy day or weekend trip from adjacent Arizona. Species of interest include Brown and Blue-footed boobies, Heermann's and **Yellow-footed gulls**, and, in winter, a good variety of other gulls that can include Glaucous, Glaucous-winged, and Western.

This site is perhaps of most interest in winter (there is often a Puerto Peñasco Christmas Bird Count), when good numbers of loons, ducks, and gulls are often present, but it offers good birding year-round. Rarities found here include Harlequin Duck, Oldsquaw, Piping Plover, and Rusty Blackbird.

Access

Puerto Peñasco is about 60 mi (100 km; 1–1.5 hours) from the Sonoita border crossing, where birds include Black Vulture. The wide, two-lane paved road to Puerto Peñasco is well signed after you have crossed the border. Stay on the road, through some nice-looking desert and scattered *ranchos* and *ejidos* (where the greenery can be worth checking for migrants during spring, at least) until you reach the edge of Puerto Peñasco.

This site also can be combined, as a long addition, with a trip to northern Baja California (Chapter 1): from Mexicali in northeast Baja, head east on Route 2 to San Luis (Río Colorado), where there is a state border crossing, and then stay on Route 2 for about 120 mi (200 km; 2–2.5 hours) through desert to Sonoita, where you fork right (south) on the well-signed road to Puerto Peñasco.

Puerto Peñasco can be reached by public buses from Sonoita, but other than the waterfront (which can be good birding), most sites nearby are not that easy to reach, although you could hitch rides.

Birding Sites

There are a number of good birding sites in and around town. As you come into town, note the right-hand junction signed "Playa Cholla 10" on the north side of Puerto Peñasco (see below) and, 1.8 mi (3 km) farther on, in "downtown," note the major left-hand junction signed to Caborca (see below). One site that's always worth a look, especially in winter, is the harbor and adjacent rocky shoreline. To reach this area, stay straight on the highway through town. Shortly past the Caborca junction, the harbor and boat masts will be visible to your right; bear right here in response to a sign that says "Malecon" (but not sharp right to "Muelle," which goes into the port) and continue around the south side of the harbor, where you can pull off easily and check for birds (loons, grebes, gulls). The road continues to the adjacent waterfront and a number of small restaurant stalls. When the fishing/shrimp boats are coming in and/or the stalls are cleaning fish, the rocky shore here can be an excellent gathering point for gulls (up to ten species), including **Yellow-footed**, and possibly Glaucous (perhaps best in February/March, when northbound migrant gulls concentrate at the head of the Gulf of California). It also is worth scanning out over the shallow waters for loons (winter), storm-petrels (windy spring and summer days), and boobies (all year). From this point you can see northwest across the Gulf to another rocky headland with tourist development (Cholla Bay). To get to other birding areas from the *malecon* (waterfront street), continue around on the one-way system back into town (turn right at the sign for Sonoita or you'll loop back in the one-way system and find yourself back on the *malecon*).

The area just to the east of Puerto Peñasco, along the road to Caborca, includes the rubbish dump and sewer ponds, both worth checking. From downtown Puerto Peñasco

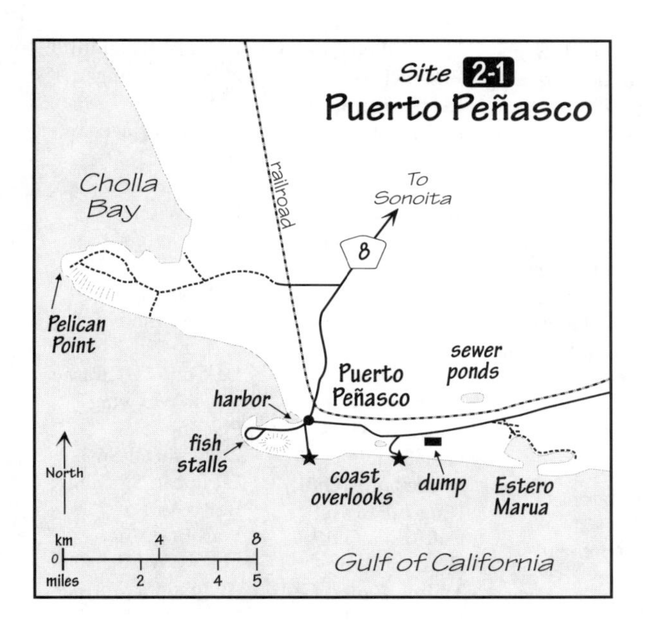

turn east on the road signed to Caborca (Route 37), and mark zero as you turn. At 1.3 mi (2.2 km) there is a right turn a short distance to the beach and a college; the seasonal pools on the right here are often good for shorebirds, and the rocky coast visible from the college parking lot can have American Oystercatcher. At 1.9 mi (3.2 km) you pass the dump excavations on the right (often good for gulls in winter), and at 2.3 mi (3.8 km) there is a sand track diagonally off to the left. A few mesquite trees 200–300 m to the north marks the site of the old sewer ponds (these trees can be good for migrants in spring). The new sewer ponds, which as yet (1997) have not had a chance to produce much vegetation, are about half a mile east of the mesquite trees and 300–500 m north of the road: you can reach them by parking off the road at around 2.9 mi (5 km) and walking north through the desert toward the first low sand hills (which mark the new ponds), crossing the railway tracks about 200 m from the road. Birds at the ponds (at any season) can include a variety of wading birds, ducks, and shorebirds. Gambel's Quail, Burrowing Owl, and Le Conte's and Curve-billed thrashers occur in the desert here.

Another area worth checking east of town is the Estero Marua: continue past the dump and sewer ponds and take the right fork at 4.7 mi (8 km) from town onto a hard but often washboarded sand road. Mark zero at the turn, go through more desert (worth checking in winter for Sage Thrashers and Sage Sparrows); at 1.2 mi (2 km) stay straight at a crossroads; at 2.0 mi (3.3 km) stay on the main road (the middle one, versus left or right) at the signs for oysters; at 2.5 mi (4.1 km) bear right; and at 2.7 mi (4.5 km) you hit the shores of the *estero* (herons, shorebirds, gulls, terns), from which point you can work left on a hard-sand track along the shore.

To reach Cholla Bay (another good area of mudflats), head back to town and turn right (north) on the highway toward Sonoita. On the far north edge of Puerto Peñasco (1.8 mi [3 km] from the Caborca road), turn left on the broad sand road signed "Playa Cholla 10" and reset to zero. You cross the railway tracks at 0.8 mi (1.3 km) and emerge from a barrage of trinket stalls into open desert at 1.1 mi (1.8 km). Bear right at 4.4 mi (7.3 km); at 5.2 mi (8.6 km) there is a right turn a short distance to the shore of Cholla Bay and some houses. At the houses you can go right (beware of soft sand) or left to view the bay, where birds can include Reddish Egret and a variety of shorebirds. Large-billed [Savannah] Sparrows are common (breeding) in the low saltmarsh vegetation around the head of the bay. This area is best at or around high tide, when the birds are closer and more concentrated; at low tide the flats are very expansive and birds can be too distant to see, let alone identify!

At 0.5 mi (0.8 km) past the turn down to the bay you enter the holiday settlement of Cholla Bay where a road loops around the headland; it is possible to see between some of the houses to the rocky coast, which can have Surfbirds (especially in March/April), Wandering Tattler, and sea ducks, and to scan the open waters for loons and boobies.

Puerto Peñasco Bird List

(including ranchos north of town along the highway to Sonoita; also, check the Christmas Bird Count issues of *American Birds/Audubon Field Notes* for more comprehensive lists of birds that can be found in winter): 133 sp. R: rare (not to be expected). †See pp. 5–7 for notes on English names.

Common Loon
Pied-billed Grebe
Eared Grebe
Western Grebe
Sooty Shearwater
Black Storm-Petrel
Least Storm-Petrel
Blue-footed Booby
Brown Booby
Brown Pelican
Double-crested Cormorant
Brandt's Cormorant
Magnificent Frigatebird
Great Blue Heron
Snowy Egret
Cattle Egret
Black-crowned Night-Heron
White-faced Ibis
Green-winged Teal
Northern Pintail
Cinnamon Teal
Northern Shoveler
Gadwall
Bufflehead
Red-breasted Merganser
Ruddy Duck
Turkey Vulture
Osprey
Red-tailed Hawk
American Kestrel
Gambel's Quail
Common Moorhen
American Coot
Black-bellied Plover
Snowy Plover
Wilson's Plover
Semipalmated Plover
Killdeer
American Oystercatcher
Black-necked Stilt
American Avocet
Greater Yellowlegs
Lesser Yellowlegs
Willet
Wandering Tattler
Spotted Sandpiper
Whimbrel
Long-billed Curlew
Ruddy Turnstone
Black Turnstone
Red Knot
Sanderling
Dunlin
Western Sandpiper
Least Sandpiper
Long-billed Dowitcher
Common Snipe
Wilson's Phalarope
Red-necked Phalarope
Laughing Gull
Bonaparte's Gull
Heermann's Gull
Ring-billed Gull
California Gull
Herring Gull
Yellow-footed Gull
Western Gull (R)
Glaucous-winged Gull (R)
Glaucous × Herring Gull (R)
Caspian Tern
Royal Tern
Elegant Tern
Common Tern
Forster's Tern
Least Tern
Black Tern
Craveri's Murrelet (R)
White-winged Dove
Mourning Dove
Inca Dove
Common Ground-Dove
Greater Roadrunner
Burrowing Owl
Lesser Nighthawk
Costa's Hummingbird
Gila Woodpecker
Western Pewee†
Willow Flycatcher
Western Flycatcher†
Say's Phoebe
Vermilion Flycatcher
Western Kingbird
N. Rough-winged Swallow
Northern Raven†
Verdin
Cactus Wren
Rock Wren
Marsh Wren
Northern Mockingbird
Curve-billed Thrasher
Le Conte's Thrasher
Cedar Waxwing
Phainopepla
Loggerhead Shrike
European Starling
Bell's Vireo
Warbling Vireo
Yellow Warbler
Audubon's Warbler†
Myrtle Warbler†
Townsend's Warbler
Hermit Warbler
MacGillivray's Warbler
Common Yellowthroat
Wilson's Warbler
Yellow-breasted Chat
Western Tanager
Lazuli Bunting
Vesper Sparrow
Large-billed Sparrow†
Lincoln's Sparrow
White-crowned Sparrow
Red-winged Blackbird
Yellow-headed Blackbird
Western Meadowlark
Brewer's Blackbird
Great-tailed Grackle
Brown-headed Cowbird
Hooded Oriole
House Finch
Lesser Goldfinch
House Sparrow

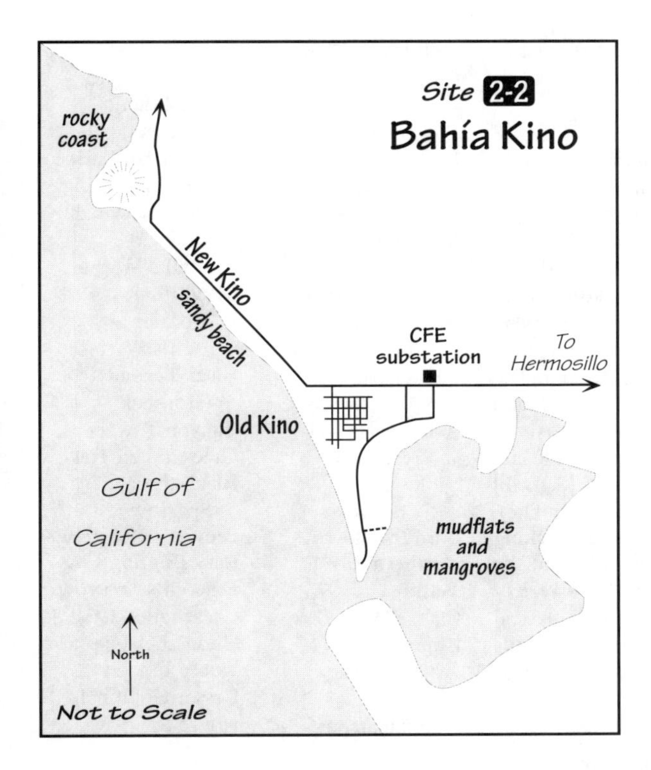

Site 2.2: Bahía Kino (Kino Bay), Sonora

(April 1982, checked 1997)

Introduction/Key Species

This site on the Gulf of California is a popular destination for sun-seeking Americans in winter, as evinced by numerous trailer parks. Bahía Kino can be visited as a weekend trip from Arizona, and birds are typical of the Sonora Desert and Gulf of California. Species of interest include Brown and Blue-footed boobies, Reddish Egret, Wilson's Plover, Heermann's and **Yellow-footed gulls**, and Gull-billed Tern. As with Puerto Peñasco, if you plan on driving farther south along the Gulf Coast (e.g., to Mazatlán, Chapter 4, and/or San Blas, Chapter 5), there is no real reason to visit Bahía Kino for birds.

Access

Bahía Kino is about 240 mi (400 km; 5.5–7 hours) south of Nogales. You can get there by taking the toll highway (Route 15D) south to Hermosillo and then following signs for Bahía Kino, which is about 65 mi (110 km; 1.5 hours) to the west, along Highway 16. This is also the road to the Hermosillo airport; reset to zero at the turn into the airport.

Much of the land between Hermosillo and Bahía Kino has been converted to agriculture, and in winter the fields can be good for migrant waterbirds, raptors, and sparrows. There are also some nice areas of desert.

Bahía Kino can be reached easily by public buses from Hermosillo, and once you are there, you can bird by walking and hitching rides.

Birding Sites

As you come into Bahía Kino there is an area of tidal mudflats visible off to the left (south), and at 60.5 mi (100 km) (after the turn into the Hermosillo airport) there is a CFE (electricity) power substation and a dirt road off to the left, signed "Cictus." Turn here and follow the dirt road around to the right and then left (thus avoiding the streets of Old Kino), where you can view the estuary and mangroves from the higher road or via a maze of tracks (beware of soft sand) that get closer to the water.

Birds here can include Reddish Egret, White Ibis, American Oystercatcher, Wilson's Plover, Gull-billed Tern, and Mangrove [Yellow] Warbler. Back on the main road, at 1.3 mi (2.2 km) after the "Cictus" junction you pass a gas station and the main left turn into Old Kino, and at 2.0 mi (3.3 km) you hit the shores of the Gulf of California and bend right, through a built-up maze of restaurants, hotels, and trailer parks ("New Kino"). The New Kino beach can be good for gulls, and Surfbirds can be common (March/April) along the rocky shore past the small volcanic hill at the north side of town (5 mi [8.4 km] north of where you first hit the shoreline). The waters off the beach are worth checking for loons, boobies, and frigatebirds, and the desert inland of the trailer parks has a variety of common desert birds, including Gila and Ladder-backed woodpeckers, Verdin, Cactus Wren, Black-tailed Gnatcatcher, and Black-throated Sparrow.

Bahía Kino Bird List

(Check Alden [1969] and the Christmas Bird Count issues of American Birds for more comprehensive lists of birds that can be found in winter): 78 sp. †See pp. 5–7 for notes on English names.

Common Loon
Eared Grebe
Brown Pelican
Blue-footed Booby
Brown Booby
Double-crested Cormorant
Magnificent Frigatebird
Great Blue Heron
Snowy Egret
Reddish Egret
Bufflehead
Common Goldeneye
Red-breasted Merganser
Turkey Vulture
Osprey
Red-tailed Hawk
American Kestrel
Black-bellied Plover
Wilson's Plover
Semipalmated Plover
Greater Yellowlegs
Lesser Yellowlegs
Spotted Sandpiper
Willet
Whimbrel
Long-billed Curlew
Marbled Godwit
Surfbird
Least Sandpiper
Western Sandpiper
Sanderling
Laughing Gull
Bonaparte's Gull
Heermann's Gull
Ring-billed Gull
California Gull
Herring Gull
Yellow-footed Gull
Gull-billed Tern
Caspian Tern
Royal Tern
Elegant Tern
Forster's Tern
Least Tern
White-winged Dove
Mourning Dove
Costa's Hummingbird
Belted Kingfisher
Gila Woodpecker
Ladder-backed Woodpecker
Ash-throated Flycatcher
Violet-green Swallow
Barn Swallow
Northern Raven†
Verdin
Cactus Wren
Blue-grey Gnatcatcher
Black-tailed Gnatcatcher
Northern Mockingbird
Curve-billed Thrasher
Loggerhead Shrike
European Starling
Orange-crowned Warbler
Nashville Warbler
MacGillivray's Warbler
Wilson's Warbler
Black-headed Grosbeak
Canyon Towhee
Green-tailed Towhee
Black-throated Sparrow
Brewer's Sparrow
Lark Bunting
Lincoln's Sparrow
Great-tailed Grackle
Hooded Oriole
House Finch
Lesser Goldfinch
House Sparrow

Site 2.3 The Sahuaripa Road, Sonora

(June 1991)

Introduction/Key Species

This site (one of many being discovered by birders in the foothills of central Sonora) provides good access to thorn forest and the associated avifauna. It may be productive at all seasons, but spring and summer are best for bird song and activity. Species common here include **Black-capped Gnatcatcher** and Five-striped Sparrow. There is also a good selection of what North American birders may think of as "Arizona specialities," such as Buff-collared Nightjar and Rose-throated Becard, plus species not found in Arizona, such as **Elegant Quail**, Military Macaw, **Colima Pygmy-Owl**, Nutting's Flycatcher, and **Sinaloa Wren**.

Access

This area is about 225–270 mi (375–450 km; 6–8 hours) from Nogales. To get here, take the toll highway (Route 15D) to Hermosillo. From the bypass on the southeastern side of Hermosillo, take the road (that should be) signed to Mazatlán and/or Novillo and Sahuaripa. This will be not long after you see a huge reservoir (often with lots of Neotropic Cormorants) inland (east) of the bypass. The first 60 mi (100 km) or so of the Sahuaripa Road alternate between overgrazed, burned grassland and brushy, semi-open desert (Bendire's Thrasher is possible; Curve-billed Thrasher is common) before getting into some areas of good thorn forest.

While there is a very basic "motel" at Novillo (take your own food), the best way to bird this area is to camp or to stay somewhere on the edge of Hermosillo and make an early morning start into the field. It should be possible to reach Novillo by public bus from Hermosillo, and good birding can be found within walking distance of Novillo.

Birding Sites

The degree of human settlement and clearing will affect birding spots, but worth checking is a small ranch with a creek near Km Post 115 (i.e., 115 km from Hermosillo), on the left (north) side of the highway. The water and animal pens here can attract a good variety of birds: species seen during a mid-afternoon visit included **Elegant Quail**, White-winged Dove, Common Ground-Dove, Broad-billed Hummingbird, Willow Flycatcher, Tropical Kingbird, Bell's Vireo, Yellow-breasted Chat, Northern Cardinal, Pyrrhuloxia, Black-headed and Blue grosbeaks, Varied Bunting, and Streak-backed Oriole.

From here on, thorn forest becomes more extensive (with many of the species listed below), and the road winds through hills before dropping down to the small settlement of Novillo (Km Post 151?), by the rushing outflow from nearby Presa Novillo (or Presa Plutarco Elias Calles), a large reservoir hidden in the hills just to the north. Birding can be good in the thorn forest along the road up to 3 mi (5 km) west of (above) Novillo, as well as around the settlement itself.

Sahuaripa Road Bird List

(Km Post 100 to Novillo): 54 sp. †See pp. 5–7 for notes on English names.

Black Vulture
Turkey Vulture
Grey Hawk
Red-tailed Hawk
Elegant Quail
White-winged Dove
Mourning Dove
Inca Dove
Common Ground-Dove
White-tipped Dove
Military Macaw
Yellow-billed Cuckoo
Colima Pygmy-Owl†
Lesser Nighthawk
Buff-collared Nightjar
Broad-billed Hummingbird
Violet-crowned Hummingbird
Elegant Trogon
Gila Woodpecker
Ladder-backed Woodpecker
N. Beardless Tyrannulet
Vermilion Flycatcher
Dusky-capped Flycatcher
Nutting's Flycatcher
Brown-crested Flycatcher
Sulphur-bellied Flycatcher
Tropical Kingbird
Cassin's Kingbird
Thick-billed Kingbird
Rose-throated Becard
N. Rough-winged Swallow
Northern Raven†
Verdin
Cactus Wren
Sinaloa Wren
Canyon Wren
Black-capped Gnatcatcher
Rufous-backed Thrush†
Bell's Vireo
Rufous-capped Warbler
Yellow-breasted Chat
Summer Tanager
Northern Cardinal
Pyrrhuloxia
Black-headed Grosbeak
Yellow Grosbeak
Varied Bunting
Canyon Towhee
Five-striped Sparrow
Bronzed Cowbird
Brown-headed Cowbird
Streak-backed Oriole
House Finch
Lesser Goldfinch

Site 2.4 Alamos, Sonora

(April 1982, August 1988, June 1991)

Introduction/Key Species

Alamos (elevation 400 m) is a small, picturesque desert town nestled near the foothills of the Sierra Madre Occidental in southern Sonora. It has been a center of ornithological exploration for many years and has hosted Christmas Bird Counts annually from 1977 to 1989, and less regularly since then. Many good birding sites can be found in the vicinity, although recent heavy rains apparently have altered areas along the Río Cuchuhaqui and perhaps have made access to the nearby mountains more problematic. If you're feeling adventurous, ask locally in Alamos about road conditions and access to the Sierra (which is not dealt with below). Even without this component to the avifauna, several readily accessible sites near Alamos offer good birding at all seasons.

The Alamos area is a meeting point of the desert, tropical, and Sierra Madrean elements of Mexico's rich avifauna. Within a few miles of town you can see birds of the Arizona deserts (e.g., Thick-billed Kingbird, Broad-billed Hummingbird, Five-striped and Rufous-winged sparrows), plus many species typical of northwestern Mexico (e.g., **Rufous-bellied Chachalaca**, **Mexican Parrotlet**, **Black-throated Magpie-Jay**, **Happy Wren**, Yellow Grosbeak [summer], Streak-backed Oriole). If you plan to continue birding to the south of Alamos, almost all of the species here can be seen more easily and/or are more common at other sites.

Access

Alamos is about 395 mi (660 km; 10–12 hours) south of Nogales, or 375 mi (620 km; 8–10 hours) north of Mazatlán, Sinaloa (Chapter 5). From Nogales, head south on the toll highway (Route 15D) to Navojoa, which is about 120 mi (200 km) south of Guaymas. From Navojoa, turn left (east) and head inland on the signed, paved road to Alamos, which is about 30 mi (50 km) distant. There are numerous forms of accommodation in the Alamos area, which is served regularly by public buses from Navojoa.

Birding Sites

The road from Navojoa to Alamos passes through a variety of desert habitats and thorn forest, cut over and grazed in places. About 9 mi (15 km) from Navojoa you'll see a prominent hill to the left (north) of the road; at 9.5 mi (16 km) a cobbled road signed "Microondas Cerro Prieto" leads about 2.2 mi (3.7 km) to the towers at the top of the hill. In late summer (June-August, when the area looks quite lush), singing Five-striped and Rufous-winged sparrows can be conspicuous on the hill. Continuing on to 22.5 mi (37.5 km), there is a trailer park/campground (El Caracol) to the right (south) of the road. Birding can be good here (ask permission if you're not staying—it's a nice spot to camp) and in the adjacent thorn forest (**Rufous-bellied Chachalaca**, **Mexican Parrotlet**, Buff-collared Nightjar, Plain-capped Starthroat, **Black-throated Magpie-Jay**, **Purplish-backed Jay**, **Sinaloa Crow**, **Black-capped Gnatcatcher**, Yellow Grosbeak).

Continuing into Alamos, birding anywhere along the road in the desert, thorn forest, and brushy fields can be good, especially early and late in the day. If you keep going through town and out to the east, you will soon cross a seasonally wet desert wash, where birding can be good, especially if there is still some water (Plain-capped Starthroat, **Happy Wren**, **Blue Mockingbird**). A dirt road (not always passable in a regular car) continues from here to the east and south and, after a few miles (and some junctions), rejoins the wash within walking distance of the Río Cuchuhaqui, which generally has water year-round and is bordered by some large trees (Montezuma Cypresses and figs) and thickets. Ask locally about road conditions and directions if you want to try this site, where birds include Bare-throated Tiger-Heron, Crane and Common Black hawks, Green Kingfisher, Squirrel Cuckoo, and Rose-throated Becard.

Alamos Bird List

(Check the Christmas Bird Count issues of American Birds/Audubon Field Notes for an idea of other species to be expected, at least in winter, remembering that birds of the higher elevation, pine-oak zone may not be easily accessible): 103 sp. †See pp. 5–7 for notes on English names.

Great Blue Heron
Cattle Egret
Green Heron
White-faced Ibis
Black Vulture
Turkey Vulture
Osprey
Cooper's Hawk
Grey Hawk
Red-tailed Hawk
Crested Caracara
American Kestrel
Rufous-bellied Chachalaca
Elegant Quail
Killdeer
Red-billed Pigeon
White-winged Dove
Mourning Dove
Inca Dove
Common Ground-Dove
White-tipped Dove
Mexican Parrotlet
White-fronted Parrot
Groove-billed Ani
Ferruginous Pygmy-Owl
Mottled Owl
Buff-collared Nightjar

Broad-billed Hummingbird
Violet-crowned Hummingbird
Plain-capped Starthroat
Elegant Trogon
Belted Kingfisher
Gila Woodpecker
Ladder-backed Woodpecker
Gilded Flicker†
N. Beardless Tyrannulet
Western Flycatcher†
Black Phoebe
Vermilion Flycatcher
Dusky-capped Flycatcher
Nutting's Flycatcher
Brown-crested Flycatcher
Sulphur-bellied Flycatcher
Tropical Kingbird
Cassin's Kingbird
Thick-billed Kingbird
Western Kingbird
N. Rough-winged Swallow
Cliff Swallow
Black-throated Magpie-Jay
Purplish-backed Jay
Sinaloa Crow
Northern Raven†
Verdin
Cactus Wren
Canyon Wren
Sinaloa Wren
Happy Wren
Northern House Wren†
Blue-grey Gnatcatcher
Black-capped Gnatcatcher
Hermit Thrush
Rufous-backed Thrush†
American Robin
Northern Mockingbird
Curve-billed Thrasher
Cedar Waxwing
Loggerhead Shrike
Bell's Vireo
Plumbeous Vireo†
Orange-crowned Warbler
Nashville Warbler
Virginia's Warbler
Lucy's Warbler
Black-throated Grey Warbler
MacGillivray's Warbler
Common Yellowthroat
Wilson's Warbler
Yellow-breasted Chat
Northern Cardinal
Pyrrhuloxia
Yellow Grosbeak
Blue Grosbeak
Lazuli Bunting
Varied Bunting
Green-tailed Towhee
Canyon Towhee
Five-striped Sparrow
Rufous-winged Sparrow
Clay-colored Sparrow
Chipping Sparrow
Brewer's Sparrow
Black-chinned Sparrow
Lark Sparrow
Grasshopper Sparrow
White-crowned Sparrow
Streak-backed Oriole
Great-tailed Grackle
Bronzed Cowbird
Brown-headed Cowbird
House Finch
Lesser Goldfinch
House Sparrow

Site 2.5 Agua Prieta, Sonora

(January 1996)

Introduction/Key Species

Agua Prieta (elevation 1200 m) is the Mexican border town opposite Douglas, Arizona. The grasslands and lakes along Route 2 east of town, toward the Chihuahua border, and west toward Cananea offer interesting birding in winter. Species are much like those of grasslands in adjacent Arizona, including Sprague's Pipit (winter), Baird's Sparrow (winter), and the resident Lilian's form of Eastern Meadowlark (Westerns occur in winter). The oak/juniper woodland in the Sierra San Luis, at the Chihuahua border, is the only known area in Mexico for Grey [Plain] Titmouse. Also see Janos, Chihuahua (Site 2.6), which can be combined in a day trip with the area east of Agua Prieta.

Access/Birding Sites

The conditions of a given area vary from year to year depending upon the extent of grazing and rainfall, so it is best to drive along and look for promising areas. If you see local farmers, it is polite to ask their permission about walking in the fields, but generally it is possible to get out almost anywhere and walk about, with due respect to fences, gates, livestock, and any No Trespassing signs.

At 39 mi (65 km) east of Agua Prieta, the grasslands around Km Post 90 can be good in winter for Sprague's Pipit and Chestnut-collared Longspur. Some 4 mi (6.5 km) farther east, the open, park-like oak woodland south of the road at Km Post 83.5 (a truck stop with plenty of room to pull off the road) has both Bridled and Grey [Plain] titmice.

The grasslands 27–30 mi (45–50 km) west of Agua Prieta, or 18–21 mi (30–35 km) east of Cananea, can be good for Baird's Sparrow (winter)—for example, around the small settlement of Ejido Cuauhtemoc (Km Post 45). Depending on water levels, the various ranch ponds in this stretch often have ducks, which can include Common Merganser (winter).

Given the seasonal nature of the habitat (except the oak woods), birding this area by public transport could be problematic but is quite possible. There is potential for accommodation and eating in both Agua Prieta and Cananea.

Agua Prieta Bird List

74 sp. †See pp. 5–7 for notes on English names

Pied-billed Grebe
Western Grebe
Clark's Grebe
Double-crested Cormorant
Great Blue Heron
Black-crowned Night-Heron
Green-winged Teal
Mallard × Mexican Duck†
Northern Pintail
Northern Shoveler
Gadwall
American Wigeon
Ring-necked Duck
Canvasback
Bufflehead
Common Merganser
Ruddy Duck
Red-tailed Hawk
American Kestrel
Merlin
American Coot
Killdeer
Mourning Dove
Acorn Woodpecker
Gila Woodpecker
Red-shafted Flicker†
Black Phoebe
Say's Phoebe
Vermilion Flycatcher
Horned Lark
Grey-breasted Jay†
Northern Raven†
Bridled Titmouse
Grey Titmouse†
Verdin
Bushtit
Cactus Wren
Bewick's Wren
Canyon Wren
Ruby-crowned Kinglet
Hermit Thrush
Curve-billed Thrasher
Sprague's Pipit
American Pipit
Phainopepla
Loggerhead Shrike
European Starling
Audubon's Warbler†
Northern Cardinal
Canyon Towhee
Spotted Towhee
Cassin's Sparrow
Chipping Sparrow
Brewer's Sparrow
Black-chinned Sparrow
Vesper Sparrow
Savannah Sparrow
Baird's Sparrow
Grasshopper Sparrow
Song Sparrow
Lincoln's Sparrow
White-crowned Sparrow
Grey-headed Junco†
Pink-sided Junco†
Chestnut-collared Longspur
Red-winged Blackbird
Yellow-headed Blackbird
Brewer's Blackbird
Eastern Meadowlark
Western Meadowlark
Great-tailed Grackle
Brown-headed Cowbird
House Finch
Lesser Goldfinch

Site 2.6 Janos, Chihuahua

(November 1986, January 1996)

Introduction/Key Species

The high desert of Chihuahua is dotted liberally with lakes of varying sizes, although access to these lakes varies considerably. Lago Colorado, northeast of the small town of Janos (elevation 1250 m), can be viewed easily from the highway. Species of interest here in winter can include Ross' Goose, Common Merganser, Bald and Golden eagles, Sandhill Crane, sparrows (including apparent Timberline [Brewer's] Sparrow), and longspurs. Based on two visits, fewer species may be expected in mid- versus early winter. This is an easy day trip from Arizona or from El Paso, Texas, and can be combined with Agua Prieta (Site 2.5) in adjacent Sonora.

Access/Birding Sites

Janos is about 100 mi (160 km; 2.5–3 hours) east of Agua Prieta, Sonora, and about 135 mi (220 km; 2.5–3 hours) from Ciudad Juárez, Chihuahua. Coming into Janos along from Agua Prieta, turn left (northeast) by the Pemex station toward Ciudad Juárez, staying on Route 2. The highway goes through some farmland, crosses a river, and winds up into some low hills. About 10.0 mi (16 km) from the Pemex junction (shortly after Km Post 188) you should see the lake off to your left (north). Continue on the highway for another 0.6 mi (1.0 km), following the road as it bends left and then right. Near the curve of the right-hand bend, about 10.6 mi (17.5 km) from Janos, you can pull off safely to the left (north) beside the highway, whence you can scan the whole lake. It is also possible from here to find a way through or over the fence and walk closer to the lake, which is about 500 m distant. As always, if you see any people, ask permission first.

The lake can be reached by second-class bus or hitching a ride from Janos, which is not replete with infrastructure; you could camp out in the desert (beware: it can be cold at night) near the lake.

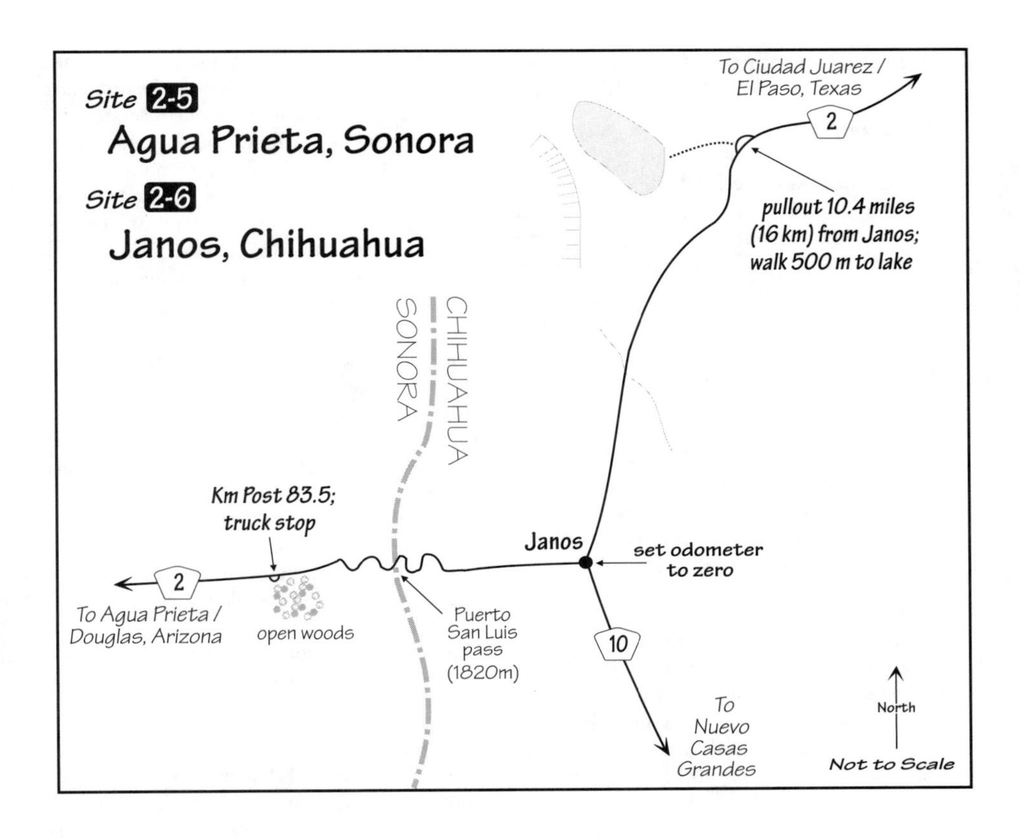

Janos (Lago Colorado) Bird List

61 sp. R: rare (not to be expected). †See pp. 5–7 for notes on English names.

Pied-billed Grebe
Eared Grebe
Western Grebe
Clark's Grebe
American White Pelican
Great Blue Heron
Great Egret
Cattle Egret
Snow Goose
Ross' Goose
Green-winged Teal
Mallard
Mexican Duck†
Northern Pintail
Northern Shoveler
Gadwall
American Wigeon
Ring-necked Duck
Common Merganser
Turkey Vulture
Northern Harrier
Cooper's Hawk
Red-tailed Hawk
Bald Eagle
Golden Eagle
American Kestrel
Prairie Falcon
Killdeer
American Avocet
Greater Yellowlegs
Spotted Sandpiper
Long-billed Curlew
Least Sandpiper
Long-billed Dowitcher
Common Snipe
Ring-billed Gull
Bonaparte's Gull (R?)
Mourning Dove
Black Phoebe
Say's Phoebe
Tree Swallow
[Northern Raven†?]
Chihuahuan Raven
Mountain Bluebird
Curve-billed Thrasher
American Pipit
Loggerhead Shrike
Audubon's Warbler†
Pyrrhuloxia
Green-tailed Towhee
Chipping Sparrow
Brewer's Sparrow
[Timberline Sparrow†?]
Black-throated Sparrow
Vesper Sparrow
Savannah Sparrow
Lincoln's Sparrow
White-crowned Sparrow
Lark Bunting
McCown's Longspur
Chestnut-collared Longspur
Yellow-headed Blackbird
Western Meadowlark

Site 2.7: Cascadas de Cusárare (Barranca del Cobre), Chihuahua

(January 1985, July 1991)

Introduction/Key Species

Often touted as Mexico's "Grand Canyon," the Barranca del Cobre (or Copper Canyon) is a truly spectacular area, and the train ride between Ciudad Chihuahua and the coast at Los Mochis, Sinaloa, is worth taking for the scenery alone. It also is possible to drive from Ciudad Chihuahua to one of the best birding areas, Cascadas de Cusárare (Cusárare Falls), which is a good site for **Eared Quetzal**. Other species of interest here include Northern Goshawk, Black Swift (summer), **Mountain Trogon**, American Dipper, Townsend's Solitaire, **Striped Sparrow**, and Evening Grosbeak.

Other areas accessible from train stops along the route offer potentially good birding, but the infrastructure tends to be either very basic or very expensive. Also, unless you plan to hike the canyons, the birds of this area can be seen more easily elsewhere, for example, along the Durango Highway (Chapter 5) or in Colima and Jalisco (Chapter 7).

Access (Train)

The train through the Barranca del Cobre goes from Ciudad Chihuahua to Los Mochis, Sinaloa, and vice versa. Ciudad Chihuahua is about 225 mi (375 km; 4.5–5 hours) by road (Route 45) south of El Paso/Ciudad Juárez U.S.A./Mexico border crossing, and has a variety of accommodations. If you have driven to Chihuahua, it is possible to leave your car safely at a hotel for the duration of your train trip and take a taxi to the railway station.

Los Mochis is about 100 mi (160 km; 2–2.5 hours) south by toll highway (Route 15D) from Navojoa and the turnoff to Alamos (Site 2.4) and about 240 mi (400 km; 5 hours) north of Mazatlán (Chapter 5). If you're doing the train trip from Los Mochis as an adjunct to a driving trip, it should also be possible to leave your vehicle safely at a hotel and take a taxi to the station.

The daily first-class (tourist) train leaves both Ciudad Chihuahua and Los Mochis in the early morning and takes about fourteen hours (sometimes longer because of delays due to weather or the derailing of cargo trains); in winter, when days are shorter and weather delays more likely, it is best to start from Los Mochis to guarantee seeing the best scenery in daylight. Check with a travel agent and/or at the railway stations for the latest schedules, which are prone to vary; generally there isn't a problem getting tickets the day before you want to travel. While food should be available on the train and also can be bought from platform vendors along the route, it isn't a bad idea to take food and drink with you, as well as toilet paper, which can be in short supply on the train. If you plan to break the journey with birding stops, be sure when you buy your ticket that it allows you to get off and on again at stations along the route.

The Cascadas de Cusárare are near the small town of Creel (2350 m elevation), which is east of the main canyon scenery. Creel offers a variety of places to stay and eat that cater to most budgets. There also is an expensive tourist lodge at Cusárare (see Birding Sites), or you could camp in the valley between the lodge and the falls.

Access (Driving)

From Ciudad Chihuahua, take the well-signed divided highway to Cuauhtemoc and from there stay on the main, paved road, following signs to Creel (about 60 mi [100 km] from Cuauhtemoc). The Pemex station in Creel has not (1996) had unleaded gasoline, so you might do well to fill up en route in La Junta, about 27 mi (45 km) west of Cuauhtemoc. In winter, the fields and grasslands between Cuauhtemoc and Creel can be good for Ferruginous Hawk, Sprague's Pipits, sparrows, and longspurs, and in summer you should watch for Swainson's Hawk.

Birding Sites

If you've reached Creel by train, the nearby Cascadas de Cusárare can be reached by taxi or hitching from Creel. There also should be public buses at least to Cusárare village, and some hotels in Creel run tours to the falls. If you're driving, at Creel continue south past town (which is off to the left, or east) on the main, paved road to Cusárare (12 mi; 20 km) and watch for a sign to the left for the village of Cusárare. Do not turn here, but stay on the main road and continue about 250 m until a sign to the right

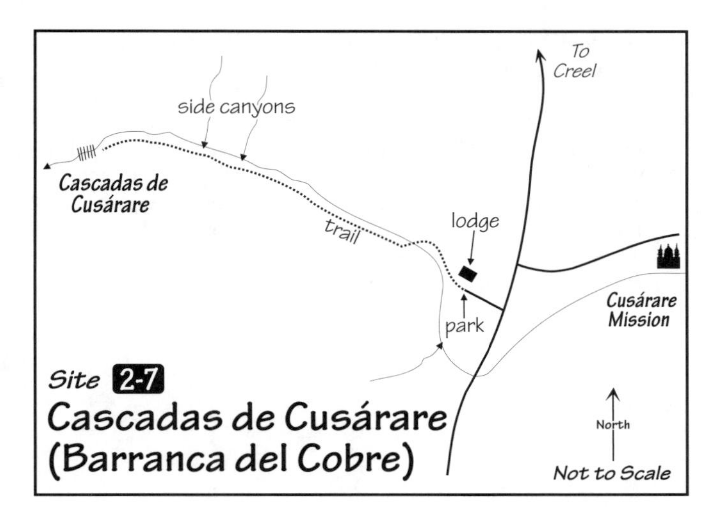

for the Copper Canyon Hiking Lodge (2400 m elevation), a large tourist lodge that could be a nice place to stay if it's within your budget.

It is possible to park by the lodge (check the hummingbird feeders if filled) and walk past it, downstream along the wide-bottomed canyon to the northwest, heading for the falls. This valley is a popular hike and picnic/camping spot. It is an easy walk, following the trail along the stream for about 2.5 km, to where the stream bends left and over the falls. One can also walk along two smaller side canyons off to the right (north) of the main canyon. **Eared Quetzals** can be fairly common here (listen for their distinctive calls, especially in the morning); Dippers live along the stream; and Mountain Pygmy-Owl, Townsend's Solitaire, and Evening Grosbeak occur in the surrounding forest. In summer, Black and Chestnut-collared swifts can be seen over the valley and may nest behind the falls, and Northern Goshawk has been reported here. **Striped Sparrows** occur in the area and can be found in fields beside (north of) the road into Cusárare village, shortly before the old church.

Migrants in the area include Red-naped and Williamson's sapsuckers, Hermit Thrush, Orange-crowned, Audubon's [Yellow-rumped], Townsend's, Hermit, and Wilson's warblers, and Grey-headed [Dark-eyed] Junco.

Cascadas de Cusárare Bird List
(including the drive from Cuauhtemoc): 80 sp. R: rare (not to be expected). †See pp. 5–7 for notes on English names.

Great Blue Heron
Green-winged Teal
Mexican Duck†
Gadwall
Black Vulture
Turkey Vulture
Northern Harrier
Sharp-shinned Hawk
Swainson's Hawk
Red-tailed Hawk
Ferruginous Hawk
American Kestrel
Prairie Falcon
Killdeer
Spotted Sandpiper
Mourning Dove
Great Horned Owl
Common Nighthawk
Mexican Whip-poor-will†
Black Swift
Chestnut-collared Swift (R)
White-throated Swift
White-eared Hummingbird
Magnificent Hummingbird
Blue-throated Hummingbird
Broad-tailed Hummingbird
Rufous Hummingbird
Mountain Trogon
Eared Quetzal†
Belted Kingfisher
Acorn Woodpecker
Hairy Woodpecker
Red-shafted Flicker†
Greater Pewee
Western Pewee†
Western Flycatcher†
Buff-breasted Flycatcher
Black Phoebe
Cassin's Kingbird
Violet-green Swallow
Barn Swallow
Horned Lark
Steller's Jay
Northern Raven†
Mexican Chickadee
White-breasted Nuthatch
Brown Creeper

Rock Wren
Canyon Wren
Brown-throated Wren†
American Dipper
Western Bluebird
Mountain Bluebird
Townsend's Solitaire
American Robin
Curve-billed Thrasher
American Pipit
Sprague's Pipit
Hutton's Vireo
Plumbeous Vireo†
Red-faced Warbler
Olive Warbler
Painted Whitestart†
Slate-throated Whitestart†
Hepatic Tanager
Canyon Towhee
Chipping Sparrow
Vesper Sparrow
Savannah Sparrow
White-crowned Sparrow
Lark Bunting
Mexican Junco†
Chestnut-collared Longspur
Western Meadowlark
Brewer's Blackbird
Great-tailed Grackle
Bronzed Cowbird
Lesser Goldfinch
Evening Grosbeak
House Sparrow

CHAPTER 3 THE MONTERREY/ SALTILLO AREA

Worthen's Sparrow

This chapter covers an area that can be reached within only a few hours of adjacent Texas, U.S.A., and that, with new toll highways, is very easy to visit for a long weekend, or even an overnight/ day trip from the U.S.A. The fastest route to Monterrey and Saltillo is via the Laredo/Nuevo Laredo U.S.A./Mexico border crossing, from which you take Route 85 direct to Monterrey, some 135 mi (225 km; 2.5–3.5 hours) to the south. Cars can be rented in Monterrey, if you're flying into the region. This area can be made into a week's trip and/or combined easily with the northeastern Gulf Slope (Chapter 4) for a longer and more varied trip.

Monterrey is the state capital of Nuevo León, and Saltillo the state capital of Coahuila. Consequently, numerous hotels and restaurants can be found in both cities. Except for Cerro El Potosí, all sites can be visited as easy day trips from a base in these two cities. With the new *au-*

topista (motorway) connecting Monterrey and Saltillo, it is possible to bird areas nearer to one while based in the other, although this will add about an hour (one-way) to your travel time.

This region marks a transition from the deserts of the Mexican Plateau to the subtropical scrub and woodland of the Gulf Slope; it also has some spectacular high mountains, at the northern end of Mexico's Sierra Madre Oriental. These sierras are home to the Maroon-fronted Parrot, probably the reason most birders visit the area; this spectacular bird is best seen between April and October and tends to wander south to relatively inaccessible mountains between November/December and February/March. Other noteworthy species include Clark's Nutcracker, Colima Warbler (summer), and Worthen's Sparrow. For those wishing to add diversity to their birding, it's worth knowing about Presa Tulillo, a small reservoir in the desert west of Saltillo.

Suggested Itineraries

Driving from Texas, it is easy to reach Monterrey and/or Saltillo in a few hours and by leaving Texas early Saturday morning, you could get to the Highrise by early afternoon and be seeing Maroon-fronted Parrots! If you reach Monterrey Friday afternoon, an early start the next morning gives a full day to Cola de Caballo and the Highrise (Site 3.1), and you can spend Sunday morning in the Tanque de Emergencia area (Site 3.3) before heading back in the afternoon. If you're based in Saltillo and arrive Friday with a few hours of daylight, Presa El Tulillo (Site 3.2) can be good at any time of day. An early-morning start from Saltillo gives a good full day to San Antonio de las Alazanas (Site 3.4) or the Highrise (Site 3.1), and the next morning you can do the Tanque area (Site 3.3) before heading back on Sunday afternoon. Cerro El Potosí can be reached in half a day's drive and deserves at least a long weekend.

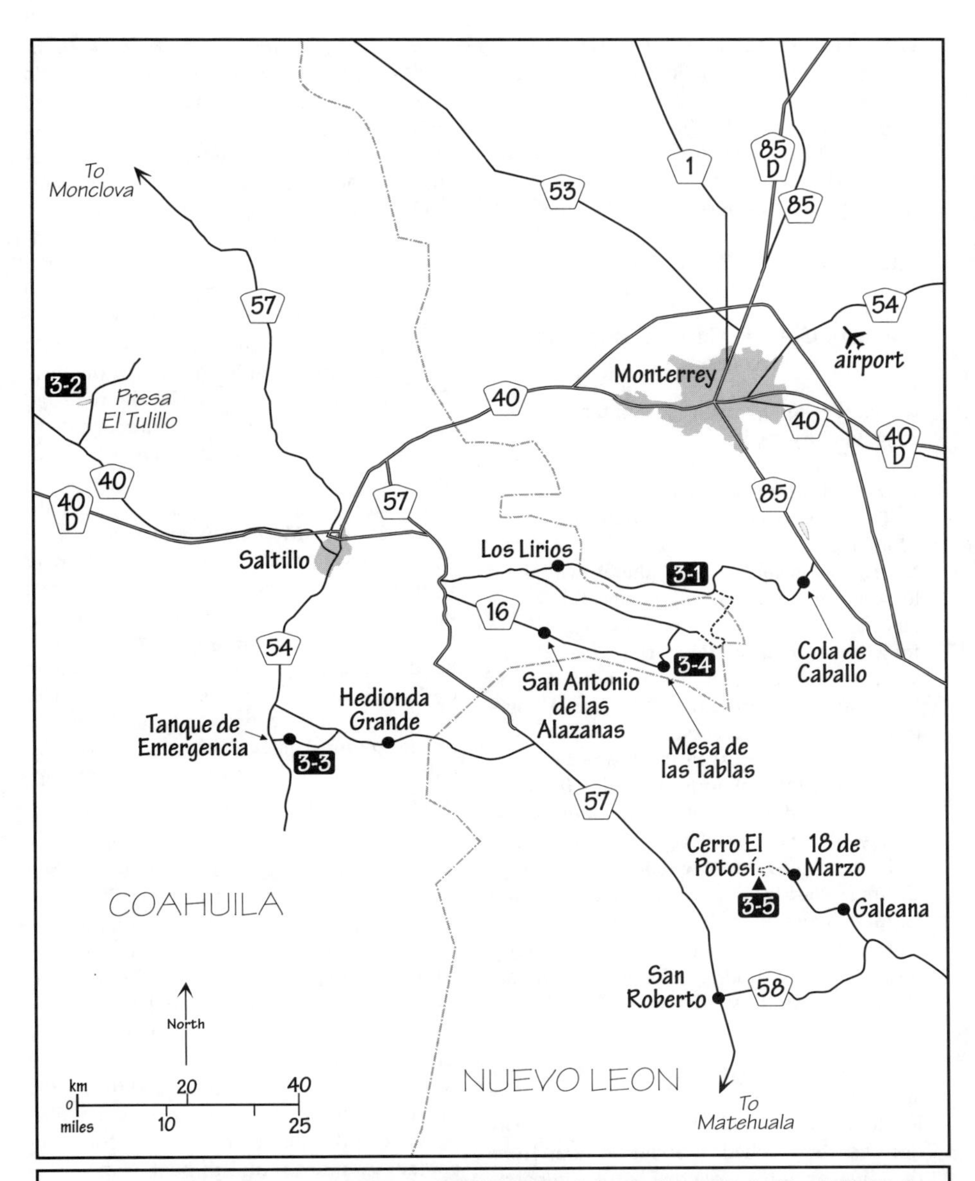

Chapter 3: The MONTERREY / SALTILLO AREA

- 3-1 Cola de Caballo / Highrise, Nuevo León
- 3-2 Presa El Tulillo, Coahuila
- 3-3 Tanque de Emergencia / Hedionda Grande, Coahuila
- 3-4 San Antonio de las Alazanas, Coahuila
- 3-5 Cerro El Potosí, Nuevo León

MONTERREY/SALTILLO AREA SITE LIST

Site 3.1 Cola de Caballo/Highrise, Nuevo León
Site 3.2 Presa El Tulillo, Coahuila
Site 3.3 Tanque de Emergencia/Hedionda Grande, Coahuila
Site 3.4 San Antonio de las Alazanas, Coahuila
Site 3.5 Cerro El Potosí, Nuevo León

Site 3.1: Cola de Caballo/Highrise, Nuevo León

(November 1986, April 1990, September 1995, January 1998)

Introduction/Key Species

This interesting area, a short distance from Monterrey, includes a variety of habitats, from subtropical semi-deciduous woodland to scrub and high-elevation pine forest (including the Highrise, a famous nesting site for **Maroon-fronted Parrot**). Increased conservation awareness in this area has led to the Highrise and surrounding area being purchased as a nature reserve.

Species of interest in the lower-elevation semi-deciduous woods include **Bronze-winged Woodpecker** and **Crimson-collared Grosbeak**, while the high-elevation pines and scrub have **Pine Flycatcher**, **Grey Silky**, Colima Warbler (summer), and **Rufous-capped Brushfinch**. Flammulated (summer) and Spotted owls occur here also. A nearby reservoir (Presa Rodrigo Gómez) adds diversity in the form of sundry waterbirds.

Access

Cola de Caballo (600 m elevation) is only about thirty minutes south of Monterrey, off Route 85, while the Highrise (1500 m elevation) can be reached in about an hour from the highway at the Cola de Caballo junction. Much of this area is a popular recreational destination for the large urban population of Monterrey; hence, it is quieter on weekdays. On weekends, traffic heading back to Monterrey can be very heavy, especially late on Sunday, and can stretch out the drive from Cola de Caballo to two hours or more. These sites also can be reached from Saltillo (about 2–2.5 hours to the Highrise, another 45 minutes to an hour to Cola de Caballo).

To reach Cola de Caballo and the Highrise from Monterrey, head south on Route 85 (toward Montemorelos and Linares) and mark zero at the intersection of Highways 85 (Avenida Garza Sada) and 40 (Avenida Constitución) on the south side of the Monterrey. Stay on the highway to El Cercado, after several miles of roadside developments and restaurants, and watch for a paved road off to the right (west) at 22.3 mi (37.3 km), well signed to Cola de Caballo ("Horsetail Falls"); this is a popular recreational spot, and there should be signs warning you in advance of the turn. Turn right onto the Cola de Caballo road and reset to zero. From here on up into the mountains there are numerous good areas for birding (see Birding Sites).

El Cercado can be reached easily via public buses from Monterrey, and you may be able to hitch a ride up to the mountains; there is at least a basic bus service into the area, but I don't have details. Other than visiting for the day from Monterrey, there are a small hotel in La Cienega (Cabañas y Hotel Rincon de Olvido, on the west edge of the village; likely to be full on weekends) and a newly opened (1997) fancy hotel at Cola de Caballo, or you can camp (take all your supplies) near the Highrise.

To reach the Highrise and Cola de Caballo from Saltillo, head out east and then south onto Route 57, as if going to San Antonio (Site 3.4), via either the old highway or the new *cuota* highway. Shortly south of Arteaga watch for a signed road east to Los Lirios. Exit onto this road, which is paved for about 14 mi (23 km) to the village of Los Lirios. As you bend into the village, you come to a T-junction: turn left and follow the main road (which soon becomes graded dirt, passable with care in a regular car) for about another 19 mi (32 km), through some spectacular scenery, to the Highrise viewpoint described under Birding Sites. Coming from Los Lirios, this viewpoint is about 6 mi (10 km) east of the (unsigned) village of San Jose de las Boquillas. The Cola de Caballo area is some 20–24 mi (35–40 km; about an hour) to the east (see Birding Sites).

Birding Sites

Coming south from Monterrey on Route 85, at about 20 mi (33 km) south of the junc-

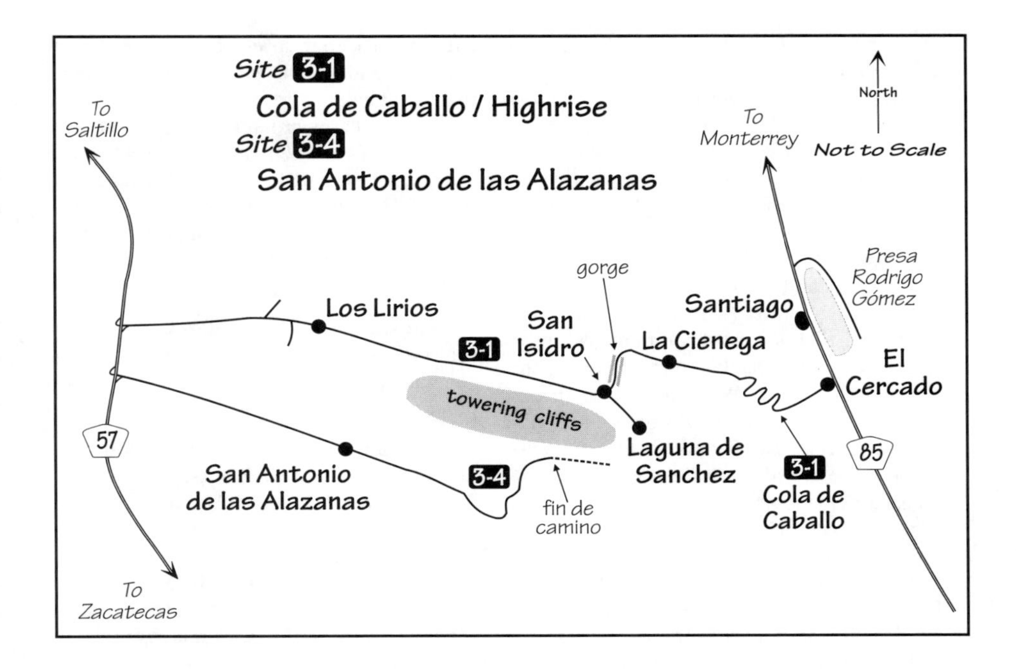

tion with Route 40, you enter Santiago (a strip of development and stalls along the highway) and can see a large reservoir (**Presa Rodrigo Gómez**) to your left. This reservoir can be reached only from the northbound lane: you can U-turn at 21.5 mi (36.3 km), at the crossroad junction signed right to El Cercado, left to El Embarcadero. From here to the north you can take any of several dirt side roads down to the reservoir and, once you're on the shore, it is possible to drive along in both directions. You can also bird the reservoir via a minor paved road about 3.6 mi (6 km) north of the U-turn and signed for Cadereyta and Presa Rodrigo Gómez: get onto the access road at the first sign (at 3 mi; 5 km) and continue a little less than a kilometer till the actual turn, which *cannot* be reached directly from the main highway. This road goes around the far (east) side of the reservoir. Birds on and beside the reservoir include Least Grebe, Double-crested and Neotropic cormorants, ducks (winter), Red-shouldered Hawk (winter), Peregrine Falcon, Plain Chachalaca, White-tipped Dove, Groove-billed Ani, Ringed Kingfisher, Long-billed Thrasher, **Crimson-collared Grosbeak**, Olive Sparrow, and Audubon's Oriole.

Once you turn off Route 85, the road winds and climbs away from the highway settlements into oak woodland, and at about 2.7 mi (4.5 km) there is a dirt road (signed inconspicuously to El Refugio) down to the right to where a stream runs through a small area of relatively lush, subtropical woodland. Above and to the right of the road here is the Cola de Caballo Hotel, whose entrance is at 3.6 mi (6.0 km) on the main road. Birds in this area of woodland and scrub (at 2.5–4 mi [4–6.5 km] along the main road) include Elegant Trogon, Blue-crowned Motmot, **Bronze-winged Woodpecker**, Green Jay, Black-crested [Tufted] Titmouse, Spot-breasted and Carolina wrens, Clay-colored Thrush, Long-billed Thrasher, Tropical Parula, Golden-crowned and **Rufous-capped warblers**, Flame-colored Tanager, **Crimson-collared Grosbeak**, Olive Sparrow, and Audubon's Oriole. The Red-crowned Parrots that can be seen here probably derive from escaped cage birds.

The main road climbs into drier woodland of oak (Bridled Titmouse, Painted Whitestart) and then pine-oak before leveling out at about 9.9 mi (16.5 km) (signed Vitroparque El Manzano) and continuing through grassy valleys, clearings with orchards, a few patches of woodland and chaparral, and a beautiful, narrow gorge (**Rufous-capped Brushfinch**, Rufous-crowned Sparrow) to the settlement of La Cienega at 13.2 mi (22 km). At 15.0 mi (25 km) you leave La Cienega and at 18.7 mi (31.3 km) bend sharply left and pass through an imposing, high-sided gorge (where Orange-billed Nightingale-Thrush has been found recently; Behrstock and Eubanks 1997), crossing the stream a few times. With care, these stream crossings usually are not a problem for ordinary cars. Shortly after the road leaves the gorge there are a few buildings comprising the small village of San Isidro; on the far side of this village, at 22.8 mi (38 km), turn right onto a graded (initially paved) road signed "Saltillo 75 km"; the road straight on goes to Laguna de Sanchez. This is now the road to Los Lirios (see Access from Saltillo).

Reset to zero as you make this turn and drive about 2.4 mi (4 km), to where you can pull off on the side of the road, shortly after a couple of farmhouses on the right, and about 25 mi (42 km) total from Route 85. Off and up to your left, if it's not hidden by clouds, you will see an impressive wall of towering cliffs—the Highrise, where **Maroon-fronted Parrots** nest. Between March/April and October/November the parrots often can be seen (and heard easily) from the road, although birds are often distant; at other seasons they are entirely or largely absent, although a few may be detected coming to roost in late afternoons. It is also possible to walk up and over the first ridge of pines and get much closer views of the cliffs, taking care not to get lost. Other birds here include **Pine Flycatcher**, and if you camp here or stay till dusk, listen for Flammulated (summer) and Spotted owls, Whiskered Screech-Owl, Mountain Pygmy-Owl, and Northern Saw-whet Owl, all of which have been reported in the area (Mexican Whip-poor-wills are also common).

If you have no luck with parrots at the Highrise, continue toward Los Lirios: parrots can be found until at least about 8.0 mi (13 km), where you enter the village of San Jose de las Boquillas (Colima Warblers in summer). Past San Jose you enter the state of Coahuila and start to get into the temperate plateau avifauna—for example, birds in the scrubby orchards around 9.7 mi (16 km) include Buff-breasted Flycatcher, Canyon Towhee, Western Bluebird, and Scott's Oriole. Birds from this point on are much like those listed for San Antonio (Site 3.4), which, as the parrot flies, is very near to the south.

Cola de Caballo/Highrise Bird List
(not including the reservoir): 81 sp.
†See pp. 5–7 for notes on English names.

Turkey Vulture
Osprey
Sharp-shinned Hawk
Red-tailed Hawk
American Kestrel
Peregrine Falcon
Band-tailed Pigeon
Mourning Dove
Inca Dove
White-tipped Dove
Maroon-fronted Parrot
White-throated Swift
Black-chinned Hummingbird
Elegant Trogon
Blue-crowned Motmot
Acorn Woodpecker
Golden-fronted Woodpecker
Yellow-bellied Sapsucker
Ladder-backed Woodpecker
Hairy Woodpecker
Red-shafted Flicker†
Greater Pewee
Hammond's Flycatcher
Pine Flycatcher
Black Phoebe
Eastern Phoebe
Green Jay
Brown Jay
Grey-breasted Jay†
Northern Raven†
Mexican Chickadee
Bridled Titmouse
Black-crested Titmouse†
Bushtit
Rock Wren
Canyon Wren
Spot-breasted Wren
Carolina Wren
Bewick's Wren
Northern House Wren†
Brown-throated Wren†
Ruby-crowned Kinglet
Townsend's Solitaire (R?)
Brown-backed Solitaire
Hermit Thrush
Clay-colored Thrush†
American Robin
Northern Mockingbird
Long-billed Thrasher
Phainopepla
Grey Silky†
White-eyed Vireo
Blue-headed Vireo†
Cassin's Vireo†
Plumbeous Vireo†
Hutton's Vireo
Orange-crowned Warbler
Nashville Warbler
Crescent-chested Warbler

Violet-green Swallow
N. Rough-winged Swallow
Cave Swallow
Barn Swallow
Chihuahuan Raven
Verdin
Rock Wren
Northern House Wren†
Marsh Wren
Ruby-crowned Kinglet
Blue-grey Gnatcatcher
Black-tailed Gnatcatcher
American Robin
Northern Mockingbird
American Pipit
Cedar Waxwing
Loggerhead Shrike
Bell's Vireo
Orange-crowned Warbler
Nashville Warbler
Northern Parula
Audubon's Warbler†
Myrtle Warbler†
Black-and-white Warbler
Common Yellowthroat
Northern Cardinal
Pyrrhuloxia
Green-tailed Towhee
Spotted Towhee
Black-throated Sparrow
Chipping Sparrow
Clay-colored Sparrow
Brewer's Sparrow
Vesper Sparrow
Lark Sparrow
Lark Bunting
Savannah Sparrow
Lincoln's Sparrow
Swamp Sparrow
White-crowned Sparrow
Western Meadowlark
Brewer's Blackbird
Great-tailed Grackle
Pine Siskin
Lesser Goldfinch
American Goldfinch

Site 3.3: Tanque de Emergencia/ Hedionda Grande, Coahuila

(November 1986, March 1988, April, June 1990, September 1995, January 1998)

Introduction/Key Species

This is the most reliable known site for the little-known **Worthen's Sparrow**, which is found most easily in winter near Tanque, where flocks may number 100+ birds, or in summer near Hedionda Grande, where a few pairs nest. Other species in this area include Ferruginous Hawk (winter), Golden Eagle, Prairie Falcon, Mountain Plover, Scaled Quail, Burrowing Owl, Western and Mountain (winter) bluebirds, Crissal Thrasher, Sprague's Pipit (winter), Cassin's Sparrow, Chestnut-collared Longspur (winter), and Western Meadowlark.

Access

Tanque de Emergencia (2100 m elevation) is about 30 mi (50 km; forty-five minutes to an hour) south of Saltillo, from which it is best visited as a day trip (mornings are best and tend to be less windy). From Saltillo, head south on Route 54 (toward Concepción del Oro and Zacatecas). At about 19 mi (32 km) from the edge of Saltillo there is a junction with a graded dirt road left (east) toward Hedionda Grande. At 6.6 mi (11 km) south of the Hedionda Grande junction, watch for an inconspicuous dirt road off to the left (east), signed "Rancho Arbolitos 2." Turn here and reset to zero; the junction is near Km Post 308 (which may be missing).

This area can be reached with public transport by getting off a second-class bus at the Tanque junction. From here you may have to walk (although buses do transit the bad dirt road) or could hitch a ride (traffic is light). There also should be public transport from Saltillo to Hedionda Grande.

Birding Sites

Having turned off the highway, in about 1.2 mi (2 km) you cross railway tracks and come to the small village of Tanque de Emergencia and a small earth-dammed cattle pond to your left. In winter, **Worthen's Sparrows** have been seen around here, and anywhere along the road from this point on. The best spot, however, has been some 3 mi (5 km) beyond the village: stay on the main dirt road (rough in places), which skirts the north side of the village, goes through some desert grassland (Cassin's and Black-throated sparrows), and, about 4.8 mi (8 km) from the highway, passes a broken, circular, concrete water tank on the left (north). There is another earth-dammed drinking pond (this and similar ponds often have a few waterbirds) a little farther ahead on the right, and the whole area to the right of the road is very short-grass, with a large prairie dog colony and Burrowing Owls. To the left, across the fence that has paralleled the road for a while, the grass is taller (in winter, Sprague's Pipit, Chestnut-collared Longspur) with scattered low bushes. **Worthen's Sparrows** have been found regularly in winter within a few hundred meters of the old water tank, on both

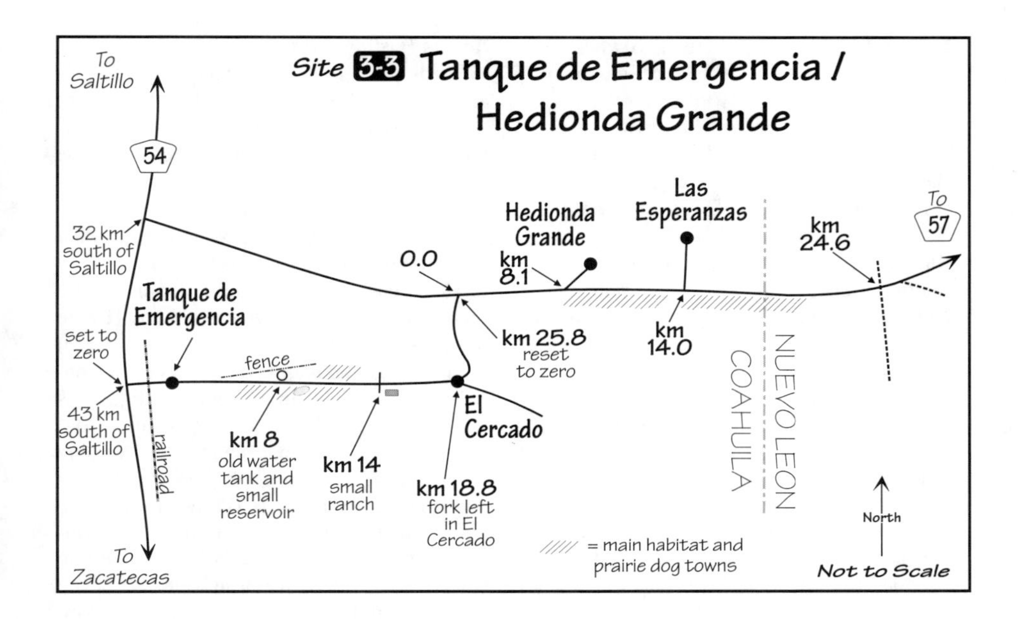

sides of the road. Flocks roam about in winter, and the few pairs that may have nested here did so on the north side of the fence.

About 3.6 mi (6 km) farther along the road beyond the water tank you pass a house and some trees on the right and another seasonal small cattle pond to the left. Beyond the house, off ahead and down to the right (south) is a slope with yuccas and scattered low trees where Crissal Thrasher has been found. If you haven't seen the sparrows by this point, continue east to a small village (El Cercado, unsigned) at about 11 mi (18 km) from the highway and turn left on the far side of town at about 11.2 mi (18.8 km). This dirt road heads north for about 4 mi (7 km) to hit the Hedionda Grande road at 15.2 mi (25.8 km), and **Worthen's Sparrows** have been seen on this stretch in winter (mixed with Chipping Sparrows and House Finches). To reach the Hedionda Grande area, turn right at this junction and reset to zero. At about 4.8 mi (8.1 km) there is a left fork signed into the village of Hedionda Grande. Bear right here on the main road and continue for several miles with prairie dog towns and short-grass habitat mainly on the right (south) side of the road (Mountain Plover, nesting Worthen's Sparrows). At 8.5 mi (14.0 km) you pass a left turn to Las Esperanzas, then cross the state line into Nuevo León, and at 14.7 mi (24.6 km) you come to a major junction. Habitat after this point appears not too good for the sparrows, but it can be good for general birding and, with some judicious navigating (take care not to get lost in the maze of dirt roads), you can reach Route 57 at about 21.5 mi (35.8 km). Alternatively, continue back past Hedionda Grande and out to Route 54 and Saltillo.

Tanque de Emergencia/Hedionda Grande Bird List

75 sp. †See pp. 5–7 for notes on English names.

Northern Pintail
Turkey Vulture
Northern Harrier
Red-tailed Hawk
Ferruginous Hawk
Golden Eagle
American Kestrel
Prairie Falcon
Scaled Quail
Killdeer
Upland Sandpiper
Baird's Sandpiper
White-winged Dove
Mourning Dove
Inca Dove
Greater Roadrunner
Burrowing Owl
Broad-tailed Hummingbird
Golden-fronted Woodpecker
Ladder-backed Woodpecker
Red-shafted Flicker†
Say's Phoebe
Cassin's Kingbird

Horned Lark
Violet-green Swallow
N. Rough-winged Swallow
Barn Swallow
Western Scrub Jay
Northern Raven†
Bushtit
Cactus Wren
Northern House Wren†
Bewick's Wren
Ruby-crowned Kinglet
Black-tailed Gnatcatcher
Western Bluebird
Mountain Bluebird
Northern Mockingbird
Curve-billed Thrasher
Crissal Thrasher
American Pipit
Sprague's Pipit
Cedar Waxwing
Phainopepla
Loggerhead Shrike
Audubon's Warbler†
Wilson's Warbler
Pyrrhuloxia
Green-tailed Towhee
Spotted Towhee
Canyon Towhee
Cassin's Sparrow
Rufous-crowned Sparrow
Black-throated Sparrow
Chipping Sparrow
Clay-colored Sparrow
Brewer's Sparrow
Worthen's Sparrow
Black-chinned Sparrow
Vesper Sparrow
Lark Sparrow
Lark Bunting
Grasshopper Sparrow
Savannah Sparrow
Lincoln's Sparrow
White-crowned Sparrow
Chestnut-collared Longspur
Red-winged Blackbird
Western Meadowlark
Brewer's Blackbird
Great-tailed Grackle
Black-vented Oriole
Scott's Oriole
House Finch
Lesser Goldfinch

Site 3.4: San Antonio de las Alazanas, Coahuila

(December 1986, April 1990, checked in part April/May 1996)

Introduction/Key Species

This is a traditional site for **Maroon-fronted Parrot**, although the Highrise, actually very near to San Antonio but more easily reached from Monterrey, is perhaps a more reliable location (see Site 3.1). A birding trip to San Antonio offers access to high-elevation chaparral and conifer habitat, with birds including **Pine Flycatcher**, **Grey Silky**, Colima Warbler (summer), and **Hooded Yellowthroat**.

Access/Birding Sites

Birding sites near San Antonio de las Alazanas are about 45–50 mi (75–80 km; 1.5–2.5 hours) from Saltillo, from which this area makes a good day trip. It should also be possible to find at least basic accommodation in or near San Antonio, and one could camp in the area (take all your supplies and be aware that it can be cold at night).

Head east and then south from Saltillo on the new four-lane divided highway (Route 57D) toward Matehuala. About 10 mi (16 km) after Arteaga (and about 18.5 mi/31 km from Saltillo), as you come through a high pass with hills on either side and begin to descend, look for an exit signed San Antonio de las Alazanas. Turn off here (exit right and then turn back left) and follow the paved road 14 mi (23 km) through farmland dominated by orchards to San Antonio (2150 m elevation). Continue on the paved road through town, and at 22.2 mi (37.1 km) the pavement ends; stay on the main road, which climbs through pine groves and meadows, with numerous summer homes along the way. At 27.8 mi (46.4 km) you crest a pass (about 2800 m elevation) marked by an inconspicuous rusty sign announcing "Puerto de las Cumbres" (Flammulated Owl, **Russet Nightingale-Thrush**). Here the road starts dropping, and it ends (to all intents and purposes) at 28.3 mi (50.6 km) and a gate, signed *fin de camino* (end of road). You can park here, and **Pine** and Buff-breasted **flycatchers** can be found in the surrounding open pines. Up to the left are imposing high cliffs, and just below you to the right is a small farmhouse. It is possible to continue on foot along the dirt road, through the farm, and along the other side of the valley (**Hooded Yellowthroat**) for several kilometers.

Maroon-fronted Parrots can be seen anywhere around the *fin de camino* and in the conifers around the pass before this point. You can simply wait by the car, listening for far-carrying, raucous and laughing calls to signal their presence. Beware that they often fly very high, and the calls can be frustratingly hard to trace—one can hear the birds for an hour or more before seeing them, or may perhaps never see them! I also have seen the parrots feeding on agaves in the valley 1–2 km beyond the *fin de camino* (April).

Birding this area by public transport would be difficult. Buses serve San Antonio, and you might be able to hitch to the *fin de camino*, but traffic may be virtually nonexistent.

San Antonio de las Alazanas Bird List (San Antonio to fin de camino): 56 sp. †See pp. 5–7 for notes on English names.

Turkey Vulture
Red-tailed Hawk
American Kestrel
Mourning Dove
Maroon-fronted Parrot
Mexican Whip-poor-will†
White-throated Swift
Broad-tailed Hummingbird
Golden-fronted Woodpecker
Red-shafted Flicker†
Greater Pewee
Pine Flycatcher
Buff-breasted Flycatcher
Say's Phoebe
Cassin's Kingbird
Horned Lark
Violet-green Swallow
Western Scrub Jay
Grey-breasted Jay†
Northern Raven†
Bushtit
Mexican Chickadee
Bewick's Wren
Brown-throated Wren†
Ruby-crowned Kinglet
Western Bluebird
Mountain Bluebird
Townsend's Solitaire
Hermit Thrush
American Robin
Northern Mockingbird
Curve-billed Thrasher
American Pipit
Grey Silky†
Loggerhead Shrike
Hutton's Vireo
Colima Warbler
Audubon's Warbler†
Townsend's Warbler
MacGillivray's Warbler
Hooded Yellowthroat
Wilson's Warbler
Slate-throated Whitestart†
Olive Warbler
Hepatic Tanager
Spotted Towhee
Canyon Towhee
Rufous-crowned Sparrow
Chipping Sparrow
Savannah Sparrow
Mexican Junco†
Western Meadowlark
Brewer's Blackbird
Great-tailed Grackle
Scott's Oriole
House Finch
Pine Siskin

Site 3.5: Cerro El Potosí, Nuevo León

(January 1985, December 1986, April, June 1990, checked April/May 1996)

Introduction/Key Species

This isolated mountain rises to about 3800 m (12,500 ft) and stands out like a beacon from the surrounding desert plains of southern Nuevo León. A graded dirt and stone road allows access to the radio towers atop El Potosí, where one can see Clark's Nutcracker (apparently a resident population). Other species of interest include **Maroon-fronted Parrot** (nomadic), Flammulated Owl (summer), **Pine Flycatcher**, **Aztec Thrush** (in winter, at least), Crissal Thrasher, Colima Warbler (summer), and **Hooded Yellowthroat**.

Access

Access to Cerro El Potosí is from Galeana (1500 m elevation), a small town in southern Nuevo León. Galeana can be reached in about 2.5–3.5 hours from Saltillo via Route 57: drive about 100 mi (160 km) south from Saltillo to Entronque San Roberto (with a large Pemex station) and turn left (east) onto Route 58. Continue about 23 mi (38 km), until the left (north) turnoff signed to Galeana, which is about 4.8 mi (8 km) from the turn. If you come from this direction, in winter look for raptors (Ferruginous Hawk, Golden Eagle) along Route 57. **Worthen's Sparrows** have been seen in the prairie dog towns a few kilometers north of and around San Roberto.

Alternatively, Galeana can be reached in about 1.5–2 hours from the Gulf Slope at Linares, via Route 58, a drive of only about 38.5 mi (64 km) to the Galeana turnoff, but a fairly long and winding climb in places, especially if you're stuck behind trucks. If you stay overnight in Linares, note that **Tawny-collared Nightjars** (summer) can be found in the thorn forest north of Route 58 around 7.4 mi (12 km) west of town.

There is a basic motel in Galeana, or you can camp (take all your supplies) on Cerro El Potosí, but beware that nights on the mountain can be cold. While public buses serve Galeana from Linares (and presumably also from Saltillo), birding Cerro El Potosí without your own vehicle is problematic: your only chance is to walk, or to hitch a ride up from 18 de Marzo with one of the infrequent vehicles.

Birding Sites

Having turned north off Route 58, at 3.3 mi (5.5 km) you enter Galeana under an arch, shortly after which the road forks into a one-way system, by a statue. Stay straight on Calle 5 de Mayo, and shortly, at the first traffic light (4.4 mi [7.3 km] from Route 58), turn left onto Cuauhtemoc in response to signs that say "Cerro El Potosí 14 km" and "18 de Marzo." Stay straight on this road through town and follow directions as below.

If you're staying in Galeana, take the street that borders the south side of the *zócalo* and follow it westward, bearing left at the fork just off the *zócalo*. Stay on this road through the dusty outskirts of Galeana and it becomes the paved road to the small village of 18 de Marzo. About 2.5 mi (4 km) from the *zócalo* is a road to the left that heads toward the base of Cerro El Potosí: this road goes a short distance to La Laguna de Labradores, a village by a small, dammed lagoon (Pied-billed Grebe, American Coot, usually little else), and beyond, through desert with low bushes (Western Scrub Jay, Crissal Thrasher) to the lower slopes of the Cerro El Potosí, before it peters out into deeply rutted logging trails. The pines and chaparral that can be reached here hold **Pine Flycatcher**, Grey-breasted Jay, Mexican Chickadee, **Hooded Yellowthroat**, and Olive Warbler.

To get to the top of Cerro El Potosí, stay on the paved road to 18 de Marzo, 9.3 mi (15.5 km) from the Galeana *zócalo*. As you enter 18 de Marzo, a dirt road goes off to the left (signed to Cerro Potosí). Mark this as zero. It is about 18.0 mi (30 km) to the top from the 18 de Marzo turnoff, and this can take 2–3 hours, even with few or no birding stops. The road can be driven with care in a regular car, but conditions vary, as does the degree of maintenance, and other vehicles tend to be few and far between.

Birding anywhere along the road can be good, and driving slowly while looking and listening for birds is the best approach. **Pine Flycatchers** can be found between 1.8 and 3.6 mi (3–6 km) (note that Hammond's occurs in winter and Western in summer), and **Aztec Thrushes** have been seen around 7.2 mi (12 km). Pygmy Nuthatches are common above 10.2 mi (17 km), Steller's Jays appear above 12 mi (20 km), and Clark's Nutcrackers can be found right at the top, in the stunted trees around the microwave station (the nutcrackers also can be missed easily!). In summer, Flammulated Owls are locally common—for instance, around 10–11 mi (20–22 km), near the first radio towers—and Colima and MacGillivray's warblers sing from the chaparral-covered slopes. **Maroon-fronted Parrots** are sporadic in their occurrence and can be seen anywhere in the zone of pines (May to July may be the best season).

Cerro El Potosí Bird List

106 sp. [a]Overlooked as being resident by Howell and Webb (1995). †See pp. 5–7 for notes on English names.

Pied-billed Grebe
Turkey Vulture
Northern Harrier
Cooper's Hawk
Sharp-shinned Hawk
Red-tailed Hawk
Golden Eagle
American Kestrel
Peregrine Falcon
Scaled Quail
American Coot
Band-tailed Pigeon
Mourning Dove
Flammulated Owl
Mountain Pygmy-Owl†
Mexican Whip-poor-will†
White-throated Swift
Magnificent Hummingbird
Broad-tailed Hummingbird
Acorn Woodpecker
Golden-fronted Woodpecker
Yellow-bellied Sapsucker
Ladder-backed Woodpecker
Hairy Woodpecker
Red-shafted Flicker†
Greater Pewee
Hammond's Flycatcher
Pine Flycatcher
Western Flycatcher†
Eastern Phoebe
Say's Phoebe
Vermilion Flycatcher
Cassin's Kingbird
Horned Lark
Violet-green Swallow
Cave Swallow
Steller's Jay
Western Scrub Jay
Grey-breasted Jay†
Clark's Nutcracker
Chihuahuan Raven
Northern Raven†
Mexican Chickadee
Bridled Titmouse
Bushtit
Pygmy Nuthatch
Brown Creeper

Cactus Wren
Canyon Wren
Bewick's Wren
Northern House Wren†
Brown-throated Wren†
Ruby-crowned Kinglet
Blue-grey Gnatcatcher
Western Bluebird
Mountain Bluebird
Townsend's Solitaire
Brown-backed Solitaire
Hermit Thrush
Aztec Thrush
American Robin
Northern Mockingbird
Curve-billed Thrasher
Crissal Thrasher
Cedar Waxwing
Grey Silky†
Loggerhead Shrike
Hutton's Vireo
Orange-crowned Warbler
Colima Warbler
Audubon's Warbler†
Black-throated Grey Warbler
Townsend's Warbler
Hermit Warbler
MacGillivray's Warbler
Hooded Yellowthroat
Painted Whitestart†
Slate-throated Whitestart†
Olive Warbler
Blue-hooded Euphonia
Hepatic Tanager
Black-headed Grosbeak
Spotted Towhee
Canyon Towhee
Rufous-crowned Sparrow
Chipping Sparrow
Clay-colored Sparrow
Brewer's Sparrow
Black-chinned Sparrow
Vesper Sparrow
Lark Sparrow
Lark Bunting
Grasshopper Sparrow
Savannah Sparrow
Lincoln's Sparrow
White-crowned Sparrow
Mexican Junco†
Western Meadowlark
Brewer's Blackbird
Great-tailed Grackle
Scott's Oriole
House Finch
Red Crossbill
Pine Siskin[a]
Lesser Goldfinch
American Goldfinch

CHAPTER 4 THE NORTHEASTERN GULF SLOPE

Hooded Grosbeak

Like Chapter 3, this chapter covers areas that can be reached within a day's drive from the U.S.A./Mexico border in Texas. The best access to sites on the northeastern Gulf Slope is via the Brownsville/Matamoros border crossing, taking Routes 180/101 south toward Ciudad Victoria, which is about 185 mi (310 km; 4–5 hours) from Brownsville. This area can be made into a trip of about a week, or some sites can be done as a long weekend. Alternatively, sites can be combined with areas around Monterrey/Saltillo (Chapter 3) for a more varied trip of up to two weeks. Car rental is possible in Matamoros and also in Ciudad Victoria, although options in the latter city are both relatively limited and expensive. (Renting a car in Monterrey allows more choices of vehicles and prices.)

The avifauna changes strikingly between South Texas and Gómez Farías (Site 4.4) or El Naranjo (Site 4.5), only 200 miles, as the antshrike doesn't fly, to the south. Bird diversity simply shoots up as you cross that magical, invisible line, the Tropic of Cancer, marking the northernmost point at which the sun can shine directly overhead. Thus, species characteristic of the lowlands and foothills of northeastern Mexico, only six hours' drive from Texas (!), include Thicket Tinamou, Bat Falcon, Singing Quail, Military Macaw, Green Parakeet, Red-crowned and Yellow-headed parrots, Tawny-collared Nightjar, Wedge-tailed Sabrewing, Mountain Trogon, Ivory-billed Woodcreeper, Grey-collared Becard, Tamaulipas Crow, Altamira Yellowthroat, Fan-tailed Warbler, Blue-hooded Euphonia, Crimson-collared and Hooded grosbeaks, and many widespread tropical species such as Red-lored Parrot, Blue-crowned Motmot, Lineated Woodpecker, Boat-billed Flycatcher, and Golden-crowned Warbler. This is just the tip of the "neotropical iceberg" that has hooked so many birders into venturing farther south.

In general, however, the native vegetation of Mexico's Atlantic Slope has been altered drastically by human activities, such that little untouched forest remains. Most of the flatlands have been converted to ranching and crops, so today it is hard to believe that the lowlands of southern Tamaulipas and northern Veracruz were once forested—and had populations of Scarlet Macaws! But the steep limestone ridges of the foothills around Gómez Farías and El Naranjo are neither easy to clear nor particularly good for agriculture and, as a consequence, impressively large tracts of subtropical forest remain, yet are easily accessible by paved road, as at El Naranjo.

The coast is characterized by vast lagoon complexes protected by barrier beaches, much like South Texas. The lagoons are fringed by mangroves and constitute important nesting areas for waterbirds in general, as well as important wintering grounds for migrant waterfowl and shorebirds.

Suggested Itineraries

The main areas of interest here are Gómez Farías (Site 4.4) and El Naranjo (Site 4.5); the latter has easier access to a greater range of habitats. Several days at least are needed to do justice to either area, but they could be visited as a long weekend trip from South Texas if you reached Ciudad Mante (or beyond) on Friday night, birded all day Saturday and Sunday morning, and headed back Sunday afternoon. La Pesca (Site 4.2) makes an easy overnight or weekend trip from South Texas. The mouth of the Río Bravo/Río Grande (Site 4.1) is an easy day trip from Texas and, with Presa Vicente Guerrero (Site 4.3), also can be visited en route to/from points south.

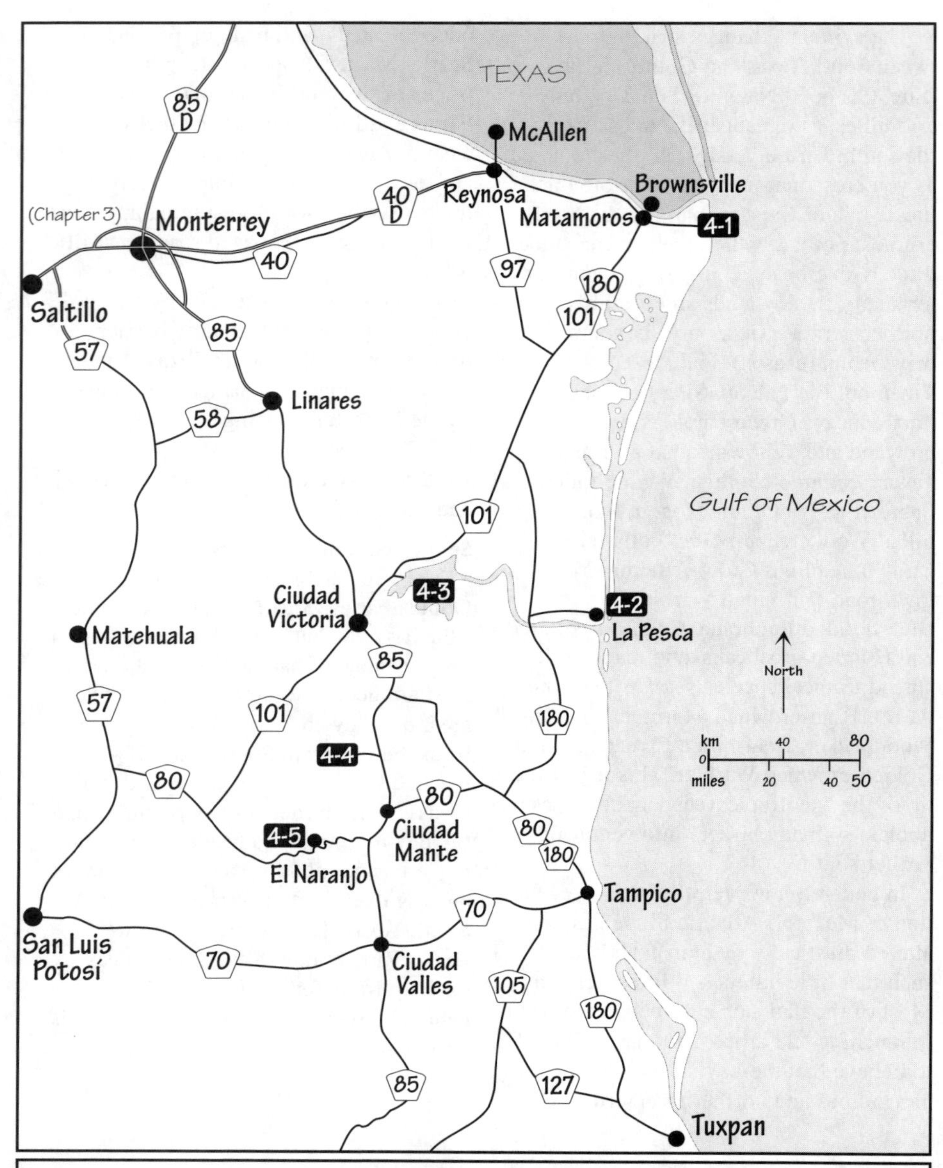

Chapter 4: The NORTHEASTERN GULF SLOPE

- **4-1** Mouth of the Río Bravo / Río Grande, Tamaulipas
- **4-2** La Pesca, Tamaulipas
- **4-3** Presa Vicente Guerrero, Tamaulipas
- **4-4** Gómez Farías, Tamaulipas
- **4-5** El Naranjo, San Luis Potosí

NORTHEASTERN GULF SLOPE SITE LIST

Site 4.1: Mouth of the Río Bravo/Río Grande, Tamaulipas

(April 1990)

Introduction/Key Species

Birders in South Texas have long known that the mouth of the Río Grande (known in Mexico as the Río Bravo) is a good place for gulls (especially Lesser Black-backed Gull) and other waterbirds; indeed, the birds are often on the Mexican side of the river! Other species of note include Reddish Egret, Mottled Duck, Wilson's Plover, Gull-billed and Least (summer) terns, Seaside Sparrow (breeds locally along the U.S. side of the river), and Nelson's Sharp-tailed Sparrow (winter).

Access/Birding Sites

The Mexican side of the Río Bravo/Río Grande is about a 33-mi drive (54 km; 1 to 1.5 hours) from Matamoros: drive east on the paved road (Highway 2) to Playa Lauro Villar, a popular recreational spot that should be signed. It is about 23 mi (38 km) to the beach, and the last few kilometers go through some nice salt marshes (nesting Willets and Forster's Terns). Mottled Ducks can be seen in roadside swales along this route. At about 10.0 mi (16 km), in the town of El Refugio (which may not be signed), a dirt road goes left (north) where the highway makes a turn to the right. This dirt road can be impassable when wet but goes through farmland with pools and marshes where Mottled Ducks also occur.

When you reach Playa Lauro Villar, to get to the mouth of the Río Bravo drive onto the beach and turn left (north). It is a 10 mi (16 km) drive along the beach (the hardest sand is often nearest the waves, so a thorough freshwater rinse of your car afterward is a good idea), and, at least on weekends, there may be a fair amount of traffic (mainly fisherman heading to the river mouth) to help in case you get stuck. The beach drive is good for gulls and shorebirds, and Lesser Black-backed Gull can be seen anywhere from here to the river mouth. When you reach the mouth there are several hard-sand places to park safely, and you can walk inland into the salt marsh (Seaside and Nelson's Sharp-tailed sparrows) along the south side of the river. Note that some tidal channels are deep and not easily crossed; consequently, there is not a huge expanse of easily accessible marsh.

Reaching Playa Lauro Villar by public transport is easy from Matamoros, and from there you'd have to try and hitch a ride north on the beach.

Mouth of the Río Grande/Río Bravo Bird List: 66 sp.

Eared Grebe
Brown Pelican
Double-crested Cormorant
Great Blue Heron
Great Egret
Snowy Egret
Little Blue Heron
Tricolored Heron
Reddish Egret
Cattle Egret
Mottled Duck
Blue-winged Teal
Northern Shoveler
Bufflehead
Red-breasted Merganser
Northern Harrier
Black-bellied Plover
Wilson's Plover
Semipalmated Plover
Killdeer
American Oystercatcher
Black-necked Stilt
American Avocet
Greater Yellowlegs
Lesser Yellowlegs
Willet
Whimbrel
Long-billed Curlew
Ruddy Turnstone
Red Knot
Sanderling
Western Sandpiper
Least Sandpiper
Pectoral Sandpiper
Dunlin
Short-billed Dowitcher
Long-billed Dowitcher
Wilson's Phalarope
Laughing Gull
Franklin's Gull
Bonaparte's Gull
Ring-billed Gull
Herring Gull
Lesser Black-backed Gull
Gull-billed Tern
Caspian Tern
Royal Tern
Sandwich Tern
Forster's Tern
Least Tern
Black Skimmer
Mourning Dove
Ruby-throated Hummingbird
Golden-fronted Woodpecker

Least Sandpiper
Pectoral Sandpiper
Laughing Gull
Franklin's Gull
Ring-billed Gull
Herring Gull
Caspian Tern
Royal Tern
Sandwich Tern
Common Tern
Forster's Tern
Least Tern
Black Skimmer
Red-billed Pigeon
White-winged Dove
Common Ground-Dove
Yellow-headed Parrot
Ferruginous Pygmy-Owl
Lesser Nighthawk
Pauraque
Chuck-will's-widow
Tawny-collared Nightjar
Chimney Swift
Ruby-throated Hummingbird
Belted Kingfisher
Golden-fronted Woodpecker
Ladder-backed Woodpecker
Ivory-billed Woodcreeper
Brown-crested Flycatcher
Scissor-tailed Flycatcher
Great Kiskadee
Tropical Kingbird (R)
Couch's Kingbird
Eastern Kingbird
Tree Swallow
Brown Jay
Chihuahuan Raven
Tamaulipas Crow†
Black-crested Titmouse†
Northern House Wren†
Marsh Wren
Ruby-crowned Kinglet
Blue-grey Gnatcatcher
Northern Mockingbird
Long-billed Thrasher
Loggerhead Shrike
White-eyed Vireo
Yellow-throated Vireo
Red-eyed Vireo
Warbling Vireo
Blue-winged Warbler
Orange-crowned Warbler
Nashville Warbler
Northern Parula
Myrtle Warbler†
Yellow-throated Warbler
Black-and-white Warbler
Worm-eating Warbler
Ovenbird
Kentucky Warbler
Common Yellowthroat
Grey-crowned Yellowthroat
Hooded Warbler
Summer Tanager
Northern Cardinal
Pyrrhuloxia
Blue Grosbeak
Indigo Bunting
Olive Sparrow
White-collared Seedeater
Clay-colored Sparrow
Vesper Sparrow
Lark Sparrow
Grasshopper Sparrow
Savannah Sparrow
Lincoln's Sparrow
Swamp Sparrow
Eastern Meadowlark
Red-winged Blackbird
Bronzed Cowbird
Great-tailed Grackle
Orchard Oriole
Altamira Oriole

Site 4.3: Presa Vicente Guerrero, Tamaulipas

(April 1990)

Introduction/Key Species

If you're driving Route 101 from Matamoros to Ciudad Victoria, it may be worth checking this large reservoir southeast of the highway about 24 mi (40 km) northeast of Victoria. Cattail-lined irrigation ditches near the reservoir appear to mark the northern range limit of **Altamira Yellowthroat**, and the reservoir holds the usual variety of waterbirds typical of such sites in northeastern Mexico. It's also a popular spot for fishermen.

Access/Birding Sites

The drive from Matamoros to Ciudad Victoria is about 185 mi (310 km). Having passed through the town of (Santander) Jimenez (about 132 mi/220 km from Matamoros), after about another 30 mi (50 km) you come to the village of Nueva Padilla (24 mi/40 km northeast of Victoria). Turn left (east) in Nueva Padilla on a paved but potholed side road, opposite a road to the right (west) signed to Barretal; mark this turn as zero. At 3.0 mi (5 km) turn left on a dirt road (probably signed for the lake) alongside a mesquite-edged dyke and reset to zero. Stay on the main road, through a gate at 2.7 mi (4.5 km), and at 6.0 mi (10 km) the reservoir will be on your left. Between 3 and 3.6 mi (5–6 km) on the dirt road you pass some cattail marsh (**Altamira Yellowthroat**, also migrant Common Yellowthroats).

Note: Some maps show another access road to the reservoir heading east off Route 101 a few kilometers south of Nueva Padilla. This may be just as good and/or easier than the route described above.

Presa Vicente Guerrero Bird List

45 sp.

American White Pelican
Double-crested Cormorant
Neotropic Cormorant
Great Blue Heron
Great Egret
Snowy Egret

Tricolored Heron
Gadwall
Black Vulture
Turkey Vulture
Harris' Hawk
Swainson's Hawk
Crested Caracara
American Kestrel
Peregrine Falcon
Common Moorhen
American Golden Plover
Killdeer
Upland Sandpiper
Whimbrel
Long-billed Curlew
Least Sandpiper
Baird's Sandpiper
Pectoral Sandpiper
Laughing Gull
Ring-billed Gull
Caspian Tern
Forster's Tern
Mourning Dove
Ruby-throated Hummingbird
Golden-fronted Woodpecker
Couch's Kingbird
Carolina Wren
Marsh Wren
White-eyed Vireo
Common Yellowthroat
Altamira Yellowthroat
Blue Grosbeak
Olive Sparrow
White-collared Seedeater
Cassin's Sparrow
Swamp Sparrow
Eastern Meadowlark
Bronzed Cowbird
Hooded Oriole

Site 4.4: Gómez Farías, Tamaulipas

(January 1985, December 1986, January 1998)

Introduction/Key Species

The Gómez Farías area, long popular with birders and ornithologists, marks an obvious point of entry into the truly neotropical avifauna of eastern Mexico, yet it is only a relatively short distance from Texas. Habitats are varied: humid tropical lowland vegetation penetrates north along the Río Sabinas, and the foothill ridges of the Sierra Madre Oriental to the west are cloaked in oak and evergreen forest.

Note: Some of the best areas (the forest on the upper slopes) are accessible only with a high-clearance and (depending on season) four-wheel-drive vehicle, or by healthy hikes. El Naranjo (Site 4.5) is about two hours by road farther south and has easier access (via a paved highway) to essentially the same avifauna, including the higher interior areas.

Gómez Farías and/or El Naranjo make an excellent introduction to neotropical birds, and persons unfamiliar with tropical avifaunas can profitably spend a week or more in either area. However, if you're planning to continue driving, or already have been, farther south in eastern Mexico, there is less reason to visit these tropical outposts—most species will be commoner farther south.

Species endemic to northeastern Mexico and found around Gómez Farías and El Naranjo are **Green Parakeet**, **Red-crowned Parrot**, **Tamaulipas Pygmy-Owl**, **Bronze-winged Woodpecker**, **Tamaulipas Crow**, and **Crimson-collared Grosbeak**. Other species of interest include Thicket Tinamou, **Singing Quail**, Military Macaw, **Grey-collared Becard**, **Fan-tailed Warbler**, and **Hooded Grosbeak**.

Access

Gómez Farías (elevation 350 m; 1150 ft) is about 260 mi (425 km; 5.5–6.5 hours) south of the Brownsville/Matamoros border crossing. From Matamoros, take Route 101 to Ciudad Victoria and, skirting town on the *periferico* (bypass), head south on Route 85 toward Ciudad Mante. About 25 mi (42 km) south of the major junction with Route 247 (some 31 mi/52 km south of Victoria), look for a paved road signed to the right (west) to Gómez Farías, about 0.6 mi (1 km) after crossing the (unsigned) Río Sabinas. This side road heads straight, through some fields, before winding up through brushy woodland and thickets to the small town of Gómez Farías, about 6.3 mi (10.5 km) from the highway.

Ciudad Mante, 24 mi (40 km; 30 minutes) south of Gómez Farías, has a number of hotels, or you could camp (be prepared for misty or rainy nights at any time of year) or try to find basic accommodation nearer Gómez Farías.

Birding Sites/Bird List

Some good birds can be found easily by walking along the road below Gómez Farías and taking narrow trails into the woodland here. The best birding is generally at 2.4–6 mi (4–10 km) from the highway. Species found below town include Thicket Tinamou, Zone-tailed Hawk, Bat Falcon, Aztec Parakeet, White-crowned, Red-lored, and **Red-crowned parrots**, Squirrel Cuckoo, Mottled Owl, **Wedge-tailed Sabrewing**, **Canivet's Emerald**, Elegant Trogon, Blue-crowned Motmot, Lineated and Pale-billed woodpeckers, Ivory-billed Woodcreeper, Barred

Antshrike, Boat-billed and Social flycatchers, Rose-throated Becard, Masked Tityra, Spot-breasted Wren, Golden-crowned and **Rufous-capped warblers**, Scrub, Yellow-throated, and Blue-hooded euphonias, Red-throated Ant-Tanager, Black-headed Saltator, **Crimson-collared Grosbeak**, **Blue Bunting**, Yellow-faced Grassquit, and Melodious Blackbird, as well as several "South Texas specialities," such as Plain Chachalaca, Buff-bellied Hummingbird, Brown Jay, Clay-colored Thrush, Tropical Parula, and Altamira and Audubon's orioles, plus numerous migrants.

It is also possible to drive through Gómez Farías to some subtropical broadleaf forest above town. Reset to zero at the *zócalo*, which is 6.2 mi (12 km) from the highway, and continue through town on the main road, which at 1.2 mi (2.0 km) becomes a broken stone road (passable with care in a regular car, depending on recent weather); fork left at 1.6 mi (2.7 km), and at 2.1 mi (3.6 km) you pass a fork: on the left is the road; on the right is a foot trail to the small *ejido* of Alta Cima, which is about 6 mi (10 km) from Gómez Farías. There is good forest along the escarpment only 4–8 km from town, and at about 5.3 mi (8.5 km) there is a small shrine beside the road and enough room to turn around if the road and your clearance don't seem to make a good match. The road continues up into the mountains from Alta Cima but is transitable only in a high-clearance vehicle.

Birds above Gómez Farías include **Singing Quail**, Blue Ground-Dove, Ruddy Quail-Dove, Military Macaw, Smoky-brown and **Bronze-winged woodpeckers**, Olivaceous and Spot-crowned woodcreepers, Green Jay, **Brown-backed Solitaire**, **Blue Mockingbird**, **Grey Silky**, Rufous-browed Peppershrike, **Crescent-chested Warbler**, **Yellow-winged**, Flame-colored, and White-winged **tanagers**, Yellow-billed Cacique, **Crimson-collared** and **Hooded** (common) **grosbeaks**, and many of the species listed for the woodlands below town. Other species to look for here include **Tamaulipas Pygmy-Owl** and **Grey-collared Becard**.

With more time, many of the species listed for El Naranjo (Site 4.5), where I have spent more time, can be found around Gómez Farías. See Arvin (1990) for an annotated checklist of species that have been recorded in the Gómez Farías region.

Site 4.5: El Naranjo, San Luis Potosí

(December 1981, March/April 1983, December 1986, April, June 1990, October 1993, January 1998)

Introduction/Key Species

The El Naranjo area, like Gómez Farías (Site 4.4), has for many years been a popular destination for birders driving down from Texas, and Christmas Bird Counts began at both sites in 1972. See the introduction/key species account under Gómez Farías for background information relating to El Naranjo (species of interest at both sites are much the same).

In particular, this is an easily accessible site for **Tamaulipas Pygmy-Owl** and is an excellent place to see (not just hear) Thicket Tinamou and **Singing Quail**, as well as Muscovy Duck, Military Macaw (nomadic), **Bronze-winged Woodpecker**, **Grey-collared Becard**, and **Crimson-collared Grosbeak**. In winter, the oak woods have large mixed-species flocks of migrants dominated by Blue-grey Gnatcatchers, Blue-headed [Solitary] Vireos, Black-throated Green and Black-and-white warblers, and, higher up, Ruby-crowned Kinglets and Townsend's Warblers. These flocks often can be attracted by imitating the slow-paced whistles of the Tamaulipas Pygmy-Owl.

Access

El Naranjo (270 m elevation) is about 300 mi (500 km; 7–8 hours) from Brownsville/Matamoros and can be reached in a day's drive from Texas, via Ciudad Victoria (see directions under Gómez Farías, Site 4.4). From Ciudad Victoria head south on Route 85, past the Gómez Farías turnoff, and on to Ciudad Mante. From Mante, head south on Route 85 for about 15.5 mi (26 km) to the small junction town of Antiguo Morelos. Turn right (west) here on Route 80 (which should be signed to El Naranjo and/or Ciudad del Maiz). After winding over two ridges of mostly cutover habitats you descend to the broad, cultivated valley of El Naranjo, about 19.2 mi (32 km) from Antiguo Morelos.

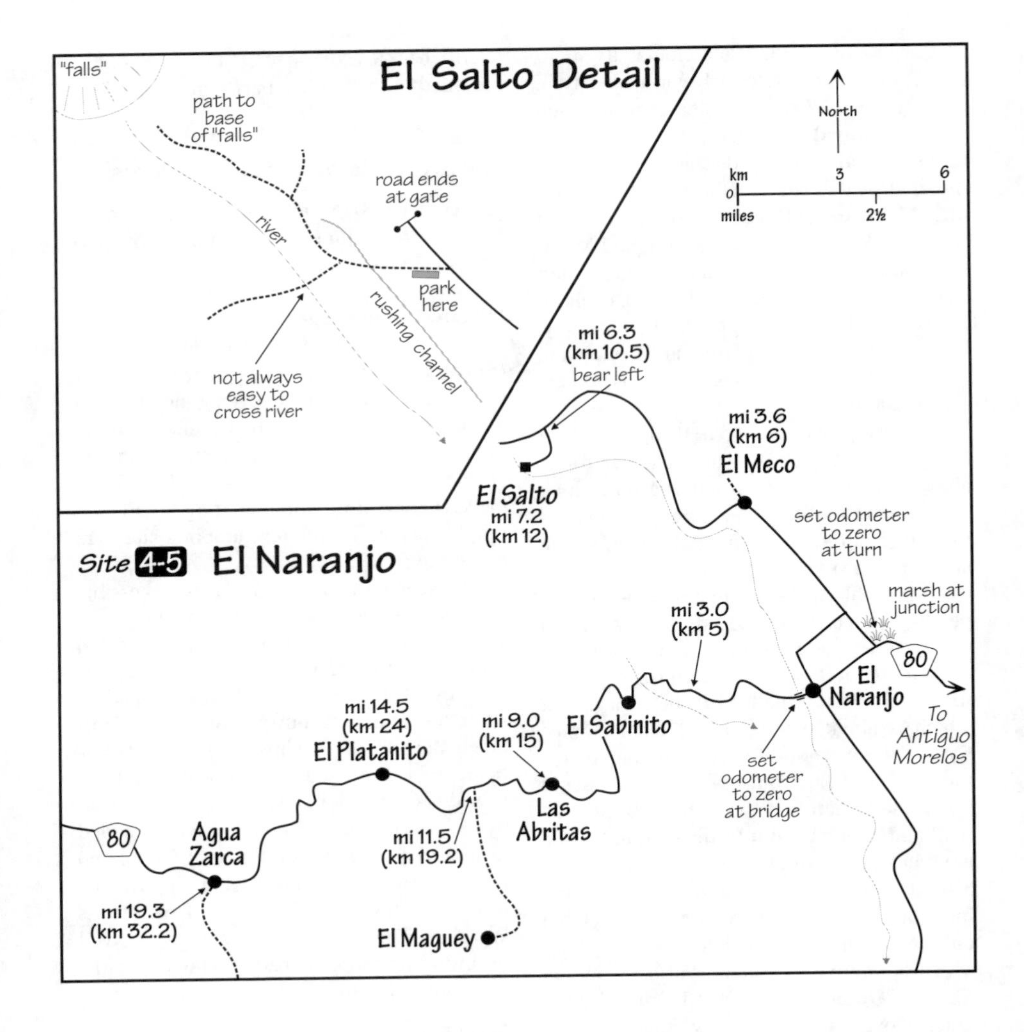

Ciudad Mante, 33 mi (55 km; an hour) north of El Naranjo, has a number of hotels, and there is a small, basic motel in El Naranjo, which also has a gas station, stores, and basic restaurants.

El Naranjo and the birding sites along Route 80 (see below) are easily reached using public buses, but note that first-class buses which ply this route aren't going to stop for you at small villages.

If you're driving on south from here and plan to visit Tlanchinol (a cloud-forest site in Hidalgo, Site 8.11), it may be best to go on to Ciudad Valles, turn east toward Tampico, and then turn off south, on Route 105, through Panuco and on to Huejutla and Tlanchinol. A more direct route would be to stay on Route 85 south from Valles to Tamazunchale and cut over to Huejutla and down to Tlanchinol. In 1986, the short stretch between Tamazunchale and Huejutla was among the worst (impassable) roads I've seen, but recent maps show this as now paved. If you can get independent confirmation of this change in road status, the route via Tamazunchale should be quicker.

If you plan to drive on to sites in Veracruz (Chapter 10), the easiest way is to head across to Tampico and from there head south on Route 180, the main "East Coast" highway.

Birding Sites

All birding sites in the El Narajo area (El Salto, El Sabinito, Las Abritas, El Platanito, Agua Zarca) correspond to small villages along and off Route 80, which runs from Antiguo Morelos on the Gulf Coast slope to Ciudad del Maiz, on the Mexican Plateau.

Set to zero at the Antiguo Morelos junction and head west on Route 80, crossing the border into San Luis Potosí at 16.2 mi (27 km). At about 19.0 mi (31.5 km), as you drop into the valley floor of El Naranjo, with the town and cane refinery visible ahead, look for a paved road to the right (north), immediately after a seasonally dry marsh (Aplomado Falcon, **Altamira Yellowthroat**) also on the right. This side road may be signed to El Meco and/or El Salto. It is paved but usually in bad condition due to the frequent passage of heavily laden cane trucks.

Birding along this road can be productive (Northern Bobwhite, **Green Parakeet**, White-crowned, Red-lored, **Red-crowned**, and **Yellow-headed parrots**, White-collared Swift). The parrots are best seen in early morning, when they fly around calling (a good site has been 1–2 km beyond El Meco); later in the day they tend to feed quietly and are easily missed. To reach El Salto, fork left at 3.6 mi (6 km) in the village of El Meco and bear left again at 6.3 mi (10.5 km) to reach, at 7.2 mi (12 km), the end of the road and a hydroelectric plant (hence, the "falls" are usually dry) with a high fence and armed guards. There is room to pull off and park safely on the left, where a dirt road leads down to the river (Muscovy Duck, Ringed, Amazon, and Green kingfishers, Black Phoebe). You can continue a short distance along a path to the right, through fields and brushy woodland to the base of the falls. Birds in this area include Thicket Tinamou, Bat Falcon, **Wedge-tailed Sabrewing**, Elegant Trogon, Blue-crowned Motmot, Red-throated Ant-Tanager, **Crimson-collared Grosbeak**, **Blue Bunting**, Yellow-faced Grassquit, and Altamira Oriole.

Back at the highway, it is about 1 km into "downtown" El Naranjo, dominated by a variably muddy or dusty main street. Just before (east of) the highway bridge over the Río El Salto on the west side of town, look for a dirt road off to the left (south). This runs through cane fields for several miles, with the river and forested hillsides to the right and may be worth driving for the chance of Aplomado Falcon. It is a good drive for other raptors (e.g., White-tailed Kite, Short-tailed Hawk), and several tracks lead to the river, where Muscovy Ducks can be found.

If you continue through town, mark zero at the bridge over the Río El Salto on the far (east) side of El Naranjo. The road winds through more woodland and fields and, after about 3 mi (5 km), runs beside a clear rushing stream to the left, bordered by large trees, and a wooded hillside to the right, just before the village of El Sabinito (4.0 mi; 6.6 km). One can park off the road here and bird from the road and/or along narrow paths into the woodland; birds include Thicket Tinamou, parrots, Ferruginous Pygmy-Owl, **Wedge-tailed Sabrewing**, Elegant Trogon, Blue-crowned Motmot, Smoky-brown and **Bronze-winged woodpeckers**, Ivory-billed Woodcreeper, Carolina Wren, Rufous-browed Peppershrike, Louisiana Waterthrush (winter), **Fan-tailed** and Golden-crowned **warblers**, **Crimson-collared Grosbeak**, and **Blue Bunting**.

From El Sabinito the road winds to the village of Chupaderos (5.0 mi; 8.3 km), after which it climbs an escarpment into subtropical oak forest. Places to pull off and bird include a side track on the left at 8 mi (13.3 km), a gated track on the right just past an inconspicuous small shrine at 9 mi (15.0 km, and 1 km before the village of Las Abritas), and the side road left (south) signed "Maguey de Oriente 6," at 11.5 mi (19.2 km). Birds that can be seen along these tracks and side roads (which go through oak forest with fairly open understory) include **Singing Quail** (listen for them scratching in the leaf litter), Blue Ground-Dove, Military Macaw, White-crowned and **Red-crowned parrots**, **Tamaulipas Pygmy-Owl**, **Azure-crowned** and **Amethyst-throated hummingbirds**, **Mountain Trogon**, **Bronze-winged Woodpecker**, Olivaceous and Spot-crowned woodcreepers, Hammond's Flycatcher (winter), **Grey-collared Becard** (often with mixed-species flocks), Grey-breasted Wood-Wren, **Brown-backed Solitaire**, **Blue Mockingbird**, **Fan-tailed Warbler**, Blue-hooded Euphonia, Red-throated Ant-Tanager, White-winged Tanager, and **Crimson-collared** and **Hooded grosbeaks**.

At about 12.4 mi (20.7 km) the road bends right, levels out along a relatively straight stretch, and emerges from the forest to clearings on the left and more extensive fields on the right, just before the village of El Platanito (14.5 mi; 24.0 km). One can park off the highway on a track that marks the start of the fields on the right (13 mi; 21.5 km) and bird from the highway (watching for traffic, as always) and along paths into the forest. Birds along this stretch, including the brushy fields and tracks into the forest, include **Singing Quail** (often seen from the highway in the open understory), Military Macaw (flying over the ridges beyond the fields to the right), **Tamaulipas Pygmy-Owl**, Long-billed Thrasher, **Grey Silky**, **Crescent-chested Warbler**, **Crimson-collared** and **Hooded grosbeaks**, Rusty Sparrow, and most of the other species listed for the lower oak woods.

From El Platanito the road winds and climbs more, and the oak woodland becomes drier, although still with many of the same bird species. At about 18.5 mi (31 km) the road emerges into an area of open ranching country around the village of Agua Zarca (19.3 mi; 32.2 km; elevation 1200 m). The oak woods and open areas shortly before (east of) Agua Zarca, where the road makes an obvious right-hand bend with a small corral to the left (18.9 mi; 31.4 km), are worth birding. Species here include Acorn Woodpecker, **Azure-crowned Hummingbird**, Say's Phoebe, Vermilion Flycatcher, Grey-breasted Jay, Bridled Titmouse, **Spotted Wren**, Eastern Bluebird, Painted Whitestart, Hepatic Tanager, Pyrrhuloxia, **Rufous-capped Brushfinch**, Canyon Towhee, House Finch, and **Black-headed Siskin**.

One can also continue on the highway beyond (east of) Agua Zarca through more oak woods and brushy fields, at least as far as a left-hand hairpin bend at 21.5 mi (36 km), after which the habitat becomes more open. Birds in this stretch include Military Macaw, **Red-crowned Parrot** (noisy and nesting in April), **Bronze-winged Woodpecker**, **Grey-collared Becard**, Northern Beardless Tyrannulet, Dusky (winter) and Buff-breasted flycatchers, **Spotted Wren**, Flame-colored Tanager, and most of the other species listed for just east of Agua Zarca.

El Naranjo Bird List

(numerous forest species are elevational migrants and can be expected at lower elevations in winter and higher elevations in summer): 254 sp. R: rare; V: vagrant (not to be expected). †See pp. 5–7 for notes on English names.

Thicket Tinamou
Least Grebe
American White Pelican
Double-crested Cormorant
Neotropic Cormorant
Bare-throated Tiger-Heron
Great Blue Heron
Great Egret
Snowy Egret
Little Blue Heron
Tricolored Heron
Green Heron
Cattle Egret
White-faced Ibis
Roseate Spoonbill
Wood Stork
Black-bellied Whistling-Duck
Muscovy Duck
Green-winged Teal
Mexican Duck†
Blue-winged Teal
Cinnamon Teal
Northern Shoveler
Gadwall
American Wigeon
Ring-necked Duck
Turkey Vulture
Black Vulture
White-tailed Kite
Northern Harrier
Sharp-shinned Hawk
Cooper's Hawk
Common Black Hawk
Great Black Hawk
Harris' Hawk
Grey Hawk
Roadside Hawk
Red-shouldered Hawk
Broad-winged Hawk
Short-tailed Hawk
Swainson's Hawk
Red-tailed Hawk
Crested Caracara
Collared Forest-Falcon
American Kestrel
Aplomado Falcon
Bat Falcon
Peregrine Falcon
Plain Chachalaca
Singing Quail
Northern Bobwhite
Sora
American Coot
Killdeer
Black-necked Stilt
Greater Yellowlegs
Solitary Sandpiper
Spotted Sandpiper
Least Sandpiper
Long-billed Dowitcher
Common Snipe
Red-billed Pigeon
White-winged Dove
Mourning Dove
Inca Dove
Common Ground-Dove
Blue Ground-Dove
White-tipped Dove
Ruddy Quail-Dove
Green Parakeet
Military Macaw
White-crowned Parrot
Red-crowned Parrot
Red-lored Parrot
Yellow-headed Parrot
Squirrel Cuckoo
Groove-billed Ani

Tamaulipas Pygmy-Owl†
Ferruginous Pygmy-Owl
Lesser Nighthawk
Pauraque
White-collared Swift
Vaux's Swift
Wedge-tailed Sabrewing
Broad-billed Hummingbird
Azure-crowned Hummingbird
Buff-bellied Hummingbird
Amethyst-throated Hummingbird
Ruby-throated Hummingbird
Mountain Trogon
Elegant Trogon
Blue-crowned Motmot
Ringed Kingfisher
Belted Kingfisher
Amazon Kingfisher
Green Kingfisher
Acorn Woodpecker
Golden-fronted Woodpecker
Yellow-bellied Sapsucker
Ladder-backed Woodpecker
Smoky-brown Woodpecker
Bronze-winged Woodpecker†
Lineated Woodpecker
Pale-billed Woodpecker
Olivaceous Woodcreeper
Ivory-billed Woodcreeper
Spot-crowned Woodcreeper
Barred Antshrike
N. Beardless Tyrannulet
Greenish Elaenia
Greater Pewee
Yellow-bellied Flycatcher
Least Flycatcher
Hammond's Flycatcher
Dusky Flycatcher
Buff-breasted Flycatcher
Eastern Phoebe
Black Phoebe
Say's Phoebe
Vermilion Flycatcher
Dusky-capped Flycatcher
Great Crested Flycatcher
Brown-crested Flycatcher
Great Kiskadee
Boat-billed Flycatcher
Social Flycatcher
Sulphur-bellied Flycatcher
Scissor-tailed Flycatcher
Tropical Kingbird
Couch's Kingbird
Cassin's Kingbird
Grey-collared Becard
Rose-throated Becard
Masked Tityra
Tree Swallow
N. Rough-winged Swallow
Barn Swallow
Green Jay
Brown Jay
Grey-breasted Jay†
Tamaulipas Crow†
Northern Raven
Bridled Titmouse
Black-crested Titmouse†
Spotted Wren
Canyon Wren
Spot-breasted Wren
Carolina Wren
Northern House Wren†
Marsh Wren
Grey-breasted Wood-Wren
Ruby-crowned Kinglet
Blue-grey Gnatcatcher
Eastern Bluebird
Brown-backed Solitaire
Orange-billed N.-Thrush
Black-headed N.-Thrush
Swainson's Thrush
Hermit Thrush
Clay-colored Thrush†
White-throated Thrush†
Grey Catbird
Blue Mockingbird
Northern Mockingbird
Long-billed Thrasher
Cedar Waxwing
Grey Silky†
Loggerhead Shrike
White-eyed Vireo
Bell's Vireo
Blue-headed Vireo†
Cassin's Vireo†
Warbling Vireo
Yellow-green Vireo
Rufous-browed Peppershrike
Blue-winged Warbler
Orange-crowned Warbler
Nashville Warbler
Crescent-chested Warbler
Northern Parula
Tropical Parula
Audubon's Warbler†
Myrtle Warbler†
Black-throated Grey Warbler
Townsend's Warbler
Black-throated Green Warbler
Hermit Warbler
Yellow-throated Warbler
Palm Warbler (V)
Black-and-white Warbler
American Redstart
Ovenbird
Louisiana Waterthrush
Kentucky Warbler
MacGillivray's Warbler
Common Yellowthroat
Altamira Yellowthroat
Grey-crowned Yellowthroat
Hooded Warbler
Wilson's Warbler
Painted Whitestart†
Fan-tailed Warbler
Golden-crowned Warbler
Rufous-capped Warbler
Yellow-breasted Chat
Scrub Euphonia
Yellow-throated Euphonia
Blue-hooded Euphonia
Yellow-winged Tanager
Blue-grey Tanager (R?)
Red-throated Ant-Tanager
Hepatic Tanager
Summer Tanager
Western Tanager
Flame-colored Tanager
White-winged Tanager
Greyish Saltator
Black-headed Saltator
Pyrrhuloxia
Crimson-collared Grosbeak
Rose-breasted Grosbeak

Black-headed Grosbeak
Blue Grosbeak
Blue Bunting
Indigo Bunting
Varied Bunting
Dickcissel
Rufous-capped Brushfinch
Canyon Towhee
Olive Sparrow
White-collared Seedeater
Blue-black Grassquit
Yellow-faced Grassquit
Rusty Sparrow
Chipping Sparrow
Clay-colored Sparrow
Lark Sparrow
Lincoln's Sparrow
Red-winged Blackbird
Eastern Meadowlark
Brewer's Blackbird
Great-tailed Grackle
Brown-headed Cowbird
Bronzed Cowbird
Melodious Blackbird
Orchard Oriole
Hooded Oriole
Altamira Oriole
Audubon's Oriole
Baltimore Oriole
Bullock's Oriole
Yellow-billed Cacique
Black-headed Siskin
Lesser Goldfinch
Hooded Grosbeak
House Sparrow

CHAPTER 5

MAZATLÁN AND THE DURANGO HIGHWAY

Tufted Jay

This chapter deals with areas in and near Mazatlán, Sinaloa, a popular resort destination on Mexico's Pacific Coast. While this region can be reached by driving from the U.S.A. (being about two days' drive from Nogales, Arizona), it is most often visited by flying into and out of Mazatlán. Needless to say, there is a wide variety of hotels and restaurants in and around Mazatlán, and vehicle rental is available from several companies with offices at the airport and in town. If you are driving down, see Chapter 2 (Northwestern Mexico) for sites to visit en route in Sonora. One could easily spend a week birding the Mazatlán/Durango Highway area and/or combine it with a visit to San Blas, Nayarit (Chapter 6), to make a one- to two-week birding trip.

Moving south from Arizona and Sonora,

Sinaloa marks a distinct jump into the Tropics (hinted at in the Alamos area of extreme southern Sonora; Site 2.4), where the Sonoran Desert gives way to thorn forest and tropical semi-deciduous woodland. The northern range limit of a host of tropical species lies in central and northern Sinaloa, and those species that do reach southern Sonora are generally commoner and easier to see in Sinaloa. Thus, species found readily around Mazatlán include Lilac-crowned Parrot, Lesser Ground-Cuckoo, Cinnamon Hummingbird, Citreoline Trogon, Russet-crowned Motmot, Red-breasted Chat, Greyish Saltator, and Yellow-winged Cacique.

In addition, the Durango Highway, the only paved road that transects Mexico's Sierra Madre Occidental, starts up from the coastal plain only a few kilometers south of Mazatlán. The Durango Highway enables easy access to highland forests with a wealth of species different from those in the tropical lowlands, not to mention a pleasant, cooler change of climate. Birds of the Durango Highway include Colima Pygmy-Owl, Bumblebee Hummingbird, White-striped Woodcreeper, Pine Flycatcher, Spotted Wren, Russet Nightingale-Thrush, Aztec Thrush, Red Warbler, Green-striped Brushfinch, and, of course, the spectacular and highly localized Tufted Jay.

Suggested Itineraries

If you have flown into Mazatlán (most flights arrive after noon), birding near town (Sites 5.1, 5.2, 5.3) could fill the late afternoon. The La Noria Road or a boat trip to the booby rocks is good for a morning, and after lunch you can drive up the Durango Highway (Site 5.4). At least a full day can be spent in the Barranca area (Site 5.6), and the Panuco Road can be birded in the morning before heading back to Mazatlán for the evening. This allows coverage of the area in a minimum of five days (including travel to and from the U.S.A.), but longer can be spent on the Durango Highway.

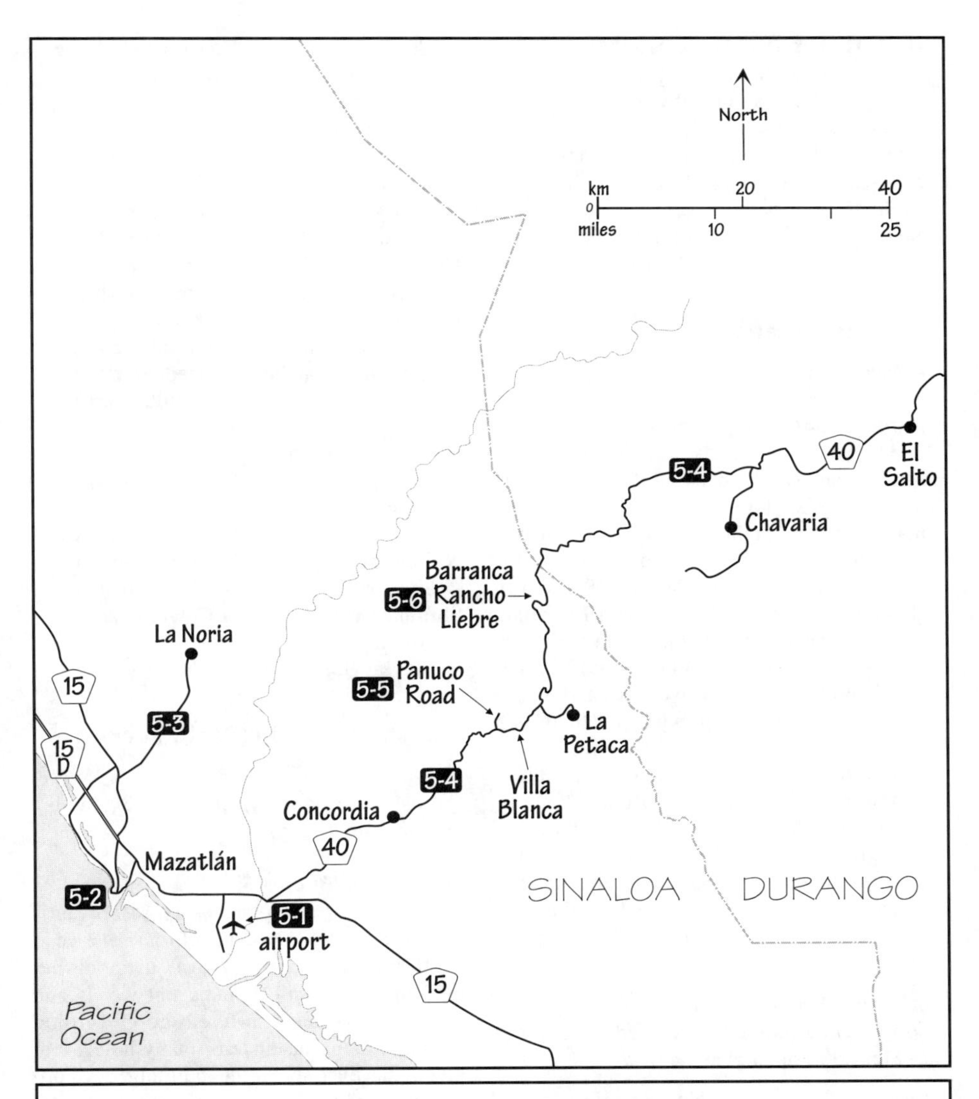

Chapter 5: MAZATLAN and the DURANGO HIGHWAY

MAZATLÁN & DURANGO HWY SITE LIST

Site 5.1 Mazatlán Airport, Sinaloa
Site 5.2 Mazatlán (Booby Rocks), Sinaloa
Site 5.3 The La Noria Roaad, Sinaloa
Site 5.4 The Durango Highway, Sinaloa/Durango
Site 5.5 The Panuco Road, Sinaloa
Site 5.6 Barranca Rancho Liebre, Sinaloa

Site 5.1: Mazatlán Airport, Sinaloa

(December 1995)

Introduction/Birding Sites

The fields and ponds near the airport can be birdy and worth checking if you have time to kill waiting for a flight or have only an hour or two of daylight left when you arrive and "have" to go birding. The airport is situated west of the main highway (Route 15), about 17 km south of downtown Mazatlán. Leave the airport on the only road out to the main highway (and Mazatlán); almost immediately after the railway overpass, and within 500 m of the airport parking lot, there is a paved road to the left (northwest) to I. de la Piedra (it may be signed only from the direction of traffic coming to the airport). You can drive along this road for several kilometers, through weedy fields lined by seasonally wet ditches, muddy pools, and some shrimp ponds. I have birded only to where the road turns sharply right and the surface becomes dirt.

Mazatlán Airport Bird List

60 sp. R: rare (not to be expected). †See pp. 5–7 for notes on English names.

Least Grebe
Neotropic Cormorant
Anhinga
Great Blue Heron
Great Egret
Snowy Egret
Little Blue Heron
Cattle Egret
White Ibis
Black-bellied Whistling-Duck
Black Vulture
Turkey Vulture
Osprey
Northern Harrier
Sharp-shinned Hawk
Harris' Hawk
Grey Hawk
Zone-tailed Hawk
Red-tailed Hawk
Crested Caracara
American Kestrel
Merlin
Common Moorhen
American Coot
Killdeer
American Avocet
Greater Yellowlegs
Willet
Spotted Sandpiper
Least Sandpiper
Western Sandpiper
Long-billed Dowitcher
Forster's Tern
White-winged Dove
Common Ground-Dove
Groove-billed Ani
Belted Kingfisher
Vermilion Flycatcher
Social Flycatcher
Great Kiskadee
Tropical Kingbird
Thick-billed Kingbird
Mangrove Swallow
N. Rough-winged Swallow
Sinaloa Crow
Northern Mockingbird
Loggerhead Shrike
Orange-crowned Warbler
Northern Waterthrush
Common Yellowthroat
Grey-crowned Yellowthroat
Wilson's Warbler
Dickcissel
Cinnamon-rumped Seedeater†
Lincoln's Sparrow
Swamp Sparrow (R?)
White-crowned Sparrow
Great-tailed Grackle
Bronzed Cowbird
Brown-headed Cowbird

Site 5.2: Mazatlán (Booby Rocks), Sinaloa

(January 1984, January/February 1985, January, December 1995)

Introduction/Birding Sites

Mazatlán is a popular beach-resort destination for many thousands of tourists a year. On the northwest edge of town, the pools behind the beach and the adjacent weedy and scrubby lots have offered good birding within walking distance of many hotels, but given the ever-spreading building developments, good birding spots are getting harder to find. If you get out into the overgrown weedy lots with marshy pools, birds that can be seen include Least Grebe, egrets, Wood Stork, Black-bellied Whistling-Duck, **Elegant Quail**, Vermilion Flycatcher, **Sinaloa Crow**, **Cinnamon-rumped** [White-collared] **Seedeater**, and sundry migrant vireos, warblers, and sparrows: plenty to look at for one or two mornings and/or afternoons.

From almost any point along the shore, especially the rocky promontories at the south end of town, you can see boobies, frigatebirds, and, depending on season and luck, loons, tropicbirds, and shearwaters. The rocky shore is also worth checking for Wandering Tattler and Surfbirds (March/April). The two small, whitewashed rocks in the south end of the bay are known as the Dos Hermanos (Two Brothers) and also as the Booby Rocks. A short boat trip around these rocks offers the chance for excellent, close-up views of both Blue-footed and Brown boobies, and there is a slight chance for Red-billed Tropicbird.

Tourist trips in the bay may (or may not) include the Booby Rocks (make sure to find this out before you go), or you can hire your own boat. Boats leave from south side of town, inside the harbor. Most hotels and tourist information stalls should know about scheduled trips and will also know where you can get boats, in case things change. The trips are short (about two hours), and if you feel adventurous and are willing to pay more, you can try going farther out into the bay in search of other birds. The bird list from two winter trips includes **Black-vented Shearwater**, Blue-footed and Brown boobies, Brown Pelican, Neotropic Cormorant, Magnificent Frigatebird, Great Blue Heron, Turkey Vulture, Wandering Tattler, Laughing, Bonaparte's, Ring-billed, Heermann's, Herring, and **Yellow-footed** (rare?) gulls, and Caspian, Royal, Forster's, and Black terns.

Site 5.3: The La Noria Road, Sinaloa

(January 1985, 1995, October 1993, September, December 1995)

Introduction/Key Species

The La Noria road is an example of the type of birding spot you can find simply by driving around and trying side roads. It goes through some brushy, cutover thorn forest and cornfields before winding up into low hills covered with fairly extensive thorn forest. Species of interest here include **Rufous-bellied Chachalaca**, **Elegant Quail**, **Lilac-crowned Parrot**, Lesser Ground-Cuckoo, Chestnut-collared and **White-naped swifts**, Nutting's Flycatcher, **Black-throated Magpie-Jay**, **Purplish-backed Jay**, **Black-capped Gnatcatcher**, and Five-striped Sparrow (winter).

Access/Birding Sites

The La Noria road is about 10 km (15 minutes) from Mazatlán and is best for birding in the morning, up to about 10 A.M., or whenever it starts to get hot and quiet. It can also be worth birding in the late afternoon, although mornings are decidedly better.

From Mazatlán, head out northwest, toward Culiacan, and stay on the old highway, avoiding the signs (*Culiacan Cuota*) for the new toll road. From the split of *libre* and *cuota* roads to the La Noria junction, the highway winds through some scrub forest where you may see **Black-throated Magpie-Jays**—traffic is fast and heavy, and it is best to resist the temptation to try and stop here. At Km Post 10 (7.5 km north from the split of *libre* and *cuota* roads), look for a paved road off to the right (east), signed to La Noria. Turn right here, and the volume of traffic will drop greatly, making it easier to stop and look for birds. There is still a fair amount of traffic, though, and pickup trucks in particular can move very quickly, so always be aware of the road.

Driving slowly and stopping where you see and/or hear birds is the best approach. It is always a good idea to keep an eye skyward for Chestnut-collared and **White-naped swifts**, flocks of which can often be seen fairly low in the early morning or just before rain showers. Initially the road is more or less straight and passes through brushy fields with hedges and patches of thorn forest. At around Km 13 it winds up over a ridge covered with good thorn forest, then down into another valley and on toward the Sierra and the small town of La Noria at about Km 22.

In early morning, groups of **Magpie-Jays** and **Purplish-backed Jays** can be seen along the road between Kms 2 and 3, and at Km 2.4/2.5 there are two tracks off to the left (north), opposite turnoffs for a village to the right (south). One can park safely off the road here and walk along these tracks, the first through fields and hedges (**Elegant Quail**, both jays noted above, and, in winter, Five-striped Sparrow), the second through thorn forest (Lesser Ground-Cuckoo, Nutting's Flycatcher). Other good pulloffs can be found between Km 13 and Km 17, in the hilly thorn-forest stretch (**Rufous-bellied Chachalaca**, **Lilac-crowned Parrot**, **Black-capped Gnat-**

catcher—usually outnumbered by Blue-greys in winter). This area can be birded using public transport by taking second-class buses from Mazatlán to La Noria and getting off when you see a good area.

The La Noria Road Bird List

122 sp. R: rare (not to be expected). †See pp. 5–7 for notes on English names.

Cattle Egret
Turkey Vulture
Hook-billed Kite
White-tailed Kite
Northern Harrier
Sharp-shinned Hawk
Cooper's Hawk
Harris' Hawk
Grey Hawk
Short-tailed Hawk
Zone-tailed Hawk
Red-tailed Hawk
Crested Caracara
Collared Forest-Falcon
American Kestrel
Merlin
Rufous-bellied Chachalaca
Elegant Quail
Red-billed Pigeon
White-winged Dove
Mourning Dove
Common Ground-Dove
Ruddy Ground-Dove (R?)
White-tipped Dove
Orange-fronted Parakeet
White-fronted Parrot
Lilac-crowned Parrot
Squirrel Cuckoo
Groove-billed Ani
Lesser Ground-Cuckoo
Ferruginous Pygmy-Owl
Mottled Owl
Lesser Nighthawk
Pauraque
Buff-collared Nightjar
Chestnut-collared Swift
White-naped Swift
Vaux's Swift
Golden-crowned Emerald
Broad-billed Hummingbird
Cinnamon Hummingbird
Violet-crowned Hummingbird
Plain-capped Starthroat
Black-chinned Hummingbird
Elegant Trogon
Belted Kingfisher
Gila Woodpecker
Golden-cheeked Woodpecker
Ladder-backed Woodpecker
Lineated Woodpecker
N. Beardless Tyrannulet
Greater Pewee
Least Flycatcher
Western Flycatcher†
Vermilion Flycatcher
Bright-rumped Attila
Dusky-capped Flycatcher
Nutting's Flycatcher
Brown-crested Flycatcher
Tropical Kingbird
Thick-billed Kingbird
Cassin's Kingbird
Violet-green Swallow
N. Rough-winged Swallow
Black-throated Magpie-Jay
Purplish-backed Jay
Sinaloa Crow
Happy Wren
Sinaloa Wren
Northern House Wren†
Blue-grey Gnatcatcher
Black-capped Gnatcatcher
Swainson's Thrush
Rufous-backed Thrush†
Northern Mockingbird
Blue Mockingbird
Curve-billed Thrasher
Loggerhead Shrike
Bell's Vireo
Black-capped Vireo
Plumbeous Vireo†
Warbling Vireo
Orange-crowned Warbler
Nashville Warbler
Lucy's Warbler
Yellow Warbler
Audubon's Warbler†
Black-throated Grey Warbler
Black-and-white Warbler
MacGillivray's Warbler
Ovenbird
Common Yellowthroat
Wilson's Warbler
Yellow-breasted Chat
Godman's Euphonia†
Summer Tanager
Western Tanager
Flame-colored Tanager
Greyish Saltator
Northern Cardinal
Pyrrhuloxia
Yellow Grosbeak
Black-headed Grosbeak
Blue Bunting
Varied Bunting
Painted Bunting
Green-tailed Towhee
Five-striped Sparrow
Black-throated Sparrow (R)
Clay-colored Sparrow
Lark Sparrow
Lark Bunting (R)
Grasshopper Sparrow
Lincoln's Sparrow
Great-tailed Grackle
Bronzed Cowbird
Orchard Oriole
Hooded Oriole
Black-vented Oriole
Streak-backed Oriole
Yellow-winged Cacique
House Finch

Site 5.4: The Durango Highway, Sinaloa/Durango

(March 1982, 1983, January 1984, February 1985, April 1987, August 1988, June 1991, October 1993, December 1995)

Introduction/Key Species

Route 40 (The Durango Highway), from the coastal plain at Villa Unión to Durango City, in the high desert of the Mexican Plateau, cuts a spectacular route through the Sierra Madre Occidental. Birding almost anywhere along the highway can be good, assuming you can find somewhere to pull off safely and park. Two popular areas dealt with separately below are the Panuco Road (Km 248.3) and the Barranca Rancho Liebre (Km 200.5); the latter is famous as a site for the handsome **Tufted Jay**. **Thick-billed Parrot** and **Eared Quetzal** have also been seen at the Barranca, but I've failed to see either on nine trips, so don't feel too badly if you miss them!

Something to be aware of is that many bird species move elevationally depending on season and the availability of food. Thus, in autumn, some birds that nest in the lowlands may move upslope into the mountains, while in winter, some species that nest in the mountains may move down to the foothills and even to the coast. In particular, hummingbirds are prone to move around and can be hard to find; the peak season for most species of hummers is November and December.

The Durango Highway has received some bad publicity in recent years, following at least one 1980s highway robbery of birders. This sort of thing can happen anywhere in the world, and sensible precautions are all you can take. I and many others have birded along the Durango Highway and camped out at the Barranca Rancho Liebre with no ill luck.

Species of interest on the Durango Highway (including the Panuco Road and Barranca Rancho Liebre, Sites 5.5 and 5.6) include **Thick-billed Parrot**, **Colima Pygmy-Owl**, Stygian Owl, **White-naped Swift**, **Sparkling-tailed Woodstar**, **Bumblebee Hummingbird**, **Eared Quetzal**, **Grey-crowned Woodpecker**, **White-striped Woodcreeper**, **Pine Flycatcher**, **Tufted Jay**, **Spotted Wren**, **Russet Nightingale-Thrush**, **Aztec Thrush**, **Red Warbler**, and **Green-striped Brushfinch**.

Access

From Mazatlán, head south on Route 15. At 7 km south of the airport junction is a major, well-signed highway junction, on the south side of Villa Unión. The Durango Highway goes left and inland (northeast), and there may be police/army checkpoints in search of drugs and arms. Stop, be polite, smile, and act like a tourist, and chances are they'll wave you on. If they do search your vehicle, stay calm and be polite and there shouldn't be a problem (unless of course you're loaded down with lethal weapons!).

While it is possible to make a day trip up the highway from Mazatlán (preferably setting out before dawn, as, for example, it's 1 to 1.5 hours to Panuco Road, and another 1.5 to 2 hours direct to the Barranca Rancho Liebre), many birders opt to base themselves at the Hotel Villa Blanca, in the village of La Capilla de Taxte (elevation 1200 m), about 61 km up the Durango Highway from its junction with Route 15 at Villa Unión.

There is a cheaper and basic hotel (with basic eating potential nearby) in Potrerillos (near Km 222), about 12 km farther up the highway from Villa Blanca, from which it is easy to walk or hitch to the nearby La Petaca Road and catch buses to the Barranca and Panuco Road. If you're using public transport, you can take buses to El Palmito, just past the Barranca Rancho Liebre, and walk/hitch back the kilometer or two to the Barranca, and also get off second-class buses anywhere along the highway, including the junction for Panuco Road and at Potrerillos.

Birding Sites

The kilometer posts along the highway count from zero at Durango to 294 at the Villa Unión junction and are used below to refer to birding spots. Coming from Villa Unión, the lowlands and first ridge or two of thorn forest have birds much like those listed for La Noria Road, including **Purplish-backed Jay** around the dirt road off to the right (south) at Km 285.1 (signed to Pantitlan). At about Km 276 you pass through the town of Concordia (elevation 150 m), after which the road starts to wind up through thorn-forested hills, then into some dry oaks and into open, dry, pine-oak woodland with brushy second growth and weedy fields (such as in the vicinity of La Capilla de Taxte, at Km 233.5). It's not a bad idea to top off your gas tank in Concordia (or Villa Unión), as this commodity can be hard to come by higher up in the mountains.

In summer, **Flammulated Flycatchers** are common, and vocal throughout much of the day, in gullies of tangled thorn forest between Km 270 and 252 (at other seasons you probably wouldn't know they were there). Other species along this stretch of the highway include **Rufous-bellied Chachalaca**, Mangrove Cuckoo (far from mangroves!), Lesser Ground-Cuckoo, **Citreoline Trogon**, and **Red-breasted Chat** (easier to find in summer, when they are singing). One spot where birders have had luck over the years is known as Cerro el Elefante, at Km 265.6. Here there is a large pullout on the right (south) for parking, whence you can walk back down the highway about 30 m to a track on the same side (south) and a stream bed, which can be walked during most of the year; birds possible include **Happy** and **Sinaloa wrens**, **Golden Vireo**, Louisiana Waterthrush (migrant), and **Fan-tailed Warbler**.

At Km 249 start to watch for an obvious, broad dirt road climbing off to the left (north) at Km 248.3, as the highway descends around a small valley. This junction may be signed to Panuco and is known to birders as the Panuco Road (Site 5.5). A full morning or two can be profitably spent birding here. From the Panuco Road to La Capilla de Taxte is 15 km, whence it is another 33 km (allow at least an hour with traffic) to the entrance for Barranca Rancho Liebre (Site 5.6). At Km 222/221 you pass through the small town of Potrerillos, where a dirt road leaves to the right (south) to the village of La Petaca (elevation 1500 m). The flower banks along this road can be good for hummingbirds (including **Sparkling-tailed Woodstar**, **Bumblebee Hummingbird**). Species along the first few kilometers of the La Petaca Road include **Mexican Parrotlet** and **Aztec Thrush**, reflecting the interesting admixture of highland and lowland avifaunas!

Tufted Jays can be seen anywhere along the Durango Highway from Km 215 to the Barranca Rancho Liebre (Site 5.6), especially in the early morning (up to 7 or 8 A.M.), but despite their large size and striking pattern, they can be elusive. At least in winter they often travel with the noisier Steller's Jays, and any time you hear Steller's it is worth watching carefully for Tufted Jays. Birding from the highway in this stretch (there are numerous pulloffs where you can park safely) can be exciting in winter, when mixed-species flocks include **Grey-crowned Woodpecker**, **White-striped Woodcreeper**, **Grey-collared Becard**, **Spotted Wren**, **Red Warbler**, **Red-headed Tanager**, and large numbers of migrant warblers (all of which can also be seen at the Barranca). It is always worth keeping an optimistic eye out for **Thick-billed Parrots** and, in spring/summer, for **Sinaloa Martin**.

Watch the kilometer posts for 202 or 201 and then, at about 200.5, the road makes a right-hand bend around a valley, marked on the left by a large pulloff and a small restaurant and house. This bend marks the entrance to the Barranca Rancho Liebre (Site 5.6). Shortly after this entrance is the small town of El Palmito, beyond which you enter the state of Durango (and change from Mountain to Central Standard Time); from here it's still 197 km to Durango City. From a birding point of view, there isn't much to be gained by continuing on up the highway above El Palmito, at least until the town of El Salto, about 100 km (2 to 2.5 hours) from El Palmito, and even then it is a long way to go for species that can be seen readily elsewhere. If you are planning to drive on to Durango, the following gives some idea of what to expect.

From Barranca Rancho Liebre, Sinaloa, birds stay much the same for another 50 km or so after entering the state of Durango (**Tufted Jay**, **Red-headed Tanager**, etc.), and some of the views are truly breathtaking. At about Km 130 (i.e., 130 km from Durango City and 70 km from Barranca Rancho Liebre) is a dirt road off to the right (south), signed for Chavaria. The first 5 km along this road can be good birding, and quieter than the highway (**Mountain Trogon**, **White-striped Woodcreeper**, Common Tufted, **Pine**, and Western (summer) **flycatchers**, **Brown-backed Solitaire**, **Grey Silky**, Red-faced (nesting) and **Red warblers**, Painted and Slate-throated whitestarts, and **Rufous-capped Brushfinch**).

Watch for Black (summer) and **White-naped swifts** and Evening Grosbeaks from the Chavaria junction to the bustling, dusty (or muddy, depending on recent weather) logging town of El Salto (Km 100; elevation 2450 m), as the road climbs into drier, generally more open pines. The pine forests (heavily logged in places) and clearings in the vicinity of El Salto (along the highway east and west of town, and to the south on the dirt logging road toward Cofradia) have a number of species not found, or less common, in the forests of the western slopes of the Sierra (Acorn Woodpecker, Cassin's Kingbird, Violet-green Swallow, Western Bluebird, Townsend's Solitaire, Canyon Towhee, **Striped Sparrow**, Mexican [Yellow-eyed] Junco, House Finch, Evening Grosbeak). White-throated Flycatchers may also be found in summer along vegetated streams through cultivated or overgrown clearings in the pine forests around El Salto: the type specimen of the northwestern race *E. a. timidus*, as well as the first described nest and eggs of the species, were collected near El Salto. Also, **Sierra Madre Sparrow** was collected (in 1951) at San Juan, 8 km west of El Salto; in June 1991, cultivation and overgrazing in the area had seemingly destroyed almost all of the native bunch grass (favored by this enigmatic species elsewhere in its range), and there was no sign of Sierra Madre Sparrows.

From El Salto to Durango City the highway runs (much straighter!) through open pine forests, clearings, and farmland, before entering high desert as one approaches the dry and dusty Durango City (elevation 1850 m).

Site 5.5: The Panuco Road, Sinaloa

(January 1984, February 1985, April 1987, June 1991, October 1993, December 1995)

Introduction/Key Species/Access

At Km 248.3 on the Durango Highway watch for an obvious dirt road climbing off to the left (north) as the highway descends around a small valley. This junction may be signed to Panuco and is known to birders as the Panuco Road (elevation 600 m). It can be reached in 1–1.5 hours from Mazatlán and is best for birds early in the day.

Drive up about 750 m to a small house (with omnipresent barking dogs) on your left, and a spectacular view out over a valley ahead to the left. You can park here or at various points along the next 1–2 km of road and bird in the brushy thorn forest along the road. Traffic is erratic, but trucks can be frequent and kick up clouds of dust. Beyond the fork in the road after a few kilometers there can be heavy truck traffic (and blasting activities!), so the first few kilometers are generally best for birding. Species of interest include Military Macaw (keep an eye and ear out over the broad valley), **Colima Pygmy-Owl**, **Sparkling-tailed Woodstar**, Calliope Hummingbird (winter), Black-capped (winter) and **Golden vireos**, and **Rusty-crowned Ground-Sparrow**.

The Panuco Road Bird List

93 sp. †See pp. 5–7 for notes on English names.

Turkey Vulture
Black Vulture
Sharp-shinned Hawk
Grey Hawk
Short-tailed Hawk
Zone-tailed Hawk
Collared Forest-Falcon
American Kestrel
Rufous-bellied Chachalaca
White-winged Dove
Inca Dove
White-tipped Dove
Orange-fronted Parakeet
Military Macaw
Mexican Parrotlet
Lilac-crowned Parrot
Squirrel Cuckoo
Lesser Ground-Cuckoo
Lesser Roadrunner
Colima Pygmy-Owl†
Ferruginous Pygmy-Owl
White-naped Swift
Golden-crowned Emerald
Broad-billed Hummingbird
Berylline Hummingbird
Cinnamon Hummingbird
Violet-crowned Hummingbird
Plain-capped Starthroat
Sparkling-tailed Woodstar
Calliope Hummingbird
Rufous/Allen's Hummingbird
Citreoline Trogon
Elegant Trogon
Russet-crowned Motmot
Gila Woodpecker
Golden-cheeked Woodpecker
Ladder-backed Woodpecker
Lineated Woodpecker
Pale-billed Woodpecker
Ivory-billed Woodcreeper
N. Beardless Tyrannulet
Greater Pewee
Least Flycatcher
Western Flycatcher†
Bright-rumped Attila
Dusky-capped Flycatcher

Ash-throated Flycatcher
Nutting's Flycatcher
Sulphur-bellied Flycatcher
Thick-billed Kingbird
Cassin's Kingbird
Western Kingbird
Masked Tityra
N. Rough-winged Swallow
Black-throated Magpie-Jay
Northern Raven†
Happy Wren
Sinaloa Wren
Northern House Wren†
Blue-grey Gnatcatcher
Black-capped Gnatcatcher
Orange-billed N.-Thrush
Swainson's Thrush
Rufous-backed Thrush†
Blue Mockingbird
Black-capped Vireo
Cassin's Vireo†
Golden Vireo
Warbling Vireo
Orange-crowned Warbler
Nashville Warbler
Tropical Parula
Audubon's Warbler†
Black-throated Grey Warbler
MacGillivray's Warbler
Wilson's Warbler
Painted Whitestart†
Slate-throated Whitestart†
Fan-tailed Warbler
Rufous-capped Warbler
Godman's Euphonia†
Western Tanager
Flame-colored Tanager
Greyish Saltator
Yellow Grosbeak
Black-headed Grosbeak
Blue Bunting
Varied Bunting
Rusty-crowned Ground-Sparrow
Lincoln's Sparrow
Hooded Oriole
Streak-backed Oriole
Yellow-winged Cacique
Lesser Goldfinch

Site 5.6: Barranca Rancho Liebre

(March 1982, 1983, January 1984, February 1985, April 1987, August 1988, June 1991, October 1993, December 1995)

Introduction/Key Species/Access

Watch the kilometer posts for 202 or 201; then, at about 200.5, the road makes a right-hand bend around a valley, marked on the left by a large pulloff and a small restaurant and house. This bend marks the entrance to the Barranca Rancho Liebre. If you reach the small town of El Palmito (with a gas station), you've gone about 1–2 km too far. Park near the restaurant (good for a cold drink) and walk up the track between the restaurant (on your left), and the small house—this is the track (formerly a road) to the Barranca, a walk of about 2 km. Birding can be good all along the track, including right at the highway, and **Tufted Jays** can be seen anywhere along the walk.

The first 200–300 m climb steeply before the track eases off to a gentle gradient. Parts of the track are not too clear, and there are one or two more major-looking tracks off to the right (don't take these, at least not to get to the Barranca). Basically you want to stay along the right-hand side of the small stream (the non-pines here are excellent for **Pine Flycatcher**; also a good stretch for **Mountain Trogon** and **Russet Nightingale-Thrush**) until you come to the edge of a clearing (an overgrown orchard), where the trail crosses the stream and continues on along the left side of the clearing to a spectacular overlook of the Barranca. There is a short, narrow trail (beware of slipping on pine needles) down to a large boulder that offers a commanding view out over the Barranca (Crested Guan can be heard and occasionally seen from here). Species found around the edge of the Barranca include Stygian Owl, **Eared Quetzal**, **Aztec Thrush**, and **Hooded Grosbeak**. From the head of the Barranca a narrow trail descends either side into the Barranca, where shady areas near water seeps are best for **Golden-browed Warbler** and **Green-striped Brushfinch**.

Barranca Rancho Liebre Bird List
(Km 212 to the Barranca): 106 sp. †See pp. 5–7 for notes on English names.

Black Vulture
Turkey Vulture
Zone-tailed Hawk
Red-tailed Hawk
American Kestrel
Crested Guan
Band-tailed Pigeon
Inca Dove
White-tipped Dove
Lilac-crowned Parrot
Whiskered Screech-Owl
Mountain Pygmy-Owl†
Stygian Owl
Northern Saw-whet Owl
Mexican Whip-poor-will†
White-naped Swift
Vaux's Swift
White-throated Swift
White-eared Hummingbird
Berylline Hummingbird
Blue-throated Hummingbird
Magnificent Hummingbird

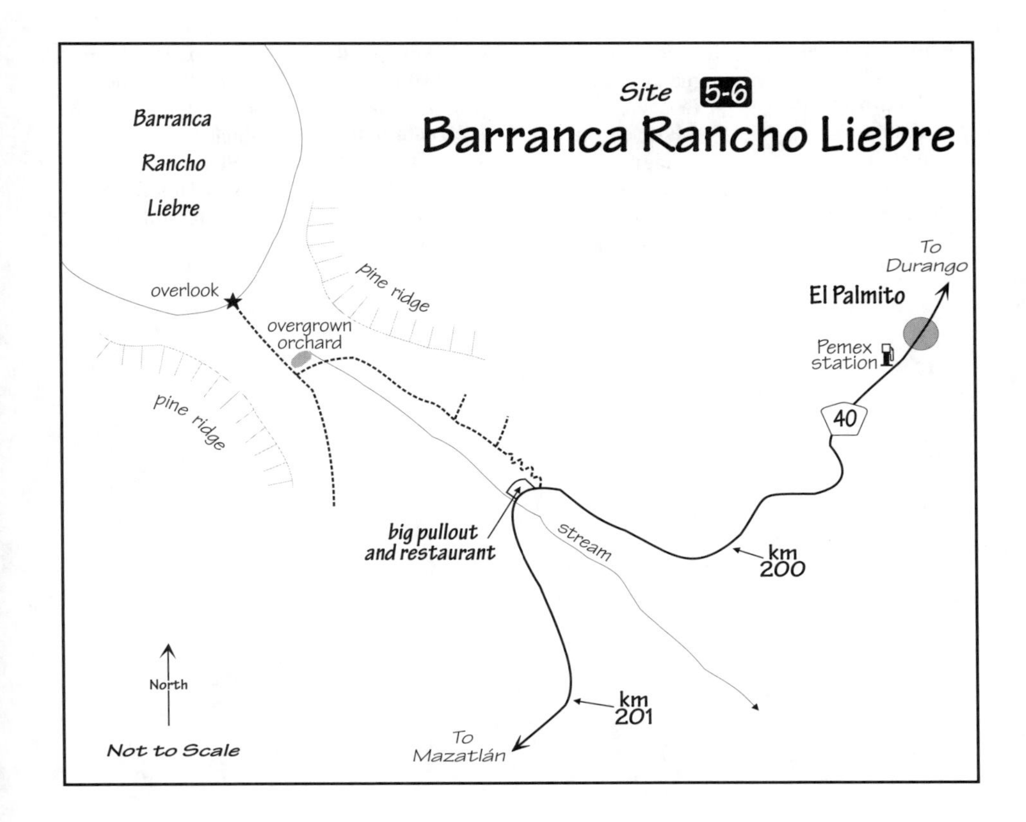

Costa's Hummingbird
Ruby-throated Hummingbird
Rufous Hummingbird
Mountain Trogon
Arizona Woodpecker†
Grey-crowned Woodpecker
Williamson's Sapsucker
Red-shafted Flicker†
White-striped Woodcreeper
Greater Pewee
Common Tufted Flycatcher†
Hammond's Flycatcher
Dusky Flycatcher
Pine Flycatcher
Western Flycatcher†
Buff-breasted Flycatcher
Dusky-capped Flycatcher
Grey-collared Becard
Violet-green Swallow
Steller's Jay
Tufted Jay
Northern Raven†
Mexican Chickadee
Bridled Titmouse
White-breasted Nuthatch
Brown Creeper
Spotted Wren
Canyon Wren
Northern House Wren†
Brown-throated Wren†
Ruby-crowned Kinglet
Blue-grey Gnatcatcher
Eastern Bluebird
Brown-backed Solitaire
Orange-billed N.-Thrush
Russet N.-Thrush
Hermit Thrush
White-throated Thrush†
Aztec Thrush
American Robin
Blue Mockingbird
Grey Silky†
Cassin's Vireo†
Plumbeous Vireo†
Hutton's Vireo
Warbling Vireo
Orange-crowned Warbler
Nashville Warbler
Virginia's Warbler
Crescent-chested Warbler
Audubon's Warbler†
Black-throated Grey Warbler
Townsend's Warbler
Hermit Warbler
Grace's Warbler
Black-and-white Warbler
MacGillivray's Warbler
Wilson's Warbler
Red-faced Warbler
Red Warbler
Painted Whitestart†
Slate-throated Whitestart†

Tree Swallow
Barn Swallow
Horned Lark
Chihuahuan Raven
Loggerhead Shrike
Cassin's Sparrow
Savannah Sparrow
Seaside Sparrow
Eastern Meadowlark
Red-winged Blackbird
Great-tailed Grackle
Bronzed Cowbird

Site 4.2: La Pesca, Tamaulipas

(April 1990)

Introduction/Key Species

This area lies just north of the Tropic of Cancer and represents a distinct step into the Neotropics from nearby South Texas. Thus, tropical species such as Thicket Tinamou, Roadside Hawk, **Yellow-headed Parrot**, **Tawny-collared Nightjar**, Ivory-billed Woodcreeper, and Grey-crowned Yellowthroat occur alongside "South Texas specialities" such as Plain Chachalaca, Pauraque, Brown Jay, and Altamira Oriole.

The Río Soto La Marina runs through the town of the same name and flows east to empty into the Gulf of Mexico, via some extensive salt marshes and lagoons, at La Pesca. The paved road between Soto la Marina and La Pesca passes through some good thorn forest, and La Pesca itself is good for waterbirds (including Piping Plover) and, much like the Texas coast, can hold fallouts of northbound migrants in spring.

Access

The town of Soto La Marina lies along Route 180, the main "East Coast" Mexican highway, about 160 mi (270 km; 4 hours) south of the Matamoros/Brownsville border crossing. From Matamoros, head south on Route 101 toward Ciudad Victoria and, about 28 mi (47 km) south of San Fernando (and about 110 mi/185 km from Matamoros), take the major left-hand fork onto Route 180, which should be signed for Tampico (and probably for Soto La Marina). From this junction it's about 51 mi (85 km) to Soto La Marina, which is right by the highway. There is an adequate motel in Soto La Marina, or you could camp along the road to, or at, La Pesca, which is 30 mi (50 km) east of Soto La Marina. This area is easily reached by public transport—for example, buses from Matamoros to Soto La Marina, and thence to La Pesca.

Birding Sites

From Soto La Marina, the paved road to La Pesca is clearly marked. The road starts out through agricultural land and then climbs over some low hills with extensive areas of thorn forest before dropping down to an area of lagoons and marshes and ending at the coast. There are several places to pull safely off the highway and find tracks to explore the thorn forest (Thicket Tinamou, **Tawny-collared Nightjar**, migrant Chuck-will's-widow calling at dawn), and **Yellow-headed Parrot** can be found from the thorn-forested hills right up to the beach scrub just before La Pesca. The beach by the breakwater in La Pesca and the adjacent flats should be checked for Piping Plover.

La Pesca Bird List

140 sp. R: rare (not to be expected). †See pp. 5–7 for notes on English names.

Thicket Tinamou
Brown Pelican
Double-crested Cormorant
Anhinga
Great Blue Heron
Great Egret
Snowy Egret
Little Blue Heron
Tricolored Heron
Reddish Egret
Cattle Egret
Green Heron
Yellow-crowned Night-Heron
White-faced Ibis
White Ibis
Black-bellied Whistling-Duck
Blue-winged Teal
Northern Shoveler
American Wigeon
Lesser Scaup
Black Vulture
Turkey Vulture
Osprey
White-tailed Kite
Northern Harrier
Sharp-shinned Hawk
Common Black Hawk
Harris' Hawk
Roadside Hawk
Broad-winged Hawk
Short-tailed Hawk
Swainson's Hawk
Red-tailed Hawk
Crested Caracara
American Kestrel
Plain Chachalaca
Northern Bobwhite
Sora
Common Moorhen
American Coot
American Golden Plover
Snowy Plover
Piping Plover
Killdeer
American Oystercatcher
American Avocet
Greater Yellowlegs
Lesser Yellowlegs
Willet
Solitary Sandpiper
Spotted Sandpiper
Upland Sandpiper
Whimbrel
Long-billed Curlew
Ruddy Turnstone
Sanderling
Semipalmated Sandpiper

Rufous-capped Warbler
Golden-browed Warbler
Yellow-breasted Chat
Olive Warbler
Blue-hooded Euphonia
Hepatic Tanager
Flame-colored Tanager
Red-headed Tanager
Blue Grosbeak
Varied Bunting
Spotted Towhee
Rufous-capped Brushfinch
Green-striped Brushfinch
Rusty Sparrow
Chipping Sparrow
Lincoln's Sparrow
Mexican Junco†
Hooded Oriole
Bullock's Oriole
Scott's Oriole
Black-headed Siskin
Lesser Goldfinch
Red Crossbill

CHAPTER 6 SAN BLAS, NAYARIT

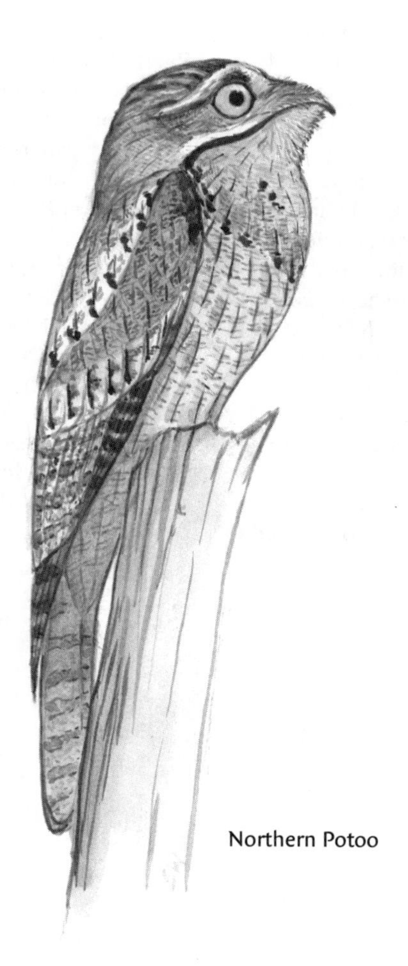

Northern Potoo

Mexico's west-coast state of Nayarit is synonymous among birders with the small coastal town of San Blas, made popular in the 1960s by pioneering birders such as Peter Alden, whose book *Finding the Birds in Western Mexico* (1969) devoted twenty-four pages to the San Blas area! As an idea of species richness, Christmas Bird Counts at San Blas (begun in 1973) have recorded up to 300 species. The varied habitats within easy reach of town are home to numerous western Mexico endemics, a good selection of widespread tropical species, and, in winter, large numbers of migrant waterbirds and landbirds.

Although somewhat of a resort, San Blas retains the character of a small coastal fishing village, unlike, for example, Mazatlán and Puerto Vallarta, which have been taken over by generic high-rise developments. Some people attribute this to the abundance of biting insects, for which the area is famous . . . But don't let this put you off—in winter insect activity is rarely a nuisance except locally in early morning and around dusk. The best time to visit San Blas is December to April; May to June is hotter, and July to November tends to be rainy and often (locally) unbearably buggy.

San Blas offers a variety of accommodations for all budgets, from luxury hotels to camping, and it would be remiss not to mention the highly recommended, family-run Hotel Garza Canela (whose name was changed in 1996 from the Hotel Las Brisas), which has many years of experience catering to birders.

Traditionally, most birders have reached San Blas via the five-hour drive from Mazatlán (Chapter 5). Ironically, the recent and exponential road improvements in northwestern Mexico have led to *autopistas* (often toll) all the way from Arizona to Guadalajara, Jalisco (indeed, to Mexico City), *except* for the stretch from Mazatlán airport to the San Blas junction off Route 15! While this situation may be remedied some day, the recent completion of a paved route from Puerto Vallarta to San Blas (a scenic but sinuous three-hour drive) makes this a more convenient option to consider if you plan to fly rather than drive.

It is easy to spend a week or longer based in San Blas, and many birders return again and again to this great spot. Also note that either Mazatlán and the Durango Highway (Chapter 5) or Colima and Jalisco (Chapter 7) can be combined with San Blas for an excellent one- to two-week birding trip. Birding sites in this chapter are divided into those of the San Blas area, reached easily by foot, bus, boat, or hitchhiking, and those of nearby areas, best reached by private car and done as day trips from San Blas.

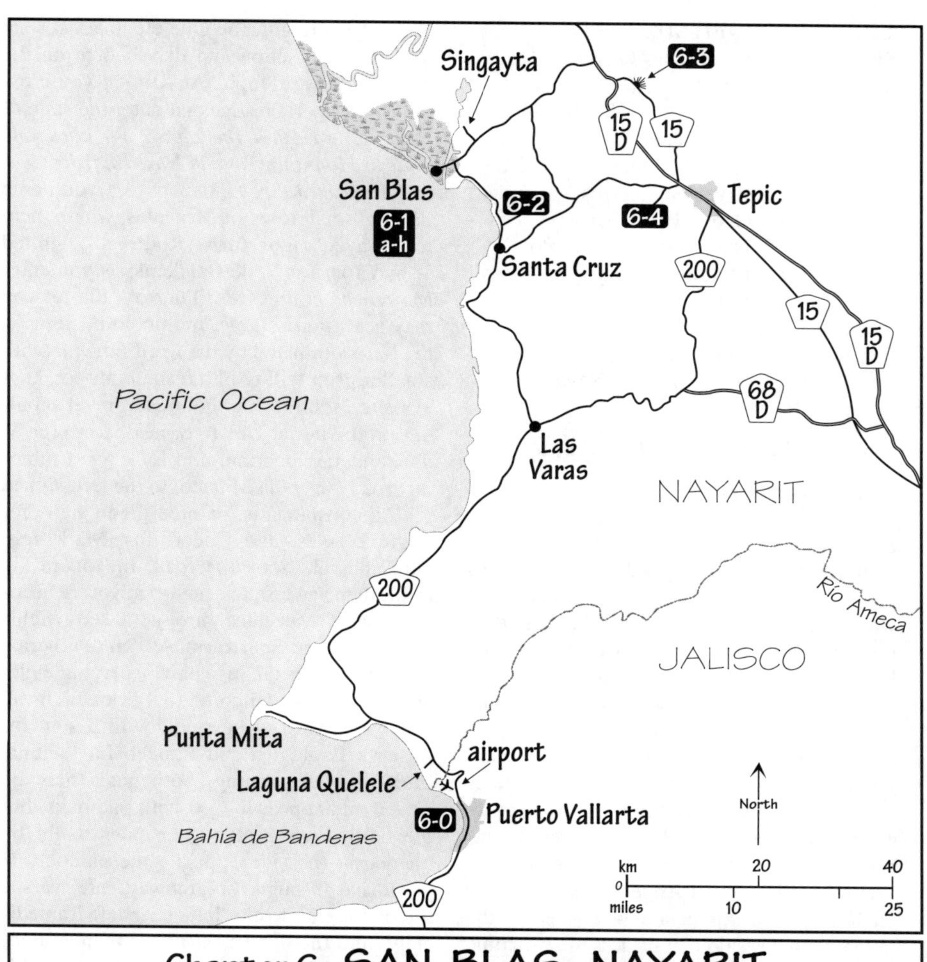

Chapter 6: SAN BLAS, NAYARIT

- **6-0** Puerto Vallarta, Jalisco
- **6-1** San Blas, Nayarit
  - **6-1a** San Blas - Sewer Ponds Trail
  - **6-1b** San Blas - Peso Island
  - **6-1c** San Blas - San Blas Fort
  - **6-1d** San Blas - River Boat Trips
  - **6-1e** San Blas - Ocean Boat Trips
  - **6-1f** San Blas - The Shrimp Ponds Road
  - **6-1g** San Blas - Matanchen Bay
  - **6-1h** San Blas - Singayta
- **6-2** La Bajada, Nayarit
- **6-3** El Mirador del Aguila, Nayarit
- **6-4** Cerro de San Juan, Nayarit

SAN BLAS, NAYARIT SITE LIST

Site 6.0 Puerto Vallarta, Jalisco
Site 6.1 San Blas, Nayarit
6.1a Sewer Ponds Trail
6.1b Peso Island
6.1c San Blas Fort
6.1d River Boat Trips
6.1e Ocean Boat Trips
6.1f The Shrimp Ponds Road
6.1g Matanchen Bay
6.1h Singayta
Site 6.2 La Bajada, Nayarit
Site 6.3 El Mirador del Aguila, Nayarit
Site 6.4 Cerro de San Juan, Nayarit

Site 6.0: Puerto Vallarta, Jalisco

(January 1997, 1998)

Birding Sites

If you fly into Puerto Vallarta, there are a couple of birding spots near town which are worth checking and which can be visited en route to San Blas: **Laguna de Quelele** and **Punta Mita**. Also see **El Tuito, Jalisco** (Site 7.0), a site for Military Macaws less than an hour south of Puerto Vallarta.

The **Laguna de Quelele** is a large, mangrove-fringed lagoon about fifteen minutes north of the Puerto Vallarta airport. Access is via a private ranch that has a restaurant-bar, which might make a good lunch spot. The owner is Miguel Angel Mateos, who is glad that people appreciate the concept of the reserve and has said that birders are welcome to visit, even in early morning when there may be nobody about. There is no official entrance fee, although a suggested donation of about $1.00 per person is requested. This is a great little birding spot, and in winter (at least) it is possible to record 100 species here in a morning! Please respect the private property and be courteous should you meet anyone.

Birds here include Roseate Spoonbill, Wood Stork, Fulvous and Black-bellied (sometimes thousands!) whistling-ducks, Crane Hawk, Rufous-necked Wood-Rail (from the second platform), **Mexican Parrotlet**, **San Blas Jay**, Bell's Vireo (winter), Painted Bunting (winter), and **Yellow-winged Cacique**.

To reach Laguna de Quelele, mark zero at the airport junction and drive north on the highway toward Tepic. At Km 5.7 you cross the Río Armeria bridge and enter the state of Nayarit: *note that the clocks go back one hour to Mountain Time in Nayarit, from Central Time in Jalisco*. At Km 10.7 you come into the small town of Mezcales, where there is a paved right turn (Route 58) signed at Km 10.9 for V. de Banderas; as you enter Mezcales coming from Puerto Vallarta, you may see a fairly large, multicolored sign on the left, dominated by the word *Laguna* (with smaller print telling of a restaurant, bar, kayaks, etc.). Pull off on the broad gravel shoulder on the left at Km 10.8, just before the V. de Banderas junction, and look for a minor dirt road (Avenida Mexico) to the left, signed with inconspicuous, painted green signs for Laguna, Aves, and Cocodrilos. Mark zero here and take this dirt road; in 300 m zig right, then zag left, and at 500 m you're heading out of Mezcales through weedy fields (Stripe-headed Sparrows). At Km 0.9 there's a right-hand bend just past a playing field, and at Km 1.5 you come to a brick archway and two gates on the right, with a sign for Reserva Ecologica and Rancho La Laguna. Early in the morning, both gates may be closed and appear locked with padlocks, but the barbed-wire cattle gate can usually be opened at the side opposite the padlock. If you drive through the archway gate, versus the cattle gate to the left, bear left immediately onto the dirt track (rather than straight on the cobbled track to the restaurant buildings) and continue 150 m to an obvious parking spot beside a small mangrove channel on your left.

From where you park there is a loop trail of 1 km or so through weedy fields and along the shore of Laguna de Quelele: start along the bank beside the mangrove channel where you park, and at the end, by the edge of the lagoon, turn right and follow a narrow path along the mangrove edge, to a wooden observation platform overlooking the lagoon; continue on to a second observation platform accessed by narrow wooden steps (watch for broken boards); from here, a broad mowed track cuts straight back to the ranch and where you parked.

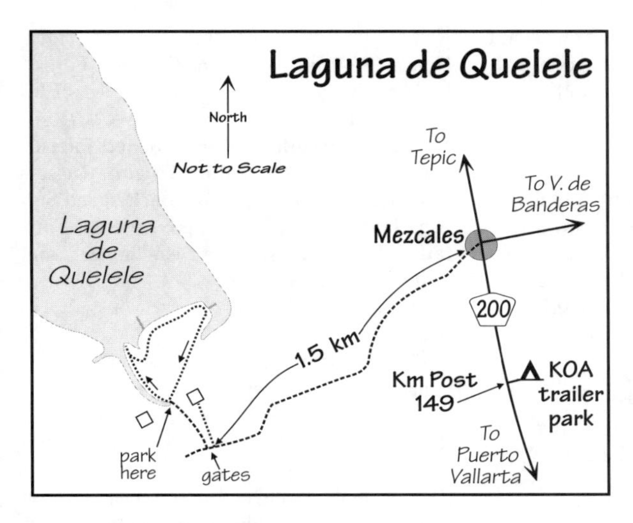

Laguna de Quelele Bird List

119 sp. †See pp. 5–7 for notes on English names.

American White Pelican
Brown Pelican
Neotropic Cormorant
Magnificent Frigatebird
Great Blue Heron
Great Egret
Snowy Egret
Little Blue Heron
Tricolored Heron
Cattle Egret
Green Heron
Black-crowned Night-Heron
Yellow-crowned Night-Heron
White Ibis
White-faced Ibis
Roseate Spoonbill
Wood Stork
Fulvous Whistling-Duck
Black-bellied Whistling-Duck
Green-winged Teal
Northern Pintail
Blue-winged Teal
Cinnamon Teal
Northern Shoveler
Lesser Scaup
Black Vulture
Turkey Vulture
Osprey
Hook-billed Kite
White-tailed Kite
Sharp-shinned Hawk
Cooper's Hawk
Crane Hawk
Common Black Hawk
Short-tailed Hawk
Zone-tailed Hawk
Collared Forest-Falcon
American Kestrel
Peregrine Falcon
Rufous-necked Wood-Rail
Common Moorhen
Black-bellied Plover
Wilson's Plover
Semipalmated Plover
Black-necked Stilt
American Avocet
Northern Jacana
Greater Yellowlegs
Lesser Yellowlegs
Willet
Spotted Sandpiper
Whimbrel
Long-billed Curlew
Marbled Godwit
Western Sandpiper
Least Sandpiper
Stilt Sandpiper
Short-billed Dowitcher
Long-billed Dowitcher
Laughing Gull
Caspian Tern
Red-billed Pigeon
White-winged Dove
Mourning Dove
Inca Dove
Common Ground-Dove
Ruddy Ground-Dove
White-tipped Dove
Mexican Parrotlet
Groove-billed Ani
Ferruginous Pygmy-Owl
Cinnamon Hummingbird
Ruby-throated Hummingbird
Belted Kingfisher
Green Kingfisher
Golden-cheeked Woodpecker
Ladder-backed Woodpecker
N. Beardless Tyrannulet
Least Flycatcher
Western Flycatcher†
Vermilion Flycatcher
Great Kiskadee
Social Flycatcher
Tropical Kingbird
Thick-billed Kingbird
N. Rough-winged Swallow
Bank Swallow
San Blas Jay
Happy Wren
Sinaloa Wren
Blue-grey Gnatcatcher
Rufous-backed Thrush†
Bell's Vireo
Warbling Vireo
Orange-crowned Warbler
Nashville Warbler
Yellow Warbler
Audubon's Warbler†
Black-and-white Warbler
American Redstart

Northern Waterthrush
MacGillivray's Warbler
Common Yellowthroat
Wilson's Warbler
Yellow-breasted Chat
Summer Tanager
Greyish Saltator
Blue Grosbeak
Painted Bunting
Dickcissel
Blue-black Grassquit
Cinnamon-rumped Seedeater†
Stripe-headed Sparrow
Lincoln's Sparrow
Great-tailed Grackle
Bronzed Cowbird
Orchard Oriole
Streak-backed Oriole
Yellow-winged Cacique

Another spot worth checking is the road out to **Punta Mita**, some 25 km (about thirty minutes) north of Puerto Vallarta. This paved road runs for about 20 km through thorn forest and ranches along the northern edge of the Bahía de Banderas. The well-signed left turn to Punta Mita is 21.3 km north of Puerto Vallarta airport, along the Tepic highway. Anywhere you can stop may be good but, frustratingly, there are few places to pull off safely and bird in the best stretches of thorn forest. However, at Km 4 there are two short cobbled roads and a large cobbled pulloff on the left, indications of an unfinished development. Species in the thorn forest here include Lesser Ground-Cuckoo, Cinnamon, Violet-crowned, and Broad-billed hummingbirds, **Citreoline Trogon**, Nutting's Flycatcher, Lucy's Warbler (winter), **Orange-breasted Bunting**, and Stripe-headed Sparrow; Flammulated Flycatcher and Red-breasted Chat also can be expected.

Site 6.1: San Blas, Nayarit

(March 1982, December/January 1982/83, 1983/84, January/February 1985, August 1988, October 1993, January, September, December 1995, January 1997, 1998)

Introduction/Key Species

San Blas is undoubtedly one of the most popular birding destinations in Mexico, and rightly so. Within easy reach of town the habitats range from mangrove lagoons to palm forest and arid scrub, and several boat trips contribute a suite of exciting birds. Species of interest in and near town include Red-billed Tropicbird, Bare-throated Tiger-Heron, Boat-billed Heron, Collared Forest-Falcon, **Rufous-bellied Chachalaca**, **Elegant Quail**, Rufous-necked Wood-Rail, **Mexican Parrotlet**, **Lilac-crowned Parrot**, Northern Potoo, **Citreoline Trogon**, **Russet-crowned Motmot**, **Purplish-backed** and **San Blas jays**, **Happy** and **Sinaloa wrens**, **Fan-tailed Warbler**, Rosy Thrush-Tanager, and **Yellow-winged Cacique**.

Access

Coming from **Mazatlán** (about a five-hour drive; see Chapter 5 for birding sites near Mazatlán), the kilometer posts along Route 15 count down to zero at Tepic; the junction for San Blas (Crucero San Blas) is at Km 34, immediately before the start of a new divided highway. While the gas station in San Blas has been renovated and now (1997) does have unleaded gas, in case of potential shortages it may be best to fill up at the Crucero San Blas. From here to San Blas it is 36 km, and birding can be good anywhere along the drive. The seasonal pond just left (south) of the road almost immediately after you turn off the highway is worth checking (at times it has hundreds of Black-bellied Whistling-Ducks) and can be viewed safely from a large pullout on the right (north) 300 m from the Crucero San Blas.

Around Km 25 the road winds down through some tropical palm forest with large trees, goes by the village of Singayta (a good birding spot; see below) at about Km 28, and then runs through a few kilometers of mangroves before crossing the bridge over the Río San Cristóbal (Km 35), which marks the entrance to San Blas. On the left-hand (south) riverbank nearer town is where boats leave for river trips, and above that is an isolated hill with the fort (see below). The cobbled one-way streets of San Blas lie ahead.

Coming from **Puerto Vallarta**, head north past the airport (mark zero here) on the highway toward Tepic. After the Punta Mita junction (see Site 6.0) the road winds through low hills and some nice tropical semi-deciduous forest and reaches the town of Las Varas at Km 87 (1.2–1.5 hours); at about Km 88, in downtown Las Varas, take the well-signed left turn for San Blas. Basically you stay on this road till Km 132 and an obvious stop sign at a T-junction (at Km 97.4 bend sharply left at the *zócalo* in Zacualpan; at Km 104.5 bear right, staying on the main highway to skirt the edge of Ixtapa de la Concepción; and at Km 130, just before the T-junction, go through El Llano).

Turn left at the T-junction, which is signed for San Blas (right goes to Tepic and is one option for reaching Cerro de San Juan [Site 6.4]) and in 800 m turn sharply right in response to a sign that says "San Blas 22" (straight on ends in the village of Santa Cruz). Stay on this road to the edge of San Blas: at Km 140 note the signed right turn for La Palma, shortly after which the road winds down to run behind the broad sweep of Matanchen Bay for several kilometers; the road bends sharply right at Km 149.5, in Matanchen, and ends at a stop sign at Km 151 (the seasonal ponds on either side of this junction should be checked for Collared Plover; see below under Matanchen Bay). Turn left here and it is 2 km to the bridge over the Río San Cristóbal, on the edge of San Blas, and another 1.5 km to the *zócalo*.

In terms of public transport, if there are not direct buses from Mazatlán, change at the Crucero San Blas (if you have a wait between buses, the pond described above is good for killing time), whence you may well be able to hitch a ride to San Blas. There is direct bus service from Puerto Vallarta to San Blas.

Birding Sites

These can be divided into those in and around town (easily accessible on foot), those reached by boat from San Blas, and those within a few kilometers of town (reached, if you are without a car, by bus, taxi, or hitching rides). A useful little booklet titled *Where to Find Birds in San Blas, Nayarit* (Novick and Wu 1994), now in its third edition, mentions locations additional to those described here.

Birding sites in and around town include the Sewer Ponds Trail (Site 6.1a), Peso Island (Site 6.1b), and San Blas Fort (Site 6.1c).

Boat trips include those on the rivers (Site 6.1d: La Tovara, Laguna de los Pájaros, Estero el Pozo) and those out to sea (Site 6.1e: Piedra de la Virgen and Roca Elefante).

Birding sites nearby include the Shrimp Pond Road (Site 6.15f), Matanchen Bay (Site 6.1g), and Singayta (Site 6.1h).

Site 6.1a: The Sewer Ponds Trail makes for a good morning's birding and runs from town through fields and scrubby woodland patches to the small sewer ponds by the mangrove-fringed Río San Cristóbal. Miscellaneous trails through the adjacent thickets and weedy fields with marshy pools are also good for birds. To find the start of this trail, walk three blocks south from the *zócalo* and turn left (east). Keep going a few blocks to the edge of town, where the road swings right by a cleared (January 1998) area (the fort is ahead of you) and then goes more or less straight for a kilometer or so to the sewer ponds and river. Once you know it, you'll likely be able to find quicker/more direct access to this trail, depending on where you are staying. This is usually a very birdy walk, particularly if some fig trees are fruiting, and species to look for include **Rufous-bellied Chachalaca**, **Elegant Quail**, **Mexican Parrotlet**, Ferruginous Pygmy-Owl, **Citreoline Trogon**, **Russet-crowned Motmot**, **Golden-cheeked** and Lineated **woodpeckers**, **Black-throated Magpie-Jay**, **Purplish-backed Jay** (infrequent), **Happy** and **Sinaloa wrens**, **Rufous-backed Thrush**, **Blue Mockingbird**, Yellow-breasted Chat (winter), Rosy Thrush-Tanager, Painted Bunting (winter), and, in the mangroves at the river, Rufous-necked Wood-Rail. The sewer ponds themselves (banked-up impoundments you might miss—if you're not downwind) are on the left (east) at the end of the trail, just before you get to the mangroves. A gated track leads up to a good overlook at the near corner of the ponds. Species to look for here include Least Grebe, Sora (winter), Northern Jacana, and Ruddy-breasted Seedeater. As the morning warms up, a good variety of raptors can be seen on this walk, including Crane, Common Black, Great Black, Grey, Short-tailed, and Zone-tailed hawks.

Site 6.1b: Peso Island is the name given by birders to the "island" across the El Pozo estuary from the town of San Blas. Launches ferry people back and forth (and traditionally cost a peso, hence the name), usually from 8 or 9 A.M. to 5 P.M., but one can arrange an

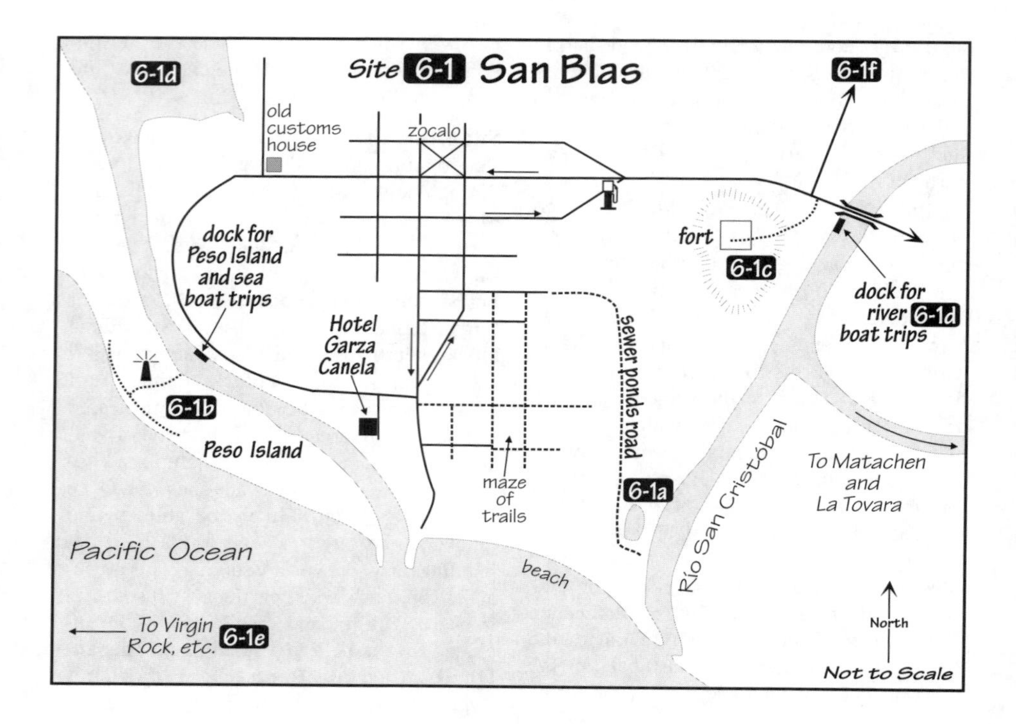

earlier or later trip. Be prepared to get your feet wet getting off or on at the far side of the estuary. The crossing in January 1998 cost 5 pesos per person, payable on the return crossing.

To get to the dock from the Hotel Garza Canela, turn left at the first corner after you turn left out of the front door and walk 600 m, until you come to the estuary dock; or, from the *zócalo*, head west 400 m from the southwest corner, past the old customs house (worth checking in the evenings for roosting Grey-breasted Martins) and curve around to the left and the estuary dock. The mudflats just north of here, at the fishing fleet harbor, can be good at low tide for shorebirds, including Wilson's Plover, Dunlin (winter), and Short-billed Dowitcher (winter). The launches run if and when there are people who want to cross, and you may have to look around a bit for the boatman.

From where the launch drops you off a trail heads into the scrubby woodland, through a couple of grassy openings and to the outer beach, with a good view across to La Piedra de la Virgen (Virgin Rock, named for the statue atop). The grassy areas with acacia scrub are good for Stripe-headed Sparrow, and in winter one can find Cassin's, Grasshopper, and Savannah sparrows. Other species include **Elegant Quail**, **Russet-crowned Motmot**, Ladder-backed Woodpecker, **Purplish-backed Jay**, Grey-crowned Yellowthroat, and a variety of wintering warblers, including Lucy's Warbler and Yellow-breasted Chat. The beach, especially near the breakwater down to your left by the mouth of the estuary, can have good roosting flocks of gulls and terns, as well as American Oystercatcher and Collared and Snowy plovers; Wandering Tattler and Surfbird (March/April) can be found on the breakwaters. Brown and Blue-footed boobies roost offshore on Virgin Rock. At times in late summer and also during El Niño winters, Black and Least storm-petrels can be seen off the beach and breakwaters.

Site 6.1c: The San Blas Fort, atop an isolated hill between town and the river, provides an excellent view of town and the surrounding area. It is worth visiting for the view and the possibility of green-flash sunsets, as well as for birds. To reach the fort, head out of town toward the bridge over the Río San

Cristóbal, and take the fork to the right (opposite the turn for the Shrimp Pond Road on the left), between rows of small restaurant stalls, toward the river and launches. In 100 m a cobblestone road goes steeply up to the right. One can walk or drive the 600 m up this road, past the cemetery to the fort. Starting in 1998, a 5-peso entry fee is charged per person. Birding here in the large trees, humid thickets, and weedy fields can be good, especially if some trees are fruiting (**Mexican Parrotlet**, **Citreoline Trogon**, **Rufous-backed Thrush**, etc.). If you stay till dusk, you have a chance of Collared Forest-Falcon, Barn and Mottled owls, Pauraque, and Buff-collared Nightjar (winter).

Site 6.1d: River Boat Trips, popular among many tourists as a "jungle boat ride," offer the chance for a number of species difficult to see for a birder based only on land. The best trips for birders are upriver from the bridge to some large lagoons, including Laguna de los Pájaros, and downriver toward the river mouth and along a connecting waterway to the spring at La Tovara. Boats for the river trips leave from the south side of the bridge on the town side of the river (1.5 km from the *zócalo*), and those for La Tovara also leave from Matanchen. Another trip good for birds is up the Estero el Pozo to Laguna de Pericos (where low tide is best for exposed mudflats), for which boats leave from the Peso Island dock. Two seagoing boat trips are also of interest for birds (Site 6.1e): to Virgin Rock (visible off the beach) and to the more distant Roca Elefante (visible from the fort, offshore to the north).

While most boatmen tend to be good bird spotters, a smaller number know the birds well and know their names in English. Recommended boatmen for river trips are Oscar Partida (phone 50324) and Chencho, and for the Estero trip and trips to Virgin Rock and Roca Elefante, Armando Santiago. They can be contacted via the Hotel Garza Canela or at the respective docks, or the hotel personnel can recommend other boatmen. Prices tend to be set depending on the number of people in a boat, and as always it is good to make sure you have the price agreed upon beforehand. Recent cost for the San Cristóbal river trips (January 1998) was 400 pesos for up to four persons, while the *estero*/ocean trips cost 500–600 pesos for up to four persons.

Birds to look for on the trip upriver include Bare-throated Tiger-Heron, Boat-billed Heron, Muscovy Duck, and sometimes spectacular numbers of waterfowl and shorebirds (including thousands of both species of whistling-ducks), raptors (including both black hawks, Crane Hawk, and Laughing Falcon), Mangrove Cuckoo, Mangrove Vireo, and Mangrove [Yellow] Warbler. Birds to look for downriver include Reddish Egret, Rufous-necked Wood-Rail, Wilson's Plover (on exposed sandbars), and, on the La Tovara trip, Bare-throated Tiger-Heron, Boat-billed Heron, Mangrove Cuckoo, and Northern Potoo. The potoos require a special evening trip, going up to La Tovara in late afternoon and coming back at dusk with a spotlight: make sure the boatman has a good flashlight or spotlight, and take your own flashlight as a backup and to watch out for mangrove branches. Also possible on this evening trip are Mottled Owl and Pauraque.

Birds to look for on the Estero el Pozo trip include Bare-throated Tiger-Heron, Reddish Egret, raptors, Rufous-necked Wood-Rail, Mangrove Cuckoo, Mangrove Vireo, Mangrove [Yellow] Warbler, and a variety of wading birds and shorebirds.

On all boat trips there is a good chance for photography, of wading birds in particular, and one also tends to see a variety of "landbirds," which can include chachalacas, parrots, woodpeckers, and jays. If you're looking for Mangrove Vireo and Mangrove Warbler, it is good to pull the boat into the mangroves, cut the engine, and imitate a Ferruginous Pygmy-Owl. This can bring in a band of mobbing warblers (mainly North American migrants), which may draw in a vireo.

Site 6.1e: Ocean Boat Trips offer a chance for good close views of boobies and, depending on venue and season, Red-billed Tropicbird and Bridled Tern. Boats leave from the Peso Island dock. As with river trips, it is good to agree upon a price beforehand and, if you feel the need, to also ask if life vests can be provided. Getting out of the mouth of the *estero* can involve going

Tropical Parula
Audubon's Warbler†
Townsend's Warbler
Black-throated Green Warbler
Black-and-white Warbler
Wilson's Warbler
Painted Whitestart†
Golden-crowned Warbler
Rufous-capped Warbler
Olive Warbler
Hepatic Tanager
Flame-colored Tanager
Rufous-capped Brushfinch
Olive Sparrow
Spotted Towhee
Canyon Towhee
Rufous-crowned Sparrow
Chipping Sparrow
Black-chinned Sparrow
Lincoln's Sparrow
Mexican Junco†
Audubon's Oriole
House Finch
Pine Siskin
Lesser Goldfinch

Site 3.2: Presa El Tulillo, Coahuila

(November 1986, March 1988, April 1990)

Introduction/Key Species

This small reservoir lies in a large area of desert and, consequently, acts as a magnet for migrants and a remarkable variety of vagrants, which have included Common Loon, Surf Scoter, and Pomarine and Parasitic jaegers! Birds typical of the surrounding desert can also be found around and en route to the reservoir.

Access/Birding Sites

Presa El Tulillo (1150 m elevation) is a small, easily accessible reservoir in the desert about 42 mi (70 km; an hour) west of Saltillo. From Saltillo, head west on the old highway, Route 40 (not the new toll highway, Route 40D), and after about 32 mi (53 km) watch for a right-hand junction (north) signed to Hipolito; this is about 7 mi (12 km) after a signed left turn to General Cepeda. Turn right onto the paved road toward Hipolito, and after 1.2 mi (2 km), you'll see the reservoir to your left (west). At 1.5 mi (2.5 km) the road crosses the dam, which offers a good view over the water as well as to your right, down into the outflow channels and riparian vegetation, where one can find paths to explore. Near 1.8 mi (3 km) there is a dirt track (with a gate) off to the left (west) that runs along near the north side of the reservoir, with sundry tracks leading down to picnic sites, with tables, by the reservoir. It also may be possible to drive along the south side, or this can be walked.

If you're using public buses, the reservoir can be reached from Saltillo by buses to Hipolito, or you could walk/hitch a ride from the Route 40 junction. Note that Tulillo is a popular weekend picnic destination, so for birding it is best visited on weekdays or in early morning on weekends.

Presa El Tulillo Bird List

110 sp. R: rare; V: vagrant (not to be expected). †See pp. 5–7 for notes on English names.

Common Loon (V)
Pied-billed Grebe
Eared Grebe
Western Grebe (R)
Great Blue Heron
Great Egret
Snowy Egret
Cattle Egret
Green Heron
Black-crowned Night-Heron
White-fronted Goose†
Green-winged Teal
Mexican Duck†
Northern Pintail
Blue-winged Teal
Cinnamon Teal
Northern Shoveler
Gadwall
American Wigeon
Canvasback
Redhead
Ring-necked Duck
Lesser Scaup
Surf Scoter (V)
Bufflehead
Hooded Merganser (R)
Ruddy Duck
Turkey Vulture
Northern Harrier
Sharp-shinned Hawk
Harris' Hawk
Red-shouldered Hawk
Swainson's Hawk
American Kestrel
American Coot
Black-bellied Plover (R)
Snowy Plover
Killdeer
Black-necked Stilt
Greater Yellowlegs
Spotted Sandpiper
Western Sandpiper
Least Sandpiper
Baird's Sandpiper
Long-billed Dowitcher
Common Snipe
Laughing Gull
Bonaparte's Gull
Ring-billed Gull
White-winged Dove
Mourning Dove
Lucifer Hummingbird
Black-chinned Hummingbird
Belted Kingfisher
Golden-fronted Woodpecker
Ladder-backed Woodpecker
Grey Flycatcher
Eastern Phoebe
Black Phoebe
Say's Phoebe
Vermilion Flycatcher
Ash-throated Flycatcher
Scissor-tailed Flycatcher
Horned Lark
Tree Swallow

through some waves and, after mid morning, the wind can pick up on the sea, so it is a good idea to protect optics and other items with plastic bags or some form of waterproofing. A light jacket/windbreaker and sunscreen are other essentials.

The trip out to Virgin Rock and back can be done in an hour or so and at any time of day (and can be combined with a trip up the Estero el Pozo), although more roosting birds are likely to be found in early morning and late afternoon. You should see Blue-footed and Brown boobies and, in summer, Bridled Terns, which nest on the rock. Migrant Sooty Terns and Brown Noddies are possible in spring and autumn. If one goes farther offshore, possibilities (mainly in winter) include **Black-vented Shearwater**, Black and **Least storm-petrels**, Red-necked and Red phalaropes, Pomarine and Parasitic jaegers, and Black Tern.

The trip to Roca Elefante (so named because from some angles it looks like an elephant's head) is longer; you should allow 4–5 hours. An early morning start is best to avoid any wind that picks up during the day. The main reason for this trip is the chance to see Red-billed Tropicbird, a few pairs of which nest at the rock and can be seen there at any season. All of the species listed for Virgin Rock may also be seen.

Site 6.1f: The Shrimp Ponds Road offers access to mangroves, lagoons, and a somewhat drier and more agricultural environment than is found elsewhere around San Blas. It is a good trip for mid to late afternoon. Take the road out of San Blas and turn left on the edge of town, 1.2 km from the *zócalo* and just before the bridge, on a straight, paved road signed "GuadalupeVictoria 16." Mark zero at this junction. Driving and stopping along this road is the best way to bird it, as water levels in the various ponds (and hence the distribution of birds) varies from year to year and season to season. The first few large ponds to the left (west) within 1 km of the junction are often good for shorebirds, including Collared Plover. The weedy roadside patches can also be very birdy in winter, for instance, around Km 4.9, where there is a track on the left to pull off from the highway. At Km 7.7 there is a dirt road off to the right (east) to Chacalilla, which passes through more ponds and agricultural fields (Harris' Hawk) and gets to some active shrimp ponds in 2.5 km. One can also stay on the main road, which gets into similar, drier habitat.

Birds to look for along the Shrimp Ponds Road include a good variety of wading birds, ducks, and shorebirds, including Wood Stork, Roseate Spoonbill, Collared Plover, raptors including White-tailed Kite, Northern Harrier (winter), Harris' Hawk, and Peregrine Falcon (migrant), plus Clapper Rail, Ringed Kingfisher, Northern Raven, Loggerhead Shrike, Mangrove Vireo, Grey-crowned Yellowthroat, Ruddy-breasted Seedeater (locally fairly common in damp weedy areas, at least in winter, but overlooked easily in basic plumage; note that **Cinnamon-rumped** [White-collared] **Seedeaters** are common), Eastern Meadowlark, and flocks of blackbirds.

Site 6.1g: Matanchen Bay is a large sweep of sandy beach around a fairly protected bay a few kilometers south of San Blas. A creek flows into the bay and, depending on local conditions of tide and silt inflow, there can be a lagoon with muddy sand flats at the northern end of the bay.

To reach Matanchen, head inland out of San Blas 2 km from the bridge and turn right (south) at the first junction (which may be signed to Puerto Vallarta); reset to zero at this turn. The ponds on either side of the road immediately as you turn here can be good for wading birds (Roseate Spoonbill early and late in the day) and, if not flooded, for shorebirds (especially Collared Plover). It is possible to pull off the road right beside the ponds and view them from your vehicle. However, as this is now the start of the new route to Puerto Vallarta, you should be aware of non-local traffic, which may be less accommodating of birders stopped beside the road. The ponds to the right (south) between the San Blas bridge and the junction can also be good for birds, particularly Roseate Spoonbills early in the morning, but are harder to see clearly, or to park near.

From the junction you drive through mangroves for a couple of kilometers and, just before the road bends sharply left along Matanchen Bay, you'll see a few houses,

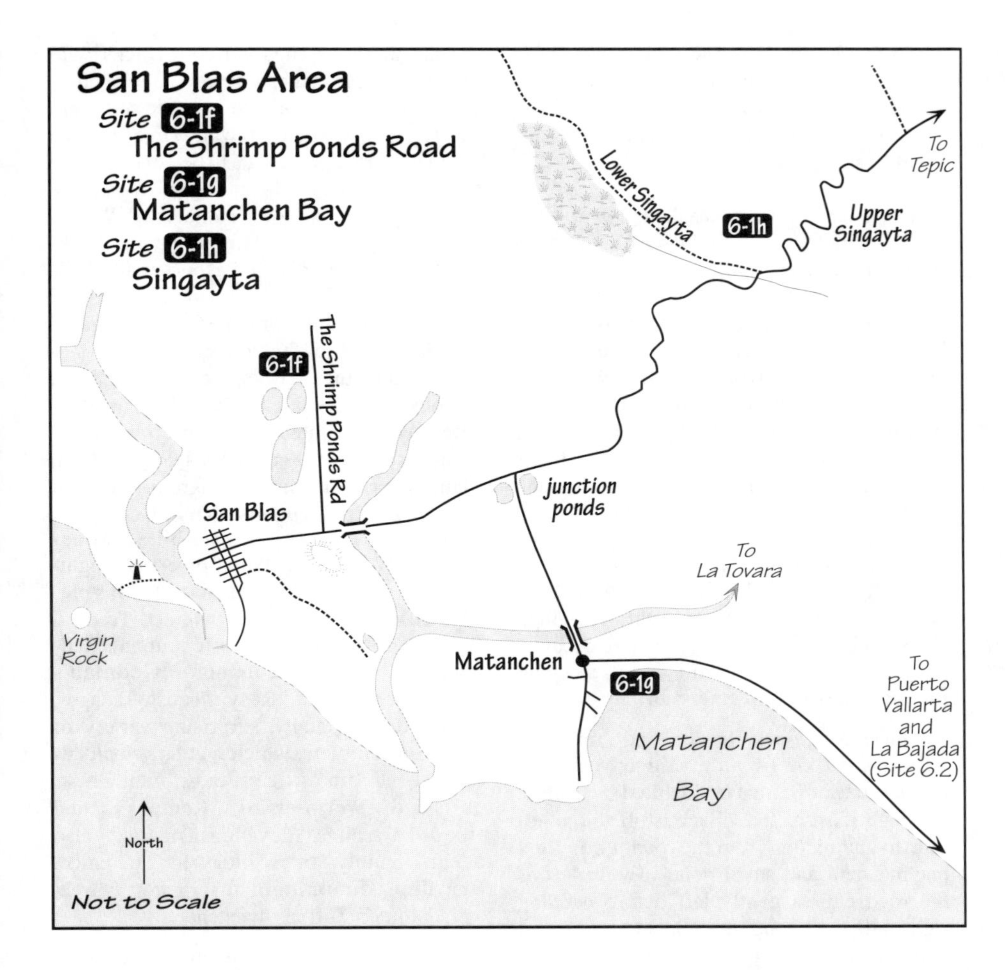

restaurants, and, to your left, a small "dock" with launches. This is Matanchen, and boats can be taken from here to La Tovara (see River Boat Trips, Site 6.1d). At the point where the main road bends sharply left, at Km 2.5, stay straight on a dirt road that runs behind the beach for 1.5 km to Las Islitas, an aggregation of *palapas* (thatched open huts) and restaurants.

Once on the dirt road, in 300 m you cross a small bridge over a creek through mangroves: the muddy channels here should be checked for Rufous-necked Wood-Rail and are also good for Yellow-crowned Night-Heron and White Ibis. The mangroves and scrub here and along the road beyond have Mangrove [Yellow] Warbler and Tropical Parula (migrant?). A hundred meters after the bridge you can take the first hard-looking sand track to the left and some *palapas* and restaurants by the beach. Park here and walk left toward the mouth of the creek to check the status of the "lagoon" behind the beach (sadly, this site has been silted up in recent years and, consequently, seems to be less attractive to shorebirds).

Flocks of gulls and terns often gather on the beach, and Common Loons (winter) and boobies can be seen out in the bay. With the right conditions this can be an excellent place for wading birds (Reddish Egret), shorebirds (Collared, Snowy, and Wilson's plovers), gulls, and terns. If you continue on to Las Islitas, the rocky promontories should be checked (March/April) for migrant Surfbirds. Birds in the grassy areas with acacia scrub behind the beach are much like those of Peso Island: Ladder-backed Woodpecker, Ash-throated Flycatcher (winter), Grey-crowned Yellowthroat, Pyrrhuloxia (migrant?), sparrows, and so on.

Site 6.1h: Singayta is a village about 9 km (10–15 minutes by car) inland from San Blas on the road back to Crucero San Blas and the main highway. This is perhaps the single best birding location in the San Blas area. Traditionally there have been two birding areas: Upper Singayta and Lower Singayta. The former involves birding from the winding main road above the town but, due to increased traffic, is less attractive (and less safe) than it was some years ago—although the birding is still good. The latter refers to a dirt road that runs through the village and continues between scrub and marsh to your left and a hillside of tropical palm/semi-deciduous forest to your right. Although areas are increasingly being cleared at Lower Singayta, this location still offers excellent birding and is best in early morning.

Lower Singayta: Head inland from San Blas and, 7.2 km from the bridge, you come to the village of Singayta (marked by a sign), although most of the village is off to the left (north) and hidden from the road. Look for a playing field and small pond to your left as the road curves gently left before bending right and winding up through palm forest and plantations. At the right-hand bend a cobbled road goes sharply left into the village of Singayta but is somewhat hidden if you are coming from San Blas; take care making this turn. You can drive through the village, about 600 m to the far side, and park off the road. Continue on foot from here along the dirt road, and birding is good for at least 3 km, to where there are mostly open fields (which is about as far as most birders manage to get!). You can also drive and park at various pull-outs in the forest—for example, a track on the left at Km 2.5.

If you arrive and park in early morning, listen for Collared Forest-Falcon, which sometimes can be seen in the canopy of an emergent tree. Past the last corral on the edge of town (on your left), check the fields for Stripe-headed Sparrows and the big fig trees beyond the pasture for perched Muscovy Ducks (wild ones!). The marsh that will be to your left after a few hundred meters can be viewed from several points, some of which may require opening barbed wire gates (be sure to close them behind you) and crossing small pastures (e.g., at Km 1.5). Good numbers of White-throated Flycatchers winter in the marsh (locate them by their distinctive burry calls; Willow and Least flycatchers also occur). Other species to look for along the Lower Singayta road include Hook-billed Kite, Crane, Zone-tailed, and other hawks, Laughing Falcon, **Rufous-bellied Chachalaca**, **Mexican Parrotlet**, White-fronted and **Lilac-crowned** (irregular) **parrots**, Ferruginous Pygmy-Owl, Plain-capped Starthroat (migrant), Elegant and **Citreoline trogons**, **Russet-crowned Motmot**, Lineated and Pale-billed woodpeckers, Ivory-billed Wood-creeper, Bright-rumped Attila, **Black-throated Magpie-Jay**, Black-capped Vireo (winter), Tropical Parula, **Fan-tailed Warbler** (common), **Red-breasted Chat** (uncommon), **Godman's** [Scrub] **Euphonia**, Rosy Thrush-Tanager, **Blue Bunting**, and a surprising variety of wintering warblers, which regularly includes vagrant "eastern" species, most often Chestnut-sided, Worm-eating, Kentucky, and Hooded warblers. Elevational migrants also occur, mainly in winter—for instance, **Berylline Hummingbird**, Greater Pewee, and Common Tufted Flycatcher.

Upper Singayta: The main road through Singayta continues inland, winding up through palm forest and plantations before leveling out 3 km or so from the Lower Singayta junction. There are several places to pull a car off, but the best way to bird this area is to get a ride to the top, where the road levels and straightens out, and walk back down to the village. The fields at the "top" and for a few kilometers farther inland along the highway can be very birdy in winter, with lots of migrant seed-eating birds (grosbeaks including Pyrrhuloxia, buntings, sparrows, Lesser Goldfinch); the roadside wires may have Cassin's Kingbirds (rarely seen around San Blas), and Lesser Ground-Cuckoo can be heard calling from the thickets. The Upper Singayta forest can birded from the road, and there are also several small side tracks and trails you can take into the forest and to weedy fields and plantations. Species to look for include Rufous-bellied Chachalaca, White-fronted and **Lilac-crowned** (irregular) **parrots**, **Mexican Hermit**, trogons, Boat-billed Flycatcher (rare), San Blas and Purplish-backed jays (fields at top), **Fan-tailed Warbler**, Red-crowned Ant-Tanager, and **Rusty-crowned Ground-Sparrow**. It is worth noting that most or all of the Upper Singayta forest birds can be found at least as easily at La Bajada, together with a number of other interesting species.

San Blas Bird List

305 sp. Rather than repeat long lists of birds for each site, the following species can be found, in habitat and season, at the birding sites in and around town, including the river boat trips, Shrimp Ponds Road, Matanchen Bay, and Singayta; this list does not include birds recorded only from ocean boat trips (Site 6.1e). Species to look for at a given site are noted in the Birding Sites. R: rare; V: vagrant (not to be expected). †See pp. 5–7 for notes on English names.

Common Loon
Least Grebe
Pied-billed Grebe
Eared Grebe
Black Storm-Petrel
Least Storm-Petrel
Blue-footed Booby
Brown Booby
American White Pelican
Brown Pelican
Neotropic Cormorant
Anhinga
Magnificent Frigatebird
Bare-throated Tiger-Heron
Least Bittern
Great Blue Heron
Great Egret
Snowy Egret
Little Blue Heron
Tricolored Heron
Reddish Egret
Cattle Egret
Green Heron
Black-crowned Night-Heron
Yellow-crowned Night-Heron
Boat-billed Heron
White Ibis
White-faced Ibis
Roseate Spoonbill
Wood Stork
Fulvous Whistling-Duck
Black-bellied Whistling-Duck
Brant (V)
Muscovy Duck
Green-winged Teal
Northern Pintail
Blue-winged Teal
Cinnamon Teal
Northern Shoveler
Gadwall
American Wigeon
Ring-necked Duck
Lesser Scaup
Ruddy Duck
Black Vulture
Turkey Vulture
Osprey
Hook-billed Kite
White-tailed Kite
Northern Harrier
Sharp-shinned Hawk
Cooper's Hawk
Crane Hawk
Common Black Hawk
Great Black Hawk
Harris' Hawk
Grey Hawk
Broad-winged Hawk
Short-tailed Hawk
Zone-tailed Hawk
Red-tailed Hawk
Crested Caracara
Laughing Falcon
Collared Forest-Falcon
American Kestrel
Merlin
Peregrine Falcon
Rufous-bellied Chachalaca
Elegant Quail
Clapper Rail
Virginia Rail
Spotted Rail (R)
Rufous-necked Wood-Rail
Sora
Purple Gallinule
Common Moorhen
American Coot
Black-bellied Plover
Collared Plover
Snowy Plover
Wilson's Plover
Semipalmated Plover
Piping Plover (V)
Killdeer
American Oystercatcher
Black-necked Stilt
American Avocet
Northern Jacana
Greater Yellowlegs
Lesser Yellowlegs
Solitary Sandpiper
Willet
Wandering Tattler
Spotted Sandpiper
Whimbrel
Long-billed Curlew
Marbled Godwit
Ruddy Turnstone
Black Turnstone (R)
Surfbird
Red Knot
Sanderling
Western Sandpiper
Least Sandpiper
Dunlin
Stilt Sandpiper
Short-billed Dowitcher
Long-billed Dowitcher
Common Snipe
Wilson's Phalarope
Red Phalarope
Laughing Gull
Franklin's Gull
Bonaparte's Gull
Heermann's Gull
Ring-billed Gull
California Gull
Herring Gull
Western Gull (V)
Black-legged Kittiwake (V)
Gull-billed Tern
Caspian Tern
Royal Tern
Elegant Tern
Common Tern
Forster's Tern
Least Tern
Black Tern
Black Skimmer
Red-billed Pigeon
White-winged Dove
Mourning Dove
Common Ground-Dove
Ruddy Ground-Dove
White-tipped Dove
Orange-fronted Parakeet
Mexican Parrotlet
White-fronted Parrot
Lilac-crowned Parrot
Yellow-billed Cuckoo

Mangrove Cuckoo
Squirrel Cuckoo
Lesser Ground-Cuckoo
Groove-billed Ani
Barn Owl
Vermiculated Screech-Owl
Colima Pygmy-Owl†
Ferruginous Pygmy-Owl
Mottled Owl
Lesser Nighthawk
Pauraque
Buff-collared Nightjar
Northern Potoo
Black Swift
Chestnut-collared Swift
White-naped Swift
Vaux's Swift
Great Swallow-tailed Swift (R)
Mexican Hermit†
Golden-crowned Emerald
Broad-billed Hummingbird
Berylline Hummingbird
Cinnamon Hummingbird
Violet-crowned Hummingbird
Plain-capped Starthroat
Ruby-throated Hummingbird
Black-chinned Hummingbird
Rufous Hummingbird
Allen's Hummingbird
Citreoline Trogon
Elegant Trogon
Russet-crowned Motmot
Ringed Kingfisher
Belted Kingfisher
Green Kingfisher
Golden-cheeked Woodpecker
Gila Woodpecker
Ladder-backed Woodpecker
Lineated Woodpecker
Pale-billed Woodpecker
Ivory-billed Woodcreeper
N. Beardless Tyrannulet
Greenish Elaenia
Greater Pewee
Common Tufted Flycatcher†
Willow Flycatcher
White-throated Flycatcher
Least Flycatcher
Western Flycatcher†
Vermilion Flycatcher
Bright-rumped Attila
Dusky-capped Flycatcher
Ash-throated Flycatcher
Brown-crested Flycatcher
Great Kiskadee
Social Flycatcher
Boat-billed Flycatcher
Sulphur-bellied Flycatcher
Tropical Kingbird
Cassin's Kingbird
Thick-billed Kingbird
Western Kingbird
Rose-throated Becard
Masked Tityra
Purple Martin
Grey-breasted Martin
Tree Swallow
Mangrove Swallow
N. Rough-winged Swallow
Bank Swallow
Cliff Swallow
Barn Swallow
Black-throated Magpie-Jay
Purplish-backed Jay
San Blas Jay
Sinaloa Crow
Northern Raven†
Happy Wren
Sinaloa Wren
Northern House Wren†
Marsh Wren
Blue-grey Gnatcatcher
Orange-billed N.-Thrush
Swainson's Thrush
White-throated Thrush†
Rufous-backed Thrush†
Northern Mockingbird
Blue Mockingbird
American Pipit
Loggerhead Shrike
Mangrove Vireo
Bell's Vireo
Black-capped Vireo
Plumbeous Vireo†
Yellow-throated Vireo (V)
Yellow-green Vireo
Golden Vireo
Warbling Vireo
Golden-winged Warbler (V)
Orange-crowned Warbler
Nashville Warbler
Lucy's Warbler
Tropical Parula
Yellow Warbler
Mangrove Warbler†
Chestnut-sided Warbler (V)
Magnolia Warbler (V)
Audubon's Warbler†
Black-throated Grey Warbler
Black-and-white Warbler
American Redstart
Worm-eating Warbler (V)
Ovenbird
Northern Waterthrush
Kentucky Warbler (V)
MacGillivray's Warbler
Common Yellowthroat
Grey-crowned Yellowthroat
Hooded Warbler (V)
Wilson's Warbler
Fan-tailed Warbler
Yellow-breasted Chat
Red-breasted Chat
Godman's Euphonia†
Red-crowned Ant-Tanager
Summer Tanager
Western Tanager
Flame-colored Tanager
Rosy Thrush-Tanager
Greyish Saltator
Pyrrhuloxia
Yellow Grosbeak
Rose-breasted Grosbeak
Black-headed Grosbeak
Blue Grosbeak
Blue Bunting
Lazuli Bunting
Indigo Bunting
Varied Bunting
Painted Bunting
Dickcissel
Rusty-crowned Ground-Sparrow
Blue-black Grassquit
Cinnamon-rumped Seedeater†
Ruddy-breasted Seedeater
Stripe-headed Sparrow

Cassin's Sparrow
Clay-colored Sparrow
Lark Sparrow
Grasshopper Sparrow
Savannah Sparrow
Lincoln's Sparrow
Eastern Meadowlark
Red-winged Blackbird
Yellow-headed Blackbird
Great-tailed Grackle
Bronzed Cowbird
Brown-headed Cowbird
Orchard Oriole
Hooded Oriole
Black-vented Oriole
Streak-backed Oriole
Bullock's Oriole
Yellow-winged Cacique
Lesser Goldfinch
House Sparrow

Site 6.2: La Bajada, Nayarit

(March 1982, December 1982, 1983, January 1985, October 1993, January, September, December 1995, January 1997, 1998)

Introduction/Key Species

La Bajada (elevation 200 m), a coffee-growing village in the hills south of Matanchen Bay, is about 18 km (30–40 minutes) from San Blas. The plantations above the village, with their tall shade trees, host a variety of species not found, or decidedly less numerous, around San Blas, including Crested Guan, Ruddy Quail-Dove, **Colima Pygmy-Owl**, **Mexican Hermit**, **Mexican Woodnymph**, **Grey-crowned Woodpecker**, Common Tufted and Boat-billed flycatchers, **San Blas Jay**, **Brown-backed Solitaire**, **Golden Vireo**, Red-crowned Ant-Tanager, Yellow Grosbeak, and **Rusty-crowned Ground-Sparrow**. A few winter migrants more typical of higher elevations can also be found here, including Greater Pewee, Common Tufted Flycatcher, Hammond's Flycatcher, Ruby-crowned Kinglet, Grey Silky, Townsend's Warbler, and Slate-throated Whitestart.

Access/Birding Sites

Head out from San Blas as if going to Matanchen Bay (Site 6.1g), and mark zero at the junction 2 km inland of the Río San Cristóbal bridge. Rather than going straight on the dirt road to Matanchen Bay (Site 6.1g), stay on the main road where it bends sharply left in Matanchen, at Km 2.5. The road runs behind the bay for about 7 km before winding up into the hills at the south end. At Km 11.1, and about 1.8 km from the bay, there is an obvious junction (the first you come to), with a paved road off to the left signed to La Palma. Turn left here (reset to zero at this junction), and in 500 m you enter the village of La Palma: at Km 1.0 stay straight rather than bearing right into town, then bear right by a church and at Km 1.4 turn right and then left in one block, which is the road out of La Palma and on to La Bajada. At Km 3.7 you enter La Bajada; at Km 4.3 stay straight on the main cobbled road (rather than bearing right) and continue through the village to the edge of the coffee plantations at Km 4.8. At this point there is a circular brick wall up to the left and an area where you can park safely off the main (cobbled) road, which continues up into the plantations. The streams in La Bajada and La Palma should be checked for Louisiana Waterthrush (winter), although human disturbance means the birds are rarely near the road.

You can walk for 10 km or more (most birds can be seen in the first 3–4 km) along the main cobbled road, and on sundry side trails, up through the plantations. The elevation (200–500 m) in combination with the tall shade trees means that birding can be good throughout the day, although morning is best. On some visits, the trees seem to be dripping with birds, especially mixed-species canopy flocks dominated in winter by Blue-grey Gnatcatchers, Warbling Vireos, and Nashville and Black-throated Grey warblers. At other times, even the next day, La Bajada seems dead, so more than one visit may be needed to convince you it is a good birding spot!

La Bajada Bird List

125 sp. R: rare; V: vagrant (not to be expected). †See pp. 5–7 for notes on English names.

Black Vulture
Turkey Vulture
Sharp-shinned Hawk
Grey Hawk
Short-tailed Hawk
Zone-tailed Hawk
Collared Forest-Falcon
Rufous-bellied Chachalaca
Crested Guan (R)
Red-billed Pigeon
Inca Dove
White-tipped Dove
Ruddy Quail-Dove
Orange-fronted Parakeet
Mexican Parrotlet
White-fronted Parrot
Lilac-crowned Parrot
Squirrel Cuckoo

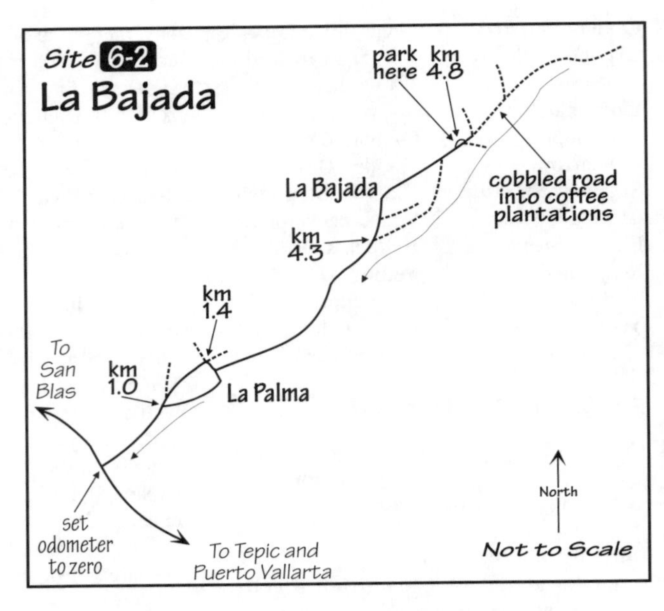

Groove-billed Ani
Colima Pygmy-Owl†
Ferruginous Pygmy-Owl
Pauraque
Chestnut-collared Swift
Vaux's Swift
Mexican Hermit†
Golden-crowned Emerald
Mexican Woodnymph
Broad-billed Hummingbird
Berylline Hummingbird
Cinnamon Hummingbird
Violet-crowned Hummingbird
Plain-capped Starthroat
Sparkling-tailed Woodstar†
Ruby-throated Hummingbird
Citreoline Trogon
Elegant Trogon
Russet-crowned Motmot
Golden-cheeked Woodpecker
Grey-crowned Woodpecker
Lineated Woodpecker
Pale-billed Woodpecker
Ivory-billed Woodcreeper
N. Beardless Tyrannulet
Greenish Elaenia
Greater Pewee
Common Tufted Flycatcher†
Yellow-bellied Flycatcher (V)
Least Flycatcher
Hammond's Flycatcher
Western Flycatcher†
Bright-rumped Attila
Dusky-capped Flycatcher
Great Kiskadee
Social Flycatcher
Boat-billed Flycatcher
Sulphur-bellied Flycatcher
Tropical Kingbird
Thick-billed Kingbird
Rose-throated Becard
Masked Tityra
N. Rough-winged Swallow
Black-throated Magpie-Jay
Green Jay (R)
San Blas Jay
Sinaloa Crow
Happy Wren
Sinaloa Wren
Northern House Wren†
Ruby-crowned Kinglet
Blue-grey Gnatcatcher
Brown-backed Solitaire
Orange-billed N.-Thrush
Swainson's Thrush
White-throated Thrush†
Rufous-backed Thrush†
Blue Mockingbird
Grey Silky† (R)
Bell's Vireo
Black-capped Vireo
Plumbeous Vireo†
Cassin's Vireo†
Golden Vireo
Warbling Vireo
Yellow-green Vireo
Orange-crowned Warbler
Nashville Warbler
Tropical Parula
Yellow Warbler
Chestnut-sided Warbler (V)
Audubon's Warbler†
Black-throated Grey Warbler
Townsend's Warbler
Black-and-white Warbler
American Redstart
Kentucky Warbler (V)
MacGillivray's Warbler
Ovenbird
Northern Waterthrush
Louisiana Waterthrush

Wilson's Warbler
Slate-throated Whitestart†
Fan-tailed Warbler
Yellow-breasted Chat
Godman's Euphonia†
Red-crowned Ant-Tanager
Summer Tanager
Western Tanager
Flame-colored Tanager
Rosy Thrush-Tanager
Greyish Saltator
Yellow Grosbeak
Rose-breasted Grosbeak
Black-headed Grosbeak
Blue Grosbeak
Varied Bunting
Painted Bunting
Rusty-crowned Ground-Sparrow
Lincoln's Sparrow
Great-tailed Grackle
Orchard Oriole
Hooded Oriole
Black-vented Oriole
Streak-backed Oriole
Bullock's Oriole
Yellow-winged Cacique

Site 6.3: El Mirador del Aguila, Nayarit

(January, December 1995, January 1997, 1998)

Introduction/Key Species

This viewpoint (*mirador*) at 550 m elevation along the old highway between Crucero San Blas and Tepic is one of the best places left in Mexico to see Military Macaws. It is about 50 km (45 minutes to an hour) from San Blas and can be combined with a trip to Cerro de San Juan (Site 6.4). The main reason to visit is for the macaws, although other species can be seen, including Zone-tailed and Red-tailed hawks, **Rufous-bellied Chachalaca,** and **Black-throated Magpie-Jay**, and **Great Swallow-tailed Swift** has been reported here.

Access/Birding Sites

From San Blas, head inland 36 km back to Route 15 (see access to San Blas from Mazatlán) and the Crucero San Blas. Turn right toward Tepic and mark zero at the junction. At Km 4.5 there is a well-signed fork, left for "Tepic libre" and straight for "Tepic cuota." Turn left onto the old (*libre*) highway, and in about 5 km you go through the village of Buenos Aires, immediately after Km Post 24. The *mirador* pulloff is on the left, 2.5 km beyond Buenos Aires at a right-hand bend, and is marked by a low, broken stone wall and a pile of garbage.

Coming from Tepic (or en route back to San Blas from Cerro de San Juan), follow the signs for Mazatlán and, 1.0 km after getting onto Route 15 on the outskirts of Tepic/going under the last overpass, watch for the split of *cuota* and *libre* roads: fork right onto the *libre* road and mark this as zero. At Km 17.5, after winding up and down, often behind noisy trucks, there is a sign on the right announcing El Mirador del Aguila (this is not it!); just after this, at Km 17.7, is Km Post 21, then a fairly sharp left-hand bend immediately after a small, blue-and-pink shrine on a rise to the right of the highway. The *mirador* is the pullout on the right just after the shrine: check carefully for traffic, signal well in advance, and take care pulling in.

Note: Turning into the *mirador* is potentially dangerous, especially if you're coming from San Blas (when you would have to cross traffic), in which case it may be best to continue on about 500 m to a wide pulloff at Km Post 21, turn around, and come back to the *mirador*. Particularly if you have a vehicle with low clearance, beware that the pulloff is not at the same level as the highway.

The macaws can be seen at any time of day and usually within 15–30 minutes of arriving at the *mirador*. Their loud and raucous cries carry over long distances (but usually not over the nearer sound of trucks!) and often indicate that the birds are flying. They are usually somewhat distant, down in the valley, but groups sometimes fly closer. A telescope is useful for getting good views of perched (and flying) birds.

Site 6.4: Cerro de San Juan, Nayarit

(December 1995, January 1997, 1998)

Introduction/Key Species

Cerro de San Juan, a massif just southwest of Tepic, is about an hour's drive from San Blas and offers access to pine-oak forest and a variety of bird species not found in the lowlands. It makes a good day trip from San Blas and can be combined with an afternoon stop at El Mirador del Aguila (Site 6.3) for Military Macaws.

Birds here include **Lesser Roadrunner**, **Colima Pygmy-Owl**, **Mexican Wood-**

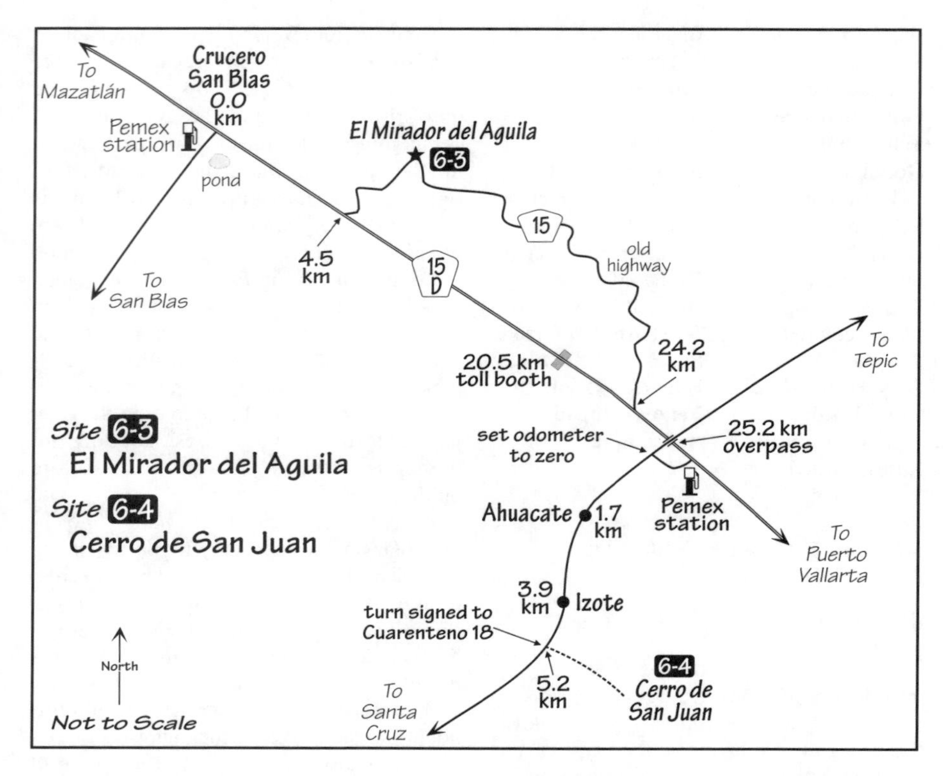

nymph, Calliope (winter) and **Bumblebee hummingbirds**, Arizona and **Grey-crowned woodpeckers**, Olivaceous and **White-striped woodcreepers**, Buff-breasted (migrant?) and Nutting's flycatchers, **Grey-collared Becard**, Green Jay, **Spotted Wren**, **Russet Nightingale-Thrush** (migrant?), **Blue Mockingbird**, Black-capped (winter) and **Golden vireos**, **Crescent-chested**, Red-faced (winter), Golden-crowned, and **Rufous-capped warblers**, Flame-colored and **Red-headed tanagers**, Yellow Grosbeak, **Rusty-crowned Ground-Sparrow**, **Dickey's** [Audubon's] **Oriole**, and **Black-headed Siskin**. Other species that should be looked for here include Eared Poorwill and Sparkling-tailed Woodstar.

Access

From San Blas to Cerro de San Juan via Tepic is about 70 km (1–1.5 hours); add another 6 km (10–15 minutes) if you take the *libre* rather than *cuota* road. Head inland 36 km to the Crucero San Blas on Route 15, the Mazatlán-Tepic highway. Turn right toward Tepic on Route 15 (mark zero at this junction): at Km 4.5 you pass the fork to the old *libre* road (see under El Mirador del Aguila, Site 6.3), and the toll booth is at Km 20.5 (19 pesos in January 1997), where there are also bathrooms; at Km 24.2 the *libre* road merges back, and the highway becomes two-lane.

At Km 25.2, you pass under a highway bridge with signs for straight on to Puerto Vallarta, right to Tepic Centro and Miramar. Turn right immediately after you go under the bridge (Km 25.3) and get into the left lane as you curve around right and up to a stop sign and the Tepic-Miramar highway (left to Miramar, right to Tepic). Turn left here and reset to zero.

At Km 1.7 you go through the small town of Ahuacate, and at Km 3.9 through the village of Izote. At Km 5.2, look for a sign for a left turn signed "Cuarenteno 18," which is also the dirt road to Cerro de San Juan. The turn itself is at Km 5.3, on a right-hand bend; remember to signal and make the left turn with care, as you will be crossing traffic.

Coming back to San Blas, turn right here onto the highway, and at Km 5.0 (after Izote and Ahuacate) there is a right fork well-signed to Mazatlán and Puerto Vallarta (straight goes into Tepic). Fork right here, and in 400 m you come to a stop sign and Route 15. Turn left, under the bridge, and you're back on the road to the Crucero San Blas; you'll reach the right-hand fork onto the *libre* road (and the Mirador del Aguila) in about 1 km.

Cerro de San Juan can also be reached from San Blas via Matanchen, which is shorter (61 km) but takes longer (1.5–2 hours) because the road is narrower and much more winding. If you want to try this route, turn right toward Matanchen at the junction 2 km inland from the Río San Cristóbal bridge (mark zero at the turn). Pass the turnoff to La Palma and La Bajada at Km 11.1, and at Km 19.4 you come to a junction where you should stop and turn left (right goes to Santa Cruz). In 800 m there is a well-signed right turn for Las Varas (and Puerto Vallarta), but stay straight. The road climbs and winds inland to Tepic, via Jalcocotan (Km 38), past the left turn to Mecatan (Km 50.7), and through the village of Platanitos (Km 58). Immediately after Platanitos, watch for Km Post 6 on the right, and just after this, at Km 58.5, watch for a right turn at a sign that says "Cuarenteno 18." The turn is immediately as the highway bends right, so slow down, signal, and make the turn with care.

Birding Sites

Once you've turned on to the Cuarenteno road, which goes up and over Cerro de San Juan, you can stop anywhere and bird along 17 km of mostly dirt road. Although traffic is generally light, this can be a dusty road, so cover your optics, and watch out if you wear contact lenses.

The Cerro de San Juan road starts at about 1000 m elevation in brushy thickets and oak-thorn forest, climbs through oak and pine-oak woodland to 1500 m at Rancho La Noria (Km 8), and then winds down through humid pine-oak forest to semi-deciduous tropical forest and coffee plantations at around 1000 m, above the village of Cuarenteno, which you enter at Km 16.5. You can encounter birds anywhere along the road, so birding is a matter of intuition and stopping when you see or hear something. Many of the more interesting birds occur with mixed-species flocks which, in winter, regularly include large numbers of migrant vireos and warblers. Species to look for in the flocks include Arizona and **Grey-crowned woodpeckers**, Olivaceous, Ivory-billed, and **White-striped woodcreepers**, **Grey-collared** and Rose-throated **becards**, Green Jay, **Golden Vireo**, **Crescent-chested** and Red-faced (winter) **warblers**, Flame-colored and **Red-headed tanagers**, and **Dickey's** [Audubon's] **Oriole**.

Some places that have been good for birding include the following: 300 m from the highway there is a large parking area on the right; birding along the road here can produce Ferruginous Pygmy-Owl, Nutting's Flycatcher, **Blue Mockingbird**, Black-capped Vireo (winter), Yellow Grosbeak, and **Rusty-crowned Ground-Sparrow.** At Km 2.3 you start into a stretch of weedy cane and cornfields with scattered pines and hedgerows; at least during December/January the flower banks here can have up to ten species of hummingbirds (including Ruby-throated, Costa's, and Calliope); other birds in this area include **Elegant Quail**, **Lesser Roadrunner**, Buff-breasted Flycatcher, **Spotted Wren**, **Blue Mockingbird**, and **Rusty-crowned Ground-Sparrow**.

From here on the road winds up through oak woodland (good for flocks) and then levels out at about Km 4.5, where there are a couple of pulloffs. Walking this stretch can be good, especially if you run into a mixed-species flock. At about Km 6 the road climbs up again, and at Km 7.7 it levels out as you come alongside the open fields of Rancho La Noria on your left. Birds along this open stretch include Eastern Bluebird, **Grey Silky**, and **Black-headed Siskin**. After the ranch, the road winds down and through some more humid canyons—for example, at Km 10.7, where birds can include **Mexican Woodnymph**, Blue-throated and **Bumblebee hummingbirds**, and Golden-crowned Warbler. Flocks can be found anywhere along this section. At Km 13.9 you start to get into coffee, and at Km 16 you hit a short paved stretch and the first banana plantations.

Cerro de San Juan Bird List
131 sp. †See pp. 5–7 for notes on English names.

Black Vulture
Turkey Vulture
Sharp-shinned Hawk
Cooper's Hawk
Grey Hawk
Broad-winged Hawk
Short-tailed Hawk
Zone-tailed Hawk
Red-tailed Hawk
Collared Forest-Falcon
American Kestrel
Rufous-bellied Chachalaca
Elegant Quail
White-winged Dove
Mourning Dove
Inca Dove
White-tipped Dove
Squirrel Cuckoo
Lesser Roadrunner
Colima Pygmy-Owl†
Ferruginous Pygmy-Owl
Vaux's Swift
Broad-billed Hummingbird
Mexican Woodnymph
White-eared Hummingbird
Berylline Hummingbird
Violet-crowned Hummingbird
Blue-throated Hummingbird
Magnificent Hummingbird
Ruby-throated Hummingbird
Costa's Hummingbird
Calliope Hummingbird
Rufous Hummingbird
Bumblebee Hummingbird
Elegant Trogon
Russet-crowned Motmot
Acorn Woodpecker
Gila Woodpecker
Yellow-bellied Sapsucker
Ladder-backed Woodpecker
Arizona Woodpecker†
Grey-crowned Woodpecker
Pale-billed Woodpecker
Olivaceous Woodcreeper
Ivory-billed Woodcreeper
White-striped Woodcreeper
Greenish Elaenia
Greater Pewee
Common Tufted Flycatcher†
Hammond's Flycatcher
Dusky Flycatcher
Western Flycatcher†
Buff-breasted Flycatcher
Dusky-capped Flycatcher
Nutting's Flycatcher
Boat-billed Flycatcher
Social Flycatcher
Cassin's Kingbird
Thick-billed Kingbird
Grey-collared Becard
Rose-throated Becard
Masked Tityra
Violet-green Swallow
N. Rough-winged Swallow
Black-throated Magpie-Jay
Green Jay
Northern Raven†
Spotted Wren
Happy Wren
Sinaloa Wren
Northern House Wren†
Ruby-crowned Kinglet
Blue-grey Gnatcatcher
Eastern Bluebird
Brown-backed Solitaire
Orange-billed N.-Thrush
Russet Nightingale-Thrush
White-throated Thrush†
Blue Mockingbird
Grey Silky†
Black-capped Vireo
Cassin's Vireo†
Plumbeous Vireo†
Hutton's Vireo
Golden Vireo
Warbling Vireo
Orange-crowned Warbler
Nashville Warbler
Crescent-chested Warbler
Audubon's Warbler†
Black-throated Grey Warbler
Townsend's Warbler
Hermit Warbler
Black-throated Green Warbler
Grace's Warbler
Black-and-white Warbler
MacGillivray's Warbler
Wilson's Warbler
Red-faced Warbler
Painted Whitestart†
Slate-throated Whitestart†
Fan-tailed Warbler
Golden-crowned Warbler
Rufous-capped Warbler
Yellow-breasted Chat
Hepatic Tanager
Summer Tanager
Western Tanager
Flame-colored Tanager
Red-headed Tanager
Greyish Saltator
Yellow Grosbeak
Rose-breasted Grosbeak
Black-headed Grosbeak
Blue Grosbeak
Varied Bunting
Rusty-crowned Ground-Sparrow
Cinnamon-rumped Seedeater†
Rusty Sparrow
Chipping Sparrow
Lincoln's Sparrow
Hooded Oriole
Black-vented Oriole
Dickey's Oriole†
Streak-backed Oriole
Bullock's Oriole
Scott's Oriole
Yellow-winged Cacique
House Finch
Pine Siskin
Black-headed Siskin
Lesser Goldfinch

CHAPTER 7 COLIMA AND JALISCO

Aztec Thrush

The tiny state of Colima and adjacent areas in the state of Jalisco have long been the subject of ornithological studies (e.g., Schaldach 1963), and with good reason. This small area has a great variety of habitats, ranging from sunny and wave-dashed Pacific beaches to cool and humid montane forests on the Volcánes de Colima, whose twin peaks dominate the region. Over 400 bird species occur here, including 40 Mexican endemics.

The avifauna of Colima and adjacent Jalisco is quite different from that of San Blas, Nayarit (Chapter 6), only a short distance to the north. In large part this is due to the effects of the central volcanic belt, the high mountains of which cut across Mexico, virtually coast-to-coast, from Jalisco in the west to Veracruz in the east. As a consequence, many species of lowland northwest Mexico are isolated from those to the south, and replacements between San Blas and Colima include, for example, Black-throated and White-throated magpie-jays and Rufous-bellied and West Mexican chachalacas. Other thorn-forest species that appear from the vicinity of Puerto Vallarta south include White-bellied Wren and the flashy Orange-breasted Bunting, while the Sinaloa Crow vanishes between San Blas (where it is common) and Puerto Vallarta. The high-elevation forests of the volcanoes also contribute a distinctive new element of species typical of the central volcanic belt but not found in the Sierra Madre Occidental to the north. In addition, the southeastern areas of the region are part of the Río Balsas drainage and have species typical of Mexico's arid southwest interior.

Birds that can be seen in Colima and Jalisco include Red-billed Tropicbird, Long-tailed Wood-Partridge, Banded Quail, Thick-billed Parrot, Mexican Parrotlet, Balsas Screech-Owl, Mountain and Colima pygmy-owls, Stygian Owl, Eared Poorwill, Great Swallow-tailed Swift, Mexican Woodnymph, Golden-cheeked and Grey-crowned woodpeckers, White-striped Woodcreeper, Flammulated Flycatcher, Sinaloa Martin (summer), Grey-barred and Spotted wrens, Aztec Thrush, Slaty, Black-capped (winter), Dwarf (winter), and Golden vireos, Red Warbler, Red-breasted Chat, Orange-breasted Bunting, Green-striped Brushfinch, Rusty-crowned Ground-Sparrow, Collared Towhee, Slate-blue Seedeater, Black-chested Sparrow, and Dickey's [Audubon's], Spot-breasted, and Abeille's (winter) orioles.

It almost goes without saying that the area is worth a one- or two-week birding trip. If you're not driving here, for example, from San Blas (Chapter 6) or Mexico City (Chapter 8), the two main points of entry are Guadalajara and Manzanillo, both served daily by several international flights from U.S. cities. Vehicles can be rented from a number of companies at both airports, and a wide range of accommodations and restaurants is available throughout most of the region. The birding sites are arranged roughly in clockwise sequence, beginning in Manzanillo. By flying into Puerto Vallarta, you could fairly easily combine Colima and Jalisco with San Blas (Chapter 6) for a productive two-week trip. From Puerto Vallarta to Barra de Navidad is about 220 km (3–4 hours) via Route 200.

Suggested Itineraries

If you arrive in Manzanillo (or Puerto Vallarta), it can be good to start along the coast (Sites 7.1, 7.2, 7.3, 7.4), based in Barra de Navidad or Manzanillo. From here you can head inland, either direct to Ciudad Colima (Sites 7.9, 7.10) and/or Ciudad Guzmán (Site 7.7) for access to the volcanoes (Sites 7.6, 7.8), or to Autlán (Site 7.5), from which you can make a loop to the volcanoes (Sites 7.6, 7.8) and on to Ciudad Colima.

If you arrive in Guadalajara, you can head to Ciudad Guzmán (Site 7.7) and/or Ciudad Colima (Sites 7.9, 7.10) for access to the volcanoes (Sites 7.6, 7.8) and head to the coast via Autlán (Sites 7.5, 7.6), looping back to Guadalajara via the coastal sites reached from Barra de Navidad and Manzanillo (Sites 7.1, 7.2, 7.3, 7.4).

These basic options can translate easily into a one- to two-week trip. If you're pushed for time, Autlán can be cut out, but it does have some great (though not always easy) birding.

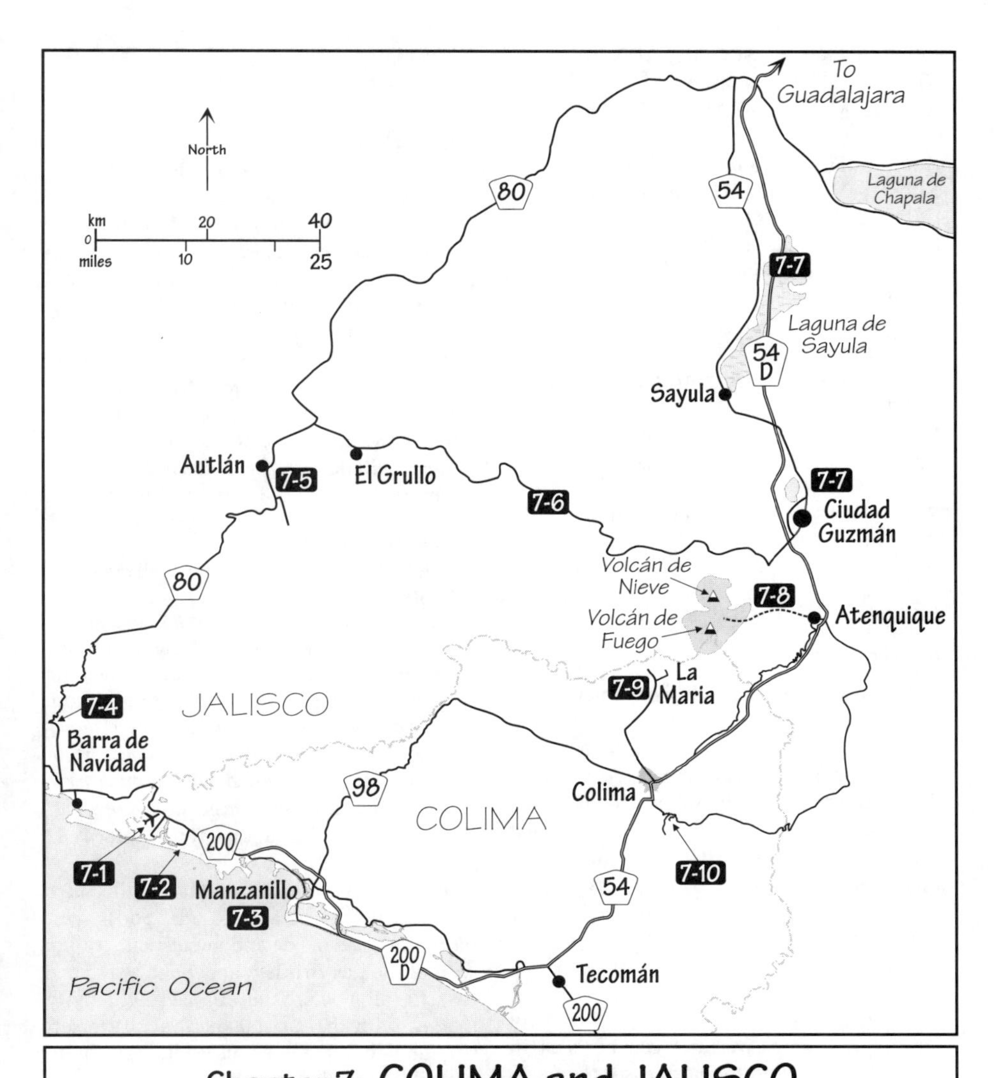

Chapter 7: COLIMA and JALISCO

- **7-1** Manzanillo Airport Marshes, Colima
- **7-2** The Playa de Oro Road, Colima
- **7-3** Manzanillo (Piedra Blanca, Power Station Outflow), Colima
- **7-4** Barranca el Choncho, Jalisco
- **7-5** Autlán / Puerto Los Mazos, Jalisco
- **7-6** Autlán to Ciudad Guzmán, Jalisco
- **7-7** Ciudad Guzmán / Laguna Sayula, Jalisco
- **7-8** Volcán de Fuego, Jalisco
- **7-9** Ciudad Colima to La Maria, Colima
- **7-10** Microondas La Cumbre, Colima

COLIMA AND JALISCO SITE LIST

Site 7.0: El Tuito, Jalisco

(checked March 1997)

If you're coming south from Puerto Vallarta, a spot worth checking is near El Tuito, 40 km (forty-five minutes to an hour) south of the Pemex station (near Km Post 214) at the south end of Puerto Vallarta along Route 200. Between Km Posts 174 and 175, and about 4.5 km north of El Tuito, a graded dirt road heads east, marked by some rusty signs. Take this road, past some abandoned quarry buildings, about 2 km uphill, where you can park and walk along the road and/or along tracks off it.

Especially in late afternoon and early morning, Military Macaws can be seen in the valley below (to the northeast); they also can be seen from the pullout on the west side of Route 200 by Km Post 176. The habitat here is mostly dry pine-oak, and other birds possible include Short-tailed Hawk, Orange-fronted Parakeet, **Lilac-crowned Parrot**, Acorn Woodpecker, Grace's and **Rufous-capped warblers**, and **Black-headed Siskin**.

Site 7.1: Manzanillo Airport Marshes, Colima

(December 1986, April 1988, March 1989, 1990, February, March 1992, 1993, 1994, February, March, December 1995, February, March 1997)

Introduction/Key Species

If you fly into Manzanillo during daylight you'll notice that the road to and from the coastal highway goes through some nice marshes: these can make a good birding start to a trip, especially if you're absolutely rabid and/or coming straight from the frozen, bird-starved north in winter.

Birds here include Least Bittern, Roseate Spoonbill, Muscovy Duck, Limpkin (a recent colonist), **Ruddy Crake**, King Rail, Collared Plover, White-throated Flycatcher (winter), Ruddy-breasted Seedeater, Spot-breasted Oriole, and a variety of waterbirds.

Access/Birding Sites

Manzanillo (or Playa de Oro) airport isn't actually *in* Manzanillo. Rather, it's some 30–35 km (30–45 minutes) along the coast to the west of the "Manzanillo" hotel and beach resort areas such as Las Brisas, which, in turn, are 8 km or so (15–20 minutes) from downtown Manzanillo. The airport is actually nearer to the smaller coastal resort of Barra de Navidad, which lies about 25 km (20–30 minutes) along the coast west of the airport and has a variety of hotels and restaurants.

If you're coming from Manzanillo, take the coastal highway (Route 200) west through the resort developments and, 12 km west of the Las Brisas roundabout, you come to a stop sign where the Colima toll highway merges with the coastal highway. Turn left here and continue west on Route 200 for 19 km, till the well-signed left-hand turn to the airport (note that you fork right and then curve back to a stop sign to cross Route 200, rather than turning across traffic).

Coming from Barra de Navidad, head east on Route 200 and, after following the one-way system through the town of Cihuatlán (as you come into Cihuatlán, turn *right* at the traffic light—straight ahead is one-way against you), you cross a long bridge over the Marabasco River (the Jalisco/Colima state line) on the east side of Cihuatlán. Watch for the well-signed right turn to the airport 6 km east of the river bridge.

The road to the airport is about 5.5 km

long, with the main area of marshes at Km 4–5 after a stretch of fields and coconut plantations. It is easy to pull off and park safely beside the road in this last kilometer before the airport. When fruiting, the line of tall fig trees on the left (south) at km 3.5 can have **Mexican Parrotlets,** and Spot-breasted Oriole can be found anywhere from here to the airport parking lot. Walking the road and looking over the marshes and aquaculture ponds (on the dikes watch for Collared Plovers and for Caimans sunning themselves) is the best way to bird this area. The edges of the reed beds, such as at the beginning of the ponds on the left, have White-throated (and Willow) Flycatcher in winter. The mangroves and thickets between the small bridge at Km 5.2 and the airport parking lot 250 m later should be checked for **San Blas Jays**. Late in the day (in February and March, at least) spectacular masses of migrant swallows and Dickcissels may swarm overhead before dropping to roost in the marsh.

This area can be birded at any time of day and makes for a good late-afternoon/evening trip after a morning looking for landbirds in the thorn forest. At dusk (after and during the mosquito "rush"!), Boat-billed Herons fly in from the east over the road at the aquaculture ponds; a strong flashlight will help you get good views.

If you're using public transport, you can either take a second-class bus and get off at the airport junction or take one of the airport *colectivo* taxis, which go direct to the airport (buses do not serve the airport).

Manzanillo Airport Marshes Bird List

148 sp. R: rare; V: vagrant (not to be expected). †See pp. 5–7 for notes on English names.

Least Grebe
Pied-billed Grebe
American White Pelican
Neotropic Cormorant
Anhinga
Magnificent Frigatebird
American Bittern
Least Bittern
Great Blue Heron
Great Egret
Snowy Egret
Little Blue Heron
Tricolored Heron
Green Heron
Reddish Egret (R)
Cattle Egret
Black-crowned Night-Heron
Yellow-crowned Night-Heron
Boat-billed Heron
White Ibis
Glossy Ibis (V)
White-faced Ibis
Roseate Spoonbill
Wood Stork
Fulvous Whistling-Duck
Black-bellied Whistling-Duck
Muscovy Duck
Green-winged Teal
Northern Pintail
Blue-winged Teal
Cinnamon Teal
Northern Shoveler
Gadwall
American Wigeon
Lesser Scaup
Black Vulture
Turkey Vulture
Osprey
White-tailed Kite
Sharp-shinned Hawk
Cooper's Hawk
Common Black Hawk
Grey Hawk
Roadside Hawk
Short-tailed Hawk
Zone-tailed Hawk
Crested Caracara
American Kestrel
Merlin
Peregrine Falcon
West Mexican Chachalaca
Ruddy Crake
King Rail
Sora
Purple Gallinule
Common Moorhen
American Coot
Limpkin
Collared Plover
Semipalmated Plover
Killdeer
Black-necked Stilt
American Avocet
Northern Jacana
Greater Yellowlegs
Lesser Yellowlegs
Solitary Sandpiper
Spotted Sandpiper
Least Sandpiper
Stilt Sandpiper
Long-billed Dowitcher
Common Snipe
Red-necked Phalarope
Gull-billed Tern
Caspian Tern
Common Tern
Forster's Tern
Red-billed Pigeon
White-winged Dove
Inca Dove
Common Ground-Dove
Ruddy Ground-Dove
Orange-fronted Parakeet
Mexican Parrotlet
Groove-billed Ani
Ferruginous Pygmy-Owl
Lesser Nighthawk
Pauraque
Ringed Kingfisher
Belted Kingfisher
Green Kingfisher
Golden-cheeked Woodpecker
Lineated Woodpecker
Willow Flycatcher
White-throated Flycatcher
Least Flycatcher
Western Flycatcher†
Vermilion Flycatcher
Dusky-capped Flycatcher
Great Kiskadee
Social Flycatcher
Tropical Kingbird
Thick-billed Kingbird
Grey-breasted Martin
N. Rough-winged Swallow
Bank Swallow
Barn Swallow
San Blas Jay
Happy Wren
Sinaloa Wren

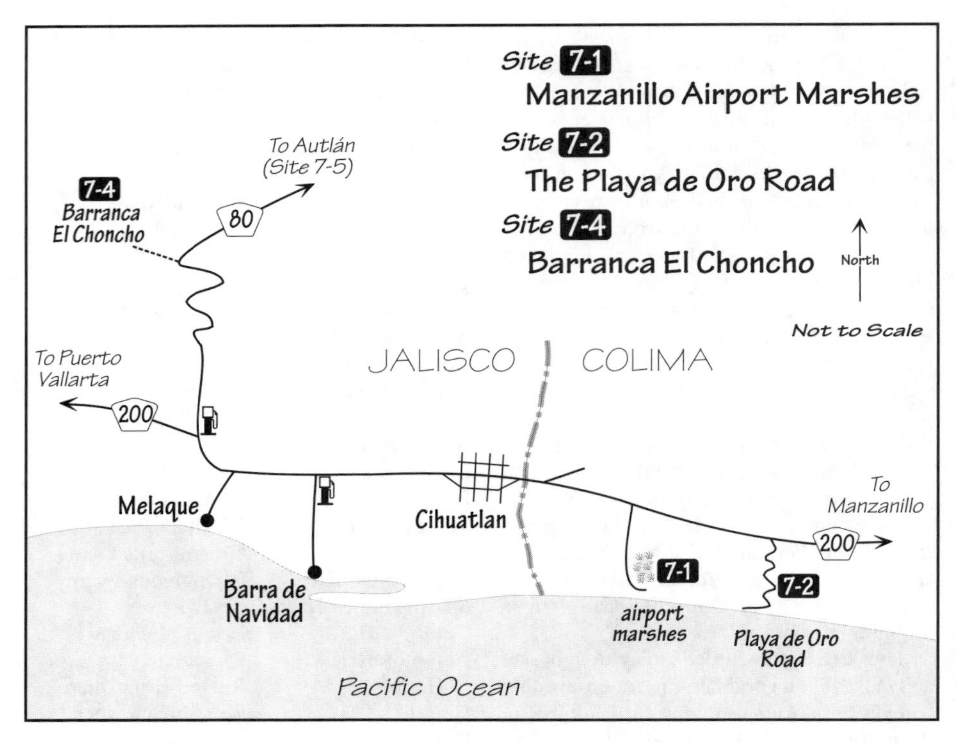

Marsh Wren
Blue-grey Gnatcatcher
Northern Mockingbird
Loggerhead Shrike
Red-throated Pipit (V)
American Pipit
Bell's Vireo
Orange-crowned Warbler
Nashville Warbler
Yellow Warbler
Palm Warbler (V)
Black-and-white Warbler
American Redstart
Northern Waterthrush
MacGillivray's Warbler
Common Yellowthroat
Wilson's Warbler
Yellow-breasted Chat
Summer Tanager
Greyish Saltator
Blue Grosbeak
Painted Bunting
Dickcissel
Blue-black Grassquit
Cinnamon-rumped Seedeater†
Ruddy-breasted Seedeater
Stripe-headed Sparrow
Lincoln's Sparrow
Red-winged Blackbird
Great-tailed Grackle
Bronzed Cowbird
Orchard Oriole
Black-vented Oriole
Streak-backed Oriole
Spot-breasted Oriole
Baltimore Oriole
Bullock's Oriole
Yellow-winged Cacique
Lesser Goldfinch

Site 7.2: The Playa de Oro Road, Colima

(December 1986/January 1987, April 1988, March 1989, 1990, 1992, February, March 1993, February 1994, February, March, December 1995, February 1997)

Introduction/Key Species

This site (between Manzanillo and Barra de Navidad) is a quiet cobbled road that runs through several kilometers of good thorn forest from the coastal highway to the beach. In the 1980s it was found to be a good location for the **Flammulated Flycatcher** (Delaney 1987).

As well as this retiring endemic, the Playa de Oro road is home to a suite of Pacific-Slope thorn-forest birds, including **West Mexican Chachalaca**, **Lilac-crowned Parrot**, Cinnamon Hummingbird, **Citreoline Trogon**, Nutting's Flycatcher, White-throated Magpie-Jay, **San Blas Jay**, **Happy**, **Sinaloa**, and **White-bellied wrens**, **Black-capped Gnatcatcher**, **Red-breasted Chat**, Rosy Thrush-Tanager, and **Orange-breasted Bunting**.

Access/Birding Sites

The Playa de Oro road leaves the coastal highway (Route 200) 7 km east of the road to the Manzanillo Airport Marshes (see Site 7.1 for directions) and is 25–40 minutes from both Manzanillo and Barra de Navidad. The junction is signed with a green and white highway sign that says "Playa de Oro 7." The first kilometer involves some care, especially if you have a vehicle with low clearance, but after that it's just cobbled.

It is 6 km from the highway to the beach, and birding can be good anywhere along the road, especially on the "coastal slopes" after you crest the main ridge at Km 2.2 (160 m elevation), when traffic noise from the highway is blocked off. Most of the thorn-forest species listed above can be found by parking off the road in any of the wide spots between Km 3 and Km 5.5 and walking for up to a kilometer along the road in either direction; there are also a few minor tracks that lead off into the thorn forest in places. Note that the **Flammulated Flycatcher** is very easy to overlook and stays within the thorn forest, rather than at the edges and in fruiting trees, where up to four species of *Myiarchus* occur together! Listening for its distinctive calls (given rarely and/or infrequently in winter) is the best way to locate it.

If you're along the road around dawn there's a good chance of hearing, and with luck seeing, both Laughing Falcon and Collared Forest-Falcon. The open acacia scrub behind the beach has Lucy's warbler (winter), Ruddy-breasted Seedeater, and Stripe-headed Sparrow, and the waters off the beach sometimes have Common Loons in winter. Offshore to the south you'll see a large white rock (called Piedra Blanca) and, with a telescope, you may be able to make out Red-billed Tropicbirds and boobies flying around it; boat trips to the rock can be arranged in Manzanillo (Site 7.3).

If you're using public transport, take a second-class bus to the junction and walk or hitch a ride (traffic is not heavy) from there along the road toward the beach. A taxi also can be rented for the morning to take you along the road, wait, and bring you back to your hotel.

The Playa de Oro Road Bird List

128 sp. V: vagrant (not to be expected). †See pp. 5–7 for notes on English names.

Common Loon
Red-billed Tropicbird
Brown Booby
Brown Pelican
Neotropic Cormorant
Anhinga
Magnificent Frigatebird
Wood Stork
Black Vulture
Turkey Vulture
Osprey
Hook-billed Kite
Sharp-shinned Hawk
Cooper's Hawk
Crane Hawk
Grey Hawk
Roadside Hawk
Short-tailed Hawk
Zone-tailed Hawk
Red-tailed Hawk
Laughing Falcon
Collared Forest-Falcon
American Kestrel
West Mexican Chachalaca
Spotted Sandpiper
Laughing Gull
Royal Tern
Elegant Tern
Red-billed Pigeon
White-winged Dove
Inca Dove
Common Ground-Dove
White-tipped Dove
Lilac-crowned Parrot
Squirrel Cuckoo
Lesser Ground-Cuckoo
Groove-billed Ani
Ferruginous Pygmy-Owl
Burrowing Owl
Mottled Owl
Pauraque
Buff-collared Nightjar
Chestnut-collared Swift
White-collared Swift
Golden-crowned Emerald
Broad-billed Hummingbird
Cinnamon Hummingbird
Black-chinned Hummingbird
Citreoline Trogon
Golden-cheeked Woodpecker
Lineated Woodpecker
Pale-billed Woodpecker
Ivory-billed Woodcreeper
N. Beardless Tyrannulet
Greenish Elaenia
Least Flycatcher
Western Flycatcher†
Vermilion Flycatcher
Bright-rumped Attila
Dusky-capped Flycatcher
Ash-throated Flycatcher
Nutting's Flycatcher
Brown-crested Flycatcher
Flammulated Flycatcher
Great Kiskadee
Social Flycatcher
Tropical Kingbird
Thick-billed Kingbird
Masked Tityra
Rose-throated Becard
Grey-breasted Martin
Tree Swallow

N. Rough-winged Swallow
Bank Swallow
Barn Swallow
White-throated Magpie-Jay
San Blas Jay
Happy Wren
Sinaloa Wren
White-bellied Wren
Blue-grey Gnatcatcher
Black-capped Gnatcatcher
Orange-billed N.-Thrush
Swainson's Thrush
White-throated Thrush†
Rufous-backed Thrush†
Northern Mockingbird
Blue Mockingbird
Bell's Vireo
Cassin's Vireo† (R)
Plumbeous Vireo†
Golden Vireo
Warbling Vireo
Orange-crowned Warbler
Nashville Warbler
Lucy's Warbler
Tropical Parula
Yellow Warbler
Magnolia Warbler (V)
Black-throated Grey Warbler
Black-and-white Warbler
American Redstart
Northern Waterthrush
MacGillivray's Warbler
Wilson's Warbler
Red-breasted Chat
Godman's Euphonia†
Red-crowned Ant-Tanager
Summer Tanager
Western Tanager
Rosy Thrush-Tanager
Greyish Saltator
Black-headed Grosbeak
Blue Grosbeak
Blue Bunting
Orange-breasted Bunting
Varied Bunting
Painted Bunting
Olive Sparrow
Stripe-headed Sparrow
Lark Sparrow
Cinnamon-rumped Seedeater†
Ruddy-breasted Seedeater
Great-tailed Grackle
Orchard Oriole
Black-vented Oriole
Streak-backed Oriole
Bullock's Oriole
Yellow-winged Cacique

Site 7.3: **Manzanillo** (Piedra Blanca, Power Station Outflow), **Colima**

(December 1983, April 1988, March 1989, 1990, 1992, 1993, 1994, March, December 1995, February, March 1997)

Introduction/Key Species

Manzanillo is Mexico's main Pacific coast port as well as a popular vacation spot, although the main beach resorts are not actually in Manzanillo but along the coast for several kilometers to the west of town. The areas of interest for most birders are those of Playa de Oro (Site 7.2) and the Manzanillo Airport Marshes (Site 7.1), although a couple of other areas can be good for a variety of waterbirds: Piedra Blanca, an offshore seabird rock with lots of nesting Red-billed Tropicbirds; and the warm-water outfall of the Manzanillo power station.

Access/Birding Sites/Bird List

Boat trips to **Piedra Blanca** can be arranged with sport-fishing boats, which are used to dealing with fishing tourists and are coming to learn that trips for birders are even easier!

The best way to arrange a trip is to stop by the office at the harbor where the fishing boats are tied up: coming from the hotel zone, follow signs for Manzanillo and, 2.7 km east of the Las Brisas *glorieta* and 400 m after the sailfish *glorieta* and Pemex station, turn right to Manzanillo (versus left for Colima); after 4 km of port and railway yards to your right the road comes alongside a stretch of waterfront where a small fleet of sport-fishing boats is tied up. At Km 4.3 there is a building on your right with a signs for sport fishing and a restaurant painted on the walls. You can park just past this building, by the right-hand fork into downtown Manzanillo, or loop left around the traffic island and park across the street on the road heading back to the highway. Boatmen hang around the office and small dock at any time of day, although 7–9 A.M. may be the best time to find someone. Let them know you want to go to the Piedra Blanca to watch birds (versus fishing), and agree upon the price beforehand. In March 1997 the smaller boats (up to 6–8 passengers) cost 800 pesos for the trip, while the larger boats (up to 12–15 passengers) cost 1400 pesos. Each boatman/captain has a card with telephone numbers, and you should expect to pay a deposit for the trip beforehand. One skipper who has done the trip a few times is Rafael Rosas R. (phone 2-06-72).

Allow five hours for the trip (roughly two hours each way—these aren't speedboats—and an hour for going around the rock and birding). An early start (around 7 A.M.) is good, in case the wind picks up in the afternoon, and you'll be back in time for lunch. You can leave from the office dock or from the opposite (west) side of the harbor at Las Brisas (more convenient if you're staying out west of town). To get to the Las Brisas dock, turn south at the Las Brisas *glorieta* and follow the road (past hotel after hotel) to its end in about 2.5 km, where it enters a naval area. At Km 2.6, where the road ends and becomes dirt, turn left to a guarded gate (say you're getting a fishing boat and you can go through) and continue a few hundred meters to the dock, where you can leave your vehicle.

Sea conditions inevitably vary: often it starts out calm and becomes gently undulating by late morning; at other times it's distinctly "lumpy" even with little or no wind. As on any pelagic trip, it can be cooler than you might think, so take a windbreaker and also Ziplock bags or some other protection for optics as well as for dry tissues to clean your lenses (don't worry, you rarely need them—the tissues, not the binoculars!).

Birds seen on the Piedra Blanca trip (January 1987, March 1989, 1990, 1992, 1993, 1994, 1995, 1997) include Common Loon, Wedge-tailed Shearwater (rare), Galapagos (rare), Black, and **Least storm-petrels,** Red-billed Tropicbird (up to 900 birds!), Masked (rare), Brown, and Blue-footed (rare) boobies, Wandering Tattler, Surfbird, Red-necked and Red phalaropes, Pomarine and Parasitic jaegers, Franklin's and Sabine's gulls, and Common and Black terns. Non-avian possibilities include Humpback Whale, Spotted Dolphins, and jumping Marlin! When you come back from the boat trip, if you're not ready to get off, it can be worth asking the boatman to cruise a few hundred meters up into the harbor, where the mudflats on the left (west) side often have lots of herons (Reddish Egret, Yellow-crowned Night-Heron, etc.), White Ibis, shorebirds (American Oystercatcher, Wilson's Plover, etc.), gulls, terns (Gull-billed, Caspian, Royal, Elegant, etc.), and Black Skimmer.

Another area worth checking in the Manzanillo area is the outflow of the large power station a few kilometers (10–15 minutes) southeast of downtown. To get there, follow signs to Manzanillo and fork right just past the sport-fishing dock (see above) and continue 500 m to the *zócalo*. Turn left at the far end of the *zócalo* and mark zero as you make the turn, which becomes a narrow one-way street with several *topes* (speed bumps). At 800 m fork right; at 1.4 km the street you are on ends/turns to dirt. Turn left here for 100 m and you hit the paved road out of town to the coast, with the power station towers visible across a large lagoon. Turn right alongside the lagoon: in 2 km you cross the intake channel; bear right 300 m later, and in another 1.5 km you cross the outflow channel opposite the entry gates for the Central Termoelectrica Manzanillo on the left. Turn right on the dirt track beside the outflow channel, and in 250 m you are at the beach and a good overlook of the outflow and any feeding or roosting terns and gulls. Coming back into Manzanillo, follow the paved road around the lagoon for 4 km, to where it hits a divided cobbled street. Turn left; the street soon becomes one-way and takes you back to the *zócalo*.

The number and variety of birds at the outflow vary considerably, and sometimes it's dead. The large lagoons beside the road on your way to the outfall should also be checked, especially for terns (Least Terns nest here in summer). Birds seen at the outfall include **Black-vented Shearwater**, **Least Storm-Petrel**, Brown Booby, Pomarine Jaeger, Franklin's and Heermann's gulls, and Gull-billed, Caspian, Royal, Elegant, Sandwich (rare), Common, Forster's, Least, and Black terns.

Site 7.4: Barranca el Choncho, Jalisco

(March 1989, 1990, 1992, 1993, February 1994, 1995, 1997)

Introduction/Key Species

This short canyon (elevation 215–350 m) contains a tongue of tropical semi-deciduous forest that cuts through the surrounding hills of thorn forest. As such it offers an interesting variety of birds, including Collared Forest-Falcon, **West Mexican Chachalaca**, **Lilac-crowned Parrot**, **Colima Pygmy-Owl**, **Mexican Hermit**, **Golden-crowned Emerald**, Ivory-billed Woodcreeper, **Flammulated Flycatcher**, **San Blas Jay**, **Golden Vireo**, **Fan-tailed Warbler**, **Red-breasted**

Chat, Rosy Thrush-Tanager, Yellow Grosbeak, and **Blue Bunting**.

Access/Birding Sites

This spot is best in the morning, although, given the high shady canopy and permanent water (if only a trickle), any time of day can be good. It may be a little far to drive here for dawn from Manzanillo (about 75 km; 1.2–1.5 hours one-way), but it's convenient (15–25 minutes) if you're staying in Barra de Navidad.

From Barra de Navidad head back to the coastal highway (Route 200) and turn left (west) toward Melaque and Puerto Vallarta: 2.2 km from the Barra de Navidad junction the highway bends sharply right (a road goes left into Melaque), and in another 1.2 km the highway forks, Route 200 heading left to Puerto Vallarta and Tomatlán, Route 80 straight to Autlán and Guadalajara. Stay straight on Route 80 and mark zero at the junction, 100 m after which there is a Pemex on the right. At Km 9.5 there are a bus stop and dirt road to the right, both signed to Lázaro Cárdenas, after which the highway starts to wind and climb inland. At Km 12.7, shortly after Km Post 246, the highway makes an acute right-hand hairpin bend, at the apex of which an inconspicuous dirt road heads off to the left (west) at Km 12.8. Turn left here (be careful crossing traffic on the bend), and in about 25 m you can park in a small open area on the left of the track, which does get some use.

The canyon can be birded by walking slowly up the track, watching for fruiting trees. In about a kilometer the track gets steeper and the habitat opens out into brushy thorn forest. All of the birds listed below occur in this first kilometer or so, making it an excellent spot for a morning's birding.

If you're using public transport, take a second-class bus from Melaque or Barra de Navidad toward Autlán and get off at the canyon (the bus won't want to stop on the bend, so you should settle for a pulloff a little farther up the highway).

Barranca el Choncho Bird List

85 sp. R: rare (not to be expected). †See pp. 5–7 for notes on English names.

Turkey Vulture
Sharp-shinned Hawk
Grey Hawk
Short-tailed Hawk
Zone-tailed Hawk
Red-tailed Hawk
Collared Forest-Falcon
West Mexican Chachalaca
Red-billed Pigeon
Inca Dove
White-tipped Dove
Orange-fronted Parakeet
Mexican Parrotlet
Lilac-crowned Parrot
Groove-billed Ani
Colima Pygmy-Owl†
Ferruginous Pygmy-Owl
Mottled Owl
Northern Potoo
Mexican Hermit†
Golden-crowned Emerald
Cinnamon Hummingbird
Plain-capped Starthroat
Sparkling-tailed Woodstar† (R)
Black-chinned Hummingbird
Citreoline Trogon
Elegant Trogon
Russet-crowned Motmot
Golden-cheeked Woodpecker
Ladder-backed Woodpecker
Lineated Woodpecker
Pale-billed Woodpecker
Ivory-billed Woodcreeper
N. Beardless Tyrannulet
Greenish Elaenia
Greater Pewee
Least Flycatcher
Western Flycatcher†
Bright-rumped Attila
Dusky-capped Flycatcher
Ash-throated Flycatcher
Nutting's Flycatcher
Brown-crested Flycatcher
Flammulated Flycatcher
Boat-billed Flycatcher
Thick-billed Kingbird
Masked Tityra
Rose-throated Becard
N. Rough-winged Swallow
San Blas Jay
Happy Wren
Sinaloa Wren
White-bellied Wren
Blue-grey Gnatcatcher
Swainson's Thrush
White-throated Thrush†
Rufous-backed Thrush†
Cassin's Vireo†
Plumbeous Vireo†
Golden Vireo
Warbling Vireo
Orange-crowned Warbler
Nashville Warbler
Tropical Parula
Black-throated Grey Warbler
Black-and-white Warbler
MacGillivray's Warbler
Louisiana Waterthrush

Wilson's Warbler
Fan-tailed Warbler
Red-breasted Chat
Godman's Euphonia†
Red-crowned Ant-Tanager
Summer Tanager
Western Tanager
Rosy Thrush-Tanager
Greyish Saltator
Yellow Grosbeak
Blue Bunting
Varied Bunting
Painted Bunting
Olive Sparrow
Black-vented Oriole
Streak-backed Oriole
Bullock's Oriole
Yellow-winged Cacique

Site 7.5: Autlán/Puerto Los Mazos, Jalisco

(December 1983, 1986, April 1988, March 1989, 1990, 1992, February 1997)

Introduction/Key Species

The town of Autlán (elevation 850 m) sits in an arid interior valley about 105 km by road north of Barra de Navidad. The highest ridge between Autlán and the coast (crossed by the highway at Puerto Los Mazos) intercepts moisture-laden winds from the Pacific and consequently supports some tropical semi-evergreen forest, the closest thing to cloud forest in northwestern Mexico. This forest supports an interesting selection of birds, including **Long-tailed Wood-Partridge**, **Singing Quail**, **Eared Poorwill** ("rediscovered" here in 1959; Schaldach and Phillips 1961), **Great Swallow-tailed Swift**, **Mexican Woodnymph**, **Amethyst-throated** and Calliope (winter) **hummingbirds**, **Sparkling-tailed Woodstar**, Smoky-brown and **Grey-crowned woodpeckers**, Olivaceous Woodcreeper, **Grey-collared Becard**, Green Jay, **Spotted Wren**, Black-capped (winter), **Dwarf** (winter), and **Golden vireos**, **Chestnut-sided Shrike-Vireo**, **Fan-tailed Warbler**, Red-crowned Ant-Tanager, **Red-headed Tanager**, **Green-striped Brushfinch**, and **Rusty-crowned Ground-Sparrow**.

In addition, the arid scrub around Autlán has **Banded Quail**, Plain-capped Starthroat, Lucifer Hummingbird (migrant), **Orange-breasted Bunting**, and **Black-chested Sparrow**.

Access

Puerto Los Mazos is visited easily from a base in Autlán, which has a number of adequate to economical hotels and restaurants. To get here from Autlán, head south on Route 80 and mark zero at the *periferico* junction on the south side of Autlán. At Km 12 (15–30 minutes) you come to the top of the first main ridge (1270 m), between Km Posts 169 and 170, with a pulloff and a couple of forestry buildings on the left (east) and a sign that says "Puerto Los Mazos 10" (although it is only 5 km to the towers). A cobbled road leads east from here, in front of the buildings and up to the radio towers (*microondas*) you may be able to see up on the greener-looking ridge top. It also has been possible to camp near the highway at Puerto Los Mazos: check with the personnel at the forestry buildings (beware, though, that traffic and barking dogs can make for a fitful sleep!).

From Barra de Navidad it is about 95 km (2–2.5 hours) to Puerto Los Mazos. If you continue inland on Route 80 (mark zero at the junction with Route 200), past the Barranca El Choncho (Site 7.4), the highway winds over ridges and through agricultural valleys (the seasonal roadside marshes around Km 57–59 can be worth checking; e.g., for both whistling-ducks) and at about Km 91 crests the top of the main ridge at Puerto Los Mazos and the forestry buildings.

If you're using public transport, take a second-class bus between Autlán (whence you also could hitch a ride) and Barra de Navidad and get off at Puerto Los Mazos, which should be a routine stop.

Birding Sites

In the past ten years the gate at the start of the cobbled road up to the microwave towers has been: unlocked, locked (but opened readily by the forestry station personnel), locked (and not opened), and (March 1997) open (unlocked) again! Clearly, you should check on this factor if you visit and perhaps should be prepared to make the 5 km walk (steep in places) up to the towers at the ridge top.

For the first kilometer or so the cobbled road to the towers goes through deciduous, oak-dominated woodland (1250–1450 m elevation) with brushy understory and weedy

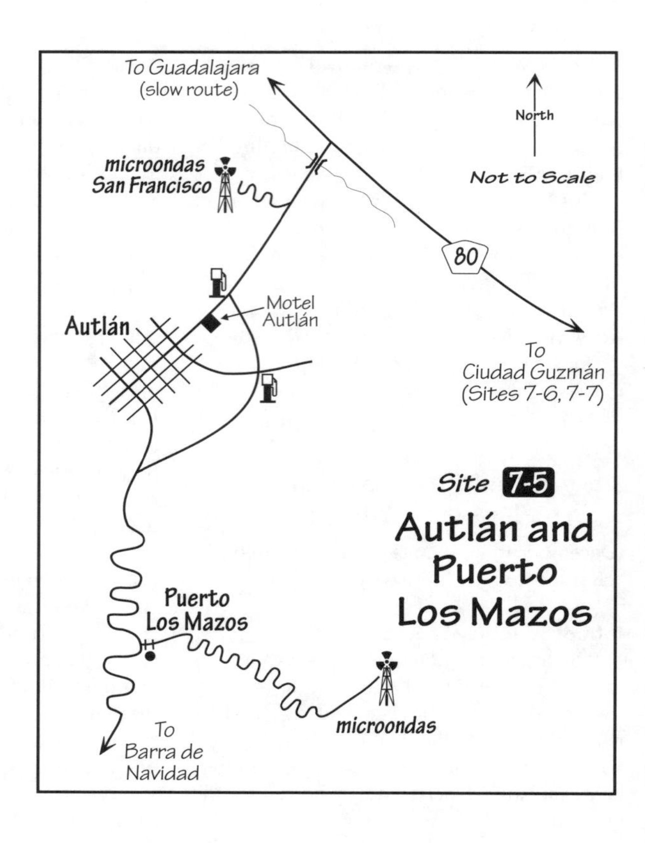

patches; this is a good stretch for Calliope Hummingbird (winter), Nutting's Flycatcher, Flame-colored Tanager, and **Rusty-crowned Ground-Sparrow**. After this the road climbs through what often seems to be a fairly lifeless belt of dry oak woods, although **Spotted Wrens** occur here and any mixed-species flocks (mainly in greener, more humid valleys) can have **Grey-collared Becard**, **Chestnut-sided Shrike-Vireo**, and **Red-headed Tanager** (all of which also occur higher).

Above the oak belt you get into more humid oak and then semi-evergreen tropical forest (1750–1950 m) near the ridge top, and this last stretch is the most interesting for birds: species here include **Long-tailed Wood-Partridge**, **Singing Quail**, **Mexican Woodnymph**, **Amethyst-throated Hummingbird**, **Sparkling-tailed Woodstar**, **Grey-crowned Woodpecker**, Olivaceous Woodcreeper, **Grey-collared Becard**, Black-capped (winter) and **Golden vireos**, **Chestnut-sided Shrike-Vireo**, **Fan-tailed Warbler**, Red-crowned Ant-Tanager, **Red-headed Tanager**, and **Green-striped Brush-finch**. **Eared Poorwills** can be found in the oaks along the *microondas* road and also along the main highway on either side of Puerto Los Mazos, although traffic noise can be distracting.

If you continue on the track beyond the towers, forest ends quickly at a barbed wire fence and gate. The brushy thickets beyond the gate should be checked for **Dwarf Vireo** (winter), and this is a good place from which to scan for swifts (including **Great Swallow-tailed**).

Another *microondas* road, this one northeast of Autlán, allows access to an area of arid scrub. From Autlán head north on Route 80 toward Guadalajara and, 7.5 km from the intersection of the *periferico* on the northeast side of Autlán, look for a cobbled road (at Km Post 145) that climbs the ridge to your left (northwest), signed to **Microondas San Francisco.** This road winds through thorn forest, and birds in the first 3 km include **Banded Quail**, Zone-tailed Hawk, Lesser Ground-Cuckoo, Ferruginous Pygmy-Owl, Buff-collared Nightjar, Broad-billed, Violet-crowned, Lucifer (migrant), and Black-chinned (winter) hummingbirds, Plain-capped Starthroat, **Golden-cheeked Woodpecker**, Northern Beardless Tyrannulet, Ash-throated (winter), Nutting's, and Brown-crested flycatchers, **San Blas Jay**, **Black-capped Gnatcatcher**, Virginia's Warbler (winter), **Orange-breasted Bunting**, **Black-chested** and Stripe-headed **sparrows**, and Streak-backed Oriole. This can be reached readily by second-class bus or hitching.

Puerto Los Mazos Bird List

123 sp. †See pp. 5–7 for notes on English names.

Black Vulture
Turkey Vulture
Hook-billed Kite
Sharp-shinned Hawk
Broad-winged Hawk
Short-tailed Hawk
Red-tailed Hawk
American Kestrel
Peregrine Falcon
West Mexican Chachalaca
Long-tailed Wood-Partridge
Singing Quail
Inca Dove
White-tipped Dove
Orange-fronted Parakeet
Mexican Parrotlet
Squirrel Cuckoo
Mottled Owl
Eared Poorwill
Vaux's Swift
Great Swallow-tailed Swift
Golden-crowned Emerald
Broad-billed Hummingbird
Mexican Woodnymph
Berylline Hummingbird
Violet-crowned Hummingbird
Amethyst-throated Hummingbird
Plain-capped Starthroat
Sparkling-tailed Woodstar†
Calliope Hummingbird
Black-chinned Hummingbird
Rufous Hummingbird
Yellow-bellied Sapsucker
Ladder-backed Woodpecker
Arizona Woodpecker†
Smoky-brown Woodpecker
Grey-crowned Woodpecker
Lineated Woodpecker
Pale-billed Woodpecker
Olivaceous Woodcreeper
Ivory-billed Woodcreeper
N. Beardless Tyrannulet
Greenish Elaenia
Greater Pewee
Common Tufted Flycatcher†
Hammond's Flycatcher
Dusky Flycatcher
Western Flycatcher†
Bright-rumped Attila
Dusky-capped Flycatcher
Nutting's Flycatcher
Western Kingbird
Masked Tityra
Grey-collared Becard
Rose-throated Becard
Violet-green Swallow
N. Rough-winged Swallow
Green Jay
Northern Raven†
Bridled Titmouse
Spotted Wren
Canyon Wren
Happy Wren
Sinaloa Wren
Northern House Wren†
Grey-breasted Wood-Wren
Ruby-crowned Kinglet
Blue-grey Gnatcatcher
Brown-backed Solitaire
Orange-billed N.-Thrush
Swainson's Thrush
Hermit Thrush
White-throated Thrush†
Blue Mockingbird
Cedar Waxwing
Grey Silky†
Black-capped Vireo
Dwarf Vireo
Cassin's Vireo†
Plumbeous Vireo†
Golden Vireo
Warbling Vireo
Chestnut-sided Shrike-Vireo
Nashville Warbler
Crescent-chested Warbler
Tropical Parula
Audubon's Warbler†
Black-throated Grey Warbler
Townsend's Warbler
Hermit Warbler
Black-throated Green Warbler
Black-and-white Warbler
Louisiana Waterthrush
MacGillivray's Warbler
Wilson's Warbler
Red-faced Warbler
Painted Whitestart†
Slate-throated Whitestart†
Fan-tailed Warbler
Golden-crowned Warbler
Rufous-capped Warbler
Blue-hooded Euphonia
Red-crowned Ant-Tanager
Hepatic Tanager
Summer Tanager
Western Tanager
Flame-colored Tanager
Red-headed Tanager
Greyish Saltator
Yellow Grosbeak
Black-headed Grosbeak
Indigo Bunting
Varied Bunting
Green-striped Brushfinch
Rufous-crowned Sparrow

Chipping Sparrow	Bullock's Oriole
Lincoln's Sparrow	**Black-headed Siskin**
Rusty-crowned Ground-Sparrow	Lesser Goldfinch
Streak-backed Oriole	

Site 7.6: Autlán to Ciudad Guzmán, Jalisco

(December 1983, 1986, April 1988, March 1989, 1990, 1992, 1993, February 1994, 1995, 1997)

Introduction/Key Species

This is a drive of about 120 km (2–3 hours without stops) that starts in arid scrub around Autlán, skirts the drier pine-oak forests on the interior slopes of the Volcán de Nieve, and ends in the dusty agricultural plains around Ciudad Guzmán. The avifauna varies accordingly, from **Mexican Parrotlet** and **Black-chested Sparrow** through **Grey-barred Wren** and **Aztec Thrush** to Canyon Towhee and House Finch.

Access/Birding Sites/Bird List

From Autlán head northeast out of town past the turn to Microondas San Francisco (Km 7.5) and to a T-junction immediately after a major river bridge (Km 14.2). Turn right here (left goes, windingly, to Guadalajara) toward El Grullo, the edge of which you reach in 7.7 km. You can either go straight through town or take the longer bypass to the south; these two routes merge again on the far (east) side of town by a Pemex station, where you can reset to zero. From this point it is 10 km to an area with a few tall fig trees and some thorny thicket vegetation with permanent water; you can pull off to the right (south) on a track and park in the shade of the large fig trees (which, when fruiting, often have **Mexican Parrotlet**). Birds here and along the track across the road, through one or two gates to a stream about 150 m from the highway, include Ferruginous Pygmy-Owl, Orange-billed Nightingale-Thrush, **Rufous-backed Thrush**, **Blue Mockingbird**, and **Rusty-crowned Ground-Sparrow**.

Continuing toward Ciudad Guzmán, another spot worth checking is by the river bridge at Km 20. Cross the bridge and you can park safely off the highway on either side. A dirt track to the left (north) follows the east bank of the river for a few hundred meters through (increasingly cleared) thorn forest and riparian willows before coming to an open field where the river bends away to the left. Birds here include **Banded Quail**, **Russet-crowned Motmot**, Ferruginous Pygmy-Owl, Rose-throated Becard, **San Blas Jay**, Orange-billed Nightingale-Thrush, **Golden Vireo**, **Rusty-crowned Ground-Sparrow**, and, in season, numerous migrants, including Bell's and Warbling vireos, Lucy's Warbler, and Varied and Painted buntings. At Km 28.5 you pass the left-hand turn into Tonaya, and the 10 km or so after this turn have some good areas of thorn forest with **Balsas Screech-Owl**, **Black-capped Gnatcatcher**, and **Black-chested Sparrow**.

After this stretch the highway heads toward the volcanoes, which are visible ahead as an impressive massif, the Volcán de Nieve on the left (north), and the Volcán de Fuego, steaming, on the right (south). Over the fields watch for White-tailed and Red-tailed hawks; birds on the roadside wires include Cassin's Kingbird and Loggerhead Shrike; the fields have Lark Sparrow (winter) and House Finch; and Acorn Woodpecker and Eastern Bluebird start to appear as you climb through cornfields into the first pines. Once you are in the pines there are numerous pulloffs and tracks into the woods. Stops anywhere here can produce species such as **Spotted Wren**, **Collared Towhee**, and, with luck, nomadic flocks of **Thick-billed Parrots** (January to April).

An area worth birding, and one which can be visited easily from Ciudad Guzmán for a morning or an afternoon, is that reached by two dirt side roads at around Km 76—that is, the pass (2270 m) before the highway drops down toward Guzmán. Coming from Autlán, the first road (to some microwave towers) is signed to the right (south) to R.M.O. Viboras; shortly thereafter you will see a wide pulloff on the right (south) of the highway. From this pulloff a second dirt road heads for a few kilometers up into the pine-oak forest. A sign announcing "Bienvenido a San Gabriel" marks this pass in the highway, which has been signed in the past as "El Floripondio."

You can park safely off the highway here and walk or drive along either road. There is a barbed wire gate at the start of the *microondas* road and a metal gate 300 m up the second road: these can be opened to allow vehicle passage. On the second road you also come to a barbed wire gate by some buildings about 750 m from the highway. You can park and ask permission (if you see anybody) to bird here (the owner, Don Jose del Toro, is glad that birders appreciate the area).

Birding along either of these roads (to 2500+ m elevation) can be excellent (although some parts are fairly steep), and large numbers of hummingbirds swarm at the flower banks between January and March. Species seen here (December 1983, 1986, April 1988, March 1989, 1990, 1992, 1993, 1994, March, December 1995, February 1997) include **Long-tailed Wood-Partridge**, **Lesser Roadrunner**, Whiskered Screech-Owl, Mountain Pygmy-Owl, Mexican Whip-poor-will, Green Violet-ear, **White-striped Woodcreeper, Pine** and Buff-breasted **flycatchers**, White-breasted and Pygmy nuthatches, **Grey-barred Wren**, **Russet Nightingale-Thrush**, **Aztec Thrush**, **Chestnut-sided Shrike-Vireo**, Colima (winter), Grace's, **Red**, and Olive **warblers**, **Rufous-capped** and **Green-striped brushfinches**, **Collared Towhee**, Scott's Oriole (winter), Red Crossbill, Pine and **Black-headed siskins**, and many of the other species listed for the Volcán de Fuego (Site 7.8).

From El Floripondio it is about 25 km (30–45 minutes) to Ciudad Guzmán, dropping out of the pines and onto the dusty plains that surround the town. Staying on the highway, you go under the *cuota* highway bridge (Km 98) before hitting the old Guzmán to Colima highway on the south side of Ciudad Guzmán at Km 99. Turn left and you'll be in town in a few kilometers.

Site 7.7: Ciudad Guzmán/Laguna Sayula, Jalisco

(December 1983, April 1988, March 1989, 1990, 1992, February, March 1993, March 1994, 1995, February, March 1997)

Introduction/Key Species

Ciudad Guzmán (formerly Zapotlán) lies at about 1450 m elevation in the extreme southwestern corner of the Mexican Plateau. The lakes and marshes north of Guzmán, along Route 54D to Guadalajara, can be excellent for waterbirds, especially in winter; species here include Clark's Grebe, American and Least bitterns, Snow Goose, Mexican Duck, King Rail, Common ("Chapala") Yellowthroat, and Yellow-headed Blackbird. These lakes and marshes can be birdy at any time of day and make a good afternoon trip following a morning birding on the Volcánes de Colima, which dominate the skyline to the southwest of town.

As a base from which to visit the volcanoes (Sites 7.6, 7.8) you can use Ciudad Colima or Ciudad Guzmán, although access to the drier interior slopes is easier, and much nearer, from Guzmán.

Access

Ciudad Guzmán is about 120 km (1–1.5 hours) south of Guadalajara (1.5–2 hours from the airport) via a new and fast *cuota* highway (Route 54D). From Guadalajara airport head north toward the city and, immediately as you go under the second highway overpass about 7 km north of the airport, turn right and curve up onto the *periferico*; once on the *periferico* (a six-lane divided highway punctuated by numerous traffic lights) look for a poorly signed exit for Colima in another 13 km. Turn right here and curve down and around onto the southbound highway, which passes back under the *periferico*. In another 26 km there is a junction for Morelia and Barra de Navidad (not, however, the fastest route), and then you are on the toll highway to Colima and Manzanillo (and the fastest route to Barra de Navidad); you reach the first toll booth in 1.5 km.

The well-signed exit for Ciudad Guzmán is 86 km south of the first toll booth, and along the intervening stretch of *cuota* highway you pass several lakes and marshy areas which can birded from the highway shoulder (see below). Exit for Guzmán here and follow signs to the center, which is a few kilometers east of the *cuota* highway. Ciudad Guzmán is about 75 km (1 to 1.2 hours) north of Ciudad Colima, whence it is reached via the same *autopista* exit or by the old (*libre*) highway (1.5 to 2 hours). See the Autlán to Ciudad Guzmán entry (Site 7.6) for access if you're coming from Autlán. While Guzmán is not a

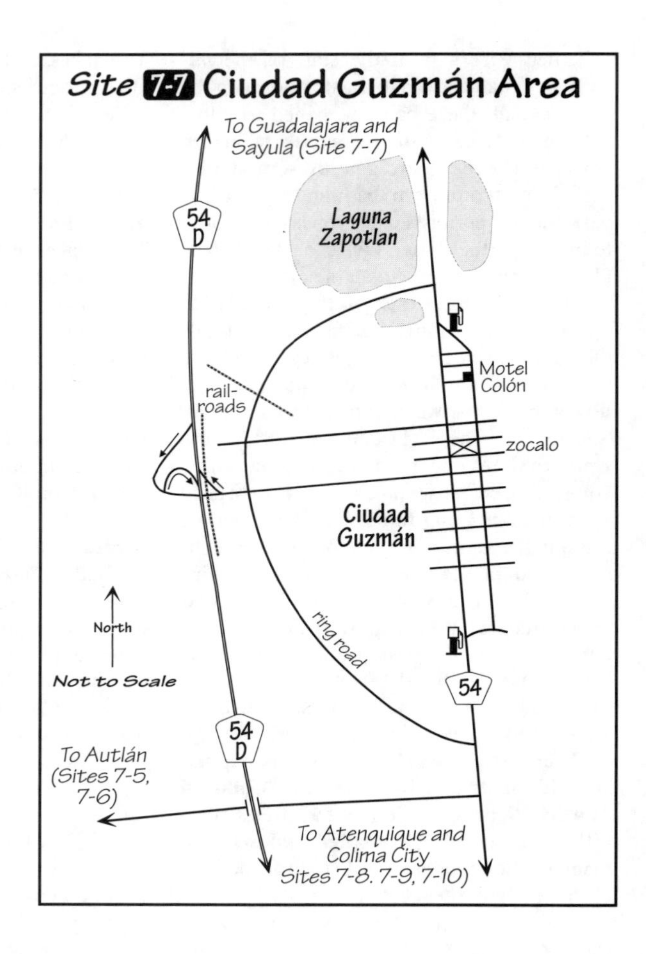

tourist town, it does have several hotels (used mostly by Mexican businessmen) and restaurants that cater to a range of budgets. The Motel Colón (née Motel Real) on the north side of town has been used by birders as a base.

Public transport to Ciudad Guzmán is easy, but getting to nearby birding areas may be problematic without your own car. To reach El Floripondio on the slopes of the Volcán de Nieve, take a second-class bus toward Autlán and get off at the pass (see Site 7.6).

Birding Sites

See Volcán de Fuego (Site 7.8) for access to the coastal slopes of the Volcánes de Colima (1–1.5 hours south of Guzmán) and Autlán to Ciudad Guzmán (Site 7.6) for access to the interior slopes of the Volcánes via El Floripondio (30–45 minutes west of Guzmán).

On the north side of Ciudad Guzmán, beside the bypass that skirts the west edge of town, is Laguna Zapotlán, a large shallow lake fringed by extensive reed beds. From downtown Guzmán, head north on the one-way street along the east side of the *zócalo*, and in 2 km you merge onto the main road north out of town, with a Pemex station on the right; 3.8 km north of here a bypass joins from the left, at the edge of a large lake. You can view this lake from the roadside straight on for a kilometer or, turning left here, from points along the first 4 km of the bypass. Beware that traffic can be heavy and sometimes fast on the bypass, so make sure you park safely off the road—for example, at 100 m,

1.3 km, and 3.0 km (all on the right) and Km 3.8 (on the left). If you want to check Laguna Zapotlán en route from Guadalajara to Colima, come off the *cuota* highway at the Guzmán exit and turn left (north) onto the bypass at a four-way stop junction 1.2 km from the *cuota*. Continue north and you come to the lake and marshes in about 3 km.

Birds at the lake here include Clark's Grebe, American White Pelican (migrant), American and Least bitterns, White-faced Ibis, Mexican Duck, King Rail, Common ("Chapala") Yellowthroat, and spectacular winter roosts of hundreds of thousands of Yellow-headed Blackbirds.

To reach another area of lakes and marshes from Guzmán, continue on the bypass 3 km past Laguna Zapotlán to the four-way stop junction (a concept not met with by universal comprehension in Mexico, so don't be surprised if few people stop), turn right, and continue (1.2 km) to the Guadalajara-Colima *autopista*, where you bear right, signed for Guadalajara. (Note that these lakes and marshes can be birded easily en route to/from from Guadalajara.)

Mark zero as you merge onto the toll highway. At around Km 25, you start to go through a plain with seasonally flooded fields; the areas of acacia-cactus scrub here (e.g., Km 34.2–35.3 on the east side of the highway) have Grey Flycatcher (winter) and Cactus Wren; the wet roadside ditches have King Rail. It is possible to pull off safely on the gravel shoulder and bird from inside or near your vehicle, and the highway makes a good vantage point for scanning the marshes and fields. The most reliably wet area is a reed-fringed lake east of the highway at Km 35.5–36.5 (beside Km Post 51), shortly before (south of) the Atoyac overpass. Depending on water conditions, there can be a large shallow lake (Laguna Sayula) at Km 42 to 46, crossed by a causeway to the east at Km 45.7; another seasonal, reed-fringed lake is at Km 54, just after a dirt road off to the east.

This whole area can have literally thousands of waterbirds in winter, mainly American White Pelicans, Great and Snowy egrets, White-faced Ibis, Snow Geese, Green-winged Teal, Northern Shovelers, American Avocets, Black-necked Stilts, and Western and Least sandpipers, with smaller numbers of sundry other species including Roseate Spoonbill, Collared and Snowy plovers, and Gull-billed Tern. Also in winter, the short-grass fields around the lakes have Sprague's Pipit.

To return to Ciudad Guzmán you can make a U-turn across the central strip of the highway at Km 45.7 (the causeway) and at Km 54 (just before the last seasonal lake); if you go beyond (north of) this point there is another possible U-turn spot at Km 59.5. Note that even on toll highways the availability of U-turns (and even exits!) is prone to change.

Ciudad Guzmán/Laguna Sayula Bird List

144 sp. V: vagrant; R: rare (not to be expected). †See pp. 5–7 for notes on English names.

Least Grebe
Pied-billed Grebe
Eared Grebe
Clark's Grebe
American White Pelican
Neotropic Cormorant
American Bittern
Least Bittern
Great Blue Heron
Great Egret
Snowy Egret
Little Blue Heron
Tricolored Heron
Cattle Egret
Green Heron
Black-crowned Night-Heron
White-faced Ibis
Roseate Spoonbill
Snow Goose
Fulvous Whistling-Duck
Green-winged Teal
Mexican Duck†
Northern Pintail
Blue-winged Teal
Cinnamon Teal
Northern Shoveler
Gadwall
Eurasian Wigeon (V)
American Wigeon
Redhead
Ring-necked Duck
Lesser Scaup
Hooded Merganser (R)
Ruddy Duck
Black Vulture
Turkey Vulture
Osprey
White-tailed Kite
Northern Harrier
Sharp-shinned Hawk
Cooper's Hawk
Harris' Hawk
Red-tailed Hawk
Crested Caracara
American Kestrel
Merlin
Peregrine Falcon
King Rail
Virginia Rail
Sora
Common Moorhen
American Coot
Black-bellied Plover
American Golden Plover
Collared Plover
Snowy Plover
Semipalmated Plover
Killdeer
Black-necked Stilt
American Avocet
Northern Jacana

Greater Yellowlegs
Lesser Yellowlegs
Willet
Ruddy Turnstone
Spotted Sandpiper
Long-billed Curlew
Marbled Godwit
Sanderling
Western Sandpiper
Least Sandpiper
Pectoral Sandpiper
Stilt Sandpiper
Long-billed Dowitcher
Common Snipe
Wilson's Phalarope
Laughing Gull
Ring-billed Gull
Gull-billed Tern
Caspian Tern
Forster's Tern
Mourning Dove
Inca Dove
Common Ground-Dove
Groove-billed Ani
Lesser Nighthawk
Chestnut-collared Swift
Vaux's Swift
Belted Kingfisher
Golden-fronted Woodpecker
Ladder-backed Woodpecker
Grey Flycatcher
Vermilion Flycatcher
Great Kiskadee
Social Flycatcher
Tropical Kingbird
Thick-billed Kingbird
Cassin's Kingbird
Tree Swallow
Violet-green Swallow
N. Rough-winged Swallow
Bank Swallow
Cliff Swallow
Barn Swallow
Northern Raven†
Cactus Wren
Marsh Wren
Blue-grey Gnatcatcher
Northern Mockingbird
American Pipit
Sprague's Pipit
Cedar Waxwing
Loggerhead Shrike
Orange-crowned Warbler
Virginia's Warbler
Yellow Warbler
Audubon's Warbler†
Northern Waterthrush
MacGillivray's Warbler
Common Yellowthroat
Wilson's Warbler
Blue Grosbeak
Indigo Bunting
Varied Bunting
Canyon Towhee
Stripe-headed Sparrow
Chipping Sparrow
Clay-colored Sparrow
Lark Sparrow
Savannah Sparrow
Lincoln's Sparrow
White-crowned Sparrow
Red-winged Blackbird
Yellow-headed Blackbird
Eastern Meadowlark
Brewer's Blackbird
Great-tailed Grackle
Bronzed Cowbird
Brown-headed Cowbird
Hooded Oriole
Streak-backed Oriole
Abeille's Oriole†
House Finch
Lesser Goldfinch

Site 7.8: Volcán de Fuego, Jalisco

(December 1986, April 1988, March 1989, 1990, 1992, 1993, 1994, March, December 1995, February 1997)

Introduction/Key Species

The twin Volcánes de Colima (which in fact lie almost entirely in the state of Jalisco!) comprise the Volcán de Fuego, still steaming, at 3820 m, and the Volcán de Nieve, often snow-capped, at 4240 m. From a birding point of view, the humid, coastal-facing slopes of the Volcán de Fuego are of the most interest (driving the variably bad dirt road up to about 4000 m on Nieve doesn't add much in the way of birds that can't be found at El Floripondio [see Site 7.6] and/or on Fuego).

Birds of interest on Fuego include **West Mexican Chachalaca**, Crested Guan, **Long-tailed Wood-Partridge**, **Banded Quail**, **Thick-billed Parrot** (winter), Stygian Owl, **Eared Poorwill**, **Amethyst-throated** and **Bumblebee hummingbirds**, **Grey-collared Becard**, **Sinaloa Martin** (summer), **Grey-barred** and **Spotted wrens**, **Russet** and Ruddy-capped **nightingale-thrushes**, **Aztec Thrush**, **Dwarf Vireo** (winter), **Chestnut-sided Shrike-Vireo**, Colima (winter) and **Red warblers**, **Red-headed Tanager**, **Green-striped Brushfinch**, **Collared Towhee,** and **Dickey's** [Audubon's] and **Abeille's** (migrant) **orioles**.

Access

Fortunately, there is a dirt road that allows access to the forested slopes of Volcán de Fuego from near the town of Atenquique. Unfortunately, this road isn't always passable, although usually you can at least get into some good habitat, if not all of the way up to the highest forests. Camping is an excellent way to bird the area, or you can visit for the day from Ciudad Guzmán or Ciudad Colima. If you decide to camp, take all of your supplies, and be aware that it can be cold at night. At least two or three full days' birding are needed to do the area justice. To reach the volcano road for dawn you will need to leave your hotel in the dark, so beware of the usual hazards: trucks without lights, bicycles, cattle.

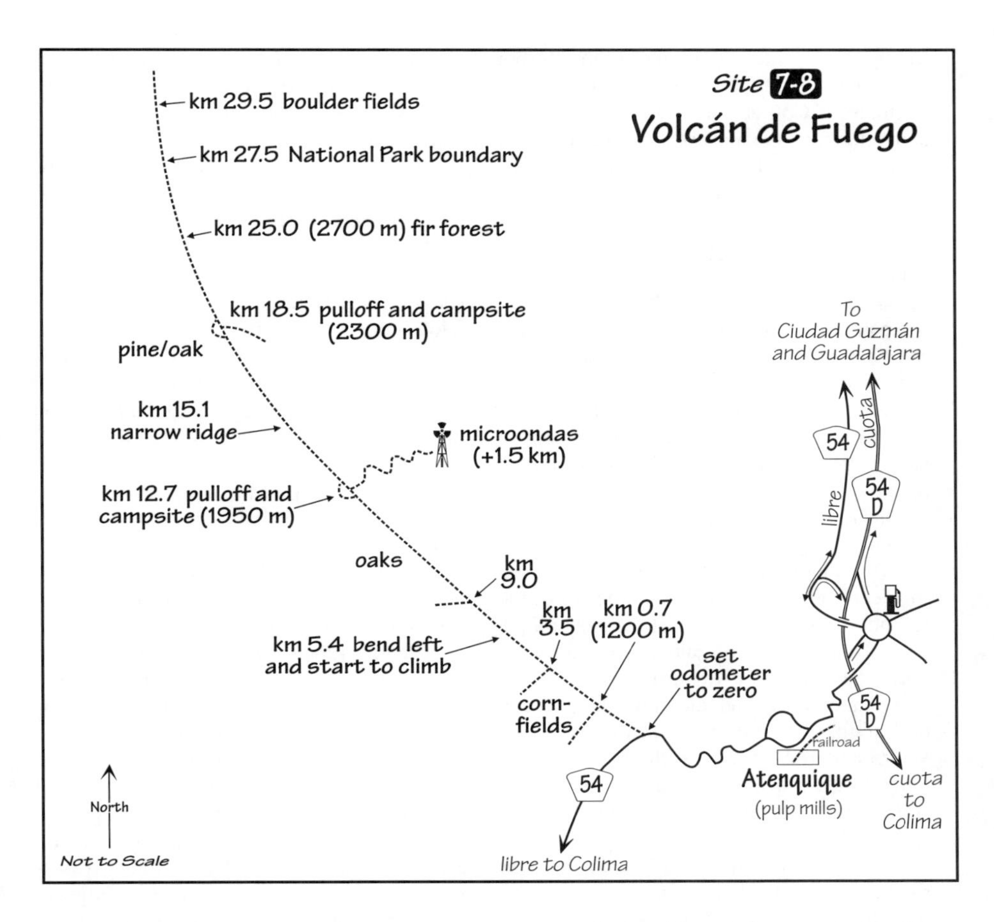

To get to Atenquique from Ciudad Guzmán, take the old highway (Route 54), because the former exit from the toll highway has been blocked off! It's 30–40 minutes to the base of the volcano road, and from there another 30–45 minutes to good forest, although the lower, agricultural slopes, near the highway, are also well worth birding.

Coming from Ciudad Guzmán, mark zero at the *zócalo*, head south on the one-way system, and merge with the highway out of town at Km 1.5 by a Pemex station on the right; you pass the *periferico* junction on the right at Km 3.0 and the right turn to Autlán and Tonaya at Km 3.8. At Km 6.5 fork right in response to signs for Colima (left goes to Jiquilpan), cross over the *periferico* at Km 7.6 so that it parallels you to the left (east), and at Km 21.8 bend right and curve back left up to a *glorieta*. Turn right and take the first exit, signed to Atenquique and "Colima Libre," marking zero as you leave the *glorieta*. The road winds down into a valley dominated by the pulp mills of Atenquique. At Km 3.4 fork left down toward the mill, crossing a narrow bridge that also is a railway bridge (trains have right of way!) and up the other side to a small *glorieta* at Km 4.1, where you take the left exit, which winds up out of the valley. At Km 6.4, the end of a straight stretch where the highway bends left, a dirt road continues on straight, toward the volcanoes, with a sign that says "R.M.O. Cerro Alto." This is the road up to Volcán de Fuego.

Coming from Ciudad Colima, you can reach Atenquique via the *autopista* or the old highway. The *autopista* (Route 54D) is distinctly quicker (40–50 minutes to the start of the volcano road, versus 1–1.5 hours). Head north from Colima on Route 54, following signs for the Guadalajara *cuota*, and you reach a toll booth 30.6 km from where the *periferico* merges with the highway on the north side of Ciudad Colima. From here it is 15 km to the Atenquique junction (though it is not signed as such), where you exit right in response to signs for "Colima Libre" and Tuxpan and come up to the *glorieta*, where the *libre* road from Guzmán also joins. Take the immediate right off the *glorieta* (the road signed to Atenquique and "Colima Libre"), reset to zero, and follow the directions above to get through Atenquique and to the start of the volcano road.

One point that can be worth stopping along the *cuota* highway between Ciudad Colima and Atenquique is the Barranca de Beltran, the deep canyon you cross 3 km north of the toll booth (or 12 km south of the Atenquique exit). There is a pulloff and viewpoint on the north side of the *barranca* (accessible from either lane). Early and late in the day this can be a good vantage point for swifts, including White-collared (at the northern limit of its range); other species here include Common Black and Red-tailed hawks and Orange-fronted Parakeet.

If you're coming from Colima on the *libre* road (Route 54), head north following signs for "Guadalajara Libre": 53.5 km from where the *periferico* merges with the highway on the north side of Colima, Route 54 bends right and down into Atenquique, and at this point the dirt road up the volcanoes goes off to the left.

There is no public transport to this site, but you may be able to hitch a ride from the highway junction to the upper ranches, from which you can walk on to good habitat.

Birding Sites

The dirt road up the volcanoes goes on for at least 30 km, with perhaps the best forest birding at Km 10 to 25. For the first few kilometers (1150–1500 m elevation) you pass through cornfields punctuated with scattered pines and bordered by brushy hedges and weedy fields. There are several side tracks and roads, but it is obvious which is the main road you should stay on. At least in winter and early spring this area can be incredibly birdy, with swarms of grosbeaks, buntings, and sparrows flushing up all along the road.

Anywhere on the first 5 km or so can be good for such birding—for instance, the track off to the left at Km 0.7 by the first small *rancho*, or around the side road to the left at Km 3.5. Birds in this first stretch include **Banded Quail**, **Lesser Roadrunner** (possible anywhere up to 2500 m, or higher), Violet-crowned, Lucifer, and Calliope hummingbirds (migrants), **Russet-crowned Motmot**, Dusky Flycatcher (winter), **Spotted** and **Happy wrens**, **Blue Mockingbird** (common up to 2500 m, or higher), **Grey Silky**, Virginia's Warbler (winter), Blue-hooded Euphonia, Lazuli, Varied, Indigo, and Painted buntings (winter), **Rusty-crowned Ground-Sparrow**, and Clay-colored and Grasshopper sparrows (winter). On and around the dead pine snags at Km 3–5 it is worth checking for **Sinaloa Martin** (summer).

At Km 5.4 the road bends left and starts to climb. From here to Km 7 a stop just before dawn can produce Great Horned Owl, Ferruginous Pygmy-Owl, **Eared Poorwill**, and Buff-collared Nightjar. **Spotted Wren** and Rusty Sparrow occur where you start to hit the first oaks. Fork right at Km 9 (left goes to a small *rancho*), shortly after which you cross a cattle guard and run along an oak-wooded valley to your left. The birding can be good anywhere along the road from this point up to the side road to the *microondas* at Km 12.7 (1950 m), and if you encounter a mixed-species flock it is worth stopping. Birds in this stretch include **Long-tailed Wood-Partridge** (easier to see higher up), Whiskered Screech-Owl, Mountain Pygmy-Owl, Mexican Whip-poor-will, Elegant Trogon, **White-striped Woodcreeper**, **Grey-collared Becard**, Grey-breasted Jay, **Grey-barred** and **Spotted wrens**, **Dwarf Vireo** (winter), **Chestnut-sided Shrike-Vireo**, **Red-headed Tanager**, **Rufous-capped Brush-finch**, **Rusty-crowned Ground-Sparrow**, **Dickey's** [Audubon's] **Oriole**, and **Black-headed Siskin**.

The cobbled side road off to the right at Km 12.7 leads about 1.5 km uphill through more nice forest to the Microondas Cerro Alto, while the clearing on the left at this junction makes a good campsite.

Staying on the main road, you wind up through relatively dry pine-oak woods until around Km 17, when the road bends back to the cooler, wetter, coastal-facing slopes and the forest becomes noticeably more humid. A good pulloff (and campsite) is a wide area on the left at Km 18.5 (2300 m), shortly after which a side track heads off to the right. From this point the road is relatively level for a few kilometers and winds around a few cool, damp canyons. Birds along this stretch include **West Mexican Chachalaca**, Crested Guan, **Long-tailed Wood-Partridge** (common), Mottled and Stygian owls, **Amethyst-throated** and **Bumblebee hummingbirds**, **Mountain Trogon**, Hammond's (winter), Western, and **Pine flycatchers**, Ruddy-capped Nightingale-Thrush, **Aztec Thrush** (seasonally in flocks of up to 100 or more birds, February to April), Colima (winter), **Red**, and **Golden-browed warblers**, **Green-striped Brushfinch**, **Collared Towhee**, **Cinnamon-bellied Flowerpiercer**, and many of the species found lower down: for example, mixed-species flocks with **Grey-collared Becard**, **Grey-barred Wrens**, **Chestnut-sided Shrike-Vireo**. Keep an ear open for the calls of **Thick-billed Parrots**, flocks of which can be encountered (usually flying) almost anywhere on the volcanoes (January to April). Spotted Owl and **Eared Quetzal** have also been reported in this area.

By about Km 24 (2700 m) you'll be in forest with an good component of firs (Golden-crowned Kinglet), and at Km 27.5 you (eventually) reach a sign announcing that you're at the park boundary. Thus, most of the good forest is outside the park! There follows an area of nice fir-pine forest before you get to boulder fields above timberline at Km 29.5.

Note: In February 1997 there was a sign near the start of the road that said "Camino Cerrado en 18 km" (Road Closed in 18 km), and about 1 km past the Km 18.5 campsite there was a large mound of dirt across the road in an apparent attempt to restrict access. However, it was possible with a high-clearance vehicle, or in a rental car, to cross this barrier and continue up to higher elevations.

Volcán de Fuego Bird List

205 sp. R: rare (not to be expected). †See pp. 5–7 for notes on English names.

Cattle Egret
Black Vulture
Turkey Vulture
Hook-billed Kite
White-tailed Kite
Sharp-shinned Hawk
Cooper's Hawk
Common Black Hawk
Grey Hawk
Broad-winged Hawk
Short-tailed Hawk
Swainson's Hawk
White-tailed Hawk
Red-tailed Hawk
Crested Caracara
American Kestrel
West Mexican Chachalaca
Crested Guan
Long-tailed Wood-Partridge
Singing Quail
Banded Quail
Band-tailed Pigeon
Mourning Dove
Inca Dove
White-tipped Dove
Common Ground-Dove
Ruddy Ground-Dove
Thick-billed Parrot
Squirrel Cuckoo
Lesser Roadrunner
Groove-billed Ani
Barn Owl
Whiskered Screech-Owl
Great Horned Owl
Mountain Pygmy-Owl†
Ferruginous Pygmy-Owl
Mottled Owl
Stygian Owl
Eared Poorwill
Buff-collared Nightjar
Mexican Whip-poor-will†
Vaux's Swift
White-throated Swift
Green Violet-ear
Golden-crowned Emerald
Broad-billed Hummingbird
White-eared Hummingbird
Berylline Hummingbird
Violet-crowned Hummingbird
Amethyst-throated Hummingbird
Blue-throated Hummingbird
Magnificent Hummingbird
Lucifer Hummingbird
Calliope Hummingbird
Broad-tailed Hummingbird
Rufous Hummingbird
Bumblebee Hummingbird
Mountain Trogon
Elegant Trogon
Russet-crowned Motmot
Acorn Woodpecker
Golden-cheeked Woodpecker
Golden-fronted Woodpecker
Yellow-bellied Sapsucker
Ladder-backed Woodpecker
Hairy Woodpecker
Arizona Woodpecker†
Red-shafted Flicker†
White-striped Woodcreeper
N. Beardless Tyrannulet
Greenish Elaenia
Greater Pewee
Common Tufted Flycatcher†

Olive-sided Flycatcher
Hammond's Flycatcher
Dusky Flycatcher
Pine Flycatcher
Western Flycatcher†
Buff-breasted Flycatcher
Vermilion Flycatcher
Bright-rumped Attila
Dusky-capped Flycatcher
Ash-throated Flycatcher
Nutting's Flycatcher
Great Kiskadee
Social Flycatcher
Cassin's Kingbird
Thick-billed Kingbird
Western Kingbird
Grey-collared Becard
Rose-throated Becard
Sinaloa Martin
Tree Swallow
Violet-green Swallow
N. Rough-winged Swallow
Cliff Swallow
Grey-breasted Jay†
Northern Raven†
Mexican Chickadee
Bridled Titmouse
Bushtit
White-breasted Nuthatch
Pygmy Nuthatch
Brown Creeper
Grey-barred Wren
Spotted Wren
Canyon Wren
Happy Wren
Sinaloa Wren
Northern House Wren†
Brown-throated Wren†
Grey-breasted Wood-Wren
Golden-crowned Kinglet
Ruby-crowned Kinglet
Blue-grey Gnat-catcher
Brown-backed Solitaire
Eastern Bluebird
Hermit Thrush
White-throated Thrush†
Rufous-backed Thrush†
Aztec Thrush
Orange-billed N.-Thrush
Russet N.-Thrush
Ruddy-capped N.-Thrush
American Robin
Northern Mockingbird
Blue Mockingbird
Cedar Waxwing
Grey Silky†
Loggerhead Shrike
Dwarf Vireo
Cassin's Vireo†
Plumbeous Vireo†
Hutton's Vireo
Golden Vireo
Warbling Vireo
Chestnut-sided Shrike-Vireo
Orange-crowned Warbler
Colima Warbler
Nashville Warbler
Virginia's Warbler
Lucy's Warbler
Crescent-chested Warbler
Audubon's Warbler†
Black-throated Grey Warbler
Townsend's Warbler
Hermit Warbler
Grace's Warbler
Black-and-white Warbler
MacGillivray's Warbler
Grey-crowned Yellowthroat
Wilson's Warbler
Red-faced Warbler
Red Warbler
Painted Whitestart†
Slate-throated Whitestart†
Rufous-capped Warbler
Golden-browed Warbler
Olive Warbler
Yellow-breasted Chat
Blue-hooded Euphonia
Hepatic Tanager
Summer Tanager
Western Tanager
Flame-colored Tanager
Red-headed Tanager
Greyish Saltator
Pyrrhuloxia (R?)
Yellow Grosbeak
Rose-breasted Grosbeak
Black-headed Grosbeak
Blue Grosbeak
Lazuli Bunting
Indigo Bunting
Varied Bunting
Painted Bunting
Dickcissel
Rufous-capped Brushfinch
Green-striped Brushfinch
Rusty-crowned Ground-Sparrow
Collared Towhee
Canyon Towhee
Blue-black Grassquit
Cinnamon-bellied Flowerpiercer
Stripe-headed Sparrow
Rusty Sparrow
Chipping Sparrow
Clay-colored Sparrow
Lark Sparrow
Grasshopper Sparrow
Lincoln's Sparrow
Mexican Junco†
Yellow-headed Blackbird
Bronzed Cowbird
Brown-headed Cowbird
Hooded Oriole
Black-vented Oriole
Dickey's Oriole†
Streak-backed Oriole
Bullock's Oriole
Abeille's Oriole†
Scott's Oriole
House Finch
Red Crossbill
Pine Siskin
Black-headed Siskin
Lesser Goldfinch

Site 7.9: Ciudad Colima to La Maria, Colima

(December 1983, April 1988, March 1989, 1990, 1992, 1993, 1994, March, December 1995, February 1997)

Introduction/Key Species

Ciudad Colima (elevation 450 m), capital of the state, sits on arid plains shadowed by the steaming peak of Volcán de Fuego. The tropical semi-deciduous forest/coffee planta-

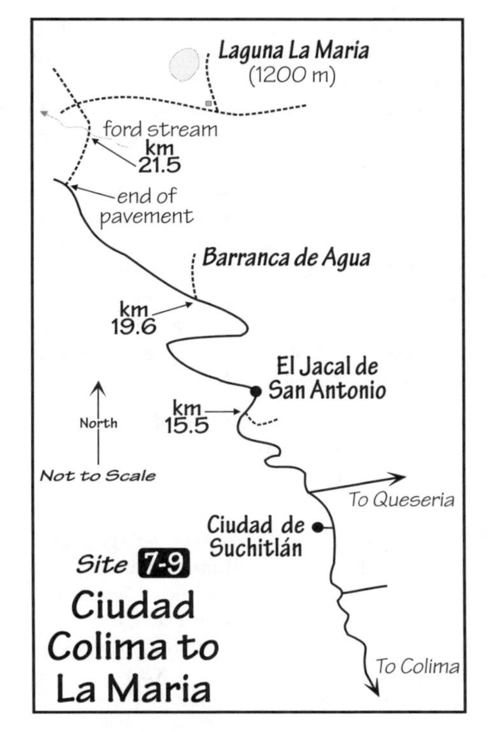

tion, brushy woodland, and weedy fields en route to, and at, Laguna La Maria, about 35 km from Colima, offer good birding. The variety of species possible includes Hook-billed and Double-toothed kites, Broad-winged Hawk (winter), **West Mexican Chachalaca**, **Banded Quail**, **Lesser Roadrunner**, **Great Swallow-tailed Swift**, **Golden-crowned Emerald**, Calliope Hummingbird (winter), **Grey-crowned Woodpecker**, **Happy** and **Sinaloa wrens**, **Blue Mockingbird**, **Slaty**, Black-capped (winter), **Dwarf** (winter), and **Golden vireos**, **Fan-tailed Warbler**, **Red-headed Tanager**, **Slate-blue Seedeater**, **Rusty-crowned Ground-Sparrow**, and Rusty Sparrow.

Access/Birding Sites

Laguna La Maria and the birding sites en route to it are twenty minutes to an hour northwest of Ciudad Colima, which has a wide range of accommodations and restaurants and can also be used as a base for birding the Volcán de Fuego (Site 7.8). The Motel Los Candiles on the north side of town is well situated and has been used by birders as a base. Colima is about 200 km (2–3 hours via the *autopista*) south of Guadalajara and about 100 km (1 to 1.5 hours via the *autopista*) from Manzanillo and the coast.

How you reach Laguna La Maria from Colima will depend on where you are staying; basically you want the road to Comalá. An easy way to find the right road (although not the most direct if you're downtown) is from the *periferico* along the east side of Ciudad Colima: 1.6 km from where the *periferico* starts on the north side (e.g., if you're coming from Ciudad Guzmán), or 5.0 km from where it starts on the south side of Colima, there is a signed exit to the west for Comalá (coming from the south, you bear right and stop before crossing traffic). Take this road, a wide four-lane street with numerous traffic lights, through the northern suburbs of Colima for 5.8 km, till you hit a *glorieta* (the first you come to), and take the first right exit here on a wide, four-lane highway. It is another 5.8 km from the *glorieta* to the small town of Comalá.

As you enter Comalá, the road bears right and continues through town on a one-way cobbled street, merging back with the paved highway on the north side of town; mark zero at this merger. Continue northwest, toward the steaming peak of Volcán de Fuego, which soon begins to dominate the skyline. Anywhere from this point onward watch along the roadside for **Banded Quail** and **Lesser Roadrunner**. Seasonally (in December, at least) the diversity of migrant hummingbirds at the roadside flowers north of Comalá can be impressive. At Km 9 you pass through the village of Nogalera, after which the road starts to wind through brushy fields and wooded patches. At Km 15 you bend right and then left around a valley where there is some (increasing) coffee understory, and at Km 15.5 (between Km Posts 16 and 17) there is a wide verge where you can pull off and park on the right, by some brick pillars at the entrance to a cobbled track up the hillside. (About 150 m ahead on the right, as the road bends left, is an amphitheater-like structure with a sign that says "El Jacal de San Antonio.") Walking the roadside (keep an eye on the traffic) in either direction from the pulloff, as well as the cobbled track, can be productive in early morning. Birds here include **West Mexican Chachalaca**, **Great Swallow-tailed Swift**, **Mexican Hermit**, **Golden-crowned Emerald**, Calliope Hummingbird (winter), Elegant Trogon, **Grey-crowned Woodpecker**, **Grey-collared Becard**, Green Jay, **Spotted**, **Happy**, and **Sinaloa wrens**, **Blue Mockingbird**, **Slaty**, Black-capped (winter), **Dwarf** (winter), and **Golden vireos**, **Red-headed Tanager**, **Rusty-crowned Ground-Sparrow**, and Rusty Sparrow.

Continuing along the highway, at Km 19 you enter a steep-sided valley with large areas of bamboo on the slopes above the road. At Km 19.6 there is room to park safely off the highway on the right by the dirt road into the village of Barranca de Agua. From here you can walk in both directions along the main road. Species are much the same as at the previous stop, and **Slate-blue Seedeater** has been found here in the seeding bamboo.

At Km 20.8 the road turns to dirt, and at Km 21.5 it fords a rushing stream, shortly after which, at Km 21.9, there is an inconspicuous right turn (signed "Las Marias 2") at a crossroads by a few houses. Turn right here; the cobbled road runs through brushy fields and hedges (**Spotted Wren**, Botteri's Sparrow). In 900 m look for a left turn through a wooden arch by some buildings. This is the entrance to Laguna La Maria (elevation 1200 m), a small crater lake whose steep walls support tropical semi-deciduous forest, much of it used as shade for growing coffee. A small entrance fee is charged (2 pesos/person), and the site can be busy on weekends, although not usually in early mornings. Conversely, Mondays can be good for birding, when Orange-billed Nightingale-Thrush and other species emerge to pick scraps from under the picnic tables. The lakeside picnic tables are about 750 m from the gate, and birding anywhere along this short stretch of road can be good.

There are second-class buses that ply this route, but I don't know what their destination is signed as—you can check at the bus station, since Laguna La Maria (near San Antonio) should be well known. It is easy to get off along the route (e.g., at El Jacal de San Antonio and Barranca de Agua), and hitching seems quite easy on this road.

Birds at La Maria include Double-toothed Kite, Collared Forest-Falcon, **West Mexican Chachalaca**, **Mexican Parrotlet**, **Lesser Roadrunner**, **Great Swallow-tailed Swift**, **Berylline Hummingbird**, Elegant Trogon, Ringed and Green kingfishers, **Grey-crowned Woodpecker**, **Spotted**, **Happy**, and **Sinaloa wrens**, **Blue Mockingbird**, **Golden Vireo**, Grey-crowned Yellowthroat, **Fan-tailed Warbler**, Flame-colored and **Red-headed tanagers**, **Rusty-crowned Ground-Sparrow**, and Botteri's Sparrow. **Sinaloa Martins** have been seen coming to drink at the lake in spring (March).

Ciudad Colima to La Maria Bird List

187 sp. R: rare (not to be expected). †See pp. 5–7 for notes on English names.

Least Grebe
Pied-billed Grebe
Eared Grebe
Neotropic Cormorant
Great Blue Heron
Great Egret
Little Blue Heron
Cattle Egret
Green Heron
Black-crowned Night-Heron
Wood Duck (R)
Black Vulture
Turkey Vulture
Osprey
Hook-billed Kite
Double-toothed Kite
White-tailed Kite
Sharp-shinned Hawk
Cooper's Hawk
Common Black Hawk
Grey Hawk
Broad-winged Hawk
Short-tailed Hawk
White-tailed Hawk
Zone-tailed Hawk
Red-tailed Hawk
Laughing Falcon
Collared Forest-Falcon
American Kestrel
West Mexican Chachalaca
Long-tailed Wood-Partridge
Banded Quail
Common Moorhen
American Coot
Spotted Sandpiper
Inca Dove
Common Ground-Dove
White-tipped Dove
Orange-fronted Parakeet
Mexican Parrotlet
Lilac-crowned Parrot
Squirrel Cuckoo
Lesser Roadrunner
Groove-billed Ani
Barn Owl
Mottled Owl
Ferruginous Pygmy-Owl
Lesser Nighthawk
Eared Poorwill
Mexican Whip-poor-will†
White-collared Swift
Vaux's Swift
White-throated Swift
Great Swallow-tailed Swift
Golden-crowned Emerald

Broad-billed Hummingbird
Berylline Hummingbird
Cinnamon Hummingbird (R?)
Amethyst-throat Hummingbird (R?)
Lucifer Hummingbird
Ruby-throated Hummingbird
Costa's Hummingbird (R?)
Calliope Hummingbird
Rufous Hummingbird
Allen's Hummingbird
Elegant Trogon
Ringed Kingfisher
Belted Kingfisher
Green Kingfisher
Golden-cheeked Woodpecker
Yellow-bellied Sapsucker
Ladder-backed Woodpecker
Smoky-brown Woodpecker
Grey-crowned Woodpecker
Olivaceous Woodcreeper
Ivory-billed Woodcreeper
N. Beardless Tyrannulet
Greenish Elaenia
Greater Pewee
Common Tufted Flycatcher†
Olive-sided Flycatcher
Hammond's Flycatcher
Dusky Flycatcher
Western Flycatcher†
Buff-breasted Flycatcher
Eastern Phoebe (R)
Vermilion Flycatcher
Bright-rumped Attila
Dusky-capped Flycatcher
Ash-throated Flycatcher
Nutting's Flycatcher
Brown-crested Flycatcher
Great Kiskadee
Social Flycatcher
Boat-billed Flycatcher
Tropical Kingbird
Cassin's Kingbird
Thick-billed Kingbird
Western Kingbird
Masked Tityra
Grey-collared Becard
Rose-throated Becard
Tree Swallow
Violet-green Swallow
N. Rough-winged Swallow
Barn Swallow
Green Jay
Northern Raven†
Spotted Wren
Happy Wren
Sinaloa Wren
Northern House Wren†
Ruby-crowned Kinglet
Blue-grey Gnatcatcher
Brown-backed Solitaire
Eastern Bluebird
Swainson's Thrush
White-throated Thrush†
Rufous-backed Thrush†
Orange-billed N.-Thrush
Northern Mockingbird
Blue Mockingbird
Cedar Waxwing
Grey Silky†
Loggerhead Shrike
Slaty Vireo
Black-capped Vireo
Dwarf Vireo
Cassin's Vireo†
Plumbeous Vireo†
Golden Vireo
Warbling Vireo
Orange-crowned Warbler
Nashville Warbler
Yellow Warbler
Audubon's Warbler†
Black-throated Grey Warbler
Townsend's Warbler
Black-throated Green Warbler
Black-and-white Warbler
American Redstart
Ovenbird
Northern Waterthrush
Louisiana Waterthrush
MacGillivray's Warbler
Grey-crowned Yellowthroat
Wilson's Warbler
Painted Whitestart†
Slate-throated Whitestart†
Fan-tailed Warbler
Rufous-capped Warbler
Yellow-breasted Chat
Godman's Euphonia†
Blue-hooded Euphonia
Red-crowned Ant-Tanager
Hepatic Tanager
Summer Tanager
Western Tanager
Flame-colored Tanager
Red-headed Tanager
Greyish Saltator
Yellow Grosbeak
Rose-breasted Grosbeak
Black-headed Grosbeak
Blue Grosbeak
Lazuli Bunting
Indigo Bunting
Varied Bunting
Painted Bunting
Rusty-crowned Ground-Sparrow
Blue-black Grassquit
Cinnamon-rumped Seedeater†
Slate-blue Seedeater†
Stripe-headed Sparrow
Rusty Sparrow
Botteri's Sparrow
Chipping Sparrow
Lark Sparrow
Grasshopper Sparrow
Lincoln's Sparrow
Great-tailed Grackle
Bronzed Cowbird
Hooded Oriole
Black-vented Oriole
Dickey's Oriole†
Streak-backed Oriole
Bullock's Oriole
Lesser Goldfinch

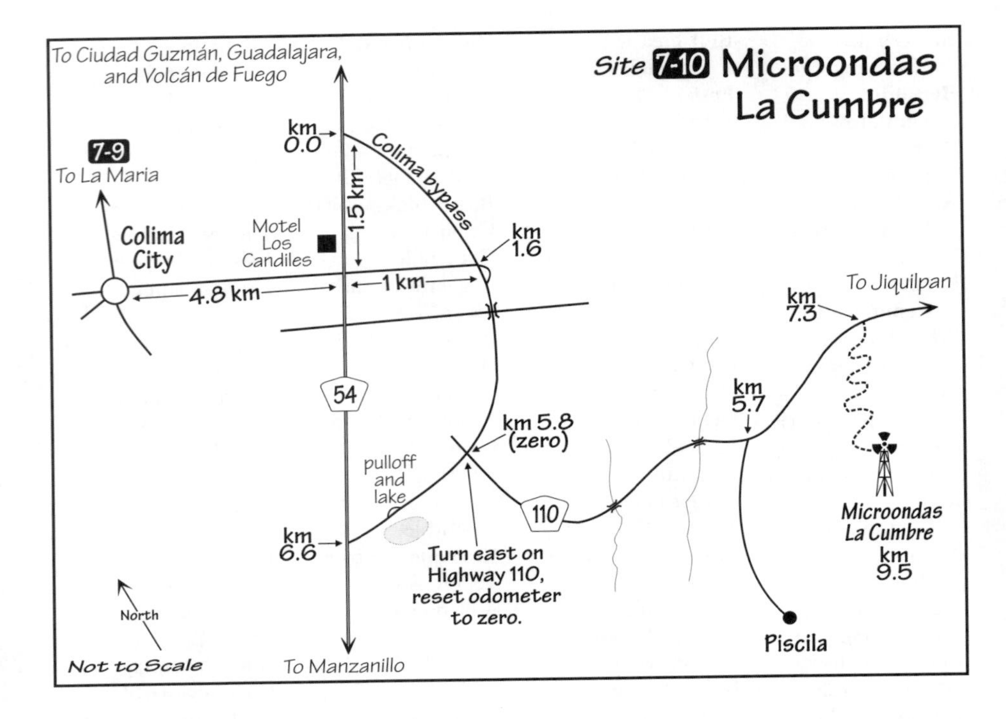

Site 7.10: Microondas La Cumbre, Colima

(April 1988, March 1989, 1990, 1992, 1993, 1994, March, December 1995, February 1997)

Introduction/Key Species

This site is off Route 110, the Jiquilpan highway, only 15–20 minutes out of Ciudad Colima, whence it makes a good late-afternoon/evening trip. Birds here include **Banded Quail**, **Lesser Roadrunner**, Lesser Ground-Cuckoo, **Balsas Screech-Owl**, **Colima Pygmy-Owl**, Buff-collared Nightjar, **Orange-breasted Bunting**, and **Black-chested Sparrow**.

Access/Birding Sites/Bird Lists

The well-signed highway to Jiquilpan (Route 110) leaves from the southeast side of the Ciudad Colima bypass (which skirts to the east of the city), at a four-way stop junction that was reinforced recently with traffic lights. This crossroads is 800 m north of where the bypass joins the highway on the south side of Colima (the roadside lake on this short stretch of bypass, easily viewed from a large, safe pullout, can be worth checking and often has Black-bellied Whistling-Ducks) and 5.8 km from where the bypass merges with the highway on the north side of Colima. Mark zero as you turn onto the Jiquilpan highway, and at Km 5.7, some 500 m after crossing a bridge across a permanent river (Ringed Kingfisher), look for the paved right turn signed for Piscila. Turning right here greatly reduces traffic, but there is still enough to warrant caution while birding along the road. Birding along the first 1–2 km of this side road (which ends in the small town of Piscila) can be good (**Banded Quail**, Lesser Ground-Cuckoo, **Lesser Roadrunner**, Broad-billed and Violet-crowned hummingbirds, White-throated Magpie-Jay, Nashville and Virginia's warblers [winter], Varied and **Orange-breasted buntings**, **Black-chested** and Stripe-headed **sparrows**, Streak-backed Oriole). The hillside to your left is covered with thorn forest, and it may be possible to see the *microondas* towers and a statue of the Virgin on top.

Continuing on Route 110 past the Piscila turnoff, you come to the signed road up to the Microondas La Cumbre on the right at Km 7.3. Turn here and mark zero. It is 3.2 km to the top, whence you have a good view over the city of Colima. The cobbled road climbs, at times steeply, through increasingly cleared thorn forest, and it is possible to pull off at several points and bird along the road. A fairly level, recently cleared side track goes off to the right at Km 2.2 and can be worth walking. A good approach for owling is to wait at the top till dusk (and watch the bats leave their roost in the statue of the Virgin) and then work slowly back down the cobbled road to the highway: Buff-collared Nightjars often sit on the road, and **Balsas Screech-Owl** can be found perched beside the road.

Birds along this *microondas* road include Short-tailed and Zone-tailed (migrant?) hawks, **West Mexican Chachalaca**, Lesser Ground-Cuckoo, **Lesser Roadrunner**, **Balsas Screech-Owl**, **Colima Pygmy-Owl**, Buff-collared Nightjar, **Russet-crowned Motmot**, Rufous-naped Wren, **Red-breasted Chat**, **Orange-breasted Bunting**, **Black-chested Sparrow**, and a selection of more common/widespread thorn-forest species.

If you're using public transport, take a second-class bus toward Jiquilpan and get off at the Piscila junction and/or the base of the microwave tower road. If you're lucky, you might be able to hitch a ride to the top of La Cumbre, but traffic is generally light.

CHAPTER 8 CENTRAL MEXICO

Chestnut-sided Shrike-Vireo

This chapter covers areas that can be reached easily from Mexico City, although many can be combined into any itinerary if you are on a driving trip from the U.S.A. If you are staying in Mexico City and driving a car, I'll assume that you're not weak of heart and will be able to use a street map to get out of the city, from wherever you're staying, and onto the relevant roads to birding sites. An early start (before 7 A.M., at the latest) is best for birding and to avoid potential rush-hour traffic. Because traffic is so unpredictable (it can take anywhere from forty-five minutes to two hours to cross Mexico City), approximate driving times are given from the toll booth exits from Mexico City: toll booths exist on roads west to Toluca and Guadalajara; south to Cuernavaca and Acapulco; east to Puebla,

Veracruz, Oaxaca, and the southeast; northeast to Pachuca; and north to Querétaro, San Luis Potosí, and Monterrey.

Note/Warning: Some years ago Mexico City implemented a "One Day without a Car" policy, in an admirable attempt to cut down on traffic and smog. Thus, during Monday to Friday, from 5 A.M. to 10 P.M., private cars (taxis and business vehicles are exempt) have been prohibited from circulating within the city limits on a given day. This system is based on the last digit of the license plate: for example, if your license plate is LXV 512 (or 512 LXV) you can't drive on Thursday (numbers ending with 1 and 2). The numbering sequence is not, to my mind, logical (it was determined by a lottery!): 1, 2 = Thursday; 3, 4 = Wednesday; 5, 6 = Monday; 7, 8 = Tuesday; 9, 0 = Friday. If you're driving into Mexico City, signs on the highways will warn you, but many an unsuspecting tourist has not understood and consequently has been an easy victim of the traffic police, who have appeared, as if by magic, at 5 A.M. on all of the on-ramps of the main Mexico City highways.

This has all changed somewhat in the last two years. Now that leaded gasoline is being phased out and newer cars have better emissions control, most newer cars (including those rented from major companies) have a sticker certifying that they can circulate on all days. However, if you are driving your own car from outside Mexico City (and even though it probably would pass the new controls), if it does not have the required sticker you could still be pulled over and fined. The fine (as of January 1998) is legally thirty days' minimum salary—about 900 pesos (a little over $100 U.S.), but police will extort whatever they can, depending upon how good you are at bargaining. Then you will have to park your car and leave it, as they will not let you continue driving, and even if you did you'd soon be stopped again! If you leave before 5 A.M. heading to the west, beware that Toluca has the same system, which at least does use the same numbering system.

If you're *really* unlucky (and at which point you might want to be outside the city anyway!), there is also a "second day without a car" policy if pollution levels get over about 250 (i.e., 250 percent of the maximum World Health Organization limits). These days are rare, thankfully, but if in doubt you can ask at the toll booths entering Mexico City.

Birding sites are divided into day trips (some sites can be combined) and longer trips that are best done as an overnight (or longer) trip and can be scheduled around the day you can't drive in Mexico City.

Birds possible in and around Mexico City (on day trips) include Banded Quail, Spotted Rail, Colima Pygmy-Owl, Black (summer), White-naped, and Great Swallow-tailed swifts, Dusky, Amethyst-throated, and Bumblebee hummingbirds, Strickland's Woodpecker, White-striped Woodcreeper, Pileated, Pine, and White-throated flycatchers, American Dipper, Grey-barred, Spotted, and Boucard's wrens, Blue Mockingbird, Ocellated Thrasher, Slaty and Golden vireos, Chestnut-sided Shrike-Vireo, Colima (winter) and Red warblers, Hooded and Black-polled yellowthroats, Red-headed Tanager, Rufous-capped and Green-striped brushfinches, Bridled, Black-chested, Striped, and Sierra Madre sparrows, Abeille's Oriole, and Hooded Grosbeak.

Additional birds that are possible on longer, overnight trips include Barred Forest-Falcon, Bearded Wood-Partridge, White-faced Quail-Dove, White-fronted Swift (summer only?), Emerald Toucanet, Ruddy Foliage-gleaner, Tawny-throated Leaftosser, Strong-billed Woodcreeper, Scaled Antpitta, Azure-hooded and Unicolored jays, Slate-colored Solitaire, Black Thrush, and White-naped Brushfinch.

Suggested Itineraries

If you're overnighting in Mexico City, you may have time to visit Sites 8.1 and/or 8.2, but note that they're on the opposite side of the city from the airport; both are best for birding in the morning. If you have a full day for which you need to be in the city some of the time, visiting at least one of these sites should be quite feasible. If you have a full day and don't need to stay in the city, several good day trips are possible, as follow.

Day trips. An early start to La Cima (Site 8.4), where it'll be cold at this hour, can be followed by swinging across to Almoloya del Río (Site 8.8), and you could be back in the city by mid afternoon at the latest. Alternatively, La Cima can be birded after a morning at Coajomulco (Site 8.5), or Coajomulco and La Cima can be birded in the afternoon (although by this time it'll probably be windy and harder to find birds) after a morning at Cañon de Lobos (Site 8.6).

If you get an early start to Laguna San Felipe, there are some birding sites en route (covered in Site 8.7), and you could (depending on local conditions) swing back via Popocatépetl (Site 8.3), although this site is better for birds (and views) in the mornings.

Longer trips might include an overnight to Temascaltepec (Site 8.9), which also can be done as a full day trip, perhaps swinging by Almoloya del Río and/or the Volcán de Toluca (covered under Site 8.9) on the way.

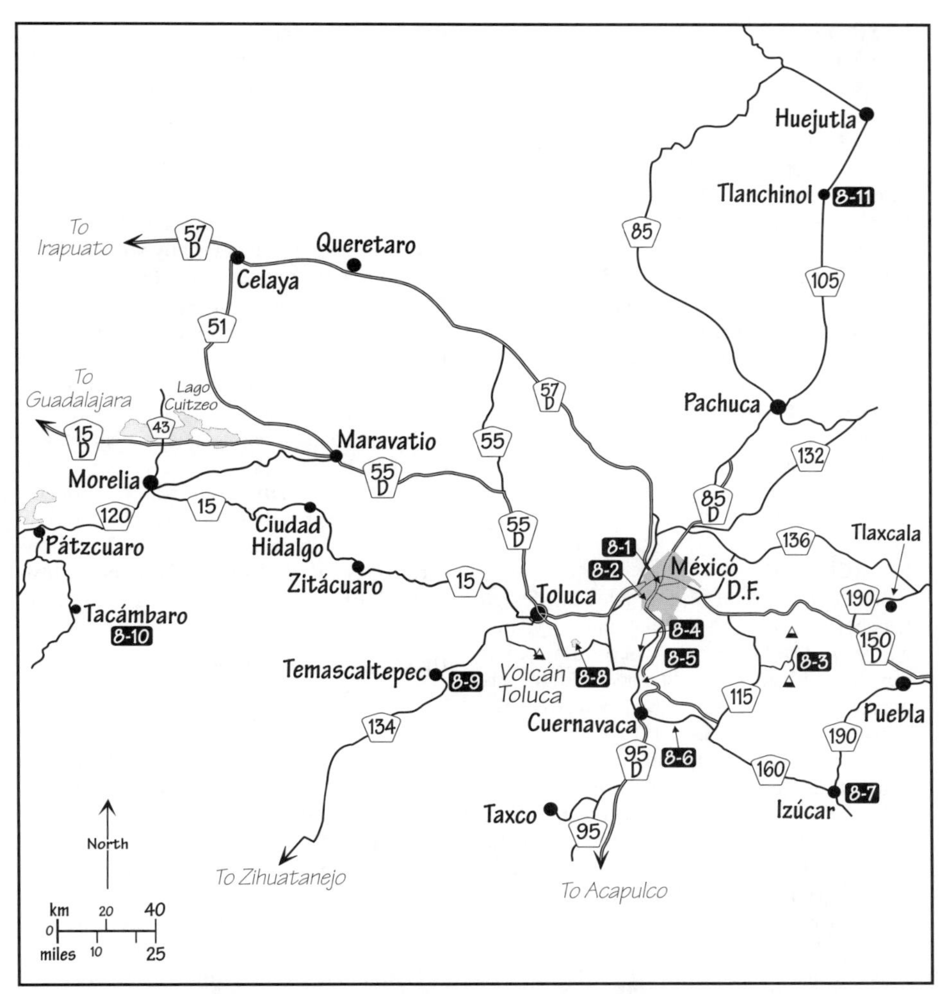

Chapter 8: CENTRAL MEXICO

8-1	UNAM Botanical Garden, D.F.
8-2	Bosque de Tlalpan (Pedregal), D.F.
8-3	Popocatepétl, México
8-4	La Cima, D.F.
8-5	Coajomulco, Morelos
8-6	Cañon de Lobos, Morelos
8-7	Laguna San Felipe, Puebla
8-8	Almoloya del Río (Lerma Marshes), México
8-9	Temascaltepec, México
8-10	Tacámbaro, Michoacán
8-11	Tlanchinol, Hidalgo

MEXICO CITY SITE LIST

Site 8.1 UNAM Botanical Garden, D.F.
Site 8.2 Bosque de Tlalpan (Pedregal), D.F.

DAY TRIPS

Site 8.3 Popocatépetl, México
Site 8.4 La Cima, D.F.
Site 8.5 Coajomulco, Morelos
Site 8.6 Cañon de Lobos, Morelos
Site 8.7 Laguna San Felipe, Puebla
Site 8.8 Almoloya del Río (Lerma Marshes), México
Site 8.9 Temascaltepec, México

OVERNIGHT TRIPS

Site 8.10 Tacámbaro (Lago de Cuitzeo), Michoacán
Site 8.11 Tlanchinol, Hidalgo

MEXICO CITY

There are several birding sites in Mexico City (elevation 2300 m) that can be visited easily in a few hours (traffic permitting!). For travelers with an overnight stopover at the airport, two sites are described here: the National University (UNAM) Botanical Garden and the Bosque de Tlalpan. Note, however, that both are on the opposite side of the city from the airport. Species of interest at these sites include Lucifer Hummingbird, **Blue Mockingbird**, **Ocellated Thrasher**, **Hooded Yellowthroat**, and **Abeille's Oriole**.

Much more extensive information on birds and birding sites in Mexico City and the adjacent areas can be found in the recommended second edition of *The Birds of Mexico City* (Wilson and Ceballos-L. 1993).

Site 8.1: UNAM Botanical Garden, D.F.

(March 1982, April 1985, May 1990, checked 1998)

Access/Birding Sites

To reach this site from downtown Mexico City, head south on Avenida Insurgentes for about 11.5 km from the intersection of Insurgentes and Paseo de la Reforma and you will see the University Stadium on your right and the main campus of the university on your left. Exit to the right here on a slip-road and drive in a loop around the south end of the stadium. After reaching the west side of the stadium, make a U-turn to the left and follow signs to the Jardin Botanico or Vivero Alto, both of which are in the far southwest of the university grounds. There is a parking lot at the entrance of the gardens, which are officially open to the public all week from 9 A.M. to 5 P.M., although usually you can get in around 8 A.M.; note, though, that access can be problematic on Sundays, when most gates to UNAM are closed. Public transport can be used to get here, but you'll have to ask how to do it, which depends upon where in the city you are staying.

Just beyond the entry gate is a cactus garden that, if the agaves are flowering, can be good for hummingbirds and orioles. From the gate, if you head to the far back left of the gardens, there is some natural arid oak scrub (**Ocellated Thrasher**, **Hooded Yellowthroat**). Elsewhere, numerous paths allow access through the gardens.

UNAM Botanical Garden Bird List

41 sp. R: rare (not to be expected). †See pp. 5–7 for notes on English names.

Inca Dove
Broad-billed Hummingbird
White-eared Hummingbird
Magnificent Hummingbird
Lucifer Hummingbird
Yellow-bellied Sapsucker
Ladder-backed Woodpecker
Greater Pewee
Cassin's Kingbird
Cliff Swallow
Western Scrub Jay
Bushtit
Canyon Wren
Bewick's Wren
Ruby-crowned Kinglet
Hermit Thrush
Rufous-backed Thrush†
American Robin
Blue Mockingbird
Ocellated Thrasher
Curve-billed Thrasher
Cedar Waxwing
Loggerhead Shrike
Hutton's Vireo
Warbling Vireo
Orange-crowned Warbler
Nashville Warbler
Black-and-white Warbler
MacGillivray's Warbler
Hooded Yellowthroat
Black-headed Grosbeak
Canyon Towhee
Rufous-crowned Sparrow
Chipping Sparrow

Black-chinned Sparrow
Bronzed Cowbird
Orchard Oriole (R)
Abeille's Oriole†
House Finch
Lesser Goldfinch
House Sparrow

Site 8.2: Bosque de Tlalpan (Pedregal), D.F.

(March 1982, January 1983, December 1986, May 1990, September 1995)

Access/Birding Sites

To reach this site from downtown Mexico City, head south on Avenida Insurgentes for about 14.5 km from the intersection of Insurgentes and Paseo de la Reforma and you will pass under the *periferico*. About 500 m after the *periferico* you go through a set of traffic signals; at the second set of lights after the *periferico*, turn right onto Camino a Sta Teresa and drive about 800 m, where you turn off left into a large parking lot (open from about 7 A.M. to 5 P.M.). Again, public transport can be used to get here, but you'll have to ask how to do it, which depends upon where in the city you are staying. Note that the entrance by the furniture and artisan market (described by Wilson and Ceballos 1993) is now walled off.

Walking almost anywhere on the rock roads that wend through the arid scrub can be good for birds (Lucifer Hummingbird, **Hooded Yellowthroat**, Varied Bunting, Black-chinned Sparrow). If you walk uphill from the parking lot and keep bearing left until you come to the end of a high stone wall, a road down to the left from here leads to Tenantongo, a picnic area with tall eucalyptus trees and a brush-covered hill, around which you can walk via narrow paths. This is a good spot for **Blue Mockingbird**, **Ocellated Thrasher**, **Hooded Yellowthroat**, **Rufous-capped Brushfinch**, and **Abeille's Oriole**. Another birding area can be reached from the main parking lot by walking uphill and keeping right, to some oak woodland in the southwest corner of the park, behind a small zoo.

Bosque de Tlalpan Bird List

55 sp. †See pp. 5–7 for notes on English names.

Sharp-shinned Hawk
Cooper's Hawk
Inca Dove
Broad-billed Hummingbird
White-eared Hummingbird
Magnificent Hummingbird
Lucifer Hummingbird
Ladder-backed Woodpecker
N. Beardless Tyrannulet
Greater Pewee
Western Pewee†
Olive-sided Flycatcher
Hammond's Flycatcher
Dusky Flycatcher
Cassin's Kingbird
Barn Swallow
Western Scrub Jay
Bushtit
Canyon Wren
Bewick's Wren
Ruby-crowned Kinglet
Blue-grey Gnatcatcher
Rufous-backed Thrush†
American Robin
Blue Mockingbird
Ocellated Thrasher
Curve-billed Thrasher
Loggerhead Shrike
Hutton's Vireo
Orange-crowned Warbler
Nashville Warbler
Audubon's Warbler†
Black-throated Grey Warbler
Townsend's Warbler
Black-and-white Warbler
Ovenbird
MacGillivray's Warbler
Hooded Yellowthroat
Wilson's Warbler
Rufous-capped Warbler
Western Tanager
Black-headed Grosbeak
Varied Bunting
Rufous-capped Brushfinch
Spotted Towhee
Canyon Towhee
Rufous-crowned Sparrow
Chipping Sparrow
Black-chinned Sparrow
Lincoln's Sparrow
Bronzed Cowbird
Bullock's Oriole
Abeille's Oriole†
House Finch
Pine Siskin
House Sparrow

DAY TRIPS

Site 8.3: Popocatépetl, México

(May 1990)

Introduction/Key Species

The transvolcanic belt of Mexico cuts east-west across the country at about the latitude of Mexico City, from the Volcánes de Colima in the west to Citlatépetl (Orizaba) in the east.

Two of the highest volcanoes in this chain, Popocatépetl and Iztaccihuatl, dominate the eastern horizon of Mexico City, assuming that the air is clear enough to see them! There is nothing here that can't be seen elsewhere, but this is a popular tourist destination that may be easier to visit than other sites as an outing with non-birding friends or family. The paved road allows easy access to high-elevation species (remember not to over-exert yourself—the air is thin) such as **Strickland's Woodpecker**, Pygmy Nuthatch, **Russet** and Ruddy-capped **nightingale-thrushes**, **Red Warbler**, and **Striped Sparrow**.

Note: Popocatépetl is presently (1998) closed due to volcanic eruptions, which have been going on for the past three years; check on the current status when you're in Mexico City; in January 1998 it was possible to get up to the pass described below.

Access

The pass (3600 m elevation) between the snow-covered peaks of Popocatépetl (5465 m) and Iztaccihuatl (5230 m) lies about 80 km by road (about an hour) southeast of Mexico City. To reach it, head out east on the toll highway (Route 190D) or old highway 190 toward Puebla and turn south, following signs for Chalco and Amecameca. About 2 km south of Amecameca center, look for a paved road to the left (east) that should be signed for Paso de Cortes. This is the road to the pass where the scenery, and the views on a clear day (mornings are best), are spectacular.

Birding Sites

The paved road to the pass goes through pine and fir forests, with bunch-grass clearings. One can stop anywhere it is safe to pull off and walk in the woods—there are several such places popular for picnics. Because of the high level of human traffic, it may be relatively easy to see the somewhat-acclimatized **Russet** and Ruddy-capped **nightingale-thrushes**; the latter, in particular, is hard to see in many other places. The bunch-grass clearings with scattered pines have **Striped Sparrow**, which is usually easy to see along the road, and these same open pine woods are good for **Strickland's Woodpecker**. Other species that have been found along the road include Mountain Pygmy-Owl, **Grey-barred Wren**, **Brown-backed Solitaire**, Golden-crowned Kinglet, **Golden-browed Warbler**, and **Abeille's Oriole**.

Note: This is a popular recreational spot, so it may be best not to visit for birds on weekends. On Mondays, however, you may find nightingale-thrushes and other birds looking for pickings at the picnic spots.

Popocatépetl Bird List

20 sp. †See pp. 5–7 for notes on English names.

Red-tailed Hawk
Strickland's Woodpecker
Olive-sided Flycatcher
Violet-green Swallow
Steller's Jay
Northern Raven†
Mexican Chickadee
Pygmy Nuthatch
Brown Creeper
Brown-throated Wren†
Russet N.-Thrush
Ruddy-capped N.-Thrush
American Robin
Red Warbler
Olive Warbler
Striped Sparrow
Chipping Sparrow
Mexican Junco†
Red Crossbill
Pine Siskin

Site 8.4: La Cima, D.F.

(March 1982, January 1983, September 1984, April 1985, December 1986, May/June 1990, July 1991, October 1993, September 1995, January 1998)

Introduction/Key Species

The **Sierra Madre Sparrow**, a Mexican endemic and a denizen of high-elevation bunch grass, is found readily at La Cima but seemingly nowhere else! Unfortunately, this site has no official protection, but the underlying volcanic rocks have kept it from being converted entirely to fields. Other species here include **Strickland's Woodpecker** and **Striped Sparrow**.

Access

La Cima (3000 m elevation) is about 25 km (20–30 minutes) south of the southern edge of Mexico City along the old highway to Cuernavaca (Route 95, signed "Cuernavaca Libre" as you leave Mexico City), not the toll highway (Route 95D, signed "Cuernavaca Cuota"). Set to zero at the fork of *cuota* and *libre* roads, after which the old highway winds up out of the valley and across open, arable fields with scattered pines. At 23.3 km you go under a railway bridge; in about another kilometer, at 24.5 km (Km Post 43.5) look for a dirt road to the right, which may be signed to La Cima. Turn right here and park off the road after about 300–400 m, amid a mosaic of ploughed fields and bunch-grass rises with scattered pines. Remember not to over-exert yourself at this elevation.

Birding Sites

In spring and autumn, **Sierra Madre Sparrows** can be easy to see (when singing or with young), and you may hear them singing before you get out of the car. At other times, such as in winter, or in summer when they are incubating, they can be hard to find, or at least to see well. If they're not immediately conspicuous, you'll probably need to do some walking: beware the very rough terrain of volcanic rocks under the bunch-grass—it's very easy to fall. It is best to start by scanning the edges of the fields where they meet the bunch grass; often the sparrows feed along these edges. If you have no luck here, pish loudly, and sparrows may perch up on grass stalks, or at least call—a quiet nasal note strongly reminiscent of Song Sparrow (which is not here). Still no luck? Set out to walk the bunch grass, first around the edges of the fields and then, if you're still unlucky, more randomly across the slopes, always scanning for birds sitting up on prominent stalks. I've never failed to see the species at La Cima, but it could be missed if you are unlucky and/or visiting the site on a windy afternoon; early to mid morning is the best time.

Return to the highway and turn right (south) for 1.5 km to where another narrow dirt road goes off to the right (west) into a mixture of pine-oak woods, fields, and brushy thickets. You can drive in and park easily about 150–200 m from the highway. Birds here include **Strickland's Woodpecker**, Buff-breasted Flycatcher, Blue-hooded Euphonia (especially at mistletoe clumps), and **Rufous-capped Brushfinch**. Evening Grosbeak (a good bird to see in Mexico) has also been found at this spot.

La Cima Bird List

50 sp. R: rare (not to be expected). †See pp. 5–7 for notes on English names.

Turkey Vulture
Red-tailed Hawk
American Kestrel
Lesser Roadrunner (R)
Broad-tailed Hummingbird
Strickland's Woodpecker
Red-shafted Flicker†
Greater Pewee
Dusky Flycatcher
Buff-breasted Flycatcher
Violet-green Swallow
N. Rough-winged Swallow
Barn Swallow
Steller's Jay
Bushtit
Mexican Chickadee
White-breasted Nuthatch
Pygmy Nuthatch
Brown Creeper
Grey-barred Wren
Brown-throated Wren†
Sedge Wren (R)
Ruby-crowned Kinglet
Blue-grey Gnatcatcher
Western Bluebird
American Robin
Curve-billed Thrasher
American Pipit
Hutton's Vireo
Audubon's Warbler†
Townsend's Warbler
Hermit Warbler
Black-and-white Warbler
MacGillivray's Warbler
Wilson's Warbler
Red Warbler
Slate-throated Whitestart†
Olive Warbler
Blue-hooded Euphonia
Rufous-capped Brushfinch
Canyon Towhee
Spotted Towhee
Striped Sparrow
Chipping Sparrow
Sierra Madre Sparrow
Mexican Junco†
Eastern Meadowlark
Abeille's Oriole†
House Finch
Pine Siskin

Site 8.5: Coajomulco, Morelos

(March 1982, January 1983, January 1984, April 1985, December 1986, checked 1997)

Introduction/Key Species

The relatively humid pine-oak forest near Coajomulco lies in the narrow transition belt between the more open, pine-dominated forest at higher elevation, and drier oaks and arid scrub lower down, nearer Cuernavaca. Species of interest here include **Long-tailed Wood-Partridge**, **White-striped Woodcreeper**, **Pine Flycatcher**, **Grey-**

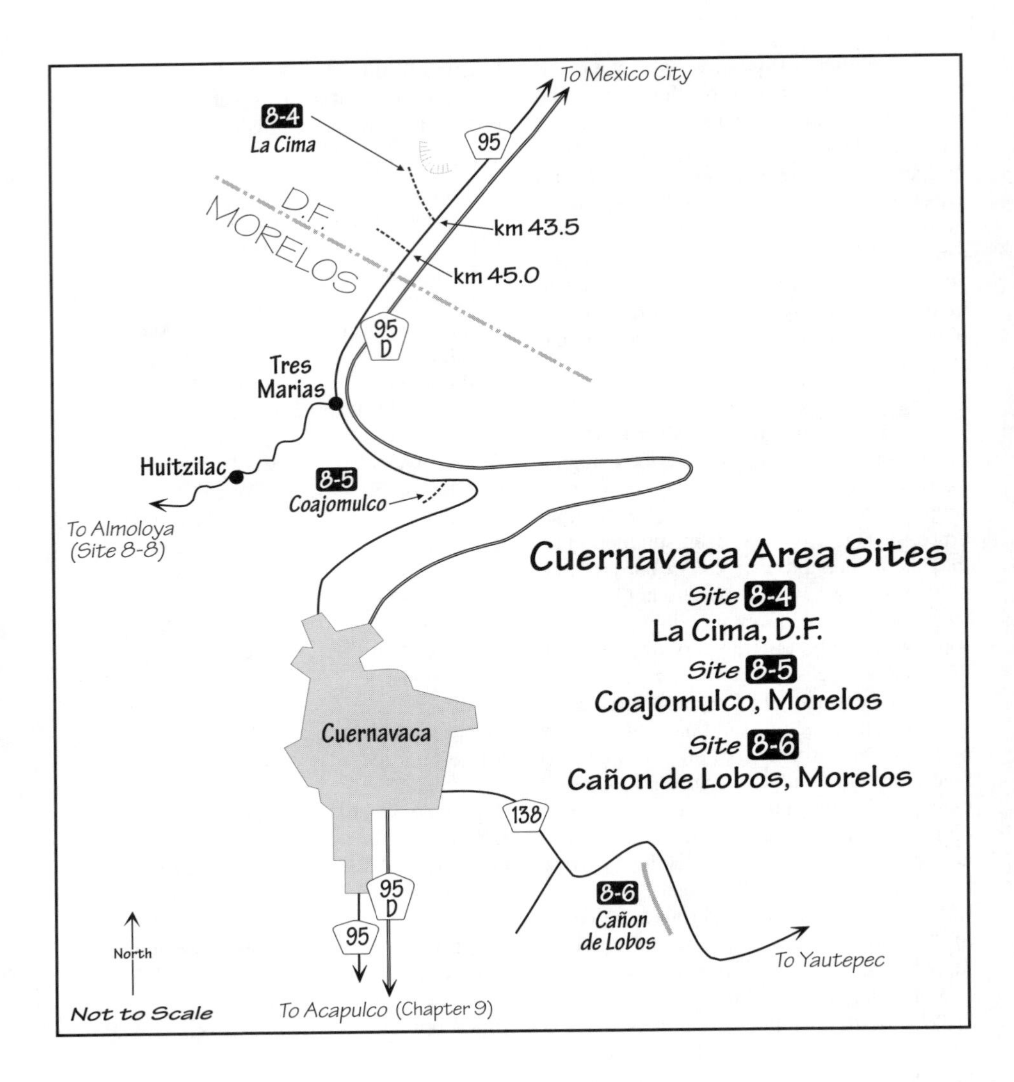

barred Wren, nightingale-thrushes, **Chestnut-sided Shrike-Vireo**, Colima Warbler (winter), **Green-striped Brushfinch**, and **Cinnamon-bellied Flowerpiercer**.

Access

Coajomulco (2600 m elevation) is about 60 km (1–1.2 hours) south of Mexico City off the old highway (Route 95) to Cuernavaca. To get there, follow the directions for La Cima and continue south on the highway, through the town of Tres Marias (numerous roadside food stalls about 7 km south of La Cima), and on for another 5 km (the highway begins to drop into humid pine-oak forest) to a left-hand turn (northeast) signed (with luck) for Coajomulco. Stay on the highway about another 100 m and look for a place to pull off on the right (south). This is between Km Posts 56 and 57. You can park here, as usual leaving nothing of value visible in the car, and walk down trails to the right into the forest, where birding can be good throughout the day. This site is also about 20 km (20–30 minutes) north of Cuernavaca.

Birding Sites

From the parking spot you can wander down and around on any of the numerous paths through the forest—you may shortly meet up with the highway again as it bends back to the right. Sooner or later you should come upon a mixed-species feeding flock, with numerous warblers. Such flocks should be checked for **Chestnut-sided Shrike-Vireo**, which is sluggish and easy to miss if it isn't singing. If the weedy and brushy flower banks have hummingbirds, they'll probably also have **Cinnamon-bellied Flowerpiercer**, and they can be good for Colima Warbler (winter) and **Green-striped Brushfinch**.

Coajomulco Bird List

78 sp. R: rare; V: vagrant (not to be expected). †See pp. 5–7 for notes on English names.

Sharp-shinned Hawk
American Kestrel
Long-tailed Wood-Partridge (R)
White-naped Swift
Green Violet-ear
White-eared Hummingbird
Blue-throated Hummingbird
Magnificent Hummingbird
Acorn Woodpecker
Yellow-bellied Sapsucker
Hairy Woodpecker
Red-shafted Flicker†
White-striped Woodcreeper
Greenish Elaenia
Greater Pewee
Common Tufted Flycatcher†
Hammond's Flycatcher
Pine Flycatcher
Western Flycatcher†
Buff-breasted Flycatcher
Dusky-capped Flycatcher
Rose-throated Becard
Steller's Jay
Grey-breasted Jay†
Mexican Chickadee
Bushtit
White-breasted Nuthatch
Brown Creeper
Grey-barred Wren
Brown-throated Wren†
Ruby-crowned Kinglet
Brown-backed Solitaire
Orange-billed N.-Thrush
Russet N.-Thrush
Ruddy-capped N.-Thrush
Black Thrush† (V)
White-throated Thrush†
American Robin
Blue Mockingbird
Grey Silky†
Cassin's Vireo†
Hutton's Vireo
Warbling Vireo
Chestnut-sided Shrike-Vireo
Nashville Warbler
Orange-crowned Warbler
Colima Warbler
Crescent-chested Warbler
Audubon's Warbler†
Townsend's Warbler
Hermit Warbler
Grace's Warbler
Black-and-white Warbler
MacGillivray's Warbler
Wilson's Warbler
Red-faced Warbler
Red Warbler
Painted Whitestart†
Slate-throated Whitestart†
Olive Warbler
Blue-hooded Euphonia
Hepatic Tanager
Western Tanager
Black-headed Grosbeak
Varied Bunting
Rufous-capped Brushfinch
Green-striped Brushfinch
Spotted Towhee
Cinnamon-bellied Flowerpiercer
Chipping Sparrow
Mexican Junco†
Bullock's Oriole
Abeille's Oriole†
Scott's Oriole
Red Crossbill
Pine Siskin
Black-headed Siskin
Lesser Goldfinch

Site 8.6: Cañon de Lobos, Morelos

(March 1982, January 1983, June 1990, June/July 1991, checked December 1997)

Introduction/Key Species

Cañon de Lobos (1200–1400 m elevation) is an area of brushy, tropical thorn forest and woodland at the upper reaches of the Río Balsas drainage. Consequently, one finds here a number of bird species characteristic of Mexico's Pacific Slope and arid southwest interior. A sporadic and easy-to-miss speciality of Cañon de Lobos is the enigmatic **Slate-blue Seedeater**, whose occurrence and abundance appear to be tied to seeding bamboo. Other species of interest are **Banded Quail**, Lesser Ground-Cuckoo, **Balsas Screech-Owl**, **Colima Pygmy-Owl**, **Golden-crowned Emerald**, **Dusky Hummingbird**, **Russet-crowned Motmot**, **Pileated Flycatcher**, **Golden Vireo**, **Rusty-crowned Ground-Sparrow**, and **Black-chested Sparrow**.

Access

Cañon de Lobos is about 110 km (1.5–2 hours) south of Mexico City, and about 15–20 km (15 minutes) southeast of Cuernavaca. From Cuernavaca head southeast on Route 138 toward Yautepec. After about 14 km you go through the town of Amador Salazar, shortly after which the road winds down through Cañon de Lobos, not really a well-marked canyon but more of a valley with wooded hillsides. While you could park beside the highway about 1 km beyond Amador Salazar, where the highway is nearest the canyon, cars have been broken into here and also have been towed by the highway police. Thus, it is best to find somewhere to park safely in Amador Salazar (preferably, arrange personally with someone to watch your car) and take a taxi the short distance down to the canyon. The taxi can wait for you, or you can arrange a time to be picked up again. It is best to be here early in the morning, as the weather tends to get hot, and bird activity wanes, by mid to late morning.

Birding Sites

The best paths lead off the highway about 1 km after Amador Salazar, at the start of the metal security barrier on the right of the highway (eastbound lane), just before the left-hand bend where the road levels out somewhat after a fairly steep descent. Paths lead to clearings adjacent to the thorn forest (**Banded Quail**, **Black-chested Sparrow**), and you can also follow cattle trails (some of them steep and narrow) up one or two of the valleys into the thorn forest (**Balsas Screech-Owl**, **Colima Pygmy-Owl**, **Pileated Flycatcher**) in search of patches of seeding bamboo (**Slate-blue Seedeater**).

Given the increased clearing and relative difficulty of access, this site may not be a high priority except for Pileated Flycatcher and Slate-blue Seedeater, both of which are best found in summer, when singing.

Cañon de Lobos Bird List

76 sp. †See pp 5–7 for notes on English names.

Black Vulture
Turkey Vulture
Grey Hawk
Red-tailed Hawk
American Kestrel
West Mexican Chachalaca
Banded Quail
Inca Dove
Common Ground-Dove
White-tipped Dove
Squirrel Cuckoo
Lesser Ground-Cuckoo
Groove-billed Ani
Balsas Screech-Owl
Colima Pygmy-Owl†
White-naped Swift
Golden-crowned Emerald
Dusky Hummingbird
Broad-billed Hummingbird
Berylline Hummingbird
Violet-crowned Hummingbird
Lucifer Hummingbird
Elegant Trogon
Russet-crowned Motmot
Golden-cheeked Woodpecker
Ladder-backed Woodpecker
N. Beardless Tyrannulet
Greenish Elaenia
Western Pewee†
Pileated Flycatcher
Dusky Flycatcher
Vermilion Flycatcher
Dusky-capped Flycatcher
Nutting's Flycatcher
Brown-crested Flycatcher
Sulphur-bellied Flycatcher
Cassin's Kingbird
Thick-billed Kingbird
Rose-throated Becard
White-throated Magpie-Jay
Canyon Wren
Happy Wren
Banded Wren
Northern House Wren†
Blue-grey Gnatcatcher
Rufous-backed Thrush†
Blue Mockingbird
Curve-billed Thrasher
Loggerhead Shrike
Solitary Vireo ssp.
Golden Vireo
Warbling Vireo
Yellow-green Vireo
Orange-crowned Warbler
Virginia's Warbler
Audubon's Warbler†
Black-and-white Warbler
Ovenbird
MacGillivray's Warbler
Grey-crowned Yellowthroat
Wilson's Warbler
Slate-throated Whitestart†
Rufous-capped Warbler
Black-headed Grosbeak
Yellow Grosbeak
Varied Bunting
Rusty-crowned Ground-Sparrow
Blue-black Grassquit
Cinnamon-rumped Seedeater†
Slate-blue Seedeater†
Black-chested Sparrow
Stripe-headed Sparrow
Rusty Sparrow
Bronzed Cowbird
Streak-backed Oriole
Lesser Goldfinch

Site 8.7: Laguna San Felipe, Puebla

(March, May 1990, January 1998)

Introduction/Key Species

The area around Izúcar de Matamoros lies in the generally arid upper reaches of the Río Balsas drainage, such that numerous Pacific Slope bird species range this far inland. A number of birds endemic to interior southwest Mexico can also be found here. Most of the country in the vicinity of the natural freshwater lake of Laguna San Felipe (also known as Laguna Epatlán) is arid and scrubby, with some agriculture, notably the extensive canefields around Izúcar de Matamoros.

Birds en route from Cuautla to Izúcar (see below), some of which are also found around the lake, include Harris' and White-tailed hawks, **Banded Quail**, **Dusky** and Violet-crowned **hummingbirds**, Plain-capped Starthroat, **Grey-breasted Woodpecker**, Thick-billed Kingbird, **Boucard's Wren**, Yellow Grosbeak, **Rusty-crowned Ground-Sparrow**, Grassland Yellow-Finch (sporadic in canefields), and **Bridled** and **Black-chested sparrows**. Species at the lake include Least Bittern (common, nesting), Spotted Rail, and **Bicolored** [Red-winged] **Blackbird**.

Access

Laguna San Felipe (1400 m elevation) is about 170 km (2.5–3 hours) southeast of Mexico City. It can be reached by taking the toll highway (Route 95D) toward Cuernavaca and turning off on a minor toll highway to Cuautla, whence you take Route 160 to the southeast for about 60 km to Izúcar de Matamoros. The lake is about 15 km northeast of Izúcar de Matamoros. If you are starting from Cuernavaca, you can reach this site easily by taking Route 138, via Yautepec, to Cuautla and on to Izúcar. See Birding Sites for directions to the lake.

Obviously, early morning is best, but bird activity at the lake can be good even at midday. In addition, a couple of places are worth stopping en route to or from the lake (see Birding Sites).

Birding Sites

Mark zero on the east side of Cuautla at the junction of Route 160 and the northern bypass to Mexico City. At about 18 km look for a crossroads shortly after a Pemex station on the left (north). This should be signed left (north) for Amayuca (which is by the highway) and/or Tlacotepec (the road opposite goes south to Jonacatepec). (There is a left turn signed for Amayuca and Tlacotepec about 2.5 km before the crossroads, which gets you to the same place in Amayuca.) At the main junction in Amayuca mark zero and continue north toward Tlacotepec, through Temoac (Km 6.5) and on to a fork in the road at Km 10.9. Turn left here and continue up another 4.5 km (15.4 km from Amayuca) to a left-hand bend in the road, just after which you can park safely on the left.

The arid oak scrub here can be birded by walking back down the road for a few hundred meters and/or taking a path off to the east (on your left as you walk back down from the parking spot). Birds found here (March 1990) include **Banded Quail**, **White-naped Swift**, **Dusky**, **White-eared**, **Berylline**, and Violet-crowned **hummingbirds**, Northern Beardless Tyrannulet, Greenish Elaenia, Bushtit, **Boucard's**, **Happy**, and Bewick's **wrens**, Orange-billed Nightingale-Thrush, **Blue Mockingbird**, **Rusty-crowned Ground-Sparrow**, **Bridled**, Rufous-crowned, and Rusty **sparrows**, and **Black-vented Oriole**. Migrants include Dusky Flycatcher, Orange-crowned, Nashville, Virginia's, Black-throated Grey, and MacGillivray's warblers, Green-tailed Towhee, Chipping, Clay-colored, Lark, Grasshopper, and Lincoln's sparrows.

Back on Route 160 reset to zero at the Amayuca junction and continue toward Izúcar. At 7.6 km (25.7 km from Cuautla) you cross the state line from Morelos into Puebla, marked by a canyon that often has water flowing through it. About 200 m after the bridge there is room to park safely on the left (north). Farther east on the highway, at 15–18 km (33–36 km from Cuautla) and shortly after the village of Tepexco, you run alongside a thorn-forested hill to the right (south); there are a few places to pull off the highway here to park, and trails lead at least a short distance into the habitat. Birds at these sites include Squirrel Cuckoo, Ferruginous Pygmy-Owl, **Dusky** and Violet-crowned **hummingbirds**, Plain-capped Starthroat, Elegant Trogon, **Grey-breasted Woodpecker**, Northern Beardless Tyrannulet, Nutting's and Brown-crested flycatchers, Canyon and Banded wrens, **Golden Vireo**, Yellow Grosbeak, Varied Bunting, **Rusty-crowned Ground-Sparrow**, **Bridled** and **Black-chested sparrows** (sympatric at the hillside), and Streak-backed Oriole.

To get to Laguna San Felipe, stay on the

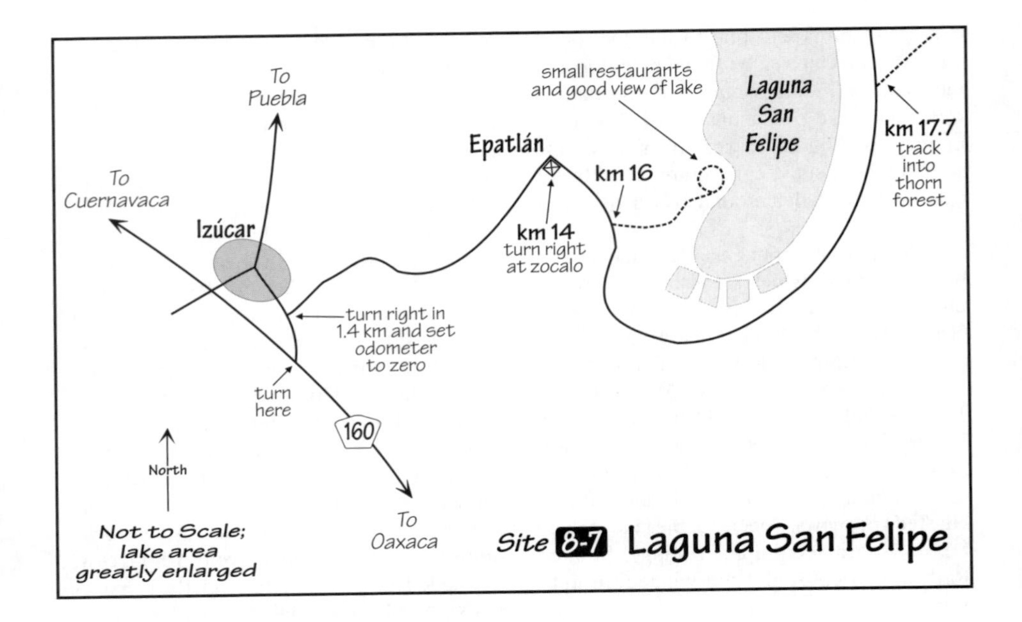

highway toward Izúcar de Matamoros, which lies off the highway to the left (north); at 39.5 km (57.6 km from Cuautla), continue *past* the main turn into Izúcar (also signed Route 190 to Puebla), and at 43 km (61.1 km), turn back sharp left (signed to Izúcar) where the highway bends gently right and is signed as going on to Oaxaca. Thus, you turn left on the entry road into Izúcar that you would take if coming from Oaxaca. Reset to zero and, at 1.4 km, turn right on the road signed (conspicuously) to Tepexi and (less obviously) to Epatlán. Continue on this for a *tope*-ridden 14 km (into Epatlán) and, just past the *zócalo*, turn sharp right, staying on the main road (signed "Huehuetlan 32"). In about 200 m fork right, staying on the main road; in another 2 km the paved road bends right, and a dirt road forks off to the left (signed to Laguna Epatlán). Turn left here, and in about 1 km the lake is on your right. If a rusty gate is closed across the road where the lake starts you should be able to push it open and continue to where the dirt road ends in 250 m by a few restaurants overlooking the lake (check the columnar cacti here for **Grey-breasted Woodpecker**). You can park here and bird on foot along the edge of the lake; the best area is perhaps to the right, where Spotted Rail has been seen along the edge of the reeds (exposed mud varies considerably with changing water levels).

If you stay on the paved road rather than turning to the restaurants, the road bends around the south end of the lake by some small impoundments and more good reed edge for rails and bitterns. At 1.7 km past the restaurant junction there is a rough dirt road to the right that leads up between the overgrazed, thorn-forest hillsides. You can park off the road here and bird the dirt road on foot (**Banded Quail**, Lesser Ground-Cuckoo, **Dusky** and Lucifer/**Beautiful hummingbirds**, **Boucard's Wren**).

Laguna San Felipe Bird List

100 sp. R: rare (note to be expected). †See pp. 5–7 for notes on English names.

Pied-billed Grebe
Eared Grebe
American Bittern
Least Bittern
Great Blue Heron
Great Egret
Snowy Egret
Little Blue Heron
Tricolored Heron
Cattle Egret
Green Heron
Black-crowned Night-Heron
White-faced Ibis
Green-winged Teal
Blue-winged Teal
Cinnamon Teal
Northern Shoveler
Gadwall
Ring-necked Duck
Lesser Scaup
Ruddy Duck

Black Vulture
Turkey Vulture
American Kestrel
Banded Quail
Virginia Rail
Spotted Rail
Sora
Common Moorhen
American Coot
Lesser Yellowlegs
Solitary Sandpiper
Spotted Sandpiper
Western Sandpiper
Least Sandpiper
Baird's Sandpiper
Pectoral Sandpiper
Long-billed Dowitcher
Common Snipe
White-winged Dove
Inca Dove
Common Ground-Dove
Squirrel Cuckoo
Lesser Ground-Cuckoo
Groove-billed Ani
Dusky Hummingbird
Violet-crowned Hummingbird
Lucifer/**Beautiful** Hummingbird
Russet-crowned Motmot
Belted Kingfisher
Green Kingfisher
Grey-breasted Woodpecker
N. Beardless Tyrannulet
Willow Flycatcher
White-throated Flycatcher
Least Flycatcher
Dusky Flycatcher
Eastern Phoebe (R)
Vermilion Flycatcher
Brown-crested Flycatcher
Great Kiskadee
Social Flycatcher
Tropical Kingbird
Thick-billed Kingbird
Cassin's Kingbird
N. Rough-winged Swallow
Barn Swallow
Boucard's Wren
Northern House Wren†
Marsh Wren
Blue-grey Gnatcatcher
Rufous-backed Thrush†
Northern Mockingbird
Curve-billed Thrasher
Loggerhead Shrike
European Starling (R?)
Orange-crowned Warbler
Virginia's Warbler
Yellow Warbler
Audubon's Warbler†
MacGillivray's Warbler
Common Yellowthroat
Wilson's Warbler
Varied Bunting
Cinnamon-rumped Seedeater†
Stripe-headed Sparrow
Clay-colored Sparrow
Lark Sparrow
Savannah Sparrow
Lincoln's Sparrow
Swamp Sparrow
Red-winged Blackbird
Bicolored Blackbird†
Great-tailed Grackle
Bronzed Cowbird
Brown-headed Cowbird
Hooded Oriole
Black-vented Oriole
Streak-backed Oriole
House Finch
Lesser Goldfinch

Site 8.8: Almoloya del Río (Lerma Marshes), México

(September 1984, December 1986, July 1988, October 1993, September 1995, February 1998)

Introduction/Key Species

The once-extensive marshes in the Lerma Valley formerly supported an interesting resident avifauna as well as large numbers of migrant waterfowl in winter. Expanding human settlement has drained and "reclaimed" all but a few areas of marsh, such as at Almoloya del Río, which now stand as isolated and still-shrinking remnants. With the extensive marshes went the **Slender-billed Grackle**, a Mexican endemic known only from the Lerma Valley and last recorded in 1910. An isolated and resident population of Yellow Rail, also endemic to the Lerma Valley, may yet survive, but no birds have been reported since 1964 . . .

One species still found in the Lerma Valley is **Black-polled Yellowthroat**, which is common at Almoloya. Other birds of interest here include King and Virginia rails and **Bicolored** [Red-winged] **Blackbird**. If you visit on weekends (at least in winter), be aware that the area is popular with duck hunters, so disturbance of waterfowl is heavy. This shouldn't affect finding the yellowthroat, however.

Access/Birding Sites

Almoloya del Río (elevation 2500 m) is a small town in the Lerma Valley (also known as the Valley of Toluca), about 60 km (an hour) southwest of Mexico City. The Lerma Valley comprises a maze of roads which, it appears, no map yet produced has managed to portray accurately, and there is good potential for getting lost in the area. Thus, while the routes described below may not be the most direct, they should get you to and from the site. The fastest route from Mexico City is via Toluca, while another recommended route is from Tres Marias, after you have birded the La Cima area (Site 8.4).

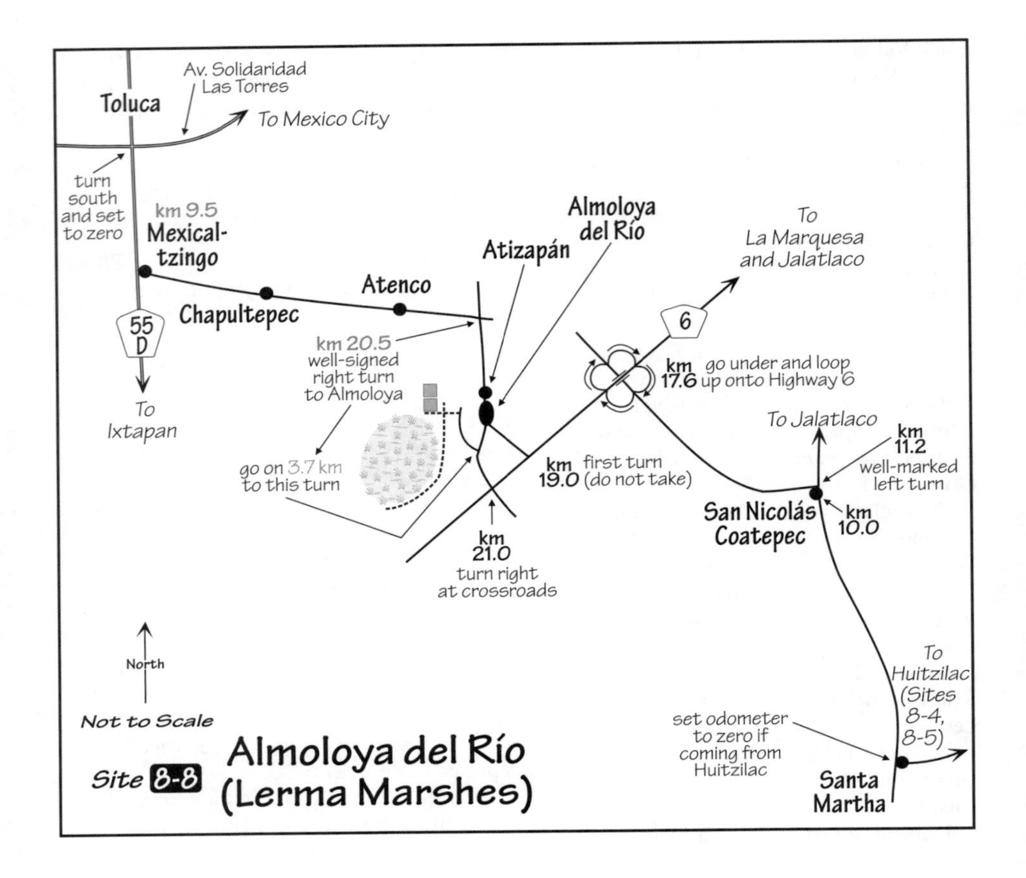

The fastest route is to take the toll highway (Route 15D) west from Mexico City to Toluca and thence take Route 55 south (signed for Ixtapan and Taxco). At 8.5 km south of the junction with Avenida Solidaridad Las Torres, which is the main (six-lane, divided) road along the south side of Toluca, turn left at Mexicaltzingo and reset to zero. Continue east on the highway (possibly signed to Tianguistenco and/or Jalatlaco) for 12 km and you come to a right turn that should be signed to Almoloya del Río. Turn right here, stay on the main road through the more or less continuous "urban" areas of Atizapán and Almoloya, and, at 3.7 km after the turn, on the south side of Almoloya, turn right on an inconspicuous, narrow paved road that skirts the southwest edge of town, with the marshes visible nearby to the left. Continue 700 m, to where you overlook a causeway and some impoundments to the left. At this point one can turn left, with care, down a short stretch of bumpy dirt road to reach the narrow, volcanic dirt road that runs to the left (south) alongside the marshes. One can continue for several kilometers along this road, although the best areas (with yellowthroats) are within 2 km. **Black-polled Yellowthroats** respond to pishing (but generally less strongly so than the resident Common Yellowthroats here, which also are common) or can be found feeding along the water's edge at the base of the reeds. In summer they sing from prominent perches up on the reeds. Early and late in the day are the best times to look for rails out at the edges of the marsh.

From Tres Marias and La Cima (Site 8.4), Almoloya is about a 1.5-hour drive. At Tres Marias, coming south from La Cima, turn right (west) on the road signed to Huitzilac and Toluca and reset to zero at the turn. Continue on the main road, forking right at 3.1 km in Huitzilac and following any signs for Toluca. A roadside stop in the humid fir forests about 2.5–4 km beyond the village of Huitzilac may produce birds such as **Amethyst-throated** and **Bumblebee hummingbirds**, Golden-crowned Kinglet, and **Hooded Grosbeak**. At 14.4 km you pass Lagunas de Zempoala, a popular recreation spot on weekends, and at 28.3 km you wind down into a broad agricultural valley and the village of Santa Martha. At 29.0 km you reach a T-junction and stop sign in Santa Martha. Turn right (signed to Tianguistenco) and stay on the main road, which at 39.0 km enters (San Nicolás) Coatepec. Wind through this little town on the main road and at 40.2 km watch for a left turn well-signed to Tianguistenco and Toluca (straight is signed to Jalatlaco and La Marquesa). Turn left here, and at 46.6 km you pass under a highway bridge and *immediately* turn right and curve up onto the highway (Route 6), which heads southwest and is signed for Almoloya del Río and Tenango. At 48.0 km *continue past* a right turn signed to Almoloya and go on to 50.0 km and a crossroads, where you turn right toward Almoloya. In 800 m look for a left turn onto an inconspicuous, narrow paved road that runs along near the marshes and in 700 m comes to the causeway and impoundments described above under the access from Toluca. See above for how to reach the marshes.

To return from Almoloya to Mexico City, follow the directions in reverse to Mexicaltzingo and the south edge of Toluca. Turn right (signed "México") at Avenida Solidaridad Las Torres and continue on this broad, divided road for about 7 km, until it ends at a T-junction; turn left, and in 1 km you hit Route 15 and turn right to Mexico City.

Almoloya del Río Bird List

70 sp. †See pp. 5–7 for notes on English names.

Pied-billed Grebe
Least Bittern
Great Blue Heron
Great Egret
Snowy Egret
Cattle Egret
Green Heron
White-faced Ibis
Green-winged Teal
Mexican Duck†
Northern Pintail
Blue-winged Teal
Cinnamon Teal
Northern Shoveler
American Wigeon
Ring-necked Duck
Ruddy Duck
Northern Harrier
Red-tailed Hawk
American Kestrel
King Rail
Virginia Rail
Sora
Common Moorhen
American Coot
Killdeer
Greater Yellowlegs
Lesser Yellowlegs
Upland Sandpiper
Least Sandpiper
Baird's Sandpiper
Long-billed Dowitcher
Common Snipe
Wilson's Phalarope
Inca Dove
Chestnut-collared Swift
Ruby-throated Hummingbird
Say's Phoebe
Vermilion Flycatcher
Cassin's Kingbird
Tree Swallow
Violet-green Swallow
Cliff Swallow
Barn Swallow
Marsh Wren
Horned Lark
American Pipit
Loggerhead Shrike
Orange-crowned Warbler
Audubon's Warbler†
Northern Waterthrush
Common Yellowthroat
Black-polled Yellowthroat
Black-headed Grosbeak
Canyon Towhee
Striped Sparrow
Lark Sparrow
Grasshopper Sparrow
Savannah Sparrow
Song Sparrow
Bicolored Blackbird†
Yellow-headed Blackbird
Brewer's Blackbird
Eastern Meadowlark
Bronzed Cowbird
Brown-headed Cowbird
Great-tailed Grackle
House Finch
Lesser Goldfinch
House Sparrow

Site 8.9: Temascaltepec, México

(May/June 1990, July 1991, February 1998)

Introduction/Key Species

Much like Tacámbaro (Site 8.10), the small town of Temascaltepec lies at the transition between temperate pine forests and the arid tropical scrub of the Río Balsas drainage. Similarly, there are several waterfalls near Temascaltepec, making it a good area for swifts. This area was visited in the 1820s by Sir William Bullock and by Ferdinand Deppe, a German naturalist, who between them collected several species then new to science, such as the American Dipper

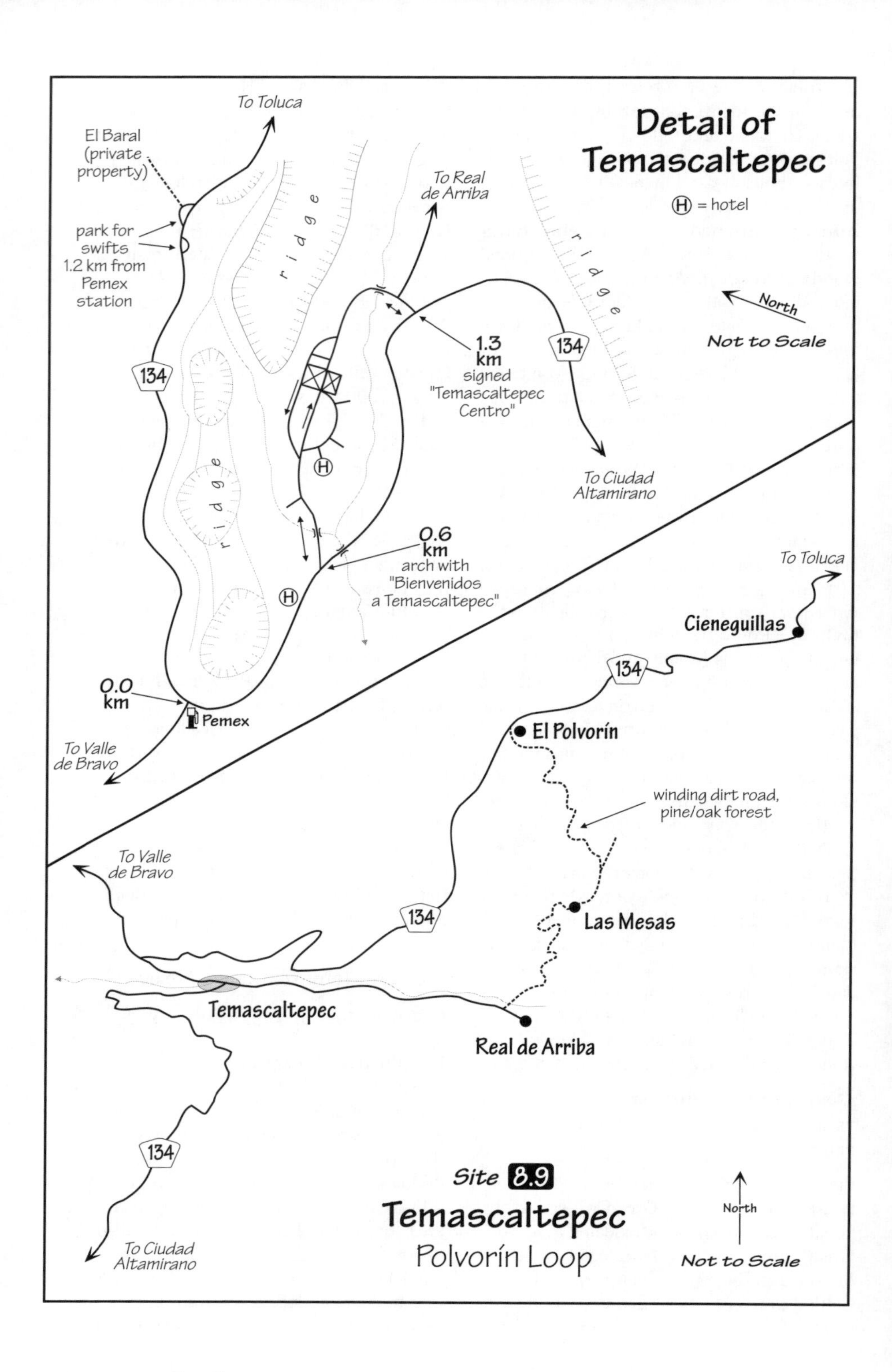

Detail of Temascaltepec
H = hotel
North
Not to Scale
To Toluca
El Baral (private property)
park for swifts 1.2 km from Pemex station
ridge
To Real de Arriba
ridge
134
1.3 km signed "Temascaltepec Centro"
134
H
To Ciudad Altamirano
ridge
0.6 km arch with "Bienvenidos a Temascaltepec"
H
0.0 km
Pemex
To Valle de Bravo
To Toluca
Cieneguillas
134
El Polvorín
winding dirt road, pine/oak forest
To Valle de Bravo
134
Las Mesas
Temascaltepec
Real de Arriba
134
Site 8.9
Temascaltepec
Polvorín Loop
North
Not to Scale
To Ciudad Altamirano

(*Cinclus mexicanus*), whose type locality is Temascaltepec!

The interesting mixture of species found here includes Solitary Eagle, Mountain Pygmy-Owl, Black (summer), Chestnut-collared, **White-naped,** and **Great Swallow-tailed swifts** (White-fronted [White-chinned] Swift should be looked for), **Amethyst-throated**, Calliope (winter), and **Bumblebee hummingbirds**, **White-striped Woodcreeper**, White-throated Flycatcher (summer), American Dipper, **Grey-barred** and **Spotted wrens**, nightingale-thrushes, **Slaty** and **Golden vireos**, Colima and Red-faced warblers (winter), **Chestnut-sided Shrike-Vireo**, **Red-headed Tanager**, **Rufous-capped** and **Green-striped brushfinches**, **Rusty-crowned Ground-Sparrow**, **Abeille's Oriole**, and **Hooded Grosbeak**.

Access

Temascaltepec (1750 m elevation) is about 140 km (2–2.5 hours) west-southwest of Mexico City and about 70 km (1.2–1.5 hours) southwest of Toluca. From Mexico City, head west on the toll highway (Route 15D) toward Toluca and follow signs to skirt the southeast edge of that city, signed initially for Valle de Bravo and Ixtapan de la Sal. Keep following signs for Valle de Bravo and you get onto Avenida Solidaridad Las Torres, which is the main (six-lane, divided) road along the south side of Toluca. Continue west on Solidaridad until a signed left turn for Valle de Bravo (the street you turn onto has large cypress trees lining it); turn left here and in 1.5 km you come to a *glorieta* with a fountain, where you turn right and onto Route 134, a divided four-lane highway heading out of Toluca (and also known as Calzada al Pacifico). Set to zero as you make the turn, and continue some 65 km to Temascaltepec (see Birding Sites).

Temascaltepec can be birded either as a day trip from Mexico City or Toluca, or can be made into an overnight trip. If you leave Mexico City in the morning, you can stop at La Cima (Site 8.4) and Almoloya del Río (Site 8.8) on the way, and at the Volcán de Toluca (see Birding Sites below). Basic to adequate accommodation and food is available in Temascaltepec, or you could camp out. The town is reached easily by public buses from Toluca, and by walking and/or hitching along the Polvorín loop road (see Birding Sites), you have easy access to good birding.

Birding Sites

If you're coming from Toluca, one place that can be worth a stop (for **Strickland's Woodpecker**) is the **Volcán** (Nevado) **de Toluca.** Having reset to zero (see Access), at about 17 km look for a paved road to the left (east), signed for Sultepec and Nevado. Turn here; the road climbs through high-elevation pine woods (remember not to over-exert yourself at this elevation). At 7.6 km a dirt road goes off left (possibly signed for the Nevado de Toluca). **Strickland's Woodpecker** can usually be found within a few hundred meters of this fork, either along the dirt road or farther along the paved road—for example, near Km Post 8, where one can pull off safely and park. In fall and winter, at least, it is often with mixed-species feeding flocks including Mexican Chickadee, White-breasted and Pygmy nuthatches, Brown Creeper, and Olive Warbler. Other species here include **White-eared** and Broad-tailed **hummingbirds**, Western Bluebird, **Russet Nightingale-Thrush**, **Blue Mockingbird**, **Red Warbler**, **Rufous-capped Brushfinch**, and **Striped Sparrow**. On weekend afternoons and holidays there can be a *lot* of traffic heading to the Nevado, but with luck the woodpecker still can be found (e.g., past Km Post 8 on the highway, where traffic is lighter).

Back at Route 134, turn left toward Temascaltepec, and the highway winds through high-elevation conifer forests (Golden-crowned Kinglet is common in the cool fir forests) with farmed clearings in and around settlements along the way. At 39 km (22 km from the Nevado junction) you pass a right turn signed to Valle de Bravo, and at 48.5 km (31.5 km) you bend right through the (unsigned) village of Cieneguillas (2300 m elevation), where the highway crosses a stream (check for Dippers), with pine-oak forest to the left and fields to the right. Toward the end of the pastures one can pull off and park on the right side of the road at 49.3 km (32.3 km), where the highway bends left by Km Post 51. Species in the fields include White-throated Flycatcher (summer), Curve-billed Thrasher, **Cinnamon-rumped** [White-collared] **Seedeater**, and Song Sparrow; the roadside trees have **Mountain Trogon** and **Red-headed Tanager**.

From Cieneguillas it is only about 10–15 minutes' drive to Temascaltepec, which you reach immediately after a left-hand bend (and another right turn to Valle de Bravo) at 65.5 km (48.5 km). There is a Pemex station on the right here, and the center of Temascaltepec lies up ahead, just off the highway. Good birding areas around town include the following:

Swift Overlook: At 1.2 km back along Route 134 toward Toluca from the Valle de Bravo turnoff and Pemex station, it is possible to pull a car off the highway with care—for instance, at the driveway to a property on the left called El Baral. From this point you can look down into a steep-sided valley, filled with waterfalls in the summer rainy season. This is a good vantage point for swifts in early to mid morning and in late afternoon. Up to 30 Black (summer), 70 Chestnut-collared, 450 **White-naped**, and 5000 Vaux's (large numbers only in winter) **swifts** have been seen from here. Note that swifts (including **Great swallow-tailed**) can be seen anywhere over the vicinity of Temascaltepec, and one should be looking up often, especially in morning, late afternoon, and just before rain showers.

The Polvorín Loop: This is a 16-km road, initially paved and then graded dirt, that loops back to Route 134 about 2.5 km south of Cieneguillas. It passes through riparian groves, orchards, gardens, brushy slopes, oaks, pine-oak woodland, clearings, and so on, and more than a hundred species can be found here in a day's birding. To get to the start of this road, continue on Route 134 past the Valle de Bravo turn and Pemex station and into Temascaltepec. At 600 m past the Pemex the highway bends right over a bridge and a cobbled road leads straight into the town center, under an arch prominently marked "Bienvenidos a Temascaltepec." Stay on the highway (check the bridge for dippers) for a further 700 m to a left turn signed to "Temascaltepec Centro." Turn left here and in 200 m, just before a small bridge, turn right up onto a narrow cobbled and then paved road that should be signed to Real de Arriba. Reset to zero here: this is the start of the Polvorín loop.

The road is paved for the first 4 km, to where it enters the small village of Real de Arriba, and one can walk along this stretch from Temascaltepec or drive and pull off at several places along the road in the brushy second growth, oak woods, and clearings beside the rushing stream (Squirrel Cuckoo, Calliope Hummingbird [winter], Greenish Elaenia, **Spotted** and **Happy wrens**, Orange-billed Nightingale-Thrush, **Rufous-backed Thrush**, **Blue Mockingbird**, **Slaty** and **Golden vireos**, Flame-colored and **Red-headed tanagers**, **Rusty-crowned Ground-Sparrow**, Rusty Sparrow).

At Km 3.8, fork left onto a dirt road (the paved road ends shortly in Real de Arriba), which descends, crosses the stream, and then winds up through oak woods, with a good overlook back down the valley (e.g., for swifts) at about 5.0 km. In the drier woods here, and in the more humid oak forest around 9–11 km, after the small, strung-out village of Las Mesas at 6.0–8.2 km (keep straight rather than forking right at 6.5 km), one can find Mountain Pygmy-Owl, **White-striped Woodcreeper**, Tufted Flycatcher, **Spotted Wren**, **Russet Nightingale-Thrush**, **Chestnut-sided Shrike-Vireo**, Grace's and Red-faced (winter) warblers, Painted Whitestart, **Rufous-capped Brushfinch**, and **Black-headed Siskin**.

Continuing on, after 11 km the road passes through several kilometers of more humid pine-oak and conifer forest with seasonal flower banks (**Amethyst-throated** and **Bumblebee hummingbirds**, **Mountain Trogon**, **Grey-barred Wren**, Grey-breasted Wood-Wren, Ruddy-capped Nightingale-Thrush, Colima Warbler (winter), **Green-striped Brushfinch**, and **Hooded Grosbeak**, plus many of the species listed above) and winds down along a stream with some pastures (White-throated Flycatcher, summer) to rejoin the highway at 16 km (called El Polvorín, although there is really nothing here!). At the highway, turn right to Toluca (and Cieneguillas) or left, back to Temascaltepec.

Temascaltepec/Polvorín Bird List

145 sp. V: vagrant; R: rare (not to be expected). †See pp. 5–7 for notes on English names.

Turkey Vulture
Sharp-shinned Hawk
Solitary Eagle (R)
Short-tailed Hawk
Red-tailed Hawk
American Kestrel
Mourning Dove
Inca Dove
White-tipped Dove
Squirrel Cuckoo
Groove-billed Ani
Mountain Pygmy-Owl†
Lesser Nighthawk
Black Swift
Chestnut-collared Swift
White-naped Swift
Vaux's Swift
White-throated Swift
Great Swallow-tailed Swift
Green Violet-ear
Broad-billed Hummingbird
White-eared Hummingbird
Berylline Hummingbird
Amethyst-throated Hummingbird
Blue-throated Hummingbird
Magnificent Hummingbird
Calliope Hummingbird
Broad-tailed Hummingbird
Bumblebee Hummingbird
Mountain Trogon
Acorn Woodpecker
Ladder-backed Woodpecker
Hairy Woodpecker
Red-shafted Flicker†
White-striped Woodcreeper
N. Beardless Tyrannulet
Greenish Elaenia
Greater Pewee
Western Pewee†
Common Tufted Flycatcher†
White-throated Flycatcher
Hammond's Flycatcher
Dusky Flycatcher
Western Flycatcher†
Buff-breasted Flycatcher
Black Phoebe
Bright-rumped Attila
Dusky-capped Flycatcher
Social Flycatcher
Thick-billed Kingbird
Cassin's Kingbird
Rose-throated Becard
Violet-green Swallow
N. Rough-winged Swallow
Cliff Swallow
Barn Swallow
Grey-breasted Jay†
Northern Raven†
Mexican Chickadee
Bridled Titmouse
Bushtit
Brown Creeper
Grey-barred Wren
Spotted Wren
Canyon Wren
Happy Wren
Northern House Wren†
Brown-throated Wren†
Grey-breasted Wood-Wren
American Dipper
Ruby-crowned Kinglet
Blue-grey Gnatcatcher
Eastern Bluebird
Brown-backed Solitaire
Orange-billed N.-Thrush
Russet N.-Thrush
Ruddy-capped N.-Thrush
Hermit Thrush
White-throated Thrush†
Rufous-backed Thrush†
American Robin
Grey Catbird (V)
Blue Mockingbird
Curve-billed Thrasher
Grey Silky†
Slaty Vireo
Hutton's Vireo
Cassin's Vireo†
Plumbeous Vireo†
Golden Vireo
Warbling Vireo
Chestnut-sided Shrike-Vireo
Orange-crowned Warbler
Colima Warbler
Nashville Warbler
Crescent-chested Warbler
Yellow Warbler
Audubon's Warbler†
Black-throated Grey Warbler
Townsend's Warbler
Hermit Warbler
Grace's Warbler
Black-and-white Warbler
Louisiana Waterthrush
MacGillivray's Warbler
Wilson's Warbler
Red-faced Warbler
Painted Whitestart†
Slate-throated Whitestart†
Rufous-capped Warbler
Golden-browed Warbler
Blue-hooded Euphonia
Hepatic Tanager
Summer Tanager
Western Tanager
Flame-colored Tanager
Red-headed Tanager
Rose-breasted Grosbeak
Black-headed Grosbeak
Lazuli Bunting
Varied Bunting
Rufous-capped Brushfinch
Green-striped Brushfinch
Rusty-crowned Ground-Sparrow
Spotted Towhee
Canyon Towhee
Cinnamon-bellied Flowerpiercer
Cinnamon-rumped Seedeater†
Rusty Sparrow
Chipping Sparrow
Song Sparrow
Lincoln's Sparrow
Mexican Junco†
Great-tailed Grackle
Bronzed Cowbird
Hooded Oriole
Black-vented Oriole
Bullock's Oriole
Abeille's Oriole†
Scott's Oriole
House Finch
Red Crossbill
Pine Siskin
Black-headed Siskin
Lesser Goldfinch
Hooded Grosbeak

OVERNIGHT TRIPS

Site 8.10: Tacámbaro, Michoacán

(March 1993, September 1995, checked in part 1997)

Introduction/Key Species

Tacámbaro (1500 m elevation) lies between the high conifer forests of the mountains in Mexico's transverse volcanic belt and the baking lowlands of the Río Balsas drainage. Thus, within a few kilometers of town, depending on which direction you drive, you can be seeing Red Warblers in cool montane forests or Orange-breasted Buntings in arid scrub. Water draining from the highlands cuts down through this area to join the Balsas and, as a consequence, there are numerous waterfalls (some seasonally dry) in the vicinity of Tacámbaro.

The first-known specimens of the recently described **White-fronted** [White-chinned] **Swift** were collected at a waterfall on the edge of Tacámbaro, and a recent sighting of this enigmatic bird was made a few kilometers south of Tacámbaro (Howell et al. 1997). Other than this swift, species of interest here include Lesser Ground-Cuckoo, **White-naped** and **Great Swallow-tailed swifts**, **Golden-crowned Emerald**, **Dusky Hummingbird** (at the western edge of its range), **Spotted** and **Happy wrens**, **Golden Vireo**, **Orange-breasted Bunting**, and **Black-chested Sparrow**. Among other birds, Banded Quail can be expected to occur.

Access

Tacámbaro is a small town about 400 km (5–6 hours) west of Mexico City and about 100 km (1.5–2 hours) south of Morelia. This is now an easy day's drive from Mexico City, given the recent completion of some toll highways: head out west to Toluca, swinging northwest to Atlacomulco and then west to Maravatio and Morelia, following signs for Morelia, for which you turn off south not long after passing Lago Cuitzeo on your right. Take the ring road around Morelia, and head out southwest toward Pátzcuaro. Perhaps the easiest route (unless you have a good map) is to continue to Pátzcuaro, whence you turn south on Route 120 toward Ario de Rosales. About 13 km south of Pátzcuaro watch for a left-hand junction (which should be signed) to Tacámbaro, which is then about 40 km away. Along this last stretch of the drive you'll notice a left-hand junction signed (probably) to Villa Madero and (possibly) for Morelia. This is a short-cut back to Morelia.

There is at least one basic, functional hotel in Tacámbaro, and a few restaurants (don't expect anything fancy), or you could camp in the area, in the warmer areas below town or in the colder forests above. Tacámbaro is reached easily from Morelia using public buses, and Salto de Santa Paula is within walking distance of town; other birding areas can be reached by second-class bus and/or hitching rides.

Birding Sites

If you're driving from Mexico City via the toll highways, it can be worth a stop alongside Lago Cuitzeo, where the new road cuts across the edge of the lake, which contains extensive reed beds. There is room to pull off safely and park here, and a brief mid-morning stop (September 1995) produced Least Bittern (common), Northern Jacana, White-throated Flycatcher (common, singing), and Common and **Black-polled yellowthroats**.

Once on the Tacámbaro road, south of Pátzcuaro, traffic is generally lighter, and one can stop and bird in the highland pine-oak and farmland. Birds along this stretch include Buff-breasted Flycatcher, Grey-breasted Jay, Mexican Chickadee, **Red** and Olive **warblers**, **Cinnamon-bellied Flowerpiercer**, and hybrid Rufous-sided × **Collared** towhees. The streams should be checked for American Dipper.

The town of Tacámbaro is off the highway to the right (south). Streets are narrow and often one-way, and it may take a few passes to get your bearings in town. One can enter town at the first turning (signed "Centro") or continue on for a kilometer or two, skirting town to the east until another road into town, uphill to the right, in the midst of several stores and a hotel-like building or two along the highway; this is the route by which buses enter town and is signed to "Centro Comercial." If you take this "bypass," 1.0 km after the first road into town ("Centro"), look left at a right-hand bend for a waterfall (season-

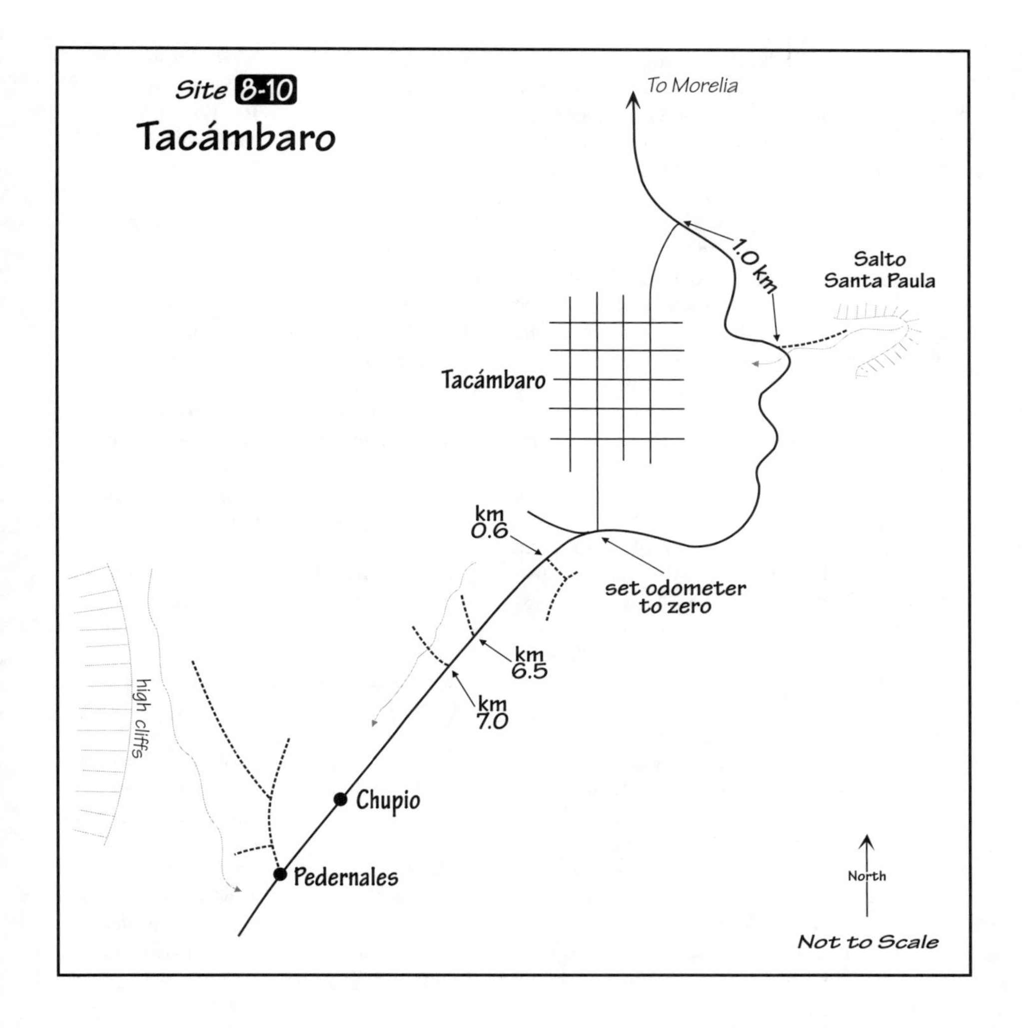

ally dry and easy to miss unless you're looking), back about 300 m from the highway, up a narrow valley. One can park, with care, off the highway here and walk up to the base of the falls (the path takes you through a few gardens); birds in the valley include **Russet-crowned Motmot** and Rusty Sparrow. This appears to be Salto Santa Paula, where the specimens of **White-fronted** [White-chinned] **Swift** were collected (in July). If they nest here, then during the summer (May to July), when they should be feeding young, might be a good time to look. At other seasons, swifts often come and go from their roosts at dusk or dawn, not allowing you much chance to see them, although groups of some species can remain in the area for up to a few hours, especially in the mornings.

To reach other birding sites, continue south from Tacámbaro, on the road to Pedernales, marking the bus entry road as zero. The first spot is at Km 0.6, where there is a dirt road up to the left (east) marked by a green-and-white sign for an electricity plant; turn left here, then turn right onto a cobbled road that winds up into pine woods (Whiskered Screech-Owl, Ferruginous Pygmy-Owl, **Spotted Wren**, Eastern Bluebird, Grace's Warbler).

Past the Km 0.6 junction, the highway winds down through fields and brushy hillsides; a well-hidden, rushing stream cuts a narrow gorge, with falls, through the fields to your right. At Km 6.5, by a left-hand bend, there is an inconspicuous, grassed-over track to a fallen-down shack on the right (east). One can pull off here and scan the slopes for swifts in the morning. At Km 7 is a narrow dirt road to the right (east) where one can drive in 100 m or so and park. At 500 m from the highway the dirt road fords the stream, and down to your left is a small waterfall that can be viewed, with caution, for roosting **White-naped Swifts** clinging to the damp, overhung cliffs. One can also cross the stream and walk up the track to look over the slopes. Birds in the hedges and brushy tangles here include **Happy Wren** and various hummingbirds and orioles (if flowers are blooming).

Continue down the highway, watching for **Lesser Roadrunners**, through the next small village, Chupio, to the slightly less small village of Pedernales (around 1000 m elevation, with Great Kiskadee, **Yellow-winged Cacique**, etc.), where, on the near side (north) of "town" one can turn right (west) on a dirt road up through the canefields toward some imposing-looking cliffs. These roads through the canefields can be in bad shape and may not always be transitable in a regular car. Once on the dirt road, shortly out of Pedernales a short track goes left to a stream. Stay right/straight here and take the next left fork, which after a few kilometers comes fairly close to the cliffs. One can park and walk through the canefields and, crossing the gorge via an old brick aqueduct, explore the slopes below the cliffs. **Great Swallow-tailed Swifts** can be seen around these cliffs, especially early and late in the day, and birds in the scrub include Lesser Ground-Cuckoo, **Orange-breasted Bunting**, and **Black-chested Sparrow**.

Tacámbaro Bird List

102 sp. †See pp. 5–7 for notes on English names.

Black Vulture
Turkey Vulture
Osprey
Sharp-shinned Hawk
Red-tailed Hawk
American Kestrel
Mourning Dove
Inca Dove
White-tipped Dove
Squirrel Cuckoo
Lesser Ground-Cuckoo
Lesser Roadrunner
Groove-billed Ani
Whiskered Screech-Owl
Ferruginous Pygmy-Owl
Buff-collared Nightjar
Black Swift
White-fronted Swift
Chestnut-collared Swift
White-naped Swift
Vaux's Swift
Great Swallow-tailed Swift
Golden-crowned Emerald
Broad-billed Hummingbird
Berylline Hummingbird
Violet-crowned Hummingbird
Lucifer Hummingbird
Russet-crowned Motmot
Golden-cheeked Woodpecker
Ladder-backed Woodpecker
N. Beardless Tyrannulet
Greater Pewee
Least Flycatcher
Hammond's Flycatcher
Black Phoebe
Vermilion Flycatcher
Dusky-capped Flycatcher
Ash-throated Flycatcher
Brown-crested Flycatcher
Great Kiskadee
Tropical Kingbird
Thick-billed Kingbird
Cassin's Kingbird
Masked Tityra
Violet-green Swallow
N. Rough-winged Swallow
Barn Swallow
Spotted Wren
Canyon Wren
Happy Wren
Banded Wren
Northern House Wren†
Blue-grey Gnatcatcher
Eastern Bluebird
Brown-backed Solitaire
Rufous-backed Thrush†
Blue Mockingbird
Curve-billed Thrasher
Cedar Waxwing
Grey Silky†
Plumbeous Vireo†
Hutton's Vireo
Golden Vireo
Warbling Vireo
Nashville Warbler
Virginia's Warbler
Yellow Warbler
Audubon's Warbler†
Black-throated Grey Warbler
Townsend's Warbler
Grace's Warbler
MacGillivray's Warbler
Louisiana Waterthrush
Grey-crowned Yellowthroat
Wilson's Warbler
Painted Whitestart†
Rufous-capped Warbler
Summer Tanager
Western Tanager
Black-headed Grosbeak
Blue Grosbeak
Indigo Bunting
Varied Bunting
Orange-breasted Bunting

Cinnamon-rumped Seedeater†
Blue-black Grassquit
Black-chested Sparrow
Rusty Sparrow
Stripe-headed Sparrow
Chipping Sparrow
Lark Sparrow
Lincoln's Sparrow
Great-tailed Grackle
Hooded Oriole
Black-vented Oriole
Streak-backed Oriole
Bullock's Oriole
Abeille's Oriole†
Scott's Oriole
Yellow-winged Cacique
House Finch
Lesser Goldfinch

Site 8.11: Tlanchinol, Hidalgo

(December 1986, June 1990)

Introduction/Key Species

The coastal-facing slopes of the Sierra Madre Oriental intercept warm, moisture-laden air as it rises from the Gulf of Mexico. In places this moisture condenses to produce a misty environment where a subtropical, humid evergreen forest (cloud forest) thrives. Much of the cloud forest in eastern Mexico has long since been cleared or reduced to a patchwork of small lots, mostly on relatively inaccessible, steep and muddy slopes. Traditional cloud-forest birding sites in the Sierra Madre Oriental have been Xilitla in San Luis Potosí (to the north) and Teziutlán in Puebla (to the south), both now greatly cutover and cleared, although with effort some good birds can be found at Teziutlán. For some reason, a fairly extensive area of readily accessible cloud forest persists near Tlanchinol.

Birds typical of the cloud forest here include Barred Forest-Falcon, **Bearded Wood-Partridge**, **White-faced Quail-Dove**, Emerald Toucanet, Ruddy Foliage-gleaner, Tawny-throated Leaftosser, Strong-billed Woodcreeper, Scaled Antpitta, Azure-hooded and **Unicolored jays**, **Slate-colored Solitaire**, **Black Thrush**, **White-naped Brushfinch**, and **Hooded Grosbeak**.

Access

Tlanchinol (1500 m elevation) is about 250 km (4–5 hours) north of Mexico City, reached via *autopista* (Route 130D) to Pachuca (1–1.5 hours) and then by Route 105 over the mountains toward Huejutla. The drive north from Pachuca can be a slow one, especially if you get stuck behind trucks or enveloped by low clouds. Allow three hours or more, including a stop or two, from Pachuca to Tlanchinol, which can also be reached, coming south, from the El Naranjo area of southeastern San Luis Potosí (Site 4.5). There is basic accommodation in the town of Tlanchinol and in Huejutla in the warmer lowlands 55 km (an hour) to the north (200 m elevation). You could also camp (which might be a cold and damp experience, given the habitat).

Public buses serve Tlanchinol from Pachuca to the south and Huejutla to the north, and it is easy to reach good habitat just north of town (see Birding Sites) by bus, hitching, or walking.

Birding Sites

Coming from Pachuca, Route 105 starts at Km 0 and climbs through oak-chaparral to plateau desert and dry farmland (Verdin, Curve-billed Thrasher, **Black-vented Oriole**) punctuated by green, irrigated, agricultural valleys, such as the one at Río Venados/Cacacome (Km 63), where birds include Broad-billed Hummingbird, Western Pewee (summer), Vermilion Flycatcher, Rose-throated Becard, Black-crested Titmouse, **Spotted Wren**, Yellow-green Vireo (summer), **Abeille's Oriole** (summer). After this, you climb into oak and humid pine-oak forest, where stops can produce **White-eared Hummingbird**, Brown Creeper, **Brown-backed Solitaire**, Orange-billed, Ruddy-capped, and **Russet nightingale-thrushes**, American Robin, and **Rufous-capped Brushfinch**.

Tlanchinol is at about Km Post 165 (?), with the center of town set off above and to the right (east) of the highway. To reach the birding areas (1200–1500 m), stay on the highway through Tlanchinol and you will find yourself driving through cloud forest, with tall tree ferns and huge, bromeliad-covered trees, 4–8 km north of town. There are places to pull off safely and park, including some nice overlooks 7–8 km from town (swifts, **White-naped Brushfinch**), and you can bird productively from the highway (albeit with noisy truck and bus traffic). There are also various trails into the forest, including a well-used one about 5 km north of Tlanchinol, marked by an area where buses pull off on the left and a road sign that says

"Lontla," although no buildings are apparent. From this point a trail goes down to the left (west), to a village in the adjacent valley, and a fork in the trail after a while seems to mean little more than people going left (steeper) and pack animals going right (less steep, and less busy). Also, immediately opposite the start of the Lontla trail, across the road, an inconspicuous narrow trail leads up into the forest. Basically, all of the cloud-forest species can be seen along these two trails.

More information on the seasonal occurrence and relative abundance of birds at Tlanchinol can be found in Howell and Webb (1992).

Tlanchinol Bird List

(many species are more vocal/conspicuous in summer than winter): 83 sp. R: rare (not to be expected). †See pp. 5–7 for notes on English names.

Red-tailed Hawk
Ornate Hawk-Eagle
Barred Forest-Falcon
Crested Guan
Bearded Wood-Partridge (R?)
Band-tailed Pigeon
Red-billed Pigeon
White-tipped Dove
White-faced Quail-Dove
Green Parakeet
White-crowned Parrot
Mottled Owl
Black Swift
Chestnut-collared Swift
White-collared Swift
Vaux's Swift
White-bellied Emerald
Amethyst-throated Hummingbird
Magnificent Hummingbird
Bumblebee Hummingbird
Collared Trogon
Mountain Trogon
Blue-crowned Motmot
Emerald Toucanet
Acorn Woodpecker
Smoky-brown Woodpecker
Bronze-winged Woodpecker†
Pale-billed Woodpecker
Yellow-bellied Sapsucker
Ruddy Foliage-gleaner
Tawny-throated Leaftosser
Olivaceous Woodcreeper
Strong-billed Woodcreeper
Spotted Woodcreeper
Spot-crowned Woodcreeper
Scaled Antpitta
Greater Pewee
Common Tufted Flycatcher†
Hammond's Flycatcher
Western Flycatcher†
Dusky-capped Flycatcher
Boat-billed Flycatcher
Grey-collared Becard
N. Rough-winged Swallow
Barn Swallow
Azure-hooded Jay
Unicolored Jay
Grey-breasted Wood-Wren
Ruby-crowned Kinglet
Eastern Bluebird
Slate-colored Solitaire
Brown-backed Solitaire
Black-headed N.-Thrush
Black Thrush†
White-throated Thrush†
American Robin
Grey Silky†
Blue-headed Vireo†
Cassin's Vireo†
Brown-capped Vireo
Chestnut-sided Shrike-Vireo
Crescent-chested Warbler
Townsend's Warbler
Hermit Warbler
Black-throated Green Warbler
Black-and-white Warbler
Wilson's Warbler
Golden-crowned Warbler
Golden-browed Warbler
Yellow-throated Euphonia
Blue-hooded Euphonia
Flame-colored Tanager
Common Bush-Tanager
White-naped Brushfinch
Chestnut-capped Brushfinch
Rusty Sparrow
Lincoln's Sparrow
Great-tailed Grackle
Bronzed Cowbird
Audubon's Oriole
Baltimore Oriole
Black-headed Siskin
Hooded Grosbeak

CHAPTER 9 THE SIERRA DE ATOYAC, GUERRERO

White-throated Jay

The state of Guerrero, in southwest Mexico, is perhaps best known for the major beach resort of Acapulco. In terms of its avifauna, Guerrero is one of the richest (although least-studied) states in Mexico. The most interesting areas, such as those in the Sierra Madre del Sur (including the Sierra de Atoyac), have traditionally been difficult to reach because of bad road conditions. Add to this a long-term history of political activism in the state and the large-scale cultivation of certain crops in the Sierra, and the limited amount of exploration by birders is perhaps more understandable.

Consequently, Guerrero has never been a major focus for organized bird tours and has been mostly a short-term destination for target-oriented birders rather than a two-week birding vacation—although

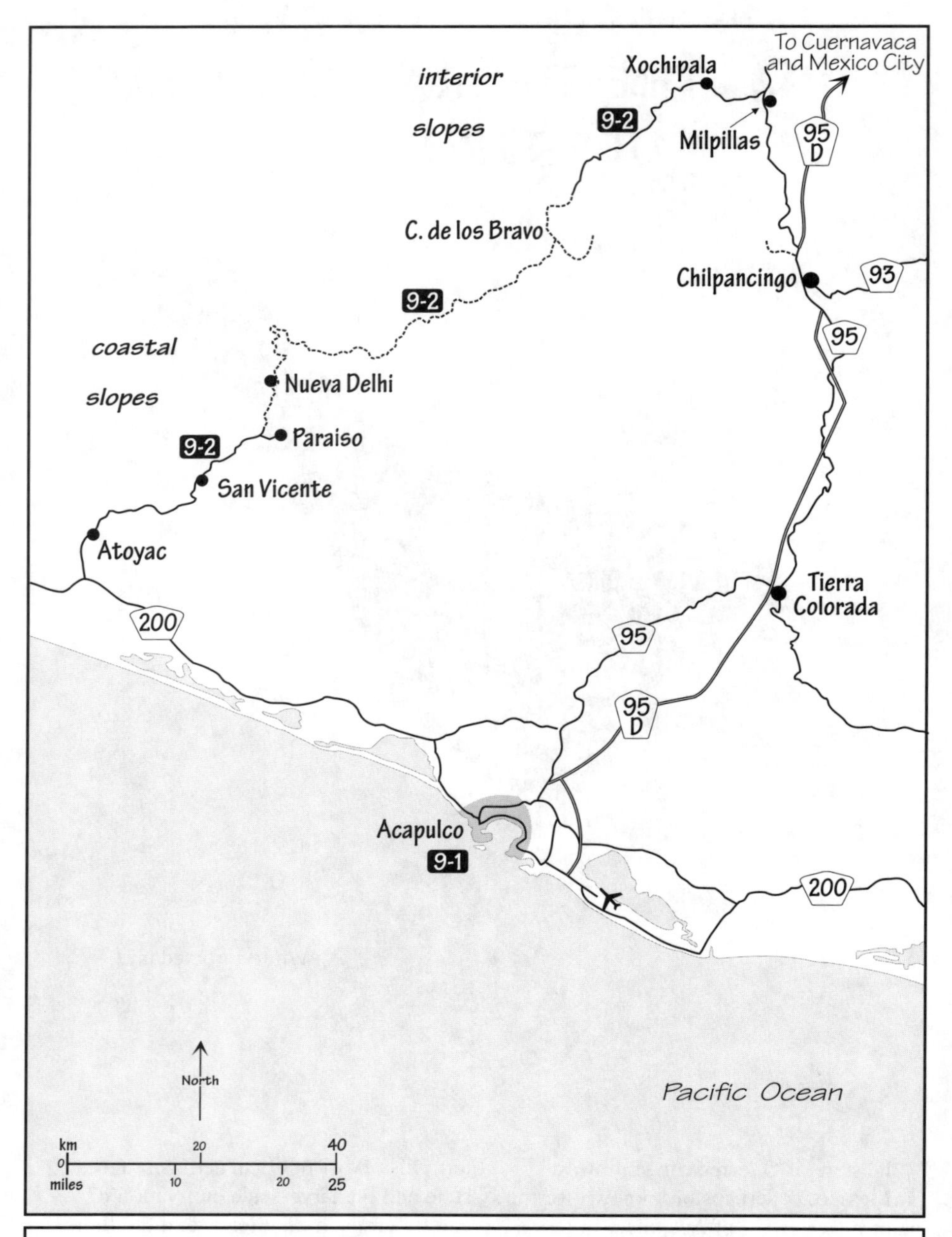

Chapter 9: THE SIERRA DE ATOYAC, GUERRERO

9-1 Acapulco, Guerrero
9-2 The Sierra de Atoyac, Guerrero

such a trip could be great. *While temerity should not be a prerequisite for birding in Guerrero, it is nonetheless common sense to check on the political climate before you visit.*

The Sierra de Atoyac is a great region for birding and is within easy reach of Acapulco. Although the most sought-after species in the Sierra de Atoyac are Short-crested Coquette, White-tailed Hummingbird, White-throated Jay, and Slate-blue Seedeater, fully one-third of the 180 species endemic to northern Middle America occur here, making it an exciting area to visit.

THE SIERRA DE ATOYAC, GUERRERO SITE LIST

Site 9.1 Acapulco, Guerrero
Site 9.2 The Sierra de Atoyac, Guerrero

Site 9.1: Acapulco, Guerrero

(January 1984, April 1988, March, May 1990, October 1993)

Access/Birding Sites

Acapulco is served daily by several flights from Mexico City as well as from a number of U.S. cities, whence package deals can often be arranged, making for an inexpensive holiday. As a major resort, Acapulco offers a wide range of accommodations and restaurants that cater to all budgets. Car rental is available through a number of companies with offices in town as well as at the airport, which is several kilometers out of town to the east.

Acapulco is about 400 km due south of Mexico City, from which, thanks to the completion of a new toll highway between these points, it can be reached in 3.5–4.5 hours. Using the older roads along this route, Acapulco is a 5.5- to 7-hour drive from Mexico City (depending on time of day/night and truck traffic), or it can be made into a full day, stopping to bird along the way (e.g., sites to the south of Mexico City, Chapter 8).

As a birding destination Acapulco offers little that cannot be seen in coastal areas of Colima (Chapter 7) or Oaxaca (Chapter 11). Getting out into the thorn forest outside town (west on Route 200 toward Pie de la Cuesta and Zihuatanejo, or east on Route 200 and/or toward the airport) should produce an interesting assortment of species typical of Mexico's Pacific Slope. Basically, any quiet side road or track into thorn forest, second growth, weedy fields, and coconut plantations could be explored productively. Birds to be expected in such habitats include Grey, Roadside, and Short-tailed hawks, Northern Bobwhite, **Banded Quail**, Inca and White-tipped doves, Orange-fronted Parakeet, Lesser Ground-Cuckoo, Ferruginous Pygmy-Owl, Cinnamon and **Doubleday's hummingbirds**, **Citreoline Trogon**, **Russet-crowned Motmot**, **Golden-cheeked Woodpecker**, Northern Beardless Tyrannulet, Dusky-capped, Nutting's, and Brown-crested flycatchers, Tropical and Thick-billed kingbirds, White-throated Magpie-Jay, Rufous-naped, **Happy**, and Banded **wrens**, **Rufous-backed Thrush**, White-lored Gnatcatcher, **Red-breasted Chat**, **Blue** and **Orange-breasted buntings**, Olive Sparrow, **Cinnamon-rumped** [White-collared] and Ruddy-breasted **seedeaters**, Altamira, Spot-breasted, and Streak-backed orioles, and **Yellow-winged Cacique**. Wintering migrants include Least and Western flycatchers, Warbling and Bell's vireos, Orange-crowned, Yellow, Black-throated Grey, Black-and-white, and MacGillivray's warblers, Western Tanager, and Orchard Oriole.

Acapulco is also a center for ocean sport fishing, and the offshore waters are good for seabirds. Birds along the waterfront include species typical of Mexico's Pacific coast (Brown Pelican, Neotropic Cormorant, Magnificent Frigatebird, Laughing Gull, etc.). Pelagic trips can be arranged at the western end of the waterfront, near the old fort, where the fishing boats and tourist launches are docked, and probably also through some of the larger hotels and tourist agencies. Scheduled or charter fishing trips that you may be able to join will probably have a fixed price, but you may be able to negotiate a "birding" rate. If you arrange your own trip (easy enough to do), make sure you agree upon the price beforehand and emphasize that you want to see birds and go as far offshore as reasonably possible in a day. An early start is worthwhile, both to beat the heat and to allow you as much time as possible before the wind picks up, which it usually, but not invariably, does in the afternoons. Species possible in a day trip include Audubon's Shearwater, Least and Black storm-petrels, Red-billed Tropicbird, Brown Booby, jaegers, Bridled and Black terns, and Brown Noddy.

Site 9.2: The Sierra de Atoyac, Guerrero

(December 1985, June 1986, April 1988, March, May 1990, October 1993)

Introduction/Key Species

The rugged and remote Sierra de Atoyac supports one of the most extensive tracts of virgin cloud forest left in Mexico. Fortunately for birders, a road crosses the Sierra: it starts in cutover thorn forest on the lower slopes, passes through coffee plantations, which dominate the middle-elevation coastal slopes, into cloud forest, pine-oak forest on the higher slopes, and then thorn forest in the Río Balsas drainage of the arid interior slopes. Rarely, however, has this road been passable to most vehicles for its entire length, although approaching the Sierra from either side should enable you to get into good habitat and find most of the endemics.

Species of interest include King Vulture, Black Hawk-Eagle, Barred Forest-Falcon, **Long-tailed Wood-Partridge**, **Singing** and **Banded quail**, **White-faced Quail-Dove**, Pheasant Cuckoo, **Balsas Screech-Owl**, **Eared Poorwill**, **White-naped** and **Great Swallow-tailed swifts**, **Mexican Hermit**, **Short-crested Coquette**, **White-tailed**, **Green-fronted**, **Amethyst-throated**, **Garnet-throated**, and **Bumblebee hummingbirds**, Long-billed and Plain-capped starthroats, **Sparkling-tailed Woodstar** (twenty-four species of hummingbirds are possible), Emerald Toucanet, **Grey-crowned Woodpecker**, Ruddy Foliage-gleaner, Scaled Antpitta, **White-throated** and **Unicolored jays**, **Boucard's Wren**, **Black** and **Aztec thrushes**, **Chestnut-sided Shrike-Vireo**, **Red** and **Fan-tailed warblers**, White-winged and **Red-headed tanagers**, **Slate-blue Seedeater**, and **Hooded Grosbeak**. In addition, the type specimen of the recently described **White-fronted** [White-chinned] **Swift** was collected in the Sierra, although it may be only a migrant or vagrant here. An article about the Short-crested Coquette and birding in the Sierra appeared in the April 1992 issue of *Birding* (Howell 1992).

Access

The Sierra de Atoyac, part of Mexico's Sierra Madre del Sur, lies a short distance inland of the coast west of Acapulco. There are two basic ways to approach it: from the inland side near Chilpancingo, or from the coastal side via Atoyac. A day trip from Acapulco to the middle-elevation coastal slopes is feasible (allow at least 2.5 hours to get to good habitat) and, at the right time of year (April/May), should produce **Short-crested Coquette** and possibly **White-tailed Hummingbird**, but not **White-throated Jay**. Unless the roads have been greatly improved, a birding day trip from Acapulco to the higher cloud forest above Paraíso is unlikely to be worthwhile.

Chilpancingo is the state capital of Guerrero and as such offers a range of accommodations and restaurants. It can be used as a base if you plan to explore the interior slopes of the Sierra and is about 265–285 km south of Mexico City (2.5–4 hours, depending on whether or not you use the new *cuota* highway), or 115–130 km (1–2 hours, likewise) north of Acapulco. This side of the Sierra seems to offer good birding at any season (e.g., June, December), although **White-throated Jay** may be easier to find in summer.

Atoyac is a town in the coastal lowlands, 6 km north of the coastal highway (Route 200) about 80 km (1–1.5 hours) west of Acapulco. The Pemex station on the south side of Atoyac has unleaded gasoline but is best found as you leave town, via the one-way system (coming into town from Acapulco you won't see it). There are a couple of basic hotels and restaurants in Atoyac, or one could camp higher up in the Sierra. There is, in fact, very basic accommodation (a cot and a shared cold shower) available in Paraíso if you ask around, although you may prefer to camp. This side of the Sierra is undoubtedly best for birding in late winter and spring (e.g., March to May) and is decidedly poorer, for hummingbirds in particular, in autumn (e.g., October). Also, by late winter the roads may have been improved from their condition after the summer and autumn rains.

Atoyac can be reached easily by public buses from Acapulco, and there are second-class buses that run between Atoyac and Paraíso and allow easy access to some good areas. Above Paraíso, public transport seems to comprise trucks, and you should enquire locally about road conditions and transport. To reach the interior slopes, there should be a bus service from Chilpancingo to Xochipala and probably beyond (ask locally).

Birding Sites

Interior Slopes. From Chilpancingo, head north on the old highway (Route 95) toward

Mexico City for about 38 km (30–45 minutes) to the "village" of Milpillas (a couple of buildings), where there is a paved road off to the left (west) that should be signed for Xochipala. If you're coming south from Mexico City on the old highway, Milpillas is 16.5 km south of where you cross the Río Balsas at Mexcala. Note that many maps show what appears to be a fairly major road heading into the Sierra to the west of Chilpancingo; this road goes to Omiltemi (the type locality for White-throated Jay), among other places, but unless it has been improved greatly, it might be best avoided.

Mark zero as you turn at Milpillas on the road to Xochipala and Carrizal de los Bravo (also shown as Corral de Bravo on some maps). This is a two-lane paved road that climbs slowly up the back side of the Sierra. If you drive through at night, look for **Balsas Screech-Owl** and Burrowing Owl, which occur along at least the first kilometer or so. From Milpillas to Xochipala (Km 11.1) and beyond, to about Km 20, the habitat is arid Balsas scrub. Stops anywhere in this habitat—for instance at Km 16/17—can produce **Banded Quail**, Orange-fronted Parakeet, Lesser Ground-Cuckoo, **Lesser Roadrunner**, **Golden-crowned Emerald**, **Dusky Hummingbird**, **Russet-crowned Motmot**, **Golden-cheeked Woodpecker**, Nutting's Flycatcher, Rufous-naped, Banded, and **Happy wrens**, White-lored Gnatcatcher, **Golden Vireo**, and **Black-chested Sparrow**. Around Km 21/22 oak scrub becomes conspicuous (Western Scrub Jay, **Boucard's Wren**, Orange-billed Nightingale-Thrush, **Fan-tailed Warbler**, Yellow Grosbeak, **Rusty-crowned Ground-Sparrow**, Rufous-crowned Sparrow, and **Black-vented Oriole**; and Dwarf Vireo seems likely to occur here), and then, for example, at Km 30, the habitat becomes dry oak woodland (Acorn Woodpecker, Greater Pewee, Buff-breasted Flycatcher, Bridled Titmouse, Painted Whitestart, Blue-hooded Euphonia, Hepatic Tanager, Rusty Sparrow, **Black-headed Siskin**).

As you climb along this road, it passes through various small settlements (stay on the main and only paved road): Xochipala at Km 11.1, La Laguna at Km 27.6, Mirabel at Km 32.2 (shortly after which pines appear), Los Morros at Km 42.6, Campo de Aviacion at 48.0. At Km 50.9 the pavement ends in the settlement of Filo de Caballo; the woods for a few kilometers before (east of) Filo have Mountain Pygmy-Owl, **Eared Poorwill**, and flocks that can include **Grey-collared Becard**. At Km 55.2 you go through the small settlement of Carrizal de los Bravo; shortly after this, at Km 58.0, look for a minor-looking side road off to the left—this is the junction with the "road" from Chilpancingo. From this point on, the humid pine-oak forest can be birded from the road or along an inconspicuous track off down to the left (southeast) at Km 60.7, a short distance before a small (possibly inactive) lumber camp. Birds here include **Long-tailed Wood-Partridge**, **Singing Quail**, **Garnet-throated** and **Bumblebee hummingbirds**, **Mountain Trogon**, Emerald Toucanet, Spotted and Spot-crowned woodcreepers, **White-throated** and **Unicolored jays**, **Brown-backed Solitaire**, **Russet** and Ruddy-capped **nightingale-thrushes**, **Black Thrush**, Colima (winter), **Red**, and **Golden-browed warblers**, Common Bush-Tanager, **Rufous-capped** and Chestnut-capped **brushfinches**, **Collared Towhee**, and **Cinnamon-bellied Flowerpiercer**. Other possibilities in the area include Strong-billed Woodcreeper, **Aztec Thrush**, and **Hooded Grosbeak**.

Coastal Slopes. Many of the above-mentioned species and a host of others can be found on the wetter, coastal-facing slopes above Atoyac. The marshy and weedy roadside fields between the main coastal highway (Route 200) and Atoyac are worth checking for Willow Flycatcher (winter) and Ruddy-breasted Seedeater. From Atoyac take the (unsigned) road toward Paraíso, which leaves Atoyac from the northeast ("far right" from where you enter) side of town (head east first, then north), marking zero as you leave the center of Atoyac. The road into the Sierra is paved, more or less, and starts by winding through cutover and burned hillsides of thorn forest and weedy fields (Lesser Ground-Cuckoo, **Doubleday's** and Cinnamon **hummingbirds**, **Citreoline Trogon**, **Golden-cheeked Woodpecker**, **Orange-breasted Bunting**, Streak-backed Oriole) before climbing into the coffee plantations which dominate the region. At Km 15 you pass through the village of Pte. San Andrés, and at Km 23 through Río Santiago. The condition of the road above this point seems highly variable: on some visits it has been badly potholed and washed away in places; a couple of months later it can be well graded and mostly paved up to Paraíso and beyond; the next time, it's in bad condition again!

At Km 32 you go through the small settlement of San Vicente (likely to be unsigned, as are most villages along this road), at 900 m elevation. From here it is about 13 km, climbing over a couple of ridges (to 1000 m) to the small town of Paraíso (800 m), shortly after which the continuous pavement, or what passes for it, ends. This whole stretch is good birding and usually can be reached in under an hour from Atoyac (or 2–2.5 hours from Acapulco). Stopping anywhere you see or hear activity is a good way to bird the road, while keeping an eye out for flowering trees such as *Inga* and *Cecropia* (both excellent for hummers). Between 1.5 and 3 km south of San Vicente and from 3.5 to 5 km north of San Vicente there are several good pulloffs and overlooks, where birds include Thicket Tinamou, King Vulture, Double-toothed and Hook-billed kites, Common Black and Great Black hawks, Black Hawk-Eagle, Barred Forest-Falcon, **West Mexican Chachalaca**, **Singing Quail**, Ruddy Quail-Dove, Pheasant Cuckoo, **Mexican Hermit**, **Short-crested Coquette**, **Green-fronted Hummingbird**, Long-billed Starthroat, Collared Trogon, **Russet-crowned Motmot**, Emerald Toucanet, **Grey-crowned Woodpecker**, Barred and Streak-headed woodcreepers, **Grey-collared Becard**, **Golden Vireo**, Tropical Parula, Red-legged Honeycreeper (summer), Red-crowned Ant-Tanager, **Rusty-crowned Ground-Sparrow**, and **Dickey's** [Audubon's] **Oriole**. At any point it is worth keeping an eye out overhead for swifts: flocks here, and especially over Paraíso, often include White-collared (common) and **White-naped** (uncommon) together, plus Chestnut-collared, Black (summer), and Vaux's. Violet Sabrewing and **White-tailed Hummingbird** occur between San Vicente and Paraíso, but both are commoner beyond Paraíso.

Above Paraíso the coffee plantations grade into cloud forest with a noticeably different avifauna. Reaching such habitat depends simply on the condition of the road. In case you're in luck and it's driveable, mark zero at the turn into Paraíso, 45 km (1.5–2 hours) from Atoyac, and bear left on the main road, rather than turning right into town. You wind along the edge of town, the center of which is off down to your right, cross a bridge over a rushing stream, and, at Km 3.0, pass a right-hand turn which may be signed "La Pintada 8" and which also heads back into Paraíso. At about Km 6.5 there is an obvious right-hand turnoff that can be taken for about 5.2 km to a minor track on the left: this leads 1.5 km, mostly through coffee plantations, to the "rancho" known as Arroyo Grande, simply a shack or two at about 1350 m. The Short-crested Coquette was rediscovered here in 1986. Birds along the Arroyo Grande track include Black Hawk-Eagle, **Long-tailed Wood-Partridge**, **Mexican Hermit**, Violet Sabrewing, **Short-crested Coquette**, **White-tailed** and **Bumblebee hummingbirds**, **Sparkling-tailed Woodstar**, Collared Trogon, Emerald Toucanet, Smoky-brown and **Grey-crowned woodpeckers**, six species of woodcreepers, Scaly-throated and Ruddy foliage-gleaners, Eye-ringed Flatbill, **Fan-tailed Warbler**, and White-winged and **Red-headed tanagers**.

Staying on the main road, rather than turning toward Arroyo Grande, you continue to climb and go through the strung-out settlement of Nueva Delhi at Km 18/21. The stretch of main road between the Arroyo Grande turn and Nueva Delhi has most if not all of the species listed for Arroyo Grande, including the coquette, as well as Bat Falcon and **Singing Quail**.

At about Km 25 you enter good cloud forest at about 1650 m, and the birding along the road for at least the next 10 km (to 2100 m) can be excellent, especially in early morning. Unless the road in the vicinity of Nueva Delhi has been improved greatly, you may need to allow up to two hours to reach the cloud forest from Paraíso, even though it's only 25 km! If the road is in bad shape, you could park in Nueva Delhi and walk or hitch a ride for a few kilometers to good habitat. Birds here include Crested Guan, **White-faced Quail-Dove** (often at the roadside in early morning), Violet Sabrewing, **Amethyst-throated** and **White-tailed hummingbirds**, **Mountain Trogon**, Scaly-throated and Ruddy foliage-gleaners, Scaled Antpitta, **White-throated** and **Unicolored jays**, American Dipper (stream crossings), Ruddy-capped Nightingale-Thrush, **Black Thrush, Chestnut-sided Shrike-Vireo**, **Golden-browed Warbler**, Chestnut-capped Brushfinch, and **Slate-blue Seedeater** (bamboo patches).

Sierra de Atoyac Bird List

250 sp. V: vagrant (not to be expected). †See pp. 5–7 for notes on English names.

Thicket Tinamou
Great Egret
Snowy Egret
Cattle Egret
Black Vulture
Turkey Vulture
King Vulture
Osprey
Double-toothed Kite
Hook-billed Kite
Sharp-shinned Hawk
Cooper's Hawk
Common Black Hawk
Great Black Hawk
Grey Hawk
Broad-winged Hawk
Short-tailed Hawk
Zone-tailed Hawk
Red-tailed Hawk
Black Hawk-Eagle
Laughing Falcon
Barred Forest-Falcon
Collared Forest-Falcon
American Kestrel
Bat Falcon
West Mexican Chachalaca
Crested Guan
Long-tailed Wood-Partridge
Singing Quail
Banded Quail
Red-billed Pigeon
Band-tailed Pigeon
Mourning Dove
Inca Dove
Common Ground-Dove
Ruddy Ground-Dove
White-tipped Dove
White-faced Quail-Dove
Ruddy Quail-Dove
Orange-fronted Parakeet
Lilac-crowned Parrot
Squirrel Cuckoo
Pheasant Cuckoo
Lesser Ground-Cuckoo
Lesser Roadrunner
Groove-billed Ani
Balsas Screech-Owl
Mountain Pygmy-Owl†
Ferruginous Pygmy-Owl
Burrowing Owl
Mottled Owl
Lesser Nighthawk
Pauraque
Eared Poorwill
Black Swift
Chestnut-collared Swift
White-collared Swift
White-naped Swift
Vaux's Swift
Great Swallow-tailed Swift
Mexican Hermit†
Violet Sabrewing
Green Violet-ear
Short-crested Coquette
Golden-crowned Emerald
Dusky Hummingbird
Doubleday's Hummingbird†
White-eared Hummingbird
Berylline Hummingbird
Cinnamon Hummingbird
Violet-crowned Hummingbird
Green-fronted Hummingbird
White-tailed Hummingbird
Amethyst-throated Hummingbird
Blue-throated Hummingbird
Garnet-throated Hummingbird
Magnificent Hummingbird
Long-billed Starthroat
Plain-capped Starthroat
Sparkling-tailed Woodstar†
Ruby-throated Hummingbird
Calliope Hummingbird
Rufous Hummingbird
Bumblebee Hummingbird
Citreoline Trogon
Mountain Trogon
Collared Trogon
Russet-crowned Motmot
Emerald Toucanet
Acorn Woodpecker
Golden-cheeked Woodpecker
Ladder-backed Woodpecker
Hairy Woodpecker
Smoky-brown Woodpecker
Grey-crowned Woodpecker
Lineated Woodpecker
Pale-billed Woodpecker
Red-shafted Flicker†
Scaly-throated Fol.-gleaner†
Ruddy Foliage-gleaner
Olivaceous Wood-creeper
Barred Woodcreeper
Ivory-billed Woodcreeper
Spotted Woodcreeper
Streak-headed Woodcreeper
Spot-crowned Woodcreeper
Scaled Antpitta
N. Beardless Tyrannulet
Greenish Elaenia
Eye-ringed Flatbill
Common Tufted Flycatcher†
Olive-sided Flycatcher
Greater Pewee
Western Pewee†
Least Flycatcher
Hammond's Flycatcher
Western Flycatcher†
Buff-breasted Flycatcher
Bright-rumped Attila
Dusky-capped Flycatcher
Nutting's Flycatcher
Great Kiskadee
Boat-billed Flycatcher
Social Flycatcher
Sulphur-bellied Flycatcher
Tropical Kingbird
Cassin's Kingbird

Thick-billed Kingbird
Western Kingbird
Rose-throated Becard
Masked Tityra
N. Rough-winged Swallow
White-throated Magpie-Jay
Steller's Jay
Green Jay
White-throated Jay
Western Scrub Jay
Unicolored Jay
Mexican Chickadee
Bridled Titmouse
Bushtit
Brown Creeper
Rufous-naped Wren
Boucard's Wren
Canyon Wren
Happy Wren
Banded Wren
Sinaloa Wren
Northern House Wren†
Brown-throated Wren†
Grey-breasted Wood-Wren
American Dipper
Ruby-crowned Kinglet
Blue-grey Gnatcatcher
White-lored Gnatcatcher
Eastern Bluebird
Brown-backed Solitaire
Orange-billed N.-Thrush
Russet N.-Thrush
Ruddy-capped N.-Thrush
Swainson's Thrush
Black Thrush†
White-throated Thrush†
Rufous-backed Thrush†
American Robin
Blue Mockingbird
Cedar Waxwing
Grey Silky†
Blue-headed Vireo†
Plumbeous Vireo†
Hutton's Vireo
Golden Vireo
Warbling Vireo
Yellow-green Vireo
Chestnut-sided Shrike-Vireo
Tennessee Warbler
Orange-crowned Warbler
Nashville Warbler
Colima Warbler
Crescent-chested Warbler
Tropical Parula
Yellow Warbler
Chestnut-sided Warbler (V)
Audubon's Warbler†
Townsend's Warbler
Hermit Warbler
Black-throated Green Warbler
Grace's Warbler
Black-and-white Warbler
American Redstart
Ovenbird
Louisiana Waterthrush
MacGillivray's Warbler
Common Yellowthroat
Wilson's Warbler
Red-faced Warbler
Red Warbler
Painted Whitestart†
Slate-throated Whitestart†
Fan-tailed Warbler
Golden-crowned Warbler
Rufous-capped Warbler
Golden-browed Warbler
Olive Warbler
Yellow-breasted Chat
Red-legged Honeycreeper
Godman's Euphonia†
Blue-hooded Euphonia
Red-crowned Ant-Tanager
Hepatic Tanager
Summer Tanager
Western Tanager
Flame-colored Tanager
White-winged Tanager
Red-headed Tanager
Common Bush-Tanager
Greyish Saltator
Black-headed Saltator
Yellow Grosbeak
Rose-breasted Grosbeak
Black-headed Grosbeak
Blue Grosbeak
Indigo Bunting
Varied Bunting
Orange-breasted Bunting
Painted Bunting
Rufous-capped Brushfinch
Chestnut-capped Brushfinch
Rusty-crowned Ground-Sparrow
Collared Towhee
Blue-black Grassquit
Cinnamon-rumped Seedeater†
Slate-blue Seedeater†
Cinnamon-bellied Flowerpiercer
Black-chested Sparrow
Rufous-crowned Sparrow
Rusty Sparrow
Lincoln's Sparrow
Mexican Junco†
Bronzed Cowbird
Orchard Oriole
Black-vented Oriole
Dickey's Oriole†
Streak-backed Oriole
Baltimore Oriole
Bullock's Oriole
Scott's Oriole
Yellow-winged Cacique
House Finch
Black-headed Siskin
Lesser Goldfinch
House Sparrow

CHAPTER 10 VERACRUZ

Sumichrast's Wren

Veracruz is a long and narrow state with 800 km of coastline bordering the southwestern edge of the Gulf of Mexico. The extremes in its physical geography, ranging from the permanently snow-capped Pico de Orizaba (at over 5600 m, the highest volcano in Mexico) to the steaming lowland rain forests of the Isthmus of Tehuantepec, are reflected in an avifauna of around 650 species. Indeed, Veracruz vies with Oaxaca and Chiapas for the highest bird diversity of any Mexican state and is home to such sought-after species as Bearded Wood-Partridge, Tuxtla [Purplish-backed] Quail-Dove, Long-tailed Sabrewing, and Sumichrast's and Nava's wrens.

The state, however, is also one of Mexico's most densely settled by people. Consequently, most of the original native forest has long since been cleared for agriculture (mostly cattle), and the tropical lowlands in particular present a rather depressing picture—finding accessible good habitat can be a challenge.

The length of the state in combination with the scattered nature of birding localities and the vast areas of "deforest" have seemingly led to Veracruz being mostly passed over by birders, other than target-oriented souls looking for a few endemics. However, the sites covered in this chapter can be made into a productive one- to two-week trip, particularly in March/April during spring migration. Alternatively, some sites could be combined into itineraries based out of other areas: Coatepec and Córdoba can be visited as overnight trips from Mexico City (Chapter 8), and Uxpanapa could be added into an itinerary covering eastern Oaxaca (Chapter 11) and/or Central and Southern Chiapas (Chapter 12). In years past, many of the sites in Veracruz (south to and including Los Tuxtlas) were popular destinations for birders driving down from the U.S.A., and this is still a very viable option (see Chapter 4).

Veracruz City is served daily by flights from Mexico City, and car rental is possible with several companies at the airport, although the choice is not as wide as you'll find at more tourist-oriented locations.

Suggested Itineraries

If you fly into Veracruz in the morning from Mexico City, your first afternoon can be used for driving to one of the more distant sites, such as Catemaco in Los Tuxtlas (Site 10.6), perhaps via Las Barrancas (Site 10.5), or north to Tecolutla (Site 10.1), perhaps via Colonia Francisco Barrios (Site 10.4).

Any number of days can be spent easily in Los Tuxtlas, from which, if you're interested in more rain-forest birding, you can head down to Uxpanapa (Site 10.7) for a day or two. Alternatively, head back north to Veracruz City and on to Tecolutla (a day's drive from Catemaco), or (a shorter drive) to Córdoba, from which you can bird Amatlán (Site 10.3) and then head north to Tecolutla.

Similarly, if you start in Tecolutla, Los Tuxtlas can be reached in a day's drive, or you could stop over in Veracruz City (for Sites 10.4, 10.5) or Córdoba (for Site 10.3).

Coatepec (Site 10.2) and Las Choapas (Site 10.8) are primarily one-species sites (Bearded Wood-Partridge and Spot-tailed Nightjar, respectively) for target-oriented birders but can easily be fitted into any itinerary.

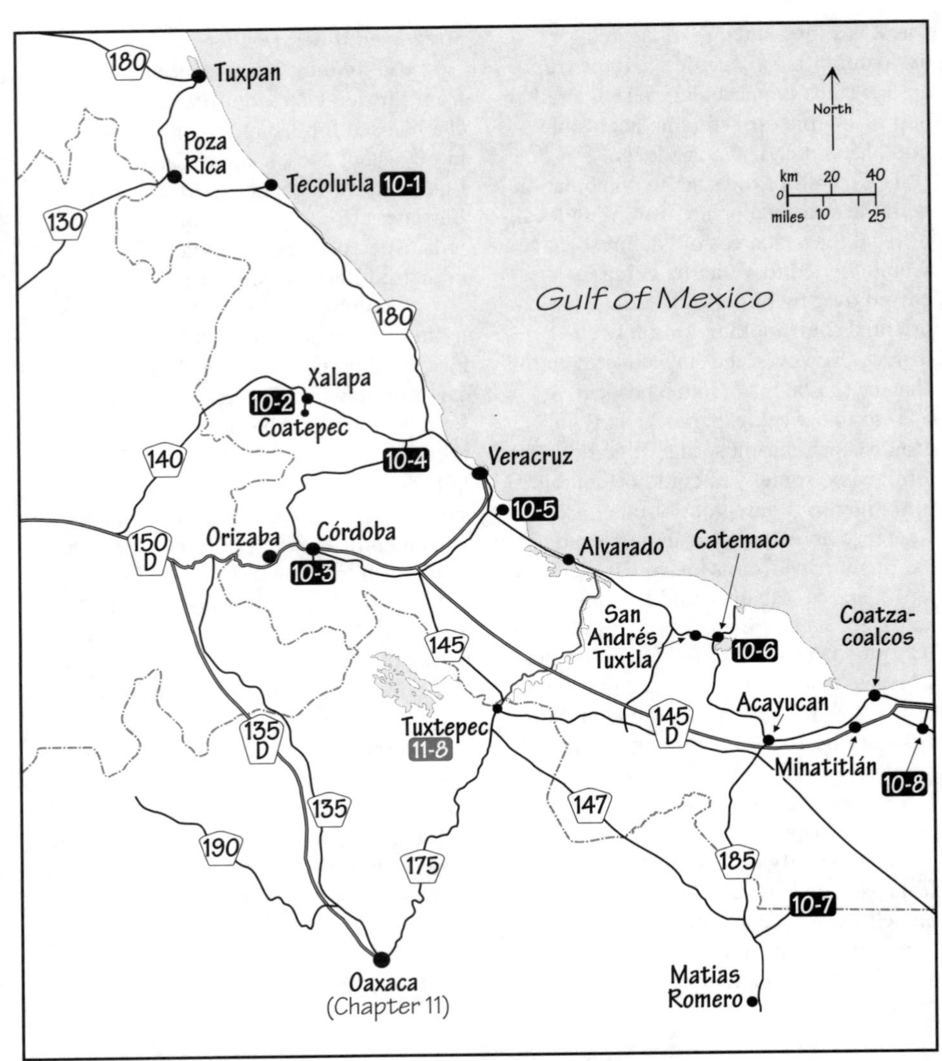

Chapter 10: VERACRUZ

VERACRUZ SITE LIST

Site 10.1: Tecolutla

(March, April 1983, April 1985, February, April 1990)

Introduction/Key Species

Tecolutla is a small coastal fishing town sited at the mouth of the Río Tecolutla in northern Veracruz. Its main attraction is the freshwater marshes just inland of town, although, inevitably, these are being drained and it may be increasingly difficult to find good, easily accessible habitat. The town also makes a good base from which to observe the spectacular northward passage of many species during late March and April: hundreds to thousands of White Pelicans, Anhingas, Franklin's Gulls, Turkey Vultures, Sharp-shinned and Broad-winged hawks, White-winged Doves (up to 14,000 per hour!), Lesser Nighthawks (at dusk), Chimney Swifts, Scissor-tailed Flycatchers, swallows, and Great-tailed Grackles stream by throughout the day, along with smaller numbers of many other species. A description of this phenomenon is given by Howell and Webb (1987).

Other birds of interest at Tecolutla include Pinnated, American (winter) and Least bitterns, Bare-throated Tiger-Heron, Boat-billed Heron, Muscovy Duck, Lesser Yellow-headed Vulture, Ruddy and Yellow-breasted crakes, King Rail, Collared Plover, Blue Ground-Dove, Sedge Wren (winter), **Altamira Yellowthroat**, Swamp Sparrow (winter), distinctive races of White-eyed Vireo (*V. g. perquisitor*) and Orchard Oriole (**Ochre Oriole**; summer), and a variety of species considered by many birders to be "South Texas specialities." Sungrebe and Spotted Rail also have been found here.

Access

Tecolutla is about 200 km (2.5–3.5 hours) north of Veracruz City, and about 300 km (4–5 hours) south of Tampico. It is reached by turning east off the main "coastal" highway, Route 180, in the town of Gutierrez Zamora. The junction is well signed, after which you follow the one-way system through Gutierrez Zamora, whence it is about 11 km via paved highway to Tecolutla.

Tecolutla is a Mexican beach resort (and can be very crowded at times, such as Easter). Consequently, there are several hotels and guest houses, which cover a range of budgets, and numerous restaurants. Public buses serve the town regularly from Gutierrez Zamora.

Birding Sites

All birding sites are within a short walk of town, or a five-minute drive or ride to the marshes 2–4 km inland. Immediately inland of Tecolutla, look for a dirt (sand) track to the north (on your left as you come into town) that cuts across fields and a marshy area for a few hundred meters to some homes and the beach on the north side of town. There is a fairly extensive freshwater marsh to the left, or west, of the track, although this is being increasingly cut into and used for cattle grazing. It is possible to get through the fence and walk about in the marsh (be prepared for wet and muddy feet, or wear wellies) in the hope of flushing Pinnated Bittern (often common), Yellow-breasted Crake (especially along the edges of channels adjacent to taller reeds; I've flushed up to seven birds in a morning), other rails, Sedge Wren (winter), and Swamp Sparrow (winter). The short length of raised dike here affords a view of the marsh (best early and late in the day for waterbirds flying around, including Muscovy Duck) and is a good vantage point for observing diurnal migration. **Altamira Yellowthroats** occur in the reed beds here but are outnumbered greatly for much of the year by migrant Commons which, frustratingly, tend to respond much more strongly to pishing. In summer (March to August) the areas of marshy acacia scrub and adjacent hedges hold **Ochre** [Orchard] **Orioles**.

The raised dike or, if you're less than excited about getting back through the marsh and fence after dark, the main track itself have been good points from which to observe a post-roost flight of Boat-billed Herons at dusk (these fly over later than both species of night-herons, which also occur); be aware that you may encounter plenty of mosquitoes if you stay out late!

The beach in town can be worth checking, especially if you get there early and make the short walk south to the river mouth (Reddish Egret, Collared Plover, terns), especially during migration. It's also possible to rent boats in town to go up the Río Tecolutla or, more productively for birds, up the mangrove-fringed tributary opposite Tecolutla, on the south bank (Bare-throated Tiger-Heron, Sungrebe, kingfishers).

One can also bird some areas of freshwater marsh just north of the road into Tecolutla, about 2–4 km back out (west) toward Gutierrez Zamora. There are a least two gated tracks where one can pull off and park safely, open and shut the gate (ask permission if you see anyone), and walk around the marshes, pools, and adjacent fields, acacia scrub, and woodlots. Birds here include Pinnated Bittern, **Green Parakeet**, **Tamaulipas Crow**, Black-crested [Tufted] Titmouse, Sprague's Pipit (winter), **Altamira Yellowthroat**, and **Ochre** [Orchard] **Oriole** (summer). The coconut plantations between the highway and the river, for a kilometer or two west of town, can also be birded but don't seem to hold too much of interest, although parakeets, Grey-breasted Martins, and Mangrove Swallows nest in the dead palm stumps.

Tecolutla Bird List

217 sp. †See pp. 5–7 for notes on English names.

Least Grebe
Pied-billed Grebe
American White Pelican
Brown Pelican
Double-crested Cormorant
Neotropic Cormorant
Anhinga
Magnificent Frigatebird
Pinnated Bittern
American Bittern
Least Bittern
Bare-throated Tiger-Heron
Great Blue Heron
Great Egret
Snowy Egret
Little Blue Heron
Tricolored Heron
Reddish Egret
Cattle Egret
Green Heron
Black-crowned Night-Heron
Yellow-crowned Night-Heron
Boat-billed Heron
White Ibis
White-faced Ibis
Roseate Spoonbill
Wood Stork
Fulvous Whistling-Duck
Black-bellied Whistling-Duck
Muscovy Duck
Green-winged Teal
Northern Pintail
Blue-winged Teal
Cinnamon Teal
Northern Shoveler
Gadwall
American Wigeon
Black Vulture
Turkey Vulture
Lesser Yellow-hd. Vulture
Osprey
White-tailed Kite
Mississippi Kite
Northern Harrier
Sharp-shinned Hawk
Cooper's Hawk
Common Black Hawk
Great Black Hawk
Harris' Hawk
Grey Hawk
Roadside Hawk
Broad-winged Hawk
Short-tailed Hawk
Swainson's Hawk
Red-tailed Hawk
Crested Caracara
Laughing Falcon
American Kestrel
Merlin
Peregrine Falcon
Plain Chachalaca
Northern Bobwhite
Limpkin
Ruddy Crake
Yellow-breasted Crake
King Rail
Virginia Rail
Sora
Purple Gallinule
Common Moorhen
American Coot
Black-bellied Plover
American Golden Plover
Collared Plover
Snowy Plover
Semipalmated Plover
Wilson's Plover
Killdeer
Black-necked Stilt
American Avocet
Northern Jacana
Greater Yellowlegs
Lesser Yellowlegs
Solitary Sandpiper
Spotted Sandpiper
Upland Sandpiper
Whimbrel
Sanderling
Semipalmated Sandpiper
Western Sandpiper
Least Sandpiper
Baird's Sandpiper
Pectoral Sandpiper
Stilt Sandpiper
Long-billed Dowitcher
Common Snipe
Wilson's Phalarope
Laughing Gull
Franklin's Gull
Ring-billed Gull
Herring Gull
Gull-billed Tern
Caspian Tern
Royal Tern
Sandwich Tern
Black Skimmer
Red-billed Pigeon
White-winged Dove
Mourning Dove
Common Ground-Dove
Ruddy Ground-Dove
Blue Ground-Dove

White-tipped Dove
Green Parakeet
Aztec Parakeet†
Red-lored Parrot
Groove-billed Ani
Barn Owl
Ferruginous Pygmy-Owl
Lesser Nighthawk
Pauraque
Chuck-will's-widow
Chimney Swift
Green-breasted Mango
Buff-bellied Hummingbird
Ruby-throated Hummingbird
Ringed Kingfisher
Belted Kingfisher
Amazon Kingfisher
Green Kingfisher
Golden-fronted Woodpecker
Yellow-bellied Sapsucker
Ladder-backed Woodpecker
Lineated Woodpecker
Willow/Alder Flycatcher
Least Flycatcher
Eastern Phoebe
Vermilion Flycatcher
Great Crested Flycatcher
Brown-crested Flycatcher
Great Kiskadee
Boat-billed Flycatcher
Social Flycatcher
Tropical Kingbird
Couch's Kingbird
Eastern Kingbird
Scissor-tailed Flycatcher
Purple Martin
Grey-breasted Martin
Mangrove Swallow
Tree Swallow
N. Rough-winged Swallow
Bank Swallow
Cliff Swallow
Barn Swallow
Green Jay
Brown Jay
Tamaulipas Crow†
Black-crested Titmouse†
Band-backed Wren
Spot-breasted Wren
Northern House Wren†
Sedge Wren
Marsh Wren
Blue-grey Gnatcatcher
Clay-colored Thrush†
Grey Catbird
Northern Mockingbird
American Pipit
Sprague's Pipit
Loggerhead Shrike
White-eyed Vireo
Blue-winged Warbler
Nashville Warbler
Orange-crowned Warbler
Yellow Warbler
Myrtle Warbler†
Black-and-white Warbler
American Redstart
Prothonotary Warbler
Northern Waterthrush
Common Yellowthroat
Altamira Yellowthroat
Grey-crowned Yellowthroat
Wilson's Warbler
Yellow-breasted Chat
Scrub Euphonia
Yellow-throated Euphonia
Blue-grey Tanager
Yellow-winged Tanager
Summer Tanager
Western Tanager
Greyish Saltator
Black-headed Saltator
Rose-breasted Grosbeak
Indigo Bunting
Olive Sparrow
Blue-black Grassquit
Yellow-faced Grassquit
White-collared Seedeater
Lark Sparrow
Grasshopper Sparrow
Savannah Sparrow
Lincoln's Sparrow
Swamp Sparrow
Red-winged Blackbird
Eastern Meadowlark
Melodious Blackbird
Great-tailed Grackle
Bronzed Cowbird
Orchard Oriole
Ochre Oriole†
Audubon's Oriole
Yellow-tailed Oriole
Altamira Oriole
Baltimore Oriole
Yellow-billed Cacique
House Sparrow

Site 10.2: Coatepec

(September 1995, January 1998)

Introduction/Key Species

Coatepec has become known recently as a site to find the poorly known **Bearded Wood-Partridge**, and PRONATURA-Veracruz, a local conservation organization, has an active program to help locate, study, and protect populations of this threatened bird. Looking for the wood-partridge should not be taken lightly, and unless you are specifically seeking this elusive endemic, all of the other species here can be seen more easily elsewhere. Trails are often steep and can be muddy, and you should be in reasonable physical shape. Also, the *maximum* group size when looking for this bird should be four or five persons (including guide). See Gomez de Silva and Aguilar (1994) for an account of finding the Bearded Wood-Partridge near Coatepec.

The habitat around Coatepec mostly comprises shaded coffee plantations, steep slopes of bracken fern, and small forest patches. Other birds of interest found in the area include **White-faced Quail-Dove**, **Wedge-tailed Sabrewing**, **Bumblebee Hummingbird**, **Bronze-winged Woodpecker**, Ruddy Foliage-gleaner, Orange-billed and Ruddy-capped nightingale-thrushes, **Blue Mockingbird**, and **White-naped Brushfinch**.

Access

The town of Coatepec (1200 m elevation) is about 10 km (20–30 minutes) southwest of Xalapa (also spelled Jalapa), the state capital of Veracruz, whence it is reached by a paved highway. Xalapa is about 330 km (4–4.5 hours using the *cuota* highways) east of Mexico City, and about 110 km (1.2–1.5 hours) from Veracruz City. There are basic hotels in Coatepec and a wide choice of accommodations in Xalapa. Public buses run regularly between Xalapa and Coatepec.

Getting to Coatepec from Xalapa. *Coming from Mexico City*, the highway into downtown Xalapa becomes Avenida 20 de Noviembre. Follow this toward the city center and fork right onto Avenida Avila Camacho for several blocks, then turn right onto Ignacio de la Llave and, after a few more blocks, right again onto the road to Coatepec. *Coming from Veracruz*, as you enter the outskirts of Xalapa, watch for the Fiesta Inn on your right; 1.5 km after this landmark, reset to zero as you turn left (well signed) onto Murillo Vidal, a wide, divided highway. Stay on this road into the city center (at 3.2 km you pass the main post office on the left and the street becomes one-way), and at 3.8 km there is a cathedral on the right and a plaza on the left. Turn left and immediately right at the end of the plaza (4.0 km), staying on the main street and following the traffic flow; at 4.8 km, just after a traffic light, the street becomes divided—take the *left-hand* side of the divide (it's still all one-way!) and at 4.9 km turn left, signed to Coatepec. At 5.0 km you pass around the right-hand edge of a *glorieta*, at the far side of which you turn right, signed to Coatepec. This becomes a newly paved divided highway out of town, and it's about 8 km to the edge of Coatepec.

Birding Sites

It is a good idea to contact PRONATURA-Veracruz in advance of your trip, as the organization may be able to help arrange a local guide. Contact details for PRONATURA-Veracruz are: phone (most personnel speak English) (011-52) (28) 128844; fax (011-52) (28) 129415; and e-mail: verpronatura@laneta.apc.org. The address is PRONATURA-Veracruz, Museo de Ciencia y Tecnología, Av. Murillo Vidal s/n, Apdo. Postal 399, Xalapa, Veracruz 91069, MEXICO. The executive director and a good field birder is Ernesto Ruelas (ruelas@compuserve.com), and the biologist in charge of the Bearded Wood-Partridge project is Sergio Aguilar (chencho@edg.net.mx), although other personnel also may be able to help you contact a guide. As always, donations to support the important conservation work of PRONATURA-Veracruz will be appreciated.

Pedro Mota, a local bird trapper, has guided many birders in search of the wood-partridge. He lives on the west edge of Coatepec at Los Carriles 103 (i.e., number 103, Los Carriles Street), which is best found by asking directions when you get to town. He speaks no English, but if your Spanish is up to it, you can contact him by telephone at (011-52) (28) 164128; it is best to ask for *Don* Pedro Mota (using this common term of respect helps greatly in interactions with Mexican landowners and elders). The local name for the **Bearded Wood-Partridge** is *chivizcoyo*, and if you say you want to look for this bird, Don Pedro will know you are a birder. You will need to start in early morning (usually while still dark), picking up Pedro Mota at his house and following directions to the site, which is best reached with a high-clearance vehicle. You may be out all day, on steep terrain, so take all your food and water.

Note, though, that even with Pedro Mota guiding you, and whistled imitations (he has whistled birds in to within a few meters!) and tape playback, seeing this elusive bird is far from guaranteed. Even if you are unsuccessful, a tip in the order of around $10–15/person should be given to appreciate your gratitude. Also, in the interests of alerting local people to the potential economic value of this bird, when in Coatepec you should make it known that looking for (but not trapping!) the *chivizcoyo* is the reason for your visit.

In case you are unable to contact Pedro Mota or PRONATURA, it may be possible to find the wood-partridges from a dirt road near Coatepec, at a site known as Guitarrero. However, you probably either will need a vehicle with high clearance or should be prepared to walk an extra kilometer or several, depending on the condition of the roads.

Entering Coatepec from Xalapa you hit a *glorieta* at the start of town, with a Pemex station on your right. Take the second exit off this *glorieta*, and in about 1 km you come to the center of Coatepec, where you should turn right so that a large church is immediately on your left and the *zócalo* is ahead on the left in the next block. Turn right again on the first street (i.e., at the near corner of the *zócalo*) and continue north until you cross a small bridge (1.5 km?) and then start to go steeply uphill; turn left about 100 m after the bridge and continue through the outskirts of town about 500 m, until a T-junction with a large church straight ahead. Reset to zero, turn left here, and follow the road around a right-hand bend and downhill, heading out of town and into coffee plantations. At 1.0 km you pass a large coffee building on the right; after this building stay straight (not right) and follow the main road up and down through more coffee plantations. At 1.9 km, pass another right turn, bend sharply right at 3.8 km and look for an area to pull off on the left at 4.3 km, where a narrow path leads away upslope. The shaded coffee plantations to your left here have **Bearded Wood-Partridge**, and late (or early) in the day is the best time to listen for their presence—for instance, as groups call before going to roost.

Coatepec Bird List

86 sp. †See pp. 5–7 for notes on English names.

Cattle Egret
Black Vulture
Turkey Vulture
Osprey
Cooper's Hawk
Grey Hawk
Roadside Hawk
Red-tailed Hawk
Plain Chachalaca
Bearded Wood-Partridge
White-tipped Dove
White-faced Quail-Dove
Squirrel Cuckoo
Mottled Owl
Pauraque
Chestnut-collared Swift
White-collared Swift
Vaux's Swift
Wedge-tailed Sabrewing
Violet Sabrewing
Azure-crowned Hummingbird
Amethyst-throated Hummingbird
Magnificent Hummingbird
Bumblebee Hummingbird
Violaceous Trogon
Collared Trogon
Blue-crowned Motmot
Acorn Woodpecker
Golden-fronted Woodpecker
Bronze-winged Woodpecker†
Ruddy Foliage-gleaner
Olivaceous Woodcreeper
Ivory-billed Woodcreeper
Spot-crowned Woodcreeper
Barred Antshrike
Greenish Elaenia
Greater Pewee
Eastern Pewee†
Western Pewee†
Common Tufted Flycatcher†
Yellow-bellied Flycatcher
Least Flycatcher
Western Flycatcher†
Dusky-capped Flycatcher
Great Crested Flycatcher
Social Flycatcher
Boat-billed Flycatcher
Tropical Kingbird
N. Rough-winged Swallow
Brown Jay
Band-backed Wren
Spot-breasted Wren
Northern House Wren†
Grey-breasted Wood-Wren
Blue-grey Gnatcatcher
Brown-backed Solitaire
Orange-billed N.-Thrush
Ruddy-capped N.-Thrush
White-throated Thrush†
Grey Catbird
Blue Mockingbird
Grey Silky†
Blue-headed Vireo†
Blue-winged Warbler
Nashville Warbler
Crescent-chested Warbler
Tropical Parula
Black-throated Green Warbler
MacGillivray's Warbler
Hooded Warbler
Wilson's Warbler
Rufous-capped Warbler
Golden-browed Warbler
Slate-throated Whitestart†
Yellow-throated Euphonia
Common Bush-Tanager
Buff-throated Saltator
Black-headed Saltator
White-naped Brushfinch
Chestnut-capped Brushfinch
Cinnamon-bellied Flowerpiercer
Rusty Sparrow
Great-tailed Grackle
Montezuma Oropendola
Baltimore Oriole
Lesser Goldfinch

Site 10.3: Amatlán

(January 1984, February 1987, April, May 1990, April 1992, October 1993, September 1995, January 1998)

Introduction/Key Species

The sharp, karst limestone outcrops around the village of Amatlán are used for growing coffee under a varied canopy of shade trees. Consequently, much of the character of the original forest has been retained, in stark contrast to other areas of Veracruz, where "sun coffee" (grown without shade trees) has created a virtual avian desert. One species among many that have benefited from the retention of surrogate forest is the very localized **Sumichrast's Wren**, endemic to forested karst outcroppings in eastern Mexico. This great bird can be found fairly readily near Amatlán, where other species of interest include Thicket Tinamou, **Singing Quail**, **Wedge-tailed Sabrewing**, three species of toucans, **Fan-tailed Warbler**, and Chestnut-headed and Montezuma oropendolas.

Access/Birding Sites

Amatlán is about 8 km (20–30 minutes) south of Córdoba, a major city and the center of Mexico's coffee-growing industry. Córdoba is about 120 km (1–1.5 hours) inland (west) from Veracruz City via a fast *cuota* highway and is about 300 km (3–4 hours by *cuota* highways, 6 or more hours via the truck-clogged *libre* highways) east of Mexico City. As one would expect, there are numerous restaurants and hotels in Córdoba.

To get to Amatlán from Veracruz there now appear to be two possible routes. **The traditional route** is to get off the *cuota* highway in response to signs for the first exit for Córdoba. In about 3 km you cross a railway line, where you should reset to zero, and the road becomes a divided four-lane. At about 3.0 km get in the left lane and at 3.5 km turn left (*not well signed*) and bear right on the main road around the edge of town; at 4.3 km there is a Pemex station (ES number 0433) on your left. *Immediately* past the Pemex, reset to zero and turn left and then left again, onto the street heading back east on the opposite side of the Pemex. Follow this for a kilometer, through a rather run-down part of town, to where it bends sharply right and crosses the railway tracks (watch your clearance on this crossing; it's not the smoothest); about 100 m after the tracks, turn sharply left on what is actually the main road (straight on looks fairly main but soon peters out). Stay on this narrow, two-lane paved road, which crosses the railway again at Km 2.0 and 2.8, passes a big sugar refinery on the left (southeast) at Km 3.0, crosses over the *cuota* highway at Km 3.7, and comes into the village of Amatlán at Km 5.0.

The new route is via an exit off the *cuota* highway, which should be signed for Amatlán. This is several kilometers past (east of) the Córdoba exit if you're coming from Veracruz and is at about Km Post 297. Exit here and head south, entering Amatlán about 2 km from the *cuota* highway. Note that in early 1998 almost all of this *cuota* exit was completed; the exception was the part leading back onto the Veracruz-bound lane. Presumably construction will soon be finished on the whole junction.

As you come into Amatlán, bear right 1.5 km from the *cuota* overpass at the first main junction (straight on is the way back out on the one-way system), then left at the next street, which comes to the *zócalo* in about two blocks. The road to the birding area leaves from the southeast corner of the *zócalo* (to the left of the large church which should be straight ahead on the far side). This road is minor, and parts of it are badly potholed and bumpy, though the worst stretch is generally within a few hundred meters of the *zócalo*, near the small cemetery on your right in 800 m. At this point, forested hills should be visible ahead: follow the road to where it ends at an open grassy area, 2 km from the Amatlán *zócalo*. There is room to park here and follow the obvious main track from the near end (where you enter) of the parking area down left and around to the right and up the other side into the coffee plantations. The small quarry here may be active (respect any blasting activities!), but the track is public and leads into the coffee, and on to some settlements, so it should always be possible to pass the quarry. The initial ascent of 100 m or so is fairly steep (the rocks can be very slippery when wet), but then the main trail levels out more for a few hundred meters.

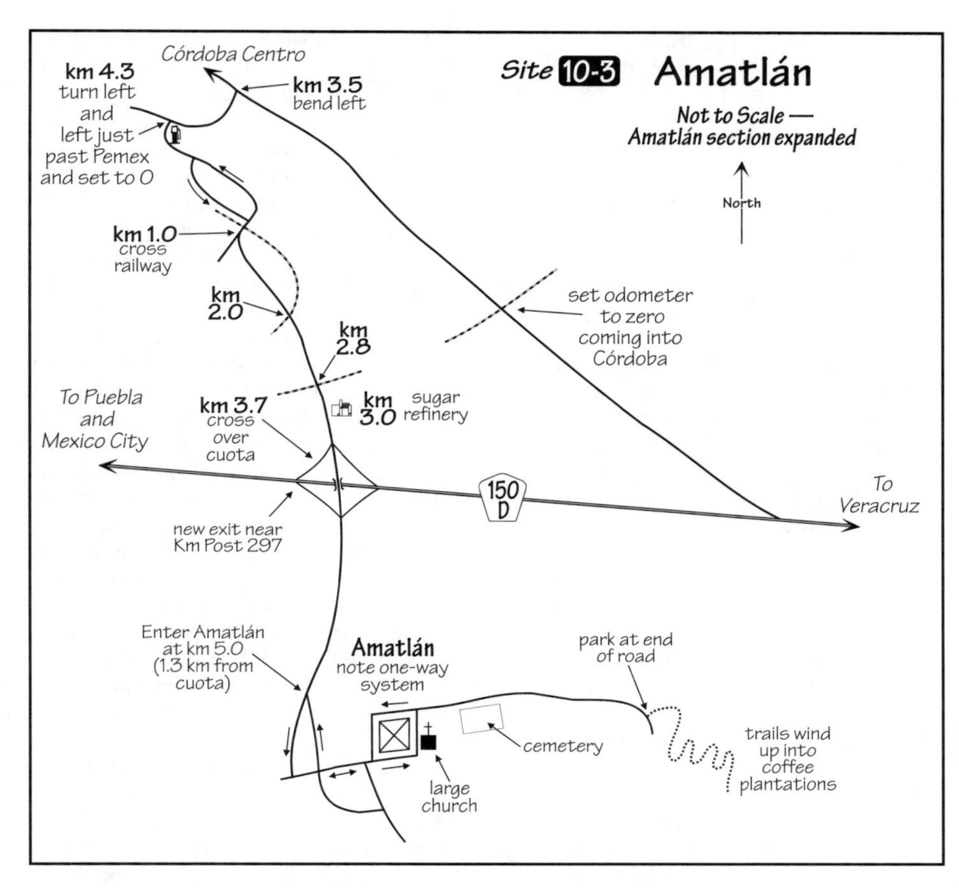

The trails through the coffee plantations split and meander, so make sure you keep an eye on your bearings. However, **Sumichrast's Wrens** occur along the first few hundred meters of trail, from where you enter the "forest," so you shouldn't need to go far enough to worry about getting lost. The wrens can be very elusive, but also very curious and confiding, and can take a while to see, even if they are singing and calling from all around.

Amatlán is served by public buses from Córdoba, though once in Amatlán you'll probably have to walk to the forest, as traffic on the dead-end road is light at best.

Amatlán Bird List

89 sp. †See pp. 5–7 for notes on English names.

Thicket Tinamou
Black Vulture
Turkey Vulture
Sharp-shinned Hawk
Short-tailed Hawk
Plain Chachalaca
Singing Quail
White-tipped Dove
Aztec Parakeet†
White-crowned Parrot
Red-lored Parrot
Squirrel Cuckoo
Ferruginous Pygmy-Owl
Mottled Owl
Chestnut-collared Swift
White-collared Swift
Vaux's Swift
Little Hermit
Wedge-tailed Sabrewing
White-bellied Emerald
Violaceous Trogon
Collared Trogon
Blue-crowned Motmot
Emerald Toucanet
Collared Aracari
Keel-billed Toucan
Golden-fronted Woodpecker
Golden-olive Woodpecker
Lineated Woodpecker
Olivaceous Woodcreeper
Ivory-billed Woodcreeper
Yellow-olive Flycatcher
Eastern Pewee†

Yellow-bellied Flycatcher
Least Flycatcher
Western Flycatcher†
Dusky-capped Flycatcher
Great Crested Flycatcher
Great Kiskadee
Boat-billed Flycatcher
Social Flycatcher
Masked Tityra
Rough-winged Swallow sp.
Green Jay
Brown Jay
Band-backed Wren
Sumichrast's Wren†
Spot-breasted Wren
Northern House Wren†
White-breasted Wood-Wren
Blue-grey Gnatcatcher
Wood Thrush
Clay-colored Thrush†
Cedar Waxwing
Blue-headed Vireo†
Warbling Vireo
Lesser Greenlet
Blue-winged Warbler
Nashville Warbler
Tropical Parula
Magnolia Warbler
Black-throated Green Warbler
Blackburnian Warbler
Black-and-white Warbler
American Redstart
Common Yellowthroat
Hooded Warbler
Wilson's Warbler
Slate-throated Whitestart†
Fan-tailed Warbler
Rufous-capped Warbler
Golden-crowned Warbler
Scrub Euphonia
Yellow-throated Euphonia
Red-crowned Ant-Tanager
Red-throated Ant-Tanager
Summer Tanager
White-winged Tanager
Common Bush-Tanager
Black-headed Saltator
Chestnut-capped Brushfinch
Blue-black Grassquit
White-collared Seedeater
Melodious Blackbird
Great-tailed Grackle
Bronzed Cowbird
Chestnut-headed Oropendola
Montezuma Oropendola
Yellow-billed Cacique

Site 10.4: Colonia Francisco Barrios

(April, May 1990, January 1998)

Introduction/Key Species

This site comprises a remnant pocket of arid thorn forest in the coastal lowlands of central Veracruz. Birds of interest here include Thicket Tinamou, Buff-collared Nightjar, **Mexican Sheartail**, Rufous-naped and **White-bellied wrens**, and Botteri's Sparrow. An account of the birding here is given by Howell and Webb (1990).

Access/Birding Sites

This site is about 60–70 km by road, depending on the route you take (forty-five minutes to an hour) northwest of Veracruz City. It can be birded as a morning or afternoon trip from Veracruz, or as a side trip if you are driving either along the coastal highway (Route 180) or along the highway inland to Xalapa (and Coatepec, Site 10.2). There is a basic hotel in Paso de Ovejas and a wide choice of accommodation in Veracruz City. Public buses transit the highway through this site regularly, but only second-class buses are likely to stop in Francisco Barrios.

To get from Veracruz to the start of the highway inland to Huatusco (Route 125), one can take two routes: either the old highway inland (slow and often bumpy) to Paso de Ovejas, where one continues north through town toward Puente Nacional and turns left (west) in the village of Conejos onto Route 125 (signed for Huatusco); or the newer and faster *cuota* highway up the coast north to Cardel, thence inland (west) on Route 140 (the highway to Xalapa [also spelled Jalapa]). In about 13 km, turn left (south) in Tamarindo (which is little more than a highway junction) and continue for about 5 km, through Puente Nacional, to Conejos, where you turn right (west) onto the Huatusco road (Route 125).

Mark zero as you turn onto the Huatusco road in Conejos. The road runs through mostly cutover, brushy fields and cattle land, and at Km 3.2 a belt of thorn forest begins on the right (north) side of the road and extends for a few kilometers, until the small village of Colonia Francisco Barrios, at about Km 6.5. A couple of tracks allow access into the habitat, which also can be birded from the highway, although be aware that traffic can be heavy and fast-moving.

At Km 5.4 look for a track off to the right (north) into the thorn forest. One can drive in on this track and park safely off the road. Continue on foot through the thorn forest, which, sadly, is a belt only about 200 m wide, and you'll come to a cleared, cutover area of low scrub past (north of) the thorn forest.

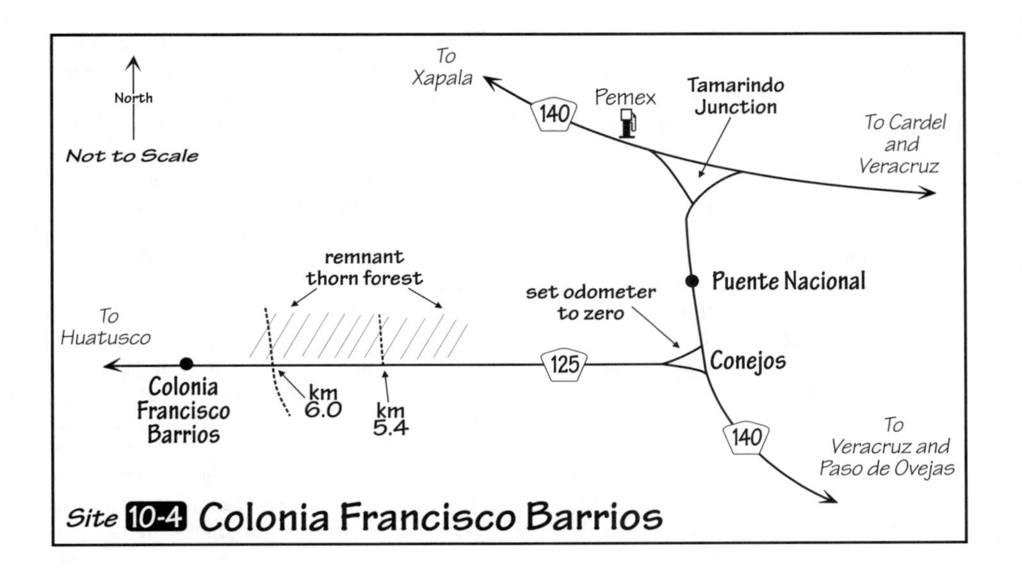

Species along this track and at the edge by the highway include Thicket Tinamou, Buff-collared Nightjar, Ferruginous Pygmy-Owl, **Canivet's Emerald**, Black-headed Trogon, Rufous-naped and **White-bellied wrens**, Varied Bunting, and Olive and Botteri's sparrows. **Mexican Sheartails** can be seen here but have been easier to find a little farther along the highway, in the scrubby fields with scattered acacias beside the road between Km 6 and 6.5, that is, immediately before (east of) Francisco Barrios. The sheartails, however, probably move around seasonally and may occur wherever there are suitable flowers.

Another track leads into the habitat on the left (south) of the highway at Km 6, and although the habitat is largely cutover, some of the species listed above can be found here (including sheartails).

Colonia Francisco Barrios Bird List
94 sp. R: rare (not to be expected). †See pp. 5–7 for notes on English names.

Thicket Tinamou
Great Egret
Cattle Egret
Black Vulture
Turkey Vulture
White-tailed Kite
Northern Harrier
Sharp-shinned Hawk
Roadside Hawk
Crested Caracara
Laughing Falcon
American Kestrel
Bat Falcon
Plain Chachalaca
Northern Bobwhite
Red-billed Pigeon
White-winged Dove
Mourning Dove
Inca Dove
Common Ground-Dove
White-tipped Dove
Aztec Parakeet†
Red-lored Parrot
Squirrel Cuckoo
Groove-billed Ani
Ferruginous Pygmy-Owl
Pauraque
Buff-collared Nightjar
White-collared Swift
Chimney Swift
Canivet's Emerald
Buff-bellied Hummingbird
Mexican Sheartail
Black-headed Trogon
Blue-crowned Motmot
Golden-fronted Woodpecker
Ladder-backed Woodpecker
N. Beardless Tyrannulet
Least Flycatcher
Willow/Alder Flycatcher
Vermilion Flycatcher
Great Crested Flycatcher
Brown-crested Flycatcher
Great Kiskadee
Boat-billed Flycatcher
Social Flycatcher
Sulphur-bellied Flycatcher
Scissor-tailed Flycatcher
Tropical Kingbird
Couch's Kingbird
Western Kingbird (R)
Eastern Kingbird
Masked Tityra
Rose-throated Becard
N. Rough-winged Swallow
Bank Swallow
Cliff Swallow
Barn Swallow
Brown Jay
Rufous-naped Wren
White-bellied Wren

Northern House Wren†
Blue-grey Gnatcatcher
White-eyed Vireo
Blue-headed Vireo†
Tennessee Warbler
Orange-crowned Warbler
Northern Parula
Yellow Warbler
Magnolia Warbler
Myrtle Warbler†
Grey-crowned Yellowthroat
Yellow-breasted Chat
Yellow-winged Tanager
Summer Tanager
Black-headed Saltator
Northern Cardinal
Rose-breasted Grosbeak
Indigo Bunting
Varied Bunting
Dickcissel
Olive Sparrow
Botteri's Sparrow
Vesper Sparrow (R?)
Grasshopper Sparrow
Blue-black Grassquit
White-collared Seedeater
Melodious Blackbird
Great-tailed Grackle
Bronzed Cowbird
Altamira Oriole
Baltimore Oriole
Montezuma Oropendola
Lesser Goldfinch

Site 10.5: Las Barrancas

(March, April 1983, January, February 1984, February, April 1985, April 1990, 1992)

Introduction/Key Species

This is an area of coastal short-grass prairie which, as well as being a site for some resident specialities, is an important stopover for long-distance grassland shorebirds and a wintering area for grassland passerines. Scrub and seasonal wetlands add to the diversity of habitats and birds found here.

Species of interest include Pinnated Bittern, Lesser Yellow-headed Vulture, Aplomado Falcon, Double-striped Thick-knee, American Golden Plover (migrant), Upland and Buff-breasted sandpipers (migrants), Striped Cuckoo, **Rufous-breasted Spinetail**, Fork-tailed Flycatcher, Sprague's Pipit (winter), Grassland Yellow-Finch, and Grasshopper Sparrow (winter).

Access/Birding Sites

This site is about 40 km (30–45 minutes) south of Veracruz City, beside the main coastal highway. Head south from Veracruz City on Route 180 toward Alvarado and Los Tuxtlas (Site 10.6) and note the junction for the highway (Route 150) inland to Córdoba, marked by a big Pemex station. Stay south on Route 180 and, 19.5 km south of the Córdoba junction (between Km Posts 19 and 20 if the highway is still labeled this way), look for a dirt road off to the left (east) through the short-grass fields. This may be signed "Las Barrancas 10." Second-class buses, such as those those running between Veracruz City and Los Tuxtlas, can drop you at this junction, from which you'll have to walk or hitch a short distance.

Turn left here (be careful turning across the traffic; it may be best to pull off on the right verge and wait for a break in the traffic) and go about 2 km to a (disused) railway crossing. The fields before and after this crossing, on both sides of the dirt road, are good for most of the grassland species. For thick-knees, park near the railway crossing and walk north along the tracks for 1–2 km, looking especially in the fields to your right (east); the scrub on your left should be checked for Striped Cuckoo, **Rufous-breasted Spinetail**, Barred Antshrike, and Yellow-bellied Elaenia. If you miss the thick-knees here (beware that they are nocturnal and spend much of the day crouched down in the grass, trying to avoid being seen), continue toward Las Barrancas along the dirt road for 2–3 km, checking the fields on either side. The seasonal ponds along here, and north along the railway tracks, can have Pinnated Bittern and Snail Kite and often act as magnets for drinking passerines, especially in late winter and spring, toward the end of the dry season.

One can continue driving out to the coast, through scrub and scattered ponds, although all of the target birds are nearer the highway. In April 1992 a large area of the grassland north of the Las Barrancas road, shortly east of the railway crossing, had been ploughed up in association with a new "ranching" settlement erected by the road; this was the traditional favored spot for the thick-knees, so you may have to work a little harder to find them if the trend of "development" continues.

Las Barrancas Bird List

(not including coast): 114 sp. R: rare; V: vagrant (not to be expected). †See pp. 5–7 for notes on English names.

Magnificent Frigatebird
Pinnated Bittern
Great Blue Heron

Great Egret
Little Blue Heron
Tricolored Heron
Cattle Egret
Green Heron
Black-crowned Night-Heron
Blue-winged Teal
Black Vulture
Turkey Vulture
Lesser Yellow-hd. Vulture
White-tailed Kite
Snail Kite
Northern Harrier
Roadside Hawk
Red-tailed Hawk
Crested Caracara
American Kestrel
Merlin
Aplomado Falcon
Peregrine Falcon
Northern Bobwhite
Double-striped Thick-knee
Black-necked Stilt
Black-bellied Plover
American Golden Plover
Killdeer
Northern Jacana
Greater Yellowlegs
Lesser Yellowlegs
Long-billed Curlew
Solitary Sandpiper
Upland Sandpiper
Semipalmated Sandpiper
Least Sandpiper
Baird's Sandpiper
Pectoral Sandpiper
Buff-breasted Sandpiper
Franklin's Gull
Gull-billed Tern
White-winged Dove
Inca Dove
Ruddy Ground-Dove
Striped Cuckoo
Groove-billed Ani
Ferruginous Pygmy-Owl
Short-eared Owl (R)
Common Nighthawk
Lesser Nighthawk
Pauraque
Chimney Swift
Buff-bellied Hummingbird
Belted Kingfisher
Ladder-backed Woodpecker
Rufous-breasted Spinetail
Barred Antshrike
Yellow-bellied Elaenia
Least Flycatcher
Vermilion Flycatcher
Great Kiskadee
Social Flycatcher
Scissor-tailed Flycatcher
Fork-tailed Flycatcher
Tropical Kingbird
Rose-throated Becard
Purple Martin
Tree Swallow
Mangrove Swallow
N. Rough-winged Swallow
Cliff Swallow
Barn Swallow
Brown Jay
Northern House Wren†
Blue-grey Gnatcatcher
Grey Catbird
Northern Mockingbird
American Pipit
Sprague's Pipit
Loggerhead Shrike
White-eyed Vireo
Orange-crowned Warbler
Northern Parula
Yellow Warbler
Magnolia Warbler
Myrtle Warbler†
Black-and-white Warbler
American Redstart
Common Yellowthroat
Wilson's Warbler
Yellow-breasted Chat
Scrub Euphonia
Blue-grey Tanager
Greyish Saltator
Northern Cardinal
Painted Bunting
Blue-black Grassquit
White-collared Seedeater
Grassland Yellow-Finch
Clay-colored Sparrow
Lark Sparrow (R?)
Grasshopper Sparrow
Savannah Sparrow
Lincoln's Sparrow
Lapland Longspur (V)
Eastern Meadowlark
Red-winged Blackbird
Yellow-headed Blackbird (R)
Melodious Blackbird
Great-tailed Grackle
Orchard Oriole
Ochre Oriole†
Altamira Oriole
Baltimore Oriole

Site 10.6: The Sierra de Los Tuxtlas

(December 1981/January 1982, January, February/March 1983, February, September, December 1984, February/March, March/April 1985, March 1987, April/May 1990, checked in part March 1998)

Introduction/Key Species

The Sierra de los Tuxtlas (also simply Los Tuxtlas, or the Tuxtlas) comprises, together with the lowlands of northern Oaxaca, the northernmost tropical rain forest in the Americas. It is a volcanic massif that rises from the Gulf Slope lowlands of southern Veracruz, on the northwestern side of the Isthmus of Tehuantepec; the highest peaks are Volcán Santa Marta in the east (1700 m) and Volcán San Martín Tuxtla in the west (1650 m). Geographic location blesses the area with a large avifauna (over 400 species), but its winter climate is dominated by frequent *nortes* (northers), winter storms that move in off the Gulf of Mexico and bring periods of cold and rain that at times last for several days. These storms, however, result in pronounced downslope migrations of species that might otherwise be difficult to see.

Habitats in the region range from mangroves to cloud forest and, on the southern slopes of Santa Marta, an area of pine-oak forest. However, seemingly little of the area has escaped the ravages of human settlement, particularly the urge to clear large areas of forest for cattle ranching, a feature branded almost inimitably upon the Tuxtlas landscape. Despite this, some areas of good rain forest are accessible (in particular at a biological research station of the National Autonomous University of Mexico [UNAM]), although the higher slopes of the volcanoes are generally not easy to reach, and most birders concentrate on the lowland habitats. The apparent designation of the San Martín and Santa Marta volcanoes as Special Biosphere Reserves reflects an awareness of the area's richness, although whether protection will go hand in hand with recognition remains to be seen.

By virtue of its isolation, the Sierra comprises a center of endemism: distinctive species and subspecies here include **Tuxtla** [Purplish-backed] **Quail-Dove**, **Long-tailed Sabrewing**, Black-headed Saltator (a buff-throated race), and **Plain-breasted** [Chestnut-capped] **Brushfinch**. Beyond the endemics, species of interest in the Tuxtlas include Slaty-breasted Tinamou, Double-toothed and Grey-headed kites, White Hawk, Barred Forest-Falcon, Aplomado Falcon, Sungrebe, Brown-hooded Parrot, Spectacled Owl, Lesser Swallow-tailed Swift, Chestnut-colored Woodpecker, Buff-throated and Scaly-throated foliage-gleaners, **Mexican Antthrush**, Yellow-bellied Tyrannulet, Lovely Cotinga, **Slate-colored Solitaire**, Green Shrike-Vireo, **Blue-crowned Chlorophonia**, **Black-throated Shrike-Tanager**, and a good variety of species typical of Mexico's Atlantic Slope lowlands. Christmas Bird Counts here in the 1970s found up to 300 species, reflecting the area's potential.

Important studies of the birds of Los Tuxtlas have been made by Wetmore and Carriker (Wetmore 1943), Edwards and Tashian (1959), Andrle (1966, 1967), and W. J. Schaldach Jr. (Schaldach and Escalante-Pliego 1997). I have spent some five months in the region, but with little time at the higher elevations.

Access

The town of Catemaco, on the shores of Lake Catemaco (350 m elevation) in the heart of the Tuxtlas, is about 170 km (2–3 hours) by road southeast from Veracruz City, beside Route 180. There are several hotels and restaurants in and near the town, which makes a good base from which to explore the area. Recently, one or two basic hotels have been built out near the coast in the vicinity of Montepio, and there are small, basic hotels in Sontecomapan and at the scenic overlook of Playa Escondida. These accommodations all allow more convenient access to the forest at the university field station (where it has also been possible to stay in the past—check on the current situation when you're there) and the coastal habitats.

Note that unless you like searching (probably in vain) for hotels and are equally fond of hordes of people swarming everywhere, Easter is not a good time to visit Los Tuxtlas (nor most other coastal areas of Mexico, for that matter).

Catemaco is served regularly by first-class and second-class buses from Veracruz City, while second-class buses regularly ply the roads from Catemaco along the lakeshore and out to Montepio on the coast, making it easy to reach most birding sites, including the UNAM biological station.

Birding Sites

While the forests have what are usually considered the most interesting birds, much can be seen in the second-growth woodland, fields, and hedges which, sadly, dominate much of the region. The waterfront in Catemaco has a nesting colony of Snowy and Cattle egrets in spring and summer, and the palm trees here often have Blue-grey Tanagers and Yellow-throated Warblers (winter). Particularly in overcast or rainy conditions, keep an eye upward anywhere in the Tuxtlas, including over Catemaco, for Chestnut-collared and Lesser Swallow-tailed swifts. Regular visitors to the lake include Magnificent Frigatebird, Laughing Gull, and Royal and Forster's terns.

Most of the birding sites are along or off the road that goes from Catemaco to Sontecomapan and on to the coast at Montepio, some 40 km (1.5–2 hours) distant. Catemaco has a variety of one-way streets, typical of any small Mexican town, and finding the Sontecomapan road can be a challenge. One way, if not the shortest, is to continue west and north along the lakeshore road in town to the last turn on the left, shortly before the prom-

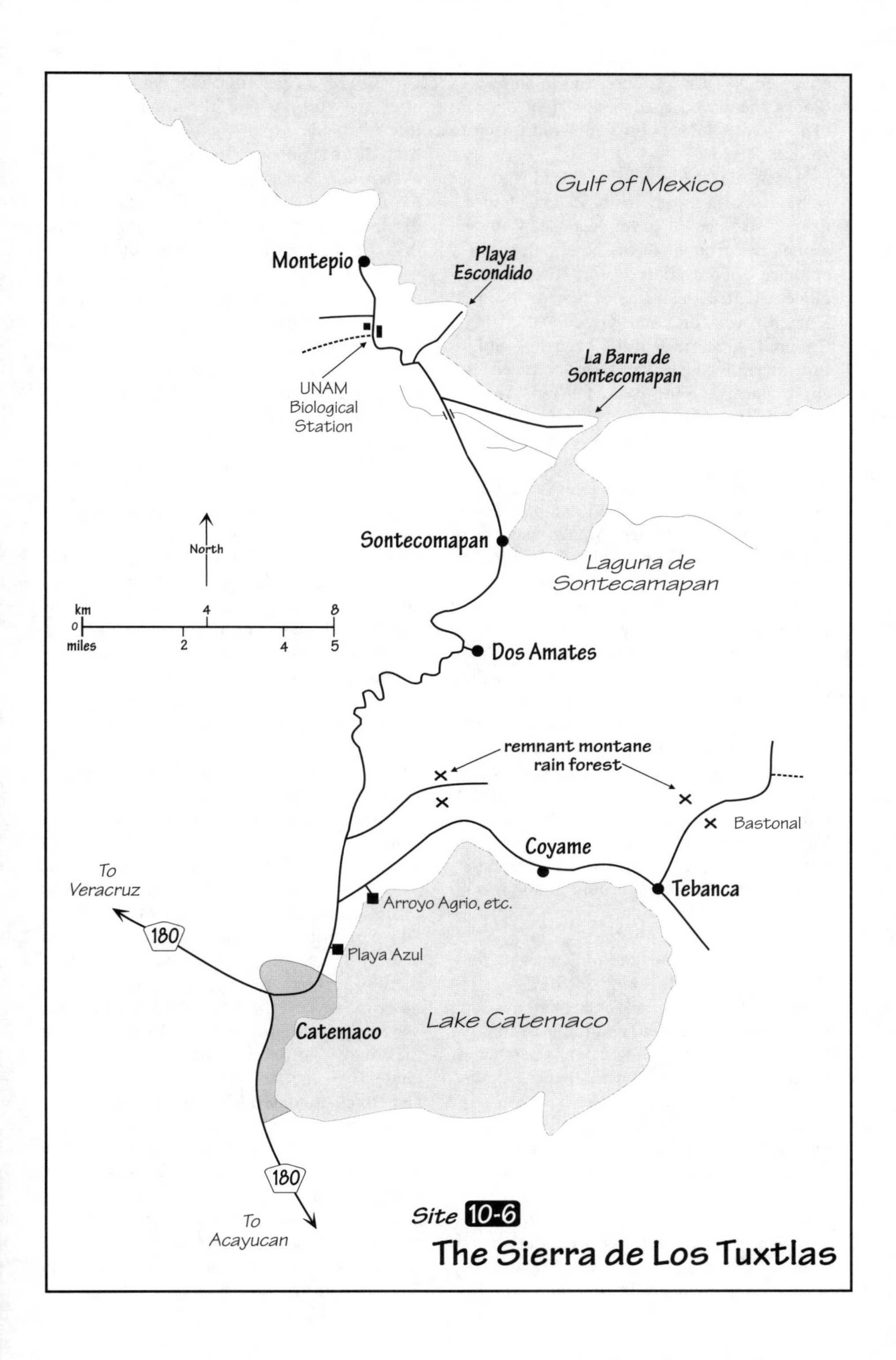
Gulf of Mexico
Montepio
Playa Escondido
UNAM Biological Station
La Barra de Sontecomapan
North
Sontecomapan
Laguna de Sontecamapan
km
0
4
8
miles
2
4
5
Dos Amates
remnant montane rain forest
Bastonal
Coyame
Tebanca
To Veracruz
180
Arroyo Agrio, etc.
Playa Azul
Catemaco
Lake Catemaco
180
To Acayucan
Site 10-6
The Sierra de Los Tuxtlas

enade ends. Turn left here, and in about 4–5 blocks the paved road bends right and out of town, at which point you should reset to zero.

At Km 2.1, the Hotel Playa Azul is to your right (east), on the shores of Lake Catemaco. Over 100 species have been seen in and around the grounds of the hotel, which birders have often used as a base. Birds here include Snail Kite, Blue-crowned Motmot, Ringed, Amazon, and Green kingfishers, Collared Aracari, **Rufous-breasted Spinetail**, Barred Antshrike, **Grey-collared Becard** (*nortes*), Band-backed Wren, **Yellow-winged Tanager**, Montezuma Oropendola, numerous wintering migrant vireos and warblers, and, depending on season, flowers, and the strength of *nortes*, an interesting selection of hummingbirds: Violet Sabrewing (winter), Green-breasted Mango (summer), **Canivet's** and **White-bellied Emeralds**, **Azure-crowned** (winter), Rufous-tailed, Buff-bellied, and Ruby-throated (migrant) **hummingbirds**. Lovely Cotinga has been seen in fruiting trees on the slopes across the road from the Playa Azul, and the weedy fields there have Northern Bobwhite and can be good for seed-eating birds, including both grassquits, migrant buntings, and orioles.

Continuing out the road toward Sontecomapan, there is a major fork at Km 4.3: right goes east around the shore of Lake Catemaco to Coyame, Tebanca, and beyond; straight (left) goes out to the coast and Montepio. There are good birding areas in both directions.

If you turn right at the fork reset to zero. Two areas worth checking for birds are the remnant lowland rain-forest patches and fields between the road and the lake shore at Km 1.4 to 2.1, and the remnant patches of montane rain forest above Tebanca, at Bastonal. The former make an easy morning or afternoon trip, while Bastonal (depending on the road conditions) is a long day trip that should be scouted in advance; alternatively, it can be done as a camping trip for one or more nights.

There are three nearby areas (all to the right of the road) worth checking shortly after forking right toward Coyame. The first is the Arroyo Agrio bottling plant at Km 1.4 after the fork, where the road levels out through fields within sight of the lake. Look for a dirt road off to the right (probably with a sign for Arroyo Agrio and/or *Embotelladora* [bottling plant]) which runs a short distance through a field to the plant buildings, on the lakeshore beside a patch of forest; there may or may not be a locked gate and guard here, and you may or may not be able to get permission to enter. If you do get permission, check the lake for herons, Least Grebe, Purple Gallinule, Northern Jacana, and, if you're lucky, Sungrebe. Scan the hedgerows and isolated trees for fruiting trees, which can have Aztec Parakeet, Keel-billed Toucan, Collared Aracari, Lovely Cotinga, and mixed-species flocks of flycatchers, Rose-throated Becard, Masked Tityra, tanagers, saltators, Rose-breasted Grosbeak (winter), and orioles. Check prominent perches for Laughing Falcon, and keep an eye and ear open for Black Hawk-Eagle overhead. It has been possible to walk past the bottling plant to the right, along the wall by the lake, into the small forest patch beyond, where birds include Collared Forest-Falcon, Black-headed and Violaceous trogons, Barred Woodcreeper, Ochre-bellied Flycatcher, and Red-throated Ant-Tanager.

At about Km 2.0 on the road there are two adjacent forest patches to the right, which are maintained as private reserves: Nanciyaga and La Jungla. In theory there is an entry fee of 5 pesos per person, but this mainly applies to the restaurants and developments by the lake. For birding, you can park on the road and walk in around the gates (paying 5 pesos if there's someone around) and walk along the dirt road toward the lake (about 1 km away) and along trails into the forest. Birds are much like Arroyo Agrio, and species seen here include Eye-ringed Flatbill, Lovely Cotinga, and **Black-throated Shrike-Tanager**.

Continuing on the road past La Jungla, you go through the town of Coyame (Km 8.1), famous for its mineral waters bottled at the source, and the pavement ends; in about another few kilometers you come to the village of Tebanca (Km ?; likely unsigned), which has a few stores and a dirt road off to the left (northeast) that leads steeply up through fields to some ranches, including Bastonal. If in doubt, you can always ask for the road to Bastonal. The dirt road up to Bastonal, the general name for a ranch on the slopes of Volcán Santa Marta, varies greatly in its condition. Frequently it has been impassable to anything other than a high-clearance and/or four-wheel-drive truck, at other times one can get up to the first forest patches in a combi (VW van), and, rarely, it can be passable with a regular car, such as a VW Rabbit. Many roads in Mexico have been improved greatly in recent years, although the Bastonal road has not been among them; check on current road conditions when you visit. In April 1990 it was possible to drive a car about 4.5 km above Tebanca, to just beyond the first ranch on the road, and walk from there for 20–30 minutes to the first remnant forest patches. In a pickup truck one could go above this, to and beyond the cleared cattle wasteland around the few ranch buildings at Bastonal.

If you want to walk from Tebanca to the first few remnant forest patches, it's about 5–7 km, an invigorating hike, much of it steeply uphill. If the road looks driveable, it is worth continuing to Bastonal and going beyond and higher on the dirt road (a walk of forty-five minutes to an hour from Bastonal) to a crossroads, where the main dirt road continues straight, over a pass and to more *ranchos* on the coastal-facing slopes. The right-hand track here goes into some excellent forest (900–1000 m elevation), much more extensive than the ever-shrinking patches below Bastonal, and **Tuxtla** [Purplish-backed] **Quail-Dove** is common. It is possible to camp along this track (take all your food and some way of purifying water).

Birding the forest patches below Bastonal can still be good (unless it's all been cleared!), especially after the passage of a *norte*. Birds at Bastonal (almost all of which I've found in the remnant patches) include Thicket and Slaty-breasted tinamous, White and Great Black hawks, Black and Ornate hawk-eagles, Barred Forest-Falcon, Blue and Maroon-chested ground-doves, **Tuxtla** [Purplish-backed] and Ruddy **quail-doves**, **Long-tailed** and Violet **sabrewings**, Collared Trogon, Emerald Toucanet, Scaly-throated and Buff-throated foliage-gleaners, Yellow-bellied Tyrannulet, Eye-ringed Flatbill, Stub-tailed Spadebill, **Slate-colored Solitaire**, Black-headed Nightingale-Thrush, White-throated Thrush, Green Shrike-Vireo, Rufous-browed Peppershrike, Tropical Parula, Slate-throated Whitestart, **Blue-crowned Chlorophonia**, Blue-hooded Euphonia, **Black-throated Shrike-Tanager**, White-winged Tanager, Common Bush-Tanager, and **Plain-breasted** [Chestnut-capped] **Brush-finch**. Many of these species, which include numerous elevational migrants, can also be found lower, for example, at the biological station, but higher elevations, such as Bastonal, are best for the Tuxtla [Purplish-backed] Quail-Dove.

If you stay straight (or left) at the fork at Km 4.3 on the main road out of Catemaco (rather than turning right to Arroyo Agrio, Tebanca, and Bastonal), the road goes out to the coast and a number of good birding sites. The first is the dirt road to the right at Km 2.0 (Km 6.3 from Catemaco), which may be signed to Ejido Vista Hermosa. Turn here and continue through 3.4 km of fields (Aplomado Falcon) to a remnant patch of montane rain forest, where **Tuxtla** [Purplish-backed] **Quail-Dove** occurs. This area has not been birded heavily, but species seem likely to be similar to Bastonal (e.g., Slaty-breasted Tinamou is quite common), and it is much easier to reach.

Continuing on the main road toward the coast, you climb gradually and crest a low pass at Km 4.2 (Km 8.5 from Catemaco) whence, on a clear day, you can look out to the coast and the mangrove-fringed Laguna de Sontecomapan. At Km 8.3 (12.6) you pass the right-hand turn into the village of Dos Amates, and at about Km 11.7 (16.0) you enter the village of Sontecomapan. At the "dock" (a couple of blocks to the right (east) of the road, near the center of Sontecomapan by the main stores), it is possible to arrange boat trips on the Laguna and along its tributary rivers. The main target bird here is Sungrebe, which can be seen along the Río Coxcuapan, the Río del Sabalo, and the Río de la Palma. Other birds around the Laguna include Pinnated Bittern, Bare-throated Tiger-

Heron, Boat-billed Heron, Grey-headed Kite, Common Black Hawk, Pygmy Kingfisher, and Yellow-tailed Oriole, and there's a chance, albeit slim, for Agami Heron. A boatman who knows the birds and has been recommended is Ismael, and he can be contacted by asking at the restaurants by the dock. To get to Sontecomapan from Catemaco, allow about thirty minutes' driving time.

Continuing through Sontecomapan on the main road, the pavement ends on the far side of town, and the road becomes dirt. You cross a bridge over the Río de la Palma at Km 21.2 (25.5), just past which there is dirt road off to the right; this side road goes for a few kilometers to La Barra de Sontecomapan, that is, to the mouth of the lagoon, where there is a small fishing settlement on the beach. The road to La Barra runs through some marshes, where birds include Muscovy Duck, Lesser Yellow-headed Vulture, White-tailed and Snail kites, Northern Harrier (winter), Common Black and Red-tailed (winter) hawks, Laughing and Aplomado falcons, Limpkin, **Ruddy Crake**, Black Rail, and Fork-tailed Flycatcher. Gulls and terns often roost on the beach near the lagoon's mouth, and shorebirds here include Collared Plover.

If you continue straight at the La Barra junction, the main road winds up into some forest patches. At Km 23.1 (27.4) there is a right turn down to the Hotel Playa Escondida (which is a couple of kilometers from the highway and is a scenic spot with some adjacent forest patches; **Long-tailed Sabrewing** can be seen on the grounds), and at Km 25.6 (29.9) you'll be in an obvious patch of good forest. At Km 26.0 (30.3) there are some modern-looking buildings on either side of the road, indicating the **UNAM Biological Station**. If you're here during office hours, you might check in at the biological station (the offices are on the right side of the road, before the more obvious dormitory buildings and laboratories on the left), which includes a trail through the forest beside the offices. There is also a published checklist of 315 species recorded at the station (Coates-Estrada and Estrada 1985). There is room to park off the road by the station, and birding can be good along the main road, on trails into the forest (ask permission if you see someone), and about 200 m past the station buildings along a newly constructed dirt road to the left, which goes to the village of Laguna Escondida, by a lake of the same name. This is a walk of about 2 km, mostly along forest edge. To get to the station from Catemaco allow about 1–1.6 hours' driving time, depending on recent weather (after and during *nortes* the muddy road beyond Sontecomapan can be slow going!).

Birding around the biological station offers a diversity of birds typical of lowland rainforest and edge: Great and Slaty-breasted tinamous, Double-toothed and Grey-headed kites, White Hawk, Black Hawk-Eagle, Barred Forest-Falcon, Short-billed Pigeon, Grey-headed Dove, Brown-hooded and Red-lored parrots, Spectacled and Mottled owls, Long-tailed and Little hermits, **Long-tailed** and Violet **sabrewings**, Long-billed Starthroat, four trogons, three toucans, seven woodpeckers, Buff-throated Foliage-gleaner, Plain Xenops, six woodcreepers, Red-capped Manakin, Lovely Cotinga, Yellow-bellied Tyrannulet, Ochre-bellied, Sepia-capped, and Sulphur-rumped flycatchers, Stub-tailed Spadebill, Long-billed Gnatwren, both greenlets, Olive-backed Euphonia, Grey-headed Tanager, **Black-throated Shrike-Tanager**, both ant-tanagers, Crimson-collared Tanager, and Black-faced and Blue-black grosbeaks. If you're here during or after a *norte*, elevational migrants (best found at fruiting trees) can include Emerald Toucanet, **Slate-colored Solitaire**, White-throated Thrush, **Blue-crowned Chlorophonia**, White-winged Tanager, and Common Bush-Tanager, all at only 200 m elevation!

Past the biological station the main road continues through fields alternating with small forest patches and reaches the coast at **Montepio** at about Km 36 (and some 40 km from Catemaco). The stretch between the biological station and Montepio can be very birdy, and numerous tracks and trails (a few of them into or beside small forest patches) can be taken to get off the main road. The river that flows beside the road to your left (west) for much of the way has Sungrebes, and the beach and small river mouth at Montepio are worth checking for waterbirds (e.g., Reddish Egret, Collared Plover), especially during migration periods.

Sierra de Los Tuxtlas Bird List

378 sp. R: rare; V: vagrant (not to be expected). †See pp. 5–7 for notes on English names.

Great Tinamou
Thicket Tinamou
Slaty-breasted Tinamou
Least Grebe
Pied-billed Grebe
Eared Grebe
American White Pelican
Brown Pelican
Double-crested Cormorant
Neotropic Cormorant
Anhinga
Magnificent Frigatebird
Bare-throated Tiger-Heron
Great Blue Heron
Great Egret
Snowy Egret
Little Blue Heron
Tricolored Heron
Reddish Egret
Cattle Egret
Green Heron
Black-crowned Night-Heron
Yellow-crowned Night-Heron
Boat-billed Heron
White Ibis
Fulvous Whistling-Duck
Black-bellied Whistling-Duck
Muscovy Duck
Wood Duck (R)
Green-winged Teal
Blue-winged Teal
Northern Shoveler
Gadwall
American Wigeon
Redhead
Lesser Scaup
Hooded Merganser (R)
Black Vulture
Turkey Vulture
Lesser Yellow-hd. Vulture
Osprey
Grey-headed Kite
Hook-billed Kite
White-tailed Kite
Double-toothed Kite
Mississippi Kite
Northern Harrier
Sharp-shinned Hawk
Bicolored Hawk
Cooper's Hawk
White Hawk
Common Black Hawk
Great Black Hawk
Grey Hawk
Roadside Hawk
Broad-winged Hawk
Short-tailed Hawk
Swainson's Hawk
Red-tailed Hawk
Zone-tailed Hawk
Black Hawk-Eagle
Ornate Hawk-Eagle
Crested Caracara
Laughing Falcon
Barred Forest-Falcon
Collared Forest-Falcon
American Kestrel
Merlin
Aplomado Falcon
Bat Falcon
Peregrine Falcon
Plain Chachalaca
Northern Bobwhite
Ruddy Crake
Black Rail
Sora
Purple Gallinule
Common Moorhen
American Coot
Sungrebe
Limpkin
Black-bellied Plover
American Golden Plover
Collared Plover
Snowy Plover
Wilson's Plover
Semipalmated Plover
Killdeer
Black-necked Stilt
Northern Jacana
Greater Yellowlegs
Lesser Yellowlegs
Solitary Sandpiper
Willet
Spotted Sandpiper
Upland Sandpiper
Whimbrel
Ruddy Turnstone
Sanderling
Semipalmated Sandpiper
Western Sandpiper
Least Sandpiper
White-rumped Sandpiper
Baird's Sandpiper
Pectoral Sandpiper
Stilt Sandpiper
Buff-breasted Sandpiper
Long-billed Dowitcher
Common Snipe
Wilson's Phalarope
Laughing Gull
Franklin's Gull
Little Gull (V)
Ring-billed Gull
Herring Gull
Caspian Tern
Royal Tern
Sandwich Tern
Common Tern
Forster's Tern
Least Tern
Black Tern
Black Skimmer
Scaled Pigeon
Red-billed Pigeon
Short-billed Pigeon
White-winged Dove
Mourning Dove
Inca Dove
Common Ground-Dove
Plain-breasted Ground-Dove
Ruddy Ground-Dove
Blue Ground-Dove
Maroon-chested Ground-Dove
White-tipped Dove
Grey-headed Dove†
Tuxtla Quail-Dove†
Ruddy Quail-Dove
Aztec Parakeet†
Brown-hooded Parrot
Red-lored Parrot
Black-billed Cuckoo
Yellow-billed Cuckoo
Squirrel Cuckoo
Groove-billed Ani
Barn Owl
Spectacled Owl
Mottled Owl
Ferruginous Pygmy-Owl
Lesser Nighthawk
Common Nighthawk
Pauraque
Northern Potoo
Chestnut-collared Swift
White-collared Swift
Chimney Swift
Vaux's Swift
Lesser Swallow-tailed Swift
Long-tailed Hermit
Little Hermit
Long-tailed Sabrewing
Violet Sabrewing
Green-breasted Mango
Black-crested Coquette
Canivet's Emerald

White-bellied Emerald
Azure-crowned Hummingbird
Rufous-tailed Hummingbird
Buff-bellied Hummingbird
Long-billed Starthroat
Ruby-throated Hummingbird
Rufous Hummingbird (V)
Black-headed Trogon
Violaceous Trogon
Collared Trogon
Slaty-tailed Trogon
Blue-crowned Motmot
Ringed Kingfisher
Belted Kingfisher
Amazon Kingfisher
Green Kingfisher
Emerald Toucanet
Collared Aracari
Keel-billed Toucan
Black-cheeked Woodpecker
Golden-fronted Woodpecker
Yellow-bellied Sapsucker
Smoky-brown Woodpecker
Golden-olive Woodpecker
Chestnut-colored Woodpecker
Lineated Woodpecker
Pale-billed Woodpecker
Rufous-breasted Spinetail
Scaly-throated Fol.-gleaner†
Buff-throated Fol.-gleaner
Plain Xenops
Tawny-winged Woodcreeper
Olivaceous Woodcreeper
Wedge-billed Woodcreeper
Barred Woodcreeper
Ivory-billed Woodcreeper
Streak-headed Woodcreeper
Barred Antshrike
Mexican Antthrush†
Yellow-bellied Tyrannulet
N. Beardless Tyrannulet
Yellow-bellied Elaenia
Ochre-bellied Flycatcher
Sepia-capped Flycatcher
Northern Bentbill
Eye-ringed Flatbill
Yellow-olive Flycatcher
Stub-tailed Spadebill
Sulphur-rumped Flycatcher
Olive-sided Flycatcher
Eastern Pewee†
Tropical Pewee
Yellow-bellied Flycatcher
Acadian Flycatcher
Alder Flycatcher
Willow Flycatcher
Least Flycatcher
Eastern Phoebe
Vermilion Flycatcher
Bright-rumped Attila
Dusky-capped Flycatcher
Great Crested Flycatcher
Brown-crested Flycatcher
Great Kiskadee
Boat-billed Flycatcher
Social Flycatcher
Sulphur-bellied Flycatcher
Piratic Flycatcher
Scissor-tailed Flycatcher
Fork-tailed Flycatcher
Tropical Kingbird
Couch's Kingbird
Eastern Kingbird
Grey-collared Becard
Rose-throated Becard
Masked Tityra
Black-crowned Tityra
Lovely Cotinga
Red-capped Manakin
Purple Martin
Grey-breasted Martin
Tree Swallow
Mangrove Swallow
N. Rough-winged Swallow
Bank Swallow
Cliff Swallow
Barn Swallow
Green Jay
Brown Jay
Band-backed Wren
Spot-breasted Wren
Northern House Wren†
Southern House Wren†
White-breasted Wood-Wren
Long-billed Gnatwren
Ruby-crowned Kinglet
Blue-grey Gnatcatcher
Slate-colored Solitaire
Black-headed N.-Thrush
Veery
Swainson's Thrush
Hermit Thrush
Wood Thrush
Clay-colored Thrush†
White-throated Thrush†
American Robin
Grey Catbird
Northern Mockingbird
Cedar Waxwing
White-eyed Vireo
Bell's Vireo
Blue-headed Vireo†
Yellow-throated Vireo
Warbling Vireo
Philadelphia Vireo
Red-eyed Vireo
Yellow-green Vireo
Tawny-crowned Greenlet
Lesser Greenlet
Green Shrike-Vireo
Rufous-browed Peppershrike
Blue-winged Warbler
Golden-winged Warbler
Tennessee Warbler
Orange-crowned Warbler
Nashville Warbler
Northern Parula
Tropical Parula
Yellow Warbler
Chestnut-sided Warbler
Magnolia Warbler
Cape May Warbler
Myrtle Warbler†
Audubon's Warbler†
Black-throated Green Warbler
Blackburnian Warbler
Yellow-throated Warbler
Palm Warbler
Bay-breasted Warbler
Cerulean Warbler
Black-and-white Warbler
American Redstart

Prothonotary Warbler
Worm-eating Warbler
Ovenbird
Northern Waterthrush
Louisiana Waterthrush
Kentucky Warbler
Mourning Warbler
Common Yellowthroat
Grey-crowned Yellowthroat
Hooded Warbler
Wilson's Warbler
Canada Warbler
Slate-throated Whitestart†
Golden-crowned Warbler
Rufous-capped Warbler
Yellow-breasted Chat
Bananaquit
Red-legged Honeycreeper
Blue-crowned Chlorophonia
Scrub Euphonia
Yellow-throated Euphonia
Blue-hooded Euphonia
Olive-backed Euphonia
Blue-grey Tanager
Yellow-winged Tanager
Grey-headed Tanager
Summer Tanager
Scarlet Tanager
Western Tanager
White-winged Tanager
Crimson-collared Tanager
Black-throated Shrike-Tanager
Red-crowned Ant-Tanager
Red-throated Ant-Tanager
Common Bush-Tanager
Greyish Saltator
Buff-throated Saltator
Black-headed Saltator
Black-faced Grosbeak
Rose-breasted Grosbeak
Blue-black Grosbeak
Blue Grosbeak
Northern Cardinal
Blue Bunting
Indigo Bunting
Varied Bunting
Painted Bunting
Dickcissel
Plain-breasted Brushfinch†
Olive Sparrow
Blue-black Grassquit
Yellow-faced Grassquit
White-collared Seedeater
Rusty Sparrow
Grasshopper Sparrow
Savannah Sparrow
Lincoln's Sparrow
Melodious Blackbird
Eastern Meadowlark
Great-tailed Grackle
Bronzed Cowbird
Giant Cowbird
Black-cowled Orchard
Hooded Oriole
Yellow-tailed Oriole
Altamira Oriole
Baltimore Oriole
Bullock's Oriole
Yellow-billed Cacique
Chestnut-headed Oropendola
Montezuma Oropendola
Lesser Goldfinch
House Sparrow

Site 10.7: Uxpanapa Road

(December 1991/January 1992, April 1992, September 1995, January 1998)

Introduction/Key Species

While the rain forests that once shadowed the Trans-Isthmus Highway (Route 185) have been all but cleared beyond the visible horizon, remnant patches remain nearby, where the underlying, jagged limestone karst makes clearing trees difficult and would leave little or no topsoil for agriculture. One such area is the Uxpanapa region in the central Isthmus of Tehuantepec, straddling the Veracruz/Oaxaca state line, where pockets of forest cling to karst beside a well-graded dirt road.

The main target bird here is the highly local and endemic **Nava's Wren**. Other species of note include Black Hawk-Eagle, White Hawk, Grey-headed Dove, **Green Parakeet**, Brown-hooded and Mealy parrots, Central American Pygmy-Owl, Black-and-white Owl, **Long-tailed Sabrewing**, Rufous-tailed Jacamar, **Rufous-breasted Spinetail**, **Mexican Antthrush**, Yellow-bellied Tyrannulet, Sepia-capped and Royal flycatchers, Slate-headed Tody-Flycatcher, Lovely Cotinga, Green Shrike-Vireo, Black-faced Grosbeak, and **Black-throated Shrike-Tanager**.

Access

The forests along the Uxpanapa Road are most easily birded from a base in or near Matias Romero, along Route 185. With the new *cuota* highways from Veracruz to Minatitlán, Matias Romero is now only about 370 km (4–5 hours) from Veracruz City: take the *cuota* highway (Route 150D) inland toward Córdoba, exiting at La Tinaja and following signs for Route 145D, which goes to Sayula, Acayucan, and on to the oil towns of Minatitlán and Coatzacoalcos. Route 145D goes through mostly cleared cattle land but also passes some nice marshes which can be worth checking. Exit this highway in response to signs for Sayula and Acayucan and turn south onto Route 185, the Trans-Isthmus Highway, through the small junction town of Sayula and on toward Matias Romero. About 109 km south of Sayula you go through the village of Boca del Monte (also known as Sarabia), whence a paved road leads off to the left (east) and may be signed "Uxpana-

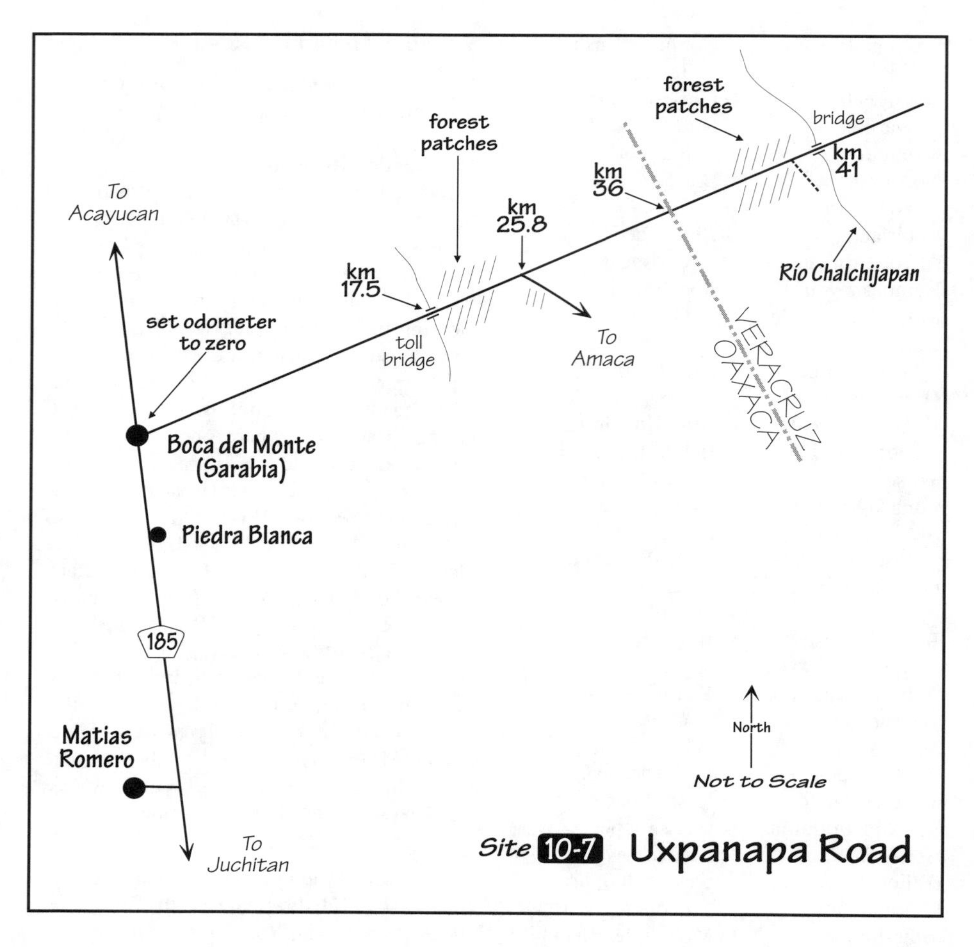

pas" on a rusty arch over the start of the road. This is the road to the forest patches and a number of villages, including government projects aimed at resettling people and "developing" the countryside.

Like most areas of lowland rain forest, this site is best visited in the early morning, and a choice of accommodations can be found in the town of Matias Romero, 22 km (20–30 minutes) south of Sarabia. There is also a basic hotel and adequate restaurant beside (east of) the highway at Piedra Blanca (which comprises little more than the Hotel Restaurant Liessa, where birders have stayed), only about 8 km (5–10 minutes) south of Sarabia; Brown Jay and **Russet-crowned Motmot** occur together in the hotel yard, reflecting the mix of Atlantic Slope and Pacific Slope avifaunas in the Isthmus.

Sarabia can be reached easily by public buses from either side of the Isthmus, and there are regular second-class buses that run along the Uxpanapa Road, allowing access to good habitat.

Birding Sites

To get to the good birding areas, allow 1.3–1.5 hours to drive from Piedra Blanca and 1.5–2 hours from Matias Romero. Mark zero as you turn onto the Uxpanapa Road in Sarabia. After about 17.5 km of potholed pavement there is a *caseta* (small building) and a chain across the road at a small bridge, where a minimal toll (5 pesos, January 1998) is charged for road maintenance. After this bridge, the bumpy dirt road usually can be transited with care in a regular car, although after heavy rain it may be impassable in places. There are a few patches of forest beside and near the road (Great Tinamou, Black Hawk-Eagle, Mealy Parrot, Keel-billed Toucan) between Km 17.5 and a right turn at Km 25.8, which may be signed to Amaca. About 500–800 m along the Amaca side road is a small patch of forest where **Nava's Wren** has been found, in the state of Oaxaca (Gomez de Silva G. and Sada 1997), although the best areas are farther out the Uxpanapa Road.

At about Km 36 you cross the state line into Veracruz, from Oaxaca, although there may be little or no sign to this effect, and a few kilometers later you go through a relatively extensive (yet still small) forest patch before coming to a major bridge over the Río Chalchijapan at Km 41.0. The forest, second growth, and edge from this bridge (where Tropical River Otter has been seen) back for 3–4 km toward Boca del Monte have a good variety of birds, and it is possible to find paths which get inside the forest, where you have a good chance of seeing Nava's Wren (versus only hearing it from the road), plus a variety of other forest-interior birds (e.g., **Mexican Antthrush** and, with flocks, **Black-throated Shrike-Tanager**). Birding from the road, along the forest edge, is also productive: watch and listen for Rufous-tailed Jacamar and for canopy flocks, which can hold Tropical Gnatcatcher, Green Shrike-Vireo, Black-faced Grosbeak, and so on; and check any suitable flowers along the roadside for hummers (**Long-tailed Sabrewing**, Black-crested Coquette). Just before (west of) the Chalchijapan bridge a track leads off to the right (south) between the forest and the river, and this trail can be walked for 1 km or more, providing access to second growth and forest edge (**Rufous-breasted Spinetail**, Slate-headed Tody-Flycatcher). In mid to late morning the bridge itself makes a good place from which to scan over the forest for raptors.

One can continue on past the Río Chalchijapan bridge for many more kilometers, but the habitat, at least near the road, doesn't seem to get any better than the forest patch west of the bridge.

Uxpanapa Road Bird List

196 sp. R: rare (not to be expected). †See pp. 5–7 for notes on English names.

Great Tinamou
Little Tinamou
Slaty-breasted Tinamou
American White Pelican
Cattle Egret
Little Blue Heron
Green Heron
Black-crowned Night-Heron
Blue-winged Teal
Black Vulture
Turkey Vulture
Osprey
White-tailed Kite
Hook-billed Kite
Sharp-shinned Hawk
Crane Hawk
White Hawk
Common Black Hawk
Grey Hawk
Roadside Hawk
Short-tailed Hawk
Black Hawk-Eagle
Crested Caracara
Laughing Falcon
American Kestrel
Bat Falcon
Plain Chachalaca
Crested Guan
Spotted Wood-Quail
Ruddy Crake
Grey-necked Wood-Rail
Spotted Sandpiper
Red-billed Pigeon
Short-billed Pigeon
Ruddy Ground-Dove
Blue Ground-Dove
Grey-headed Dove†
Green Parakeet
Aztec Parakeet†
Brown-hooded Parrot
White-crowned Parrot
Red-lored Parrot
Mealy Parrot
Squirrel Cuckoo
Striped Cuckoo
Groove-billed Ani
Central American Pygmy-Owl†
Ferruginous Pygmy-Owl
Mottled Owl
Black-and-white Owl
Pauraque
White-collared Swift
Vaux's Swift
Lesser Swallow-tailed Swift
Long-tailed Hermit
Little Hermit
Long-tailed Sabrewing
White-bellied Emerald
Rufous-tailed Hummingbird
Stripe-tailed Hummingbird
Long-billed Starthroat
Black-headed Trogon
Violaceous Trogon
Slaty-tailed Trogon
Blue-crowned Motmot
Ringed Kingfisher
Belted Kingfisher
Rufous-tailed Jacamar
Collared Aracari
Keel-billed Toucan
Black-cheeked Woodpecker
Golden-fronted Woodpecker
Smoky-brown Woodpecker
Golden-olive Woodpecker
Chestnut-colored Woodpecker
Lineated Woodpecker
Pale-billed Woodpecker
Rufous-breasted Spinetail
Buff-throated Fol.-gleaner

Tawny-winged Woodcreeper
Olivaceous Woodcreeper
Wedge-billed Woodcreeper
Ivory-billed Woodcreeper
Streak-headed Woodcreeper
Great Antshrike
Barred Antshrike
Dusky Antbird
Dot-winged Antwren
Mexican Antthrush†
Yellow-bellied Tyrannulet
N. Beardless Tyrannulet
Greenish Elaenia
Yellow-bellied Elaenia
Sepia-capped Flycatcher
Northern Bentbill
Slate-headed Tody-Flycatcher
Common Tody-Flycatcher
Eye-ringed Flatbill
Yellow-olive Flycatcher
Stub-tailed Spadebill
Royal Flycatcher
Sulphur-rumped Flycatcher
Tropical Pewee
Yellow-bellied Flycatcher
Least Flycatcher
Black Phoebe
Vermilion Flycatcher
Bright-rumped Attila
Dusky-capped Flycatcher
Great Crested Flycatcher
Great Kiskadee
Boat-billed Flycatcher
Social Flycatcher
Scissor-tailed Flycatcher
Tropical Kingbird
Eastern Kingbird
Rose-throated Becard
Masked Tityra
Rufous Piha
Lovely Cotinga
Grey-breasted Martin
Mangrove Swallow
N. Rough-winged Swallow
Ridgway's Rough-w. Swallow†
Brown Jay
Band-backed Wren
Nava's Wren†
Spot-breasted Wren
Southern House Wren†
White-breasted Wood-Wren
Long-billed Gnatwren
Blue-grey Gnat-catcher
Tropical Gnatcatcher
Wood Thrush
Clay-colored Thrush†
Grey Catbird
White-eyed Vireo
Blue-headed Vireo†
Yellow-throated Vireo
Lesser Greenlet
Green Shrike-Vireo
Rufous-browed Peppershrike
Blue-winged Warbler
Northern Parula
Tropical Parula
Yellow Warbler
Chestnut-sided Warbler
Magnolia Warbler
Black-throated Green Warbler
Blackburnian Warbler
Black-and-white Warbler
American Redstart
Worm-eating Warbler
Ovenbird
Northern Waterthrush
Louisiana Waterthrush
Kentucky Warbler
MacGillivray's Warbler (R?)
Common Yellowthroat
Grey-crowned Yellowthroat
Hooded Warbler
Wilson's Warbler
Canada Warbler
Golden-crowned Warbler
Yellow-breasted Chat
Bananaquit
Red-legged Honeycreeper
Yellow-throated Euphonia
Olive-backed Euphonia
Blue-grey Tanager
Yellow-winged Tanager
Black-throated Shrike-Tanager
Red-crowned Ant-Tanager
Red-throated Ant-Tanager
Summer Tanager
White-winged Tanager
Scarlet-rumped Tanager
Crimson-collared Tanager
Greyish Saltator
Buff-throated Saltator
Black-headed Saltator
Black-faced Grosbeak
Rose-breasted Grosbeak
Blue-black Grosbeak
Indigo Bunting
Orange-billed Sparrow
Variable Seedeater
White-collared Seedeater
Lincoln's Sparrow
Melodious Blackbird
Great-tailed Grackle
Orchard Oriole
Yellow-tailed Oriole
Baltimore Oriole
Yellow-billed Cacique
Montezuma Oropendola

Site 10.8: Las Choapas

(May 1990)

Introduction/Key Species

The natural savannas around the small town of Las Choapas, on the Veracruz/Tabasco border, are a site for the little-known Spot-tailed Nightjar, which is present here at least from late March through June; whether this species winters in Mexico has yet to be resolved satisfactorily. Other species of interest here include Pale-vented Pigeon, Plain-

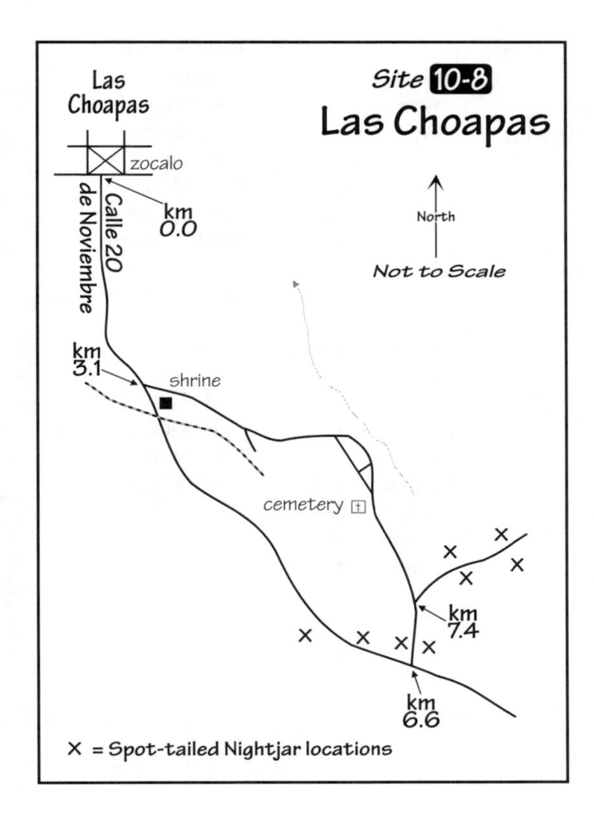

breasted Ground-Dove, Common Nighthawk (summer), and Botteri's Sparrow. More information (including habitat photos) about the nightjars and other birds at Las Choapas is given by Zimmerman (1957).

Access

Las Choapas is about 200 km (2–3 hours, depending on whether or not you make use of the *cuota* highway) southeast of Catemaco and about 150 km (1.5–2 hours) west of Villahermosa. There are a couple of very basic hotels in town, and a few restaurants along the same lines. One could also camp out in the savannas. Las Choapas can be done as an overnight trip from Los Tuxtlas (Site 10.6) or as an afternoon/evening trip from Villahermosa, as an adjunct to Northern Chiapas (Chapter 13).

The town is served by regular public buses, for example, from nearby Minatitlán and Coatzacoalcos. Once at Las Choapas, ask about public transport along the relevant roads (see Birding Sites), which could be reached easily by taxi and/or hitching.

Birding Sites

As you come into Las Choapas from the northwest, continue into the center and the *zócalo*, which will be on your left. Mark zero as you turn right (south) from the *zócalo* on the main road (Calle 20 de Noviembre) south through "downtown" Las Choapas. At Km 3.1, just out of town, there is a fork in the road with an obvious shrine in the center of the fork. Bear right at this fork, cross a railway line, and at Km 6.6 (3.5 km from the shrine fork) turn left onto a dirt road, continue 800 m, and fork right onto another dirt road. This road runs through savanna ridges, with scattered oaks and palms, interspersed with forest patches along the streams. Spot-tailed Nightjars occur in the savannas on the first and second rises, 600–700 m and 900–1000 m along this road, or 8.0 and 8.3 km from the *zócalo*.

While the nightjars can be found widely in this area, traffic on the road described above may be lighter than on the other roads. This area can also be reached by turning left at the shrine fork, although the route is slightly longer and less straightforward.

Las Choapas Bird List

43 sp. †See pp. 5–7 for notes on English names.

Muscovy Duck
Roadside Hawk
Plain Chachalaca
Pale-vented Pigeon
Plain-breasted Ground-Dove
White-tipped Dove
Aztec Parakeet†
Red-lored Parrot
Mottled Owl
Common Nighthawk
Pauraque
Chuck-will's-widow
Spot-tailed Nightjar
Little Hermit
Black-headed Trogon
Golden-fronted Woodpecker
Eastern Pewee†
Alder Flycatcher
Least Flycatcher
Brown-crested Flycatcher
Great Crested Flycatcher
Social Flycatcher
Tropical Kingbird
Eastern Kingbird
Bank Swallow
Brown Jay
Swainson's Thrush
Red-eyed Vireo
Yellow Warbler
Chestnut-sided Warbler
Magnolia Warbler
Canada Warbler
Grey-crowned Yellowthroat
Rufous-capped Warbler
Red-throated Ant-Tanager
Indigo Bunting
Blue-black Grassquit
White-collared Seedeater
Botteri's Sparrow
Eastern Meadowlark
Melodious Blackbird
Yellow-tailed Oriole
Montezuma Oropendola

CHAPTER 11 OAXACA

Bridled Sparrow

Oaxaca may have more species (almost 700) than any other Mexican state (Binford 1989). This is due to a combination of factors: the state is very mountainous; it extends to both the Pacific and Atlantic coastal slopes; and it lies at the juncture of the Nearctic and Neotropical zones. From a birding point of view there are four main regions that should be visited: the Valley of Oaxaca and surrounding mountains; the Sierra Madre del Sur (Sierra de Miahuatlán) and Pacific Coast; the Isthmus of Tehuantepec; and the cloud forests and lowlands of the Atlantic Slope.

A concentrated two-week trip could take in all of the main sites fairly easily. Alternatively, two or three of the regions could be covered more thoroughly in the same time, or a leisurely week could be based in Oaxaca City, perhaps with an overnight trip to Tehuantepec or Tuxtepec. It is also quite easy to combine Oaxaca with sites in Central and Southern Chiapas (Chapter 12) and/or with Uxpanapa, Veracruz (Site 10.7).

Oaxaca City, capital of the state, is a good starting point for a birding trip. There are daily flights from Mexico City, and car rental is straightforward at the airport (unless it's Christmas or some other fiesta, in which case you should make reservations well in advance and reconfirm them at least once). From Mexico City, Oaxaca used to be a full day's drive (and you could still do that), but with the recent completion of new *cuota* highways it is now only 4.5–5.5 hours (about 450 km) via Puebla and Tehuacán.

At 1500 m elevation, Oaxaca City lies in a dry valley ringed by mountains. It is a popular destination for tourists and birders alike, so hotels and restaurants can be found for all budgets. The city has a pleasant climate and is a convenient base from which to explore a good variety of habitats. The following sites comprise easy day trips from Oaxaca City: Monte Alban, Teotitlán del Valle, Yagul, Route 175 North, and Cerro San Felipe (La Cumbre). The other sites are arranged in an arbitrary geographic sequence, starting with the Atlantic Slope, then the Sierra Madre del Sur/Pacific Coast, and last the Isthmus of Tehuantepec.

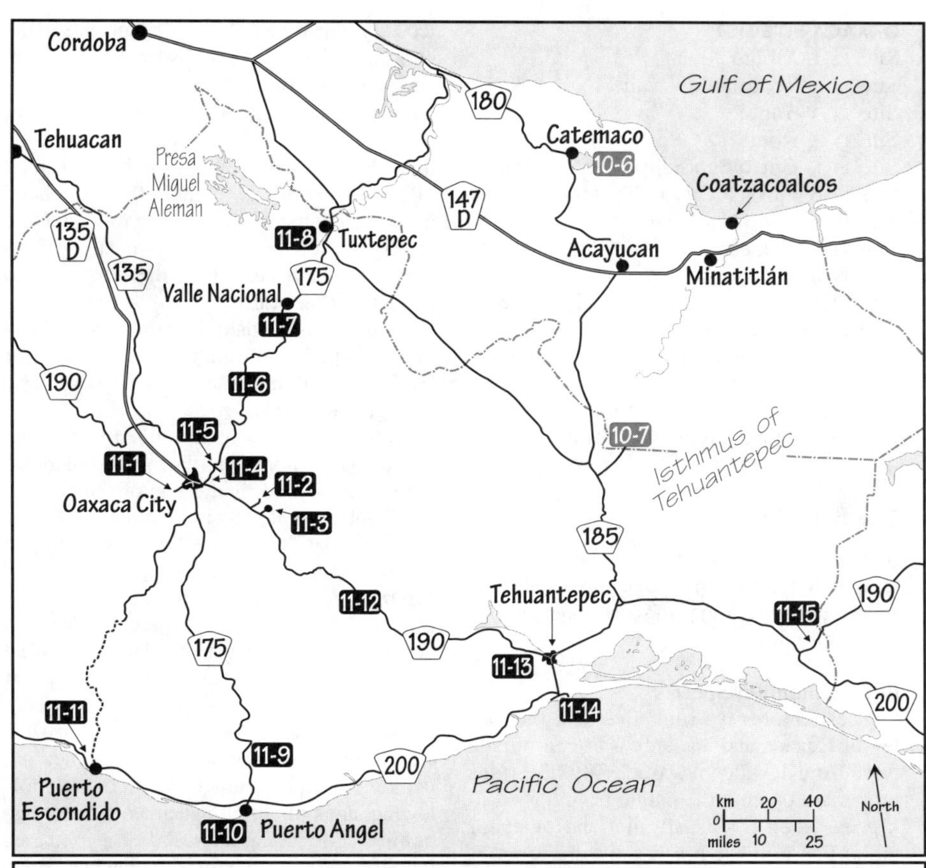

Chapter 11: OAXACA

- **11-1** Monte Alban
- **11-2** Teotitlán del Valle
- **11-3** Yagul
- **11-4** Route 175 North
- **11-5** Cerro San Felipe (La Cumbre)
- **11-6** Oaxaca City to Tuxtepec
- **11-7** Valle Nacional
- **11-8** Tuxtepec
- **11-9** La Soledad
- **11-10** Puerto Angel / Playa Zipolite
- **11-11** Puerto Escondido
- **11-12** Oaxaca City to Tehuantepec
- **11-13** Tehuantepec
- **11-14** La Ventosa Lagoon
- **11-15** Tapanatepec Foothills

OAXACA SITE LIST

Site 11.1 Monte Alban
Site 11.2 Teotitlán del Valle
Site 11.3 Yagul
Site 11.4 Route 175 North
Site 11.5 Cerro San Felipe (La Cumbre)
Site 11.6 Oaxaca City to Tuxtepec
Site 11.7 Valle Nacional
Site 11.8 Tuxtepec
Site 11.9 La Soledad
Site 11.10 Puerto Angel/Playa Zipolite
Site 11.11 Puerto Escondido
Site 11.12 Oaxaca City to Tehuantepec
Site 11.13 Tehuantepec
Site 11.14 La Ventosa Lagoon
Site 11.15 Tapanatepec Foothills

Site 11.1: Monte Alban

(January, February 1983, November 1985, January 1987, April 1988, December 1989, 1990, 1991, 1992, 1993, 1995; checked December 1996)

Introduction/Key Species

These impressive ruins overlook the Valley of Oaxaca and are only a fifteen-minute drive from downtown Oaxaca City. The ruins proper are open to the public from 9 A.M. to 5 P.M., when it's usually hot and quiet for birds. For birding, it's best to drive out at dawn to the parking lot, near which bird activity can be good as the sun's rays warm up the scrub-covered slopes. Many sought-after endemics occur here, but some can be hard to see when they're not singing: the difference between a visit in December (usually fairly quiet) and April (usually much song) is striking. But, with patience, most species can be seen at any season. Late afternoon visits are hit-and-miss for bird activity but can be good, especially in late winter and spring.

Access/Birding Sites

How you get here depends on where you are staying in Oaxaca City. Monte Alban is a major tourist site, so the route is well marked, and a new, wide, and very direct access road has now been built (versus a rather tortuous and winding narrow road that visitors prior to 1995 may recall), starting on the western edge of the city as one exits on the highway to Mexico City. The road to Monte Alban ends in the parking lot. Usually this is not chained off, and one can drive to the far end and park. Although the parking lot is empty at first light, it may be packed full when you return—it helps to park facing out! If the lot is chained off, you can park at the pulloff on the bend immediately below the parking lot; if you do so, come back and move your car by 8:30 A.M., when tour buses need the extra room to make this sharp turn.

Tourist buses visit the ruins throughout the day, but if you want to get here early for birding you'll need to take a taxi (easily arranged, e.g., at the *zócalo*).

Before the ruins open officially, one can bird along the numerous narrow footpaths on the slopes beyond the far end of the parking lot. Tumba (Tomb) 107 is the mound at the far end of the parking lot. Waiting here as the sun hits the scrub in early morning is a good bet for **Dwarf Vireo** (listen for its chatter call, *very* like Ruby-crowned Kinglet, which also occurs but is rare). **Beautiful Hummingbird** is fairly common November to February, at least (nesting November 1995), and **Dusky Hummingbird** and **White-throated Towhee** are common. As the sun warms things up, work along trails a few hundred meters, following signs of activity such as calls, songs, or birds at flowering trees. This is a good area for **Slaty Vireo, Ocellated Thrasher**, **Boucard's Wren**, and, in spring (April to May, at least), **Pileated Flycatcher**, which sings loudly and is easy to find (but seems rare or absent in winter). The main ruins are worth seeing but do not have any particular birds (Rock Wren is often easy to see around the main plaza, and, if fruiting, the large fig tree as you enter the main ruins can be full of birds, which can include Blue-hooded Euphonia). Although Oaxaca Sparrow has been reported from Monte Alban, I've never found it there.

Note: Oaxaca City has an active cage-bird trade: species noted for sale range from Long-tailed Wood-Partridge to European Starling and Blue-crowned Chlorophonia! Consequently, oddities occur around town, including Monte Alban, and Tropical Mockingbirds at the ruins may derive from escapees.

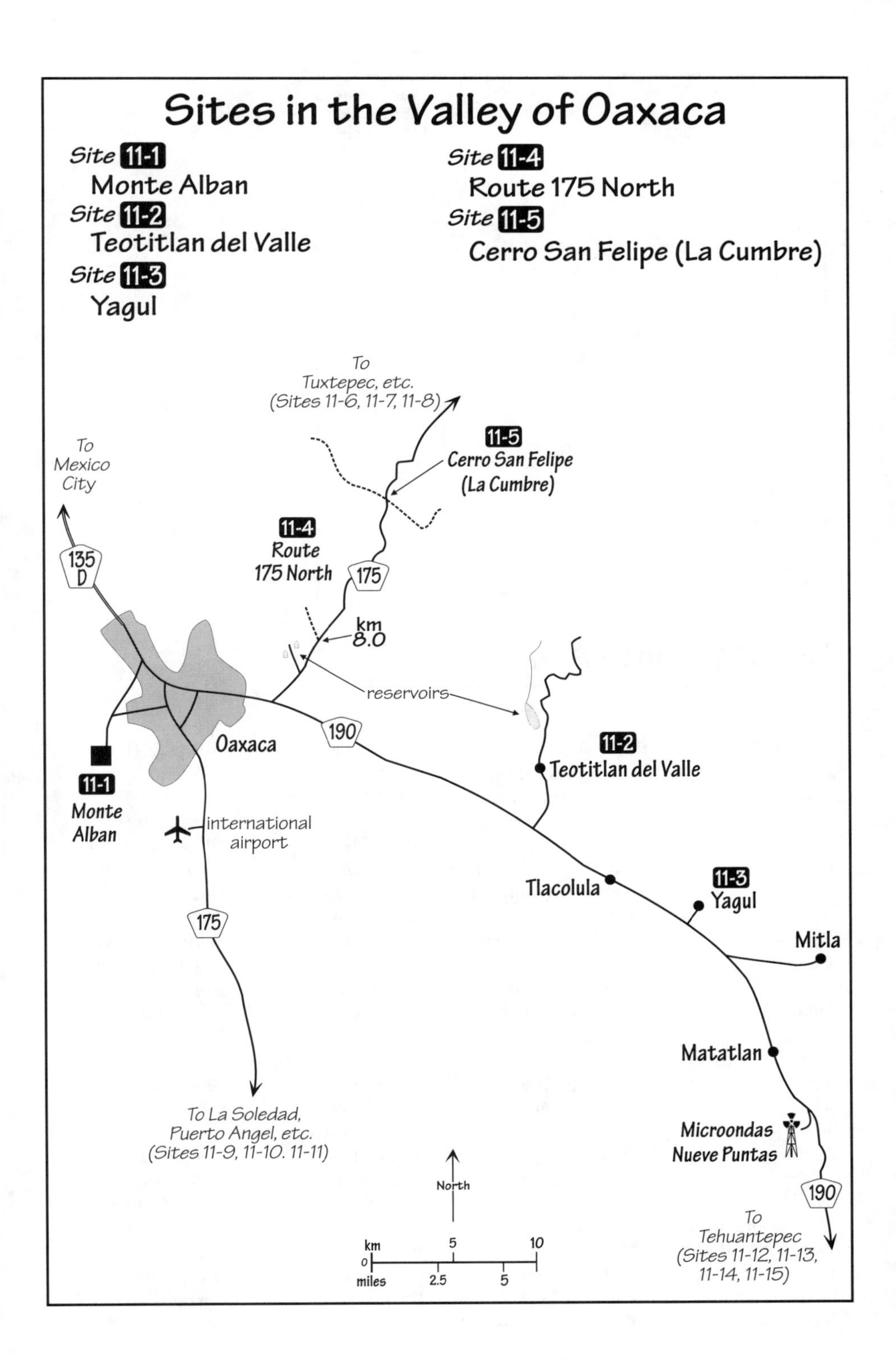
Sites in the Valley of Oaxaca
Site 11-1
Monte Alban
Site 11-2
Teotitlan del Valle
Site 11-3
Yagul
Site 11-4
Route 175 North
Site 11-5
Cerro San Felipe (La Cumbre)
To
Tuxtepec, etc.
(Sites 11-6, 11-7, 11-8)
11-5
Cerro San Felipe
(La Cumbre)
To
Mexico
City
135
D
11-4
Route
175 North
175
km
8.0
reservoirs
190
Oaxaca
11-2
Teotitlan del Valle
11-1
Monte
Alban
international
airport
Tlacolula
11-3
Yagul
175
Mitla
Matatlan
Microondas
Nueve Puntas
To La Soledad,
Puerto Angel, etc.
(Sites 11-9, 11-10. 11-11)
North
km
0
5
10
miles
2.5
5
190
To
Tehuantepec
(Sites 11-12, 11-13,
11-14, 11-15)

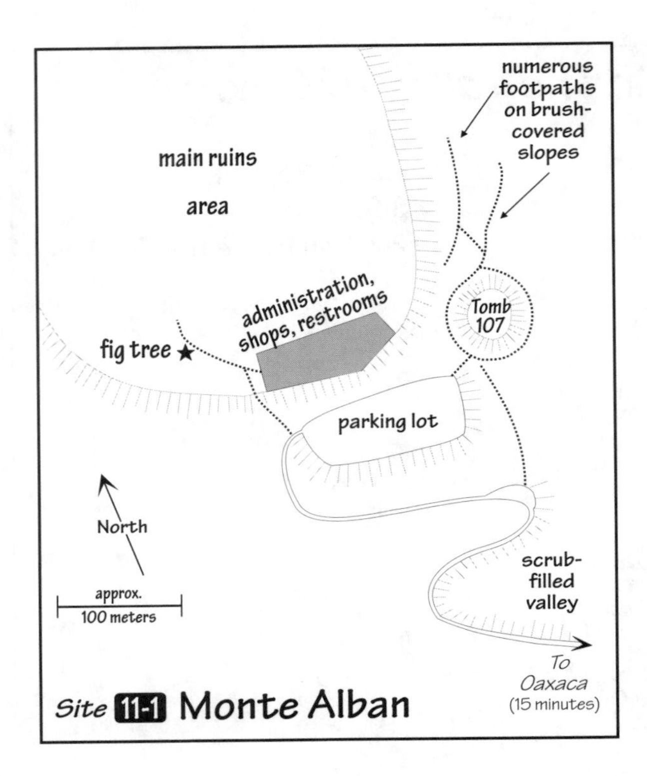

Monte Alban Bird List
(many hummingbirds and orioles are seasonal): 93 sp. R: rare (not to be expected). †See pp. 5–7 for notes on English names.

Cattle Egret
Turkey Vulture
White-tailed Kite
Sharp-shinned Hawk
Cooper's Hawk
Short-tailed Hawk
White-tailed Hawk
Red-tailed Hawk
Crested Caracara
American Kestrel
Mourning Dove
Inca Dove
Common Ground-Dove
White-tipped Dove
Groove-billed Ani
Lesser Nighthawk
Chestnut-collared Swift
Vaux's Swift
Broad-billed Hummingbird (R)
Dusky Hummingbird
Berylline Hummingbird
White-eared Hummingbird (R)
Beautiful Hummingbird
Ruby-throated Hummingbird
Rufous Hummingbird
Grey-breasted Woodpecker
N. Beardless Tyrannulet
Greenish Elaenia
Pileated Flycatcher
Greater Pewee
Western Pewee†
Least Flycatcher (R)
Dusky Flycatcher
Say's Phoebe
Vermilion Flycatcher
Ash-throated Flycatcher
Nutting's Flycatcher
Cassin's Kingbird
Thick-billed Kingbird
Western Kingbird
Violet-green Swallow
N. Rough-winged Swallow
Cliff Swallow
Barn Swallow
Northern Raven†
Boucard's Wren
Canyon Wren
Rock Wren
Bewick's Wren
Northern House Wren†
Ruby-crowned Kinglet (R)
Blue-grey Gnatcatcher
Orange-billed N.-Thrush
Hermit Thrush
American Robin
Northern Mockingbird
Tropical Mockingbird
Blue Mockingbird
Curve-billed Thrasher
Ocellated Thrasher
American Pipit
Loggerhead Shrike
Slaty Vireo
Dwarf Vireo
Blue-headed Vireo†
Cassin's Vireo†
Warbling Vireo

Orange-crowned Warbler
Nashville Warbler
Virginia's Warbler
Audubon's Warbler†
Townsend's Warbler
MacGillivray's Warbler
Wilson's Warbler
Rufous-capped Warbler
Blue-hooded Euphonia
Western Tanager
Rose-breasted Grosbeak
Black-headed Grosbeak
Blue Grosbeak
Varied Bunting
White-throated Towhee
Rufous-crowned Sparrow
Black-chinned Sparrow
Grasshopper Sparrow
Lincoln's Sparrow
Eastern Meadowlark
Bronzed Cowbird
Black-vented Oriole
Streak-backed Oriole
Scott's Oriole
House Finch
Lesser Goldfinch

Site 11.2: Teotitlán del Valle

(December 1989, 1990, 1991, 1992, 1993, 1995, checked December 1996)

Introduction/Key Species

This area of arid temperate scrub, about forty-five minutes from Oaxaca City, has **Ocellated Thrasher**, **Dwarf Vireo**, **Oaxaca** and **Bridled sparrows**, and so on. A small reservoir adds diversity in the form of a few waterbirds. The town of Teotitlán del Valle is famous for its weavings and is a well-known tourist attraction. It is easy to combine a morning's birding with a stop at the weaving market, centered on the town square.

Access/Birding Sites

Take Route 190 (the main Tehuantepec highway) out of Oaxaca City to the east-southeast and, 18.8 km from the junction with Route 175, look for the well-signed left turn to Teotitlán del Valle. Mark zero at the turn and follow this paved road into town, where it becomes a dirt road at Km 4.6, just beyond the rug market; continue straight here (eventually the road goes to the settlement of Benito Juarez up in the mountains). At Km 5.9 (1.3 km from where the pavement ends) you come to the dam of a small reservoir (Presa Piedra Azul) on the left; the large pulloff at the far end by the inflow (Km 6.8) is a good picnic spot and is often very birdy in the early mornings. From here you can continue on foot along the road for a few hundred meters, along a line of *Erythrina* (coral bean) trees which, if flowering (December to January, at least), can be excellent for hummingbirds (including **Green-fronted**) and orioles. Continue on north a few kilometers, and the dirt road winds up into arid scrub and then oak scrub. Park at the first sharp right-hand bend, at Km 9 (2.2 km from the head of the reservoir), and bird along the road in either direction. The scrub is densest along the (seasonally dry) valleys, more open on the hillsides. Here, and at the next two or three bends, **Dwarf Vireos** occur in the scrub-filled valleys; **Oaxaca** and **Bridled sparrows** are fairly common along the road, and both (especially Bridled) respond to pishing; **Ocellated Thrasher** can be found singing on the slopes above this first bend.

Teotitlán is served by public buses, and from here there are second-class buses that run up into the mountains, passing through the birding areas; unless you stay in or near Teotitlán, however, it's unlikely that you'd get to the good spots before bird activity wanes and the heat kicks in.

Teotitlán del Valle Bird List

(reservoir up into the first oaks): 116 sp. R: rare (not to be expected). †See pp. 5–7 for notes on English names.

Least Grebe
Pied-billed Grebe
Eared Grebe
Great Blue Heron
Great Egret
Cattle Egret
Green Heron
Blue-winged Teal
Turkey Vulture
Northern Harrier
Sharp-shinned Hawk
Cooper's Hawk
Red-tailed Hawk
American Kestrel
West Mexican Chachalaca
American Coot
Spotted Sandpiper
Least Sandpiper
Long-billed Dowitcher
Common Snipe
White-winged Dove
Mourning Dove
Inca Dove
Common Ground-Dove
Dusky Hummingbird
Berylline Hummingbird
Green-fronted Hummingbird
Magnificent Hummingbird
Beautiful Hummingbird
Ruby-throated Hummingbird
Belted Kingfisher
Green Kingfisher
Grey-breasted Woodpecker

Yellow-bellied Sapsucker
Ladder-backed Woodpecker
N. Beardless Tyrannulet
Greenish Elaenia
Greater Pewee
Common Tufted Flycatcher†
Least Flycatcher (R)
Dusky Flycatcher
Eastern Phoebe (R)
Black Phoebe
Say's Phoebe
Vermilion Flycatcher
Dusky-capped Flycatcher
Ash-throated Flycatcher
Nutting's Flycatcher
Great Kiskadee
Tropical Kingbird
Thick-billed Kingbird
Cassin's Kingbird
Violet-green Swallow
N. Rough-winged Swallow
Western Scrub Jay
Northern Raven†
Bridled Titmouse
Bushtit
Boucard's Wren
Bewick's Wren
Northern House Wren†
Ruby-crowned Kinglet
Blue-grey Gnatcatcher
Hermit Thrush
Orange-billed N.-Thrush
American Robin
Northern Mockingbird
Blue Mockingbird
Curve-billed Thrasher
Ocellated Thrasher
American Pipit
Grey Silky†
Loggerhead Shrike
Dwarf Vireo
Plumbeous Vireo†
Warbling Vireo
Tennessee Warbler (R)
Orange-crowned Warbler
Nashville Warbler
Virginia's Warbler
Myrtle Warbler†
Audubon's Warbler†
Black-throated Grey Warbler
Townsend's Warbler
Black-and-white Warbler
MacGillivray's Warbler
Common Yellowthroat
Wilson's Warbler
Painted Whitestart†
Slate-throated Whitestart†
Rufous-capped Warbler
Blue-hooded Euphonia
Hepatic Tanager
Summer Tanager
Western Tanager
Rose-breasted Grosbeak
Black-headed Grosbeak
Blue Grosbeak
Indigo Bunting
Spotted Towhee
White-throated Towhee
Cinnamon-rumped Seedeater†
Oaxaca Sparrow
Bridled Sparrow
Chipping Sparrow
Clay-colored Sparrow
Lark Sparrow
Lincoln's Sparrow
Great-tailed Grackle
Hooded Oriole
Black-vented Oriole
Dickey's Oriole†
Baltimore Oriole
Bullock's Oriole
Baltimore × Bullock's Oriole
Scott's Oriole
House Finch
Lesser Goldfinch

Site 11.3: Yagul

(November 1985, April 1988, December 1989, 1990, 1991, 1992, 1993, 1995, checked December 1996)

Introduction/Key Species

These ruins are surrounded by open, desert-like scrub in a spectacular, elevated setting. They tend to be quiet in the morning, as Yagul is a "minor" site, not reached by busloads of tourists much before midday. It is an excellent site for **Beautiful Hummingbird** (November to January, at least), **Grey-breasted Woodpecker**, **Boucard's Wren**, and **Bridled Sparrow**.

Access/Birding Sites

Take Route 190 (the main Tehuantepec highway) east-southeast from Oaxaca City and, 28.2 km from the junction with Route 175 (i.e., about 17.5 km past the Teotitlán del Valle turning), look for a signed turn to the left (north) for Yagul. This paved road ends in 1.5 km at the parking lot for the ruins, which are open from 9 A.M. to 5 P.M. (and are forty-five minutes to an hour from Oaxaca). If you get here early, for first light, there will probably be a cable across the road about 250 m below the parking lot; however, on the left just before the cable there is a pulloff where one can park a car (just after a track that leads off to the left/north).

Along the road here, as the morning sun hits the big cacti, is a good place for **Boucard's Wren**, **Bridled Sparrow**, **Beautiful Hummingbirds** (including males, which often are outnumbered by more than five to one by female-plumaged birds at other sites), and **Grey-breasted Woodpecker**. The open scrub here is also a good spot for Grey Flycatcher (at the southern limit of its winter range). The track to the north, and another opposite this to the south, skirt between the base of the hills and agricultural land, with tall trees along irrigation ditches, which can be good in winter for seed-eating birds (including Lazuli Bunting, at the southern limit of its winter range). The ruins themselves don't have any different birds, but if you miss **Bridled Sparrow** along the entrance road, look in the scrub below the far north edge of the ruins, where **Pileated Flycatcher** also can be found, at least in spring.

Yagul is reached easily by second-class public buses running between Oaxaca City and Mitla; ask to be dropped at the Yagul junction, from which you walk the short distance to the hills and good birding.

Yagul Bird List

83 sp. R: rare (not to be expected). †See pp. 5–7 for notes on English names.

Great Blue Heron
Great Egret
Cattle Egret
Turkey Vulture
White-tailed Kite
Northern Harrier
Sharp-shinned Hawk
Cooper's Hawk
Red-tailed Hawk
White-tailed Hawk
Crested Caracara
American Kestrel
Common Snipe
White-winged Dove
Mourning Dove
Inca Dove
Common Ground-Dove
White-throated Swift
Dusky Hummingbird
Berylline Hummingbird
Beautiful Hummingbird
Ruby-throated Hummingbird
Rufous Hummingbird
Belted Kingfisher
Grey-breasted Woodpecker
Ladder-backed Woodpecker
N. Beardless Tyrannulet
Pileated Flycatcher
Least Flycatcher (R)
Dusky Flycatcher
Grey Flycatcher (R)
Black Flycatcher
Say's Phoebe
Vermilion Flycatcher
Ash-throated Flycatcher
Nutting's Flycatcher
Great Kiskadee
Cassin's Kingbird
Thick-billed Kingbird
Western Kingbird
N. Rough-winged Swallow
Barn Swallow
Northern Raven†
Boucard's Wren
Rock Wren
Canyon Wren
Bewick's Wren
Northern House Wren†
Marsh Wren (R)
Blue-grey Gnatcatcher
Rufous-backed Thrush†
Northern Mockingbird
Curve-billed Thrasher
American Pipit
Loggerhead Shrike
Solitary Vireo ssp.
Warbling Vireo
Orange-crowned Warbler
Nashville Warbler
Virginia's Warbler
Audubon's Warbler†
MacGillivray's Warbler
Wilson's Warbler
Rufous-capped Warbler
Western Tanager
Summer Tanager
Blue Grosbeak
Lazuli Bunting (R)
Indigo Bunting
Varied Bunting
White-throated Towhee
Bridled Sparrow
Lark Sparrow
Clay-colored Sparrow
Lincoln's Sparrow
Eastern Meadowlark
Red-winged Blackbird
Great-tailed Grackle
Bronzed Cowbird
Black-vented Oriole
House Finch
Lesser Goldfinch

Site 11.4: Route 175 North

(January, February 1983, November 1985, January 1987, April 1988, December 1989, 1990, 1991, 1992, 1993, 1995, checked December 1996)

Introduction/Key Species

From its junction with Route 190 on the eastern edge of Oaxaca City, Route 175 heads north into the mountains and eventually over to Tuxtepec in the Atlantic Slope lowlands (see Sites 11.6, 11.7, 11.8). The arid scrub on the lower slopes holds a number of endemics, including **Pileated Flycatcher**, **Ocellated Thrasher**, **Slaty** and **Dwarf vireos**, and **Oaxaca** and **Bridled sparrows**. Two small reservoirs (which can be dry in spring) are also worth checking. Note that Rusty Sparrow is not known to occur in the Valley of Oaxaca and that Oaxaca Sparrow is quite common (not rare, as often suggested) but local.

Access/Birding Sites

Head east-southeast out of Oaxaca on Route 190 toward Tehuantepec. At 3.5 km east of the Calle Juarez Pemex station there is a major turn to the left (north), signed Route 175 to Tuxtepec. The Km Posts mark distance from zero at Tuxtepec to about Km 215 at the junction with Route 190. Mark zero as you turn north onto Route 175.

At Km 4.9 there is a wide paved fork to the left and a banked-up rock slope that looks like (and is) the dam for a reservoir. To check the reservoirs, take the left-hand fork to the top of the rise and turn left or right onto the dirt roads that lead down to the shores of the reservoirs on either side of the road. Because open water is scarce in the Valley of Oaxaca, lost waterbirds tend to show up here and have included Brown Pelican and Reddish Egret! The surrounding fields host a variety of species, including wintering Clay-colored, Lark, and Grasshopper sparrows.

To get to the scrub, stay on Route 175 toward the hills (noting that the left-hand fork to the reservoirs is wider than the main highway). After 1 km or so of fields the road winds up through hillsides which, in places, are not too badly overgrazed by goats and have some scrub-filled gulleys. A rushing stream parallels the right side of the highway, lined in places with tall trees. Stopping at almost any bend or pulloff between Km 7.5 and 12 (i.e., between Km Posts 207 and 202) should produce **Oaxaca Sparrow** (especially conspicuous in spring when singing); also in spring, **Pileated Flycatcher** and **Dwarf Vireo** are often conspicuous here by virtue of their songs.

A well-known spot is the right-hand bend at Km 8 (Km Post 206.5), marked by a garbage-filled pullout on the left, where there's room to park a car. From here a narrow footpath leads up the scrub-filled gulley to the left (north); continuing up this path about 200 m brings you out to a fairly open, overgrazed spur (check the upper slopes for **Bridled Sparrow**) that provides an overlook into narrow, scrub-filled valleys on either side (**Slaty** and **Dwarf vireos**). This site can be good for the last hour or so of sun in late afternoon, as well as in early to mid morning. Across the road, tracks lead down to the stream, where birding can be good in the taller trees.

Second-class buses from Oaxaca to Gueletao or Tuxtepec can drop you anywhere along this road, and hitching rides is also possible.

Route 175 North Bird List

(Km 207/202): 98 sp. R: rare (not to be expected). †See pp. 5–7 for notes on English names.

Cattle Egret
Turkey Vulture
White-tailed Kite
Sharp-shinned Hawk
Cooper's Hawk
Red-tailed Hawk
West Mexican Chachalaca
White-winged Dove
Mourning Dove
Inca Dove
White-tipped Dove
Lesser Roadrunner
Dusky Hummingbird
Berylline Hummingbird
Magnificent Hummingbird
Beautiful Hummingbird
Ruby-throated Hummingbird
Rufous Hummingbird
Elegant Trogon
Grey-breasted Woodpecker
Ladder-backed Woodpecker
N. Beardless Tyrannulet
Greenish Elaenia
Pileated Flycatcher
Greater Pewee
Western Pewee†
Common Tufted Flycatcher†
Dusky Flycatcher
Black Phoebe
Vermilion Flycatcher
Dusky-capped Flycatcher
Ash-throated Flycatcher
Nutting's Flycatcher
Cassin's Kingbird
Thick-billed Kingbird
Western Kingbird
Rose-throated Becard
Violet-green Swallow
N. Rough-winged Swallow
Northern Raven†
Western Scrub Jay
Bushtit
Boucard's Wren
Canyon Wren
Bewick's Wren
Northern House Wren†
Blue-grey Gnatcatcher
Hermit Thrush
Clay-colored Thrush†
Orange-billed N.-Thrush
Blue Mockingbird
Ocellated Thrasher
Grey Silky†
Slaty Vireo
Dwarf Vireo
Cassin's Vireo†
Golden Vireo
Warbling Vireo
Orange-crowned Warbler
Nashville Warbler
Virginia's Warbler
Audubon's Warbler†
Townsend's Warbler
Black-throated Grey Warbler
Black-and-white Warbler
Louisiana Waterthrush
MacGillivray's Warbler
Wilson's Warbler
Slate-throated Whitestart†
Rufous-capped Warbler
Blue-hooded Euphonia
Hepatic Tanager
Western Tanager
Black-headed Grosbeak
Blue Grosbeak
Lazuli Bunting (R)
Indigo Bunting
Varied Bunting
Green-tailed Towhee (R)
Spotted Towhee

White-throated Towhee
Blue-black Grassquit
Bridled Sparrow
Rufous-crowned Sparrow
Oaxaca Sparrow
Chipping Sparrow
Lark Sparrow
Grasshopper Sparrow
Lincoln's Sparrow
Bronzed Cowbird
Brown-headed Cowbird
Hooded Oriole
Dickey's Oriole†
Black-vented Oriole
Bullock's Oriole
Scott's Oriole
House Finch
Lesser Goldfinch

Site 11.5: Cerro San Felipe (La Cumbre)

(March 1982, January, February 1983, November 1985, January 1987, April 1988, December 1989, 1990, 1991, 1992, 1993, 1995, checked December 1996)

Introduction/Key Species

Cerro San Felipe, in the Sierra de Aloapaneca north of Oaxaca City, is famous as a readily accessible site for the Mexican endemic **Dwarf Jay**. It also hosts a good variety of birds typical of central Mexico's humid pine-oak highlands, as well as great mixed-species flocks of wintering warblers from western North America. Species of interest here include **Long-tailed Wood-Partridge**, **Amethyst-throated**, **Garnet-throated**, and **Bumblebee hummingbirds**, **Mountain Trogon**, **Pine Flycatcher**, **Grey-barred Wren**, **Russet** and Ruddy-capped **nightingale-thrushes**, **Chestnut-sided Shrike-Vireo**, **Red Warbler**, **Rufous-capped Brushfinch**, and **Collared Towhee**.

Access/Birding Sites

Head east-southeast out of Oaxaca City on Route 190 and turn north onto Route 175 (see Site 11.4), setting to zero at the junction. Continue north, climbing into the mountains, for 22.5 km to the top of the first main ridge at around 2700 m elevation. This is at Km Post 192 and is marked by a small group of roadside huts called La Cumbre (i.e., The Summit). Given the climb, the potential to be stuck behind slow trucks, and the need to find someone to open the "gate" (see below), it is best to allow an hour's driving time from downtown Oaxaca.

This site can be reached easily by second-class buses running between Oaxaca and Gueletao or Tuxtepec. From La Cumbre (where you'd get off the bus), dirt logging roads on either side of the highway offer excellent birding. Traffic along the logging roads (see below) is generally light, but good habitat is not far from the highway and can be reached easily by walking.

There may be a low chain across the entrance to the logging roads, purportedly to control the illegal extraction of lumber. If these chains are truly padlocked (often they look to be so but can be slackened to allow passage), stand around and whistle or ask at the nearest hut, where there is usually someone with a key. There should be no problem having the chains unlocked—there are settlements along both roads, and they are basically public roads; recently the "gatekeepers" have taken to charging a small "*propina*," or "tip" (try your bargaining skills) to unlock the gates—they may have you sign a book to give this procedure an air of legality.

Although the Valley of Oaxaca is usually hot and sunny (although often cool at first light), Cerro San Felipe can be cold, foggy, and windy at any time of day. Conversely, it can be clear, sunny, still, and beautiful. It is best to be prepared for the worst rather than get chilled.

The Northwest Side. Birding is potentially good anywhere along this road: drive slowly, watching and listening for flocks. Although I have seen both a **Long-tailed Wood-Partridge** (running across the road) and a flock with **Dwarf Jays** less than 1 km from Route 175, the consistently best areas seem to be 2–9 km from the highway. At Km 2 there is a notably sharp right-hand bend uphill, with a large pulloff on the left where one can park safely. After this the road climbs a little and then bends back left and more or less levels out for a kilometer or two, with a track leading steeply down to the left at Km 3. The entrance to this track is another good parking spot. Birding around the Km 2 pulloff and walking along the road for 1–2 km beyond the Km 3 track, and down the steep side road there, should produce a good variety of species, including **Dwarf Jay**.

In winter **Dwarf Jays** travel with flocks of

Steller's Jays and **Grey-barred Wrens**, and it is best to listen for the noisy Steller's Jay and wren calls to find the flocks. The Dwarf Jays call infrequently (if at all) and can be overlooked easily as they forage in dense oak foliage. Be patient and stay with the flocks as long as you can and you should see small blue jays flying from tree to tree, often in close association with the Steller's and wrens; they can be quite confiding when found. In spring and summer, Dwarf Jays tend to be in pairs, not with flocks, and can be harder to see, although they tend to be more vocal than in winter.

If you continue along the road, at Km 7.4 there is a junction, right to Nuevo Zooquiapan, left to Corral de Piedras, a few hundred meters past some (disused?) forestry buildings on the left. Bear left toward Corral de Piedras; the road goes through more humid forest, where **Garnet-throated Hummingbird** and **Aztec Thrush** have been found (at Km 9–10). The road toward Nuevo Zooquiapan soon gets into drier pine-oak and oak woodland on the interior slopes.

The Southeast Side. The first 2 km or so wind through fields, brush, and hedges before coming into a good patch of forest, the start of which is marked by a cobbled road going off and up to the right, to a microwave tower. There is plenty of room to park safely at this junction. The field edges and roadside can have good patches of flowers (November to February, at least) where **Bumblebee Hummingbird** is often seen. Walking is fairly easy along the road into the forest and, although one can drive on for many kilometers and get to other patches of forest, this first patch seems to have most, if not all, of the more interesting species. As on the northwest side, the trick is to find a mixed-species flock with jays and wrens.

Cerro San Felipe Bird List

99 sp. R: rare (not to be expected). †See pp. 5–7 for notes on English names.

Turkey Vulture
Sharp-shinned Hawk
Broad-winged Hawk
Short-tailed Hawk
Zone-tailed Hawk
Red-tailed Hawk
American Kestrel
Long-tailed Wood-Partridge
Band-tailed Pigeon
Lesser Roadrunner
Mountain Pygmy-Owl†
Mexican Whip-poor-will†
White-throated Swift
Great Swallow-tailed Swift (R)
Green Violet-ear
White-eared Hummingbird
Amethyst-throated Hummingbird
Blue-throated Hummingbird
Garnet-throated Hummingbird
Magnificent Hummingbird
Broad-tailed Hummingbird
Rufous Hummingbird
Bumblebee Hummingbird
Mountain Trogon
Acorn Woodpecker
Yellow-bellied Sapsucker
Hairy Woodpecker
Red-shafted Flicker†
Strong-billed Woodcreeper (R)
Spot-crowned Woodcreeper
Greenish Elaenia
Greater Pewee
Common Tufted Flycatcher†
Hammond's Flycatcher
Dusky Flycatcher
Pine Flycatcher
Western Flycatcher†
Rose-throated Becard
Steller's Jay
Dwarf Jay
Western Scrub Jay
Northern Raven†
Mexican Chickadee
Bushtit
Grey-barred Wren
Brown-throated Wren†
Grey-breasted Wood-Wren
Brown Creeper
Ruby-crowned Kinglet
Brown-backed Solitaire
Russet N.-Thrush
Ruddy-capped N.-Thrush
Hermit Thrush
Black Thrush†
Aztec Thrush
American Robin
Blue Mockingbird
Cedar Waxwing
Grey Silky†
Cassin's Vireo†
Hutton's Vireo
Warbling Vireo
Chestnut-sided Shrike-Vireo
Orange-crowned Warbler
Nashville Warbler
Crescent-chested Warbler
Audubon's Warbler†
Townsend's Warbler
Hermit Warbler
Black-throated Green Warbler
Black-and-white Warbler
Louisiana Waterthrush
MacGillivray's Warbler
Wilson's Warbler
Red-faced Warbler
Red Warbler
Slate-throated Whitestart†
Rufous-capped Warbler
Golden-browed Warbler
Olive Warbler
Blue-hooded Euphonia
Hepatic Tanager
Black-headed Grosbeak

White-naped Brushfinch (R?)
Chestnut-capped Brushfinch
Rufous-capped Brushfinch
Cinnamon-bellied Flowerpiercer
Collared Towhee
Spotted Towhee
Chipping Sparrow
Lincoln's Sparrow
Mexican Junco†
Bullock's Oriole
Abeille's Oriole†
Scott's Oriole
House Finch
Red Crossbill
Black-headed Siskin
Lesser Goldfinch

Site 11.6: Oaxaca City to Tuxtepec

(April 1988, December 1989, 1990, 1991, January, December, 1992, December 1993, 1995, checked December 1996)

Introduction/Key Species

The drive from Oaxaca City to Tuxtepec, in the Atlantic Slope lowlands, can take a full day and involves some elevational changes: you start in Oaxaca at about 1500 m, go over the Sierra de Aloapaneca (La Cumbre) at about 2700 m, drop into another dry interior valley (again at about 1500 m), climb up and over the Sierra de Juarez to about 2900 m, and then drop into the Atlantic Slope foothills, ending at Tuxtepec, only 30 m above sea level . . . but at least the road is sinuous and potholed! The main reason to subject yourself to this 220 km drive is the cloud forest above Valle Nacional (Site 11.7), although birding near Tuxtepec (Site 11.8) also can also be productive.

A day trip to the Valle Nacional cloud forest from Oaxaca City is possible but tiring at best, especially if you plan to be there at dawn! The drive to Tuxtepec, where there are a couple of adequate hotels, allows you to scout the route and birding spots, and then from Tuxtepec to good cloud forest is only about an hour's drive (versus almost three hours from Oaxaca).

Access/Birding Sites/Bird Lists

Note that the km posts along the highway mark from zero in Tuxtepec to 215 at the junction with Route 190 on the edge of Oaxaca City. An early start is best from Oaxaca, getting across Cerro San Felipe before first light and starting the day's birding in the dry interior valley below Gueletao de Juarez, about 60 km from Oaxaca (allow a good hour and a half for the drive). Near Km Post 161 the road crosses a rushing river at a noisy metal bridge. Birding in the brushy scrub along the road on either side of the bridge can be good in the early morning, including a track up to the right (east), about 200 m before the bridge. Interestingly, this river flows north, into the Balsas drainage, which accounts for some species not found in the Valley of Oaxaca, for example, Violet-crowned Hummingbird and **Happy Wren**.

Birds around the bridge include Red-tailed Hawk, Peregrine Falcon, Spotted Sandpiper (winter), Violet-crowned Hummingbird, **Grey-breasted Woodpecker**, Northern Beardless Tyrannulet, Greater Pewee, Black Phoebe, Vermilion and Nutting's flycatchers, Cassin's and Thick-billed kingbirds, Western Scrub Jay, **Boucard's**, **Happy**, and Northern House (winter) **wrens**, Ruby-crowned Kinglet (winter), Blue-grey Gnatcatcher (migrant?), **Slaty Vireo**, Orange-crowned, Nashville, Virginia's, Audubon's, Townsend's, Black-and-white, MacGillivray's, Red-faced, and Wilson's warblers (winter), Louisiana Waterthrush (winter), **Rufous-capped Warbler**, Hepatic Tanager, **White-throated Towhee**, Yellow-faced Grassquit, **Cinnamon-rumped** [White-collared] **Seedeater**, Lincoln's Sparrow (winter), Hooded Oriole (winter), House Finch, and Lesser Goldfinch.

From Gueletao the highway climbs up into humid pine-oak forest and clearings as the road wends along through the Sierra de Juarez. Birds here are much the same as those listed for Cerro San Felipe (**including Dwarf Jay**), but the impressive flower banks (November to January, at least) along the highway are better for **Garnet-throated Hummingbird** (locally fairly common), and Colima Warbler (winter) has also been found here. Brushy clearings should be checked for **Hooded Yellowthroat**. At Km Post 108.5 there is a signed *mirador* on the right of the road: in the rare event that the Atlantic Slope lowlands are not at least partly cloudy, you may see the shores of the Gulf of Mexico! From here you drop into cloud forest and then remnant patches of rain forest before reaching the town of Valle Nacional at Km Post 40 (see Site 10.7).

From Valle Nacional to Tuxtepec is about

40 km of mostly straight and level highway, a relief after the mountains. The road passes through numerous villages and mostly cleared land with scattered forest patches, and the Río Valle Nacional flows north to your right. Birds are typical of the cutover, humid Atlantic Slope lowlands. A stop almost anywhere will seem birdy after the cloud forest, but you will be experiencing quantity versus quality. Roadside birds include Cattle Egret, Black Vulture, Roadside Hawk, Red-billed Pigeon, Ruddy Ground-Dove, Aztec Parakeet, Ringed and Amazon kingfishers, Golden-fronted Woodpecker, Great Kiskadee, Social Flycatcher, Tropical and Couch's kingbirds, Brown Jay, Melodious Blackbird, and Montezuma Oropendola. Species to keep an eye out for include Laughing Falcon, **Ridgway's Rough-winged Swallow**, and Chestnut-headed Oropendola (sometimes in orchards at Valle Nacional).

Site 11.7: Valle Nacional

(April 1988, December 1989, 1990, 1991, January, April, December 1992, December 1993, 1995, checked December 1996)

Introduction/Key Species

Valle Nacional (60 m elevation) is a small town in the Atlantic Slope lowlands of Oaxaca, between Oaxaca City and Tuxtepec. It is also generally the name given by birders to the forested slopes above town, along Route 175 to Oaxaca City. The highway climbs from Valle Nacional through cutover patches of rain forest, into montane rain forest, and then through some excellent cloud forest before getting up into humid pine-oak forest along the top of the ridge (at around 2700 m), in the Sierra de Juarez, where the avifauna is similar to that noted for Cerro San Felipe (Site 11.5). This area offers an excellent variety of Atlantic Slope cloud-forest birds, plus a good selection of lowland species nearer town.

Species of interest include Crested Guan, Ornate Hawk-Eagle, White Hawk, **White-faced Quail-Dove**, Barred Parakeet, Pheasant Cuckoo, Central American Pygmy-Owl, **Emerald-chinned** and **Bumblebee hummingbirds**, Black-crested Coquette, Scaly-throated and Ruddy foliage-gleaners, Lovely Cotinga, Azure-hooded and **Unicolored jays**, **Slate-colored Solitaire**, **Black Thrush**, **Blue-crowned Chlorophonia**, and **White-naped Brushfinch**.

Weather is very important in the cloud forest: clear and sunny days, while beautiful, tend to be fairly dead for birds after the early morning; cold and drizzly or rainy days tend to have bird activity all day but are not much fun; cloud and mist moving in and out, with alternating cool and sunny spells, tend to be best for both birds and birding. Season is also very important. Thus, while the majority of species are resident, a dawn chorus in April may reveal many species you never thought were here based on numerous December trips.

Access/Birding Sites

All sites are along Route 175 between Oaxaca City and Valle Nacional. The nearest place with reasonable accommodation is Tuxtepec (about an hour from good birding above Valle Nacional), although there is a basic hotel in Valle Nacional. A day trip from Oaxaca City is possible but tiring (a three-hour drive one-way); see Site 11.6. Given the nature of the birding, working this area via public transport is problematic. However, a potentially productive approach would be to get off the bus at La Esperanza, near Km Post 80, and walk the highway back down to Km Post 71 (see below).

From Valle Nacional (Km Post 40), the first 30 km or so wind through cutover forest patches. There are numerous pullouts where you can park safely: stopping and walking along the highway or driving slowly and looking/listening are the best way to find birds, as the locations of flocks and fruiting trees are ever-changing. Spots that have been good include Km Posts 51 to Km 54 (White Hawk, Red-lored, White-crowned and Brown-hooded parrots, Little Hermit, Black-crested Coquette, Keel-billed Toucan, White-collared Manakin, Golden-hooded and Crimson-collared tanagers), and from there to the overlook at Km Post 71, where one starts to enter cloud forest. At Km Post 71 (about 1200 m) the birds are a mixture of lowland and highland: Short-billed Pigeon, Pheasant Cuckoo, Central American Pygmy-Owl, Ruddy Foliage-gleaner, Tawny-throated Leaftosser,

Slate-colored Solitaire, Green Honeycreeper, **Yellow-winged Tanager**, Black-headed Saltator, **White-naped Brushfinch**. From here on you are mostly in cloud forest (1500–2400 m) until around Km Post 100, when the road starts to climb into humid pine-oak. Good spots can be the north side of the small village of La Esperanza, at Km Post 79–80 (check the roadside flowers for **Emerald-chinned** and **Amethyst-throated hummingbirds**), and any stop from here to Km Post 99 (Crested Guan, **White-faced Quail-Dove**, Barred Parakeet, **Bumblebee Hummingbird**, Emerald Toucanet, Azure-hooded and **Unicolored jays**, **Blue-crowned Chlorophonia**), including several good pulloffs for picnic spots, a couple of them with shelters, which can be good in the rain!

From here you climb steeply into humid pine-oak (watch for **Garnet-throated Hummingbird** at roadside flowers and for **Dwarf Jays** and **Grey-barred Wrens** with flocks of **Unicolored Jays** near Km Post 102). At the top of the last steep climb is the *mirador* (Km Post 108.5) noted in the Oaxaca City to Tuxtepec entry (Site 11.6).

Note that the Km Posts noted above are only ideas; play it by ear depending on weather and where you see/hear bird activity. The best spot one day can be dead the next. And remember, birding in cloud forest can be very frustrating: you may hear lots of birds but see nothing or, if it's sunny, not even hear anything. In this case it may be best to head to the highlands or lowlands and come back another time.

Valle Nacional Bird List

192 sp. This is divided roughly into highlands, or cloud forest (H; generally above 1200/1500 m; Km Posts 71–100), and lowlands (L; generally below 1200–1500 m; Km Posts 40–71), to indicate what may be seen in the different habitats; absence of a code letter indicates that the species is found in both habitats. Some highland species are elevational migrants, marked E in the list, and can be found at lower elevations in winter. †See pp. 5–7 for notes on English names.

Great Tinamou[L]
Little Tinamou[L]
Black Vulture[L]
Turkey Vulture[L]
Sharp-shinned Hawk
White Hawk[L]
Roadside Hawk[L]
Short-tailed Hawk[L]
Red-tailed Hawk[L]
Ornate Hawk-Eagle[H]
Barred Forest-Falcon[H]
Collared Forest-Falcon[L]
American Kestrel[L]
Bat Falcon[L]
Plain Chachalaca[L]
Crested Guan[H]
Long-tailed Wood-Partridge[H]
Spotted Wood-Quail[L]
Ruddy Crake[L]
Scaled Pigeon[L]
Short-billed Pigeon[L]
Ruddy Ground-Dove[L]
White-tipped Dove[L]
White-faced Quail-Dove[H]
Aztec Parakeet†[L]
Barred Parakeet[H]
Brown-hooded Parrot[L]
White-crowned Parrot[L]
Red-lored Parrot[L]
Squirrel Cuckoo[L]
Pheasant Cuckoo[H]
Central Am. Pygmy-Owl†[L]
Ferruginous Pygmy-Owl[L]
Chestnut-collared Swift
White-collared Swift
Vaux's Swift[L]
Long-tailed Hermit[L]
Little Hermit[L]
Wedge-tailed Sabrewing[L]
Violet Sabrewing[H]
Emerald-chinned Hummingbird[H]
Black-crested Coquette[L]
Canivet's Emerald[L]
White-bellied Emerald[L]
Azure-crowned Hummingbird[L]
Berylline Hummingbird[L]
Rufous-tailed Hummingbird[L]
Amethyst-throated Hummingbird[H]
Garnet-throated Hummingbird[H]
Ruby-throated Hummingbird[L]
Bumblebee Hummingbird[H]
Violaceous Trogon[L]
Collared Trogon
Blue-crowned Motmot[L]
Rufous-tailed Jacamar[L]
Emerald Toucanet[E]
Collared Aracari[L]
Keel-billed Toucan[L]
Golden-fronted Woodpecker[L]
Smoky-brown Woodpecker[L]
Golden-olive Woodpecker[L]
Lineated Woodpecker[L]
Pale-billed Woodpecker[L]
Rufous-breasted Spinetail[L]
Scaly-throated Fol.-gleaner†[H]
Ruddy Fol.-gleaner[H]
Buff-throated Fol.-gleaner[L]
Tawny-throated Leaftosser[H]
Olivaceous Woodcreeper[L]
Ivory-billed Woodcreeper[L]
Spotted Woodcreeper[H]
Spot-crowned Woodcreeper[H]
Barred Antshrike[L]
Dusky Antbird[L]
Greenish Elaenia[L]
Yellow-bellied Elaenia[L]

Ochre-bellied Flycatcher[L]
Sepia-capped Flycatcher[L]
Northern Bentbill[L]
Eye-ringed Flatbill[E]
Yellow-olive Flycatcher[L]
Common Tufted Flycatcher†[E]
Greater Pewee[E]
Olive-sided Flycatcher
Yellow-bellied Flycatcher[L]
Least Flycatcher[L]
Hammond's Flycatcher[H]
Western Flycatcher†[E?]
Bright-rumped Attila[L]
Dusky-capped Flycatcher
Great Kiskadee[L]
Boat-billed Flycatcher[L]
Social Flycatcher[L]
Tropical Kingbird[L]
Couch's Kingbird[L]
Masked Tityra[L]
Rose-throated Becard[L]
Lovely Cotinga[L]
White-collared Manakin[L]
N. Rough-winged Swallow
Green Jay[L]
Brown Jay[L]
Azure-hooded Jay[H]
Unicolored Jay[H]
Band-backed Wren[L]
Spot-breasted Wren[L]
Northern House Wren†[L]
Southern House Wren†[L]
White-breasted Wood-Wren[L]
Grey-breasted Wood-Wren[H]
Ruby-crowned Kinglet[H]
Long-billed Gnatwren[L]
Blue-grey Gnatcatcher
Slate-colored Solitaire[E]
Orange-billed N.-Thrush[E]
Ruddy-capped N.-Thrush[H]
Black-headed N.-Thrush[E]
Swainson's Thrush
Hermit Thrush[H]
Wood Thrush[L]
Black Thrush†[H]
Clay-colored Thrush†[L]
White-throated Thrush†[E]
Grey Catbird[L]
Cedar Waxwing
White-eyed Vireo[L]
Blue-headed Vireo†[L]
Cassin's Vireo†
Warbling Vireo
Brown-capped Vireo[H]
Lesser Greenlet[L]
Blue-winged Warbler[L]
Nashville Warbler
Tropical Parula[L]
Magnolia Warbler[L]
Townsend's Warbler[H]
Black-throated Green Warbler
Black-and-white Warbler
Ovenbird[L]
Northern Waterthrush[L]
Louisiana Waterthrush
Kentucky Warbler[L]
MacGillivray's Warbler[L]
Common Yellowthroat[L]
Hooded Warbler[L]
Wilson's Warbler
Slate-throated Whitestart†[E]
Rufous-capped Warbler
Golden-browed Warbler[H]
Yellow-breasted Chat[L]
Bananaquit[L]
Golden-hooded Tanager†[L]
Green Honeycreeper[L]
Red-legged Honeycreeper[L]
Blue-crowned Chlorophonia[E]
Scrub Euphonia[L]
Yellow-throated Euphonia[L]
Blue-hooded Euphonia[E]
Olive-backed Euphonia[L]
Yellow-winged Tanager[L]
Red-crowned Ant-Tanager[L]
Red-throated Ant-Tanager[L]
Summer Tanager
Western Tanager
White-winged Tanager[E]
Crimson-collared Tanager[L]
Common Bush-Tanager[E]
Buff-throated Saltator[L]
Black-headed Saltator[L]
Black-faced Grosbeak[L]
Blue-black Grosbeak[L]
Rose-breasted Grosbeak[L]
Blue Grosbeak[L]
Indigo Bunting
White-naped Brushfinch[H]
Chestnut-capped Brushfinch[H]
Olive Sparrow[L]
Blue-black Grassquit[L]
White-collared Seedeater[L]
Thick-billed Seedfinch[L]
Cinnamon-bellied Flowerpiercer[H]
Rusty Sparrow[L]
Lincoln's Sparrow
Melodious Blackbird[L]
Great-tailed Grackle[L]
Bronzed Cowbird[L]
Black-cowled Oriole[L]
Orchard Oriole[L]
Yellow-tailed Oriole[L]
Baltimore Oriole[L]
Yellow-billed Cacique[L]
Chestnut-headed Oropendola[L]
Montezuma Oropendola[L]

Site 11.8: Tuxtepec

(December 1989, 1990, 1991, January, December 1992, December 1993, 1995, checked December 1996)

Introduction/Key Species

The bustling little town of Tuxtepec is typical of the humid Atlantic Slope lowlands and is a far cry from the arid interior of the Oaxaca Valley. The working day for many of the inhabitants begins before first light, and the streets change quickly from mud to dust and back again with the whims of rain and sun. Most of the surrounding land has been cleared for cattle and crops but a few remnant forest patches remain, and the roadside second growth and hedges tend to be very birdy, especially in winter when North American migrants are present. The areas described below are prone to change from year to year, with an overall trend to less and less forest, but still can provide some exciting birding, especially for those on their first trip to eastern Mexico.

Birds of interest near Tuxtepec include Snail Kite, White Hawk, Black Hawk-Eagle, **Ruddy Crake**, Striped Owl, **Rufous-breasted Spinetail**, Slate-headed Tody-Flycatcher, **Sumichrast's Wren**, Crimson-collared Tanager, and Thick-billed Seedfinch.

Access/Birding Sites

In Tuxtepec there are two *glorietas*. From the southern *glorieta* (nearer downtown and about 250 m north of the Hotel Hacienda, where birders have stayed) mark zero and head west for 2.0 km out of town to a three-way fork where the main road (signed to Jalapa del Diaz) goes right; turn left instead (signed to Camelia Roja) and mark this junction as zero (the middle road goes to some houses). At about Km 10 a broad river with gravel bars flows beside the road to your left, and some small dwellings along the road comprise the unsigned village of Camelia Roja. At Km 10.2, as the road bends left, there is an unsigned dirt road off to the right. This road runs between two increasingly deforested hillsides and is bordered by a changing patchwork of fields and second-growth thickets (**Rufous-breasted Spinetail**, Slate-headed Tody-Flycatcher, Thick-billed Seedfinch). Drive about 500 m and pull off to the side of this dirt road. Birding by foot along the road in both directions (onward as far as the top of the first rise, i.e., only another 500 m) can be good in early morning, with mixed-species flocks of trogons, woodpeckers, Green Jays, Band-backed Wrens, saltators, and orioles. Depending on weather (e.g., if it's overcast with light drizzle), birding can stay good to midday.

If birding quiets down by mid morning, head back to the paved road and turn right. In 500 m you cross a small bridge where a stream flows into the main river (this is a good point from which to scan for herons, shorebirds, kingfishers, and Mangrove Swallows; and Sungrebe has been seen here); stay on the road, following it another 6.5 km through a couple of small villages, to the outflow of Presa Miguel Aleman, a large reservoir that can be viewed by following the road up to the top of the dam. Don't expect much in the way of waterbirds (Eared Grebe, American White Pelican), but in winter the outflow can be good for Ospreys, gulls, and Caspian Terns. The roadside second-growth thickets just before the outflow can be birdy, especially if you imitate a Ferruginous Pygmy-Owl and stir up a mobbing band of warblers and tanagers; Thick-billed Seedfinch can often be found here. The forested limestone hill to the north, overlooking the outflow, has **Sumichrast's Wren**, but access to the forest is not easy.

One spot that may be worth checking for Striped Owl, although a newly constructed Pemex station (December 1995) may have changed this, is the roadside phone wires immediately north of the junction with Route 147 (to Playa Vicente and Palomares), on the south side of Tuxtepec.

If you're using public transport, take a second-class bus from Tuxtepec to Camelia Roja and walk from there along the dirt road described above; buses along this route continue out to near the reservoir outflow.

Tuxtepec Bird List

199 sp. R: rare (not to be expected). †See p. 5–7 for notes on English names.

Little Tinamou
Thicket Tinamou
Least Grebe
Pied-billed Grebe
Eared Grebe
American White Pelican
Neotropic Cormorant
Great Blue Heron
Great Egret
Snowy Egret
Little Blue Heron
Tricolored Heron
Reddish Egret (R)
Cattle Egret
Green Heron
Black-crowned Night-Heron
Yellow-crowned Night-Heron
White-faced Ibis
Black-bellied Whistling-Duck
Blue-winged Teal
Lesser Scaup
Black Vulture
Turkey Vulture
Osprey

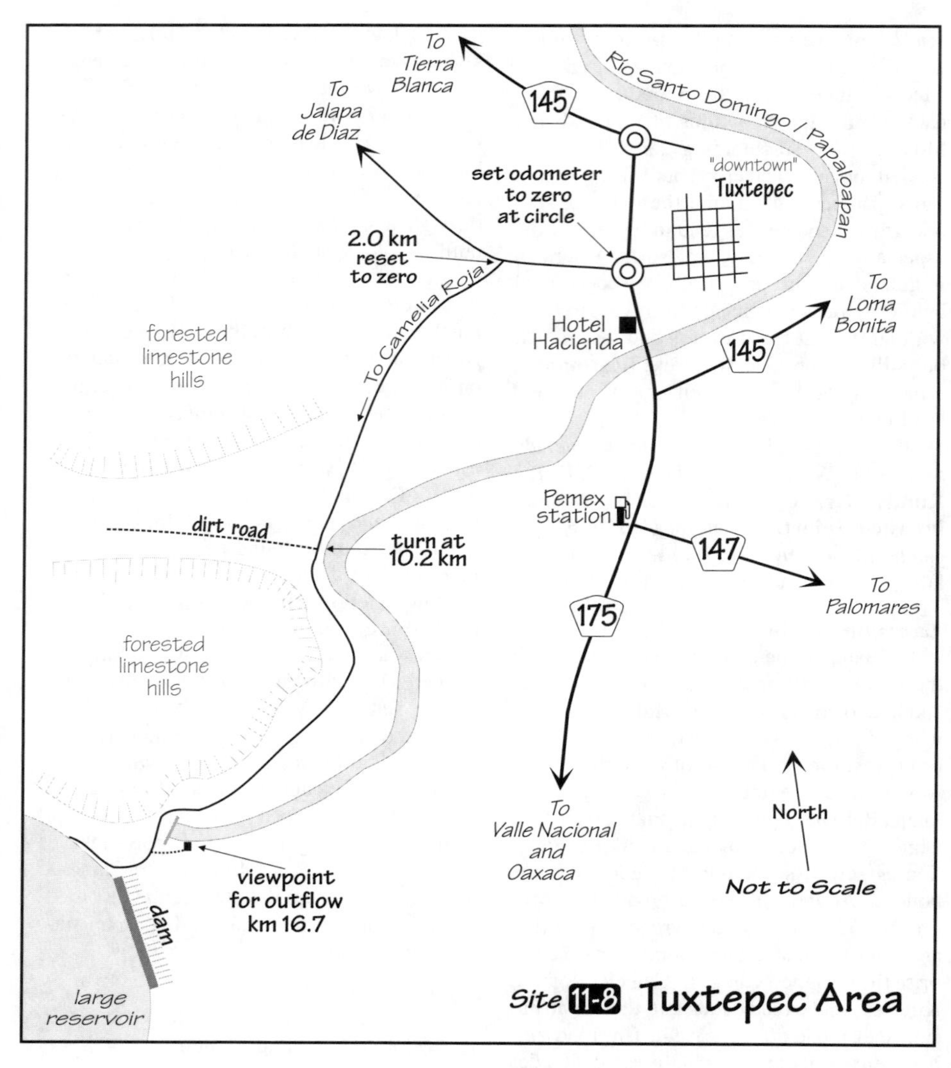

White-tailed Kite
Snail Kite
Sharp-shinned Hawk
Crane Hawk
White Hawk
Grey Hawk
Roadside Hawk
Short-tailed Hawk
Red-tailed Hawk
Black Hawk-Eagle
Crested Caracara
Collared Forest-Falcon
Laughing Falcon
Bat Falcon
American Kestrel
Plain Chachalaca
Spotted Wood-Quail
Ruddy Crake
Grey-necked Wood-Rail
Collared Plover
Killdeer
Black-necked Stilt
Northern Jacana
Greater Yellowlegs
Lesser Yellowlegs
Solitary Sandpiper
Spotted Sandpiper
Least Sandpiper
Stilt Sandpiper
Long-billed Dowitcher
Common Snipe
Laughing Gull
Franklin's Gull
Herring Gull
Caspian Tern
Red-billed Pigeon
Ruddy Ground-Dove
Inca Dove
White-tipped Dove
Grey-headed Dove†
Aztec Parakeet†
White-crowned Parrot
Red-lored Parrot
Squirrel Cuckoo
Striped Cuckoo
Groove-billed Ani
Ferruginous Pygmy-Owl
Mottled Owl
Striped Owl
Lesser Nighthawk
Pauraque

White-collared Swift
Vaux's Swift
Long-tailed Hermit
Little Hermit
Green-breasted Mango
Canivet's Emerald
White-bellied Emerald
Rufous-tailed Hummingbird
Ruby-throated Hummingbird
Black-headed Trogon
Violaceous Trogon
Collared Trogon
Blue-crowned Motmot
Ringed Kingfisher
Belted Kingfisher
Amazon Kingfisher
Green Kingfisher
Collared Aracari
Keel-billed Toucan
Golden-fronted Woodpecker
Golden-olive Woodpecker
Lineated Woodpecker
Pale-billed Woodpecker
Rufous-breasted Spinetail
Buff-throated Fol.-gleaner
Barred Woodcreeper
Ivory-billed Woodcreeper
Barred Antshrike
Dusky Antbird
Yellow-bellied Elaenia
Ochre-bellied Flycatcher
Slate-headed Tody-Flycatcher
Common Tody-Flycatcher
Yellow-olive Flycatcher
Yellow-bellied Flycatcher
Least Flycatcher
Black Phoebe
Vermilion Flycatcher
Bright-rumped Attila
Dusky-capped Flycatcher
Great Crested Flycatcher
Brown-crested Flycatcher
Great Kiskadee
Boat-billed Flycatcher
Social Flycatcher
Tropical Kingbird
Couch's Kingbird
Masked Tityra
Rose-throated Becard
Tree Swallow
Mangrove Swallow
N. Rough-winged Swallow
Ridgway's Rough-w. Swallow†
Green Jay
Brown Jay
Band-backed Wren
Sumichrast's Wren†
Spot-breasted Wren
White-bellied Wren
Northern House Wren†
White-breasted Wood-Wren
Long-billed Gnatwren
Blue-grey Gnatcatcher
Swainson's Thrush
Wood Thrush
Clay-colored Thrush†
White-throated Thrush†
Grey Catbird
White-eyed Vireo
Blue-headed Vireo†
Yellow-throated Vireo
Blue-winged Warbler
Orange-crowned Warbler
Nashville Warbler
Northern Parula
Yellow Warbler
Magnolia Warbler
Myrtle Warbler†
Black-throated Green Warbler
Black-and-white Warbler
American Redstart
Worm-eating Warbler
Ovenbird
Northern Waterthrush
Kentucky Warbler
Common Yellowthroat
Grey-crowned Yellowthroat
Hooded Warbler
Wilson's Warbler
Rufous-capped Warbler
Yellow-breasted Chat
Bananaquit
Red-legged Honeycreeper
Yellow-throated Euphonia
Yellow-winged Tanager
Blue-grey Tanager
Summer Tanager
Crimson-collared Tanager
Red-throated Ant-Tanager
Greyish Saltator
Buff-throated Saltator
Black-headed Saltator
Northern Cardinal
Black-faced Grosbeak
Blue-black Grosbeak
Rose-breasted Grosbeak
Blue Grosbeak
Indigo Bunting
Painted Bunting
Olive Sparrow
Lincoln's Sparrow
Blue-black Grassquit
White-collared Seedeater
Thick-billed Seedfinch
Melodious Blackbird
Great-tailed Grackle
Bronzed Cowbird
Black-cowled Oriole
Orchard Oriole
Yellow-tailed Oriole
Altamira Oriole
Baltimore Oriole
Yellow-billed Cacique
Chestnut-headed Oropendola
Montezuma Oropendola
Lesser Goldfinch

Site 11.9: La Soledad

(February 1983, January 1987, April 1988, checked March, April 1996)

Introduction/Key Species

The valley of La Soledad is at the transition between tropical semi-deciduous forest and pine-oak woodland in the coastal-slope foothills of the Sierra de Miahuatlán, part of Mexico's Sierra Madre del Sur. It is a site for **Blue-capped Hummingbird**, and other species of interest include **White-faced Quail-Dove**, **Colima Pygmy-Owl**, **Mexican Hermit**, **Cinnamon-sided** and **Bumblebee hummingbirds**, **Grey-crowned Woodpecker**, Ruddy Foliage-gleaner, **Chestnut-sided Shrike-Vireo**, and **Red-headed Tanager**.

Stops in suitable habitat between here and the coast should produce an assortment of Pacific Slope dry forest birds (such as noted for the Puerto Escondido area, Site 11.11).

Access/Birding Sites

La Soledad (1450 m elevation) is on Route 175 South and can be reached in about 1.5 hours from the coast at Puerto Angel/ Pochutla or in about 3–4 hours from Oaxaca City to the north. The nearest place with reasonable hotels is Puerto Angel, although there is basic accommodation in Pochutla. Second-class buses can drop you anywhere along this route, and starting early from Pochutla enables some good birding around La Soledad.

If you're coming from Oaxaca, head south past the airport on Route 175; about 15 km from the center of Oaxaca, the road forks (right to Puerto Escondido, left to Puerto Angel). Take this point as zero and go left on Route 175 toward Puerto Angel (also signed to Miahuatlán). Stay on Route 175 through a number of small towns, and you come to Miahuatlán in about 84 km (1.5–2 hours from Oaxaca). It is not necessary to enter town; rather, skirt to the left (east) on Route 175, which climbs into the Sierra de Miahuatlán and, after about 80 km, reaches La Soledad at Km Post 184 (1.5–2 hours from Miahuatlán). At around 25 km north of La Soledad, White-throated Jay has been found recently (April 1993) in pine-oak forest along a logging track running west from the highway.

Alternatively, to reach La Soledad from the coastal highway (Route 200), head north through Pochutla on Route 175 toward Oaxaca City, marking zero at the Pochutla Bus Station on the north side of town. Continue about 50 km through thorn forest and into taller, semi-deciduous woodland and oaks at the top of the coastal-slope ridge (1500 m) and a small restaurant named El Mirador. Watch for swifts (including **Great Swallow-tailed**) from the ridge 3.3 km before (south of) El Mirador and from the stretch of road between El Mirador and another small restaurant (Linda Vista), 1 km farther north. In February 1983 a **Cinnamon-sided [Green-fronted] Hummingbird** nested at El Mirador and often perched on the roadside phone wires. **Blue-capped Hummingbirds** can be seen along the road here, although La Soledad seems more reliable: continue north for 4 km from Linda Vista into the valley and small village of La Soledad (a few houses and a small store at Km Post 184).

From La Soledad, go 250 m north (i.e., toward Oaxaca) on the highway and look for an obvious track on the right (east), where there's enough room to park a car safely off the highway. In 1983 this was a narrow track, but by 1988 it had been enlarged, possibly to allow trucks to enter. One can walk down this track a few hundred meters and along subsidiary tracks into the forest. After a few hundred meters, before reaching the stream at the bottom of the valley, there is a small house on the left, and a hidden track angles steeply down to the right; this track crosses the stream to more coffee plantations. **Blue-capped Hummingbird** can be seen near the highway or anywhere here in the forest, including across the stream, and is best located by its slightly liquid, "pebbly" trills, which may suggest a Costa's Hummingbird. Its occurrence and abundance almost certainly vary seasonally, and in some months it may be absent from this site; it is present and fairly easy to find from January to April, at least. (**Blue-capped Hummingbird** has also been reported from the side road to Pluma Hidalgo, about 12 km nearer the coast.) Birding along the highway into the pines just north of La Soledad is also good, as is the entire roadside between La Soledad and El Mirador.

La Soledad to El Mirador Bird List

118 sp. R: rare (not to be expected). †See pp. 5–7 for notes on English names.

Black Vulture
Turkey Vulture
Sharp-shinned Hawk
Cooper's Hawk
Common Black Hawk
Broad-winged Hawk
Short-tailed Hawk
Zone-tailed Hawk
Red-tailed Hawk
Barred Forest-Falcon
American Kestrel

Red-billed Pigeon
Inca Dove
White-tipped Dove
White-faced Quail-Dove
Squirrel Cuckoo
Groove-billed Ani
Ferruginous Pygmy-Owl
Colima Pygmy-Owl†
Chestnut-collared Swift
White-collared Swift
Vaux's Swift
Great Swallow-tailed Swift
Mexican Hermit†
Golden-crowned Emerald
White-eared Hummingbird
Berylline Hummingbird
Cinnamon-sided Hummingbird†
Blue-capped Hummingbird
Sparkling-tailed Woodstar†
Ruby-throated Hummingbird
Broad-tailed Hummingbird (R)
Bumblebee Hummingbird
Mountain Trogon
Collared Trogon
Emerald Toucanet
Yellow-bellied Sapsucker
Ladder-backed Woodpecker
Grey-crowned Woodpecker
Lineated Woodpecker
Ruddy Foliage-gleaner
Olivaceous Woodcreeper
Ivory-billed Woodcreeper
Spot-crowned Woodcreeper
Greenish Elaenia
Eye-ringed Flatbill
Common Tufted Flycatcher†
Olive-sided Flycatcher
Greater Pewee
Yellow-bellied Flycatcher
Hammond's Flycatcher
Dusky Flycatcher
Western Flycatcher†
Bright-rumped Attila
Dusky-capped Flycatcher
Social Flycatcher
Boat-billed Flycatcher
Thick-billed Kingbird
Masked Tityra
Rose-throated Becard
Violet-green Swallow
Green Jay
Happy Wren
Banded Wren
Northern House Wren†
Grey-breasted Wood-Wren
Ruby-crowned Kinglet
Blue-grey Gnatcatcher
Brown-backed Solitaire
Swainson's Thrush
Hermit Thrush
White-throated Thrush†
Blue Mockingbird
Cedar Waxwing
Grey Silky†
Black-capped Vireo
Cassin's Vireo†
Golden Vireo
Warbling Vireo
Chestnut-sided Shrike-Vireo
Tennessee Warbler
Nashville Warbler
Orange-crowned Warbler
Audubon's Warbler†
Townsend's Warbler
Hermit Warbler
Black-throated Green Warbler
Grace's Warbler
Black-and-white Warbler
MacGillivray's Warbler
Wilson's Warbler
Slate-throated Whitestart†
Golden-crowned Warbler
Rufous-capped Warbler
Golden-browed Warbler
Yellow-breasted Chat
Red-legged Honeycreeper
Blue-hooded Euphonia
Red-crowned Ant-Tanager
Hepatic Tanager
Summer Tanager
Western Tanager
Flame-colored Tanager
Red-headed Tanager
Common Bush-Tanager
Greyish Saltator
Black-headed Saltator
Black-headed Grosbeak
Rose-breasted Grosbeak
Cinnamon-bellied Flowerpiercer
Rusty Sparrow
Chipping Sparrow
Lincoln's Sparrow
Dickey's Oriole†
Bullock's Oriole
Abeille's Oriole†
Baltimore Oriole
Black-headed Siskin

Site 11.10: Puerto Angel/Playa Zipolite

(January, February 1983, checked March 1996)

Introduction/Key Species

The thorn forest in the vicinity of Puerto Angel and the nearby beach of Playa Zipolite has a good assortment of Pacific Slope thorn-forest birds, including Lesser Ground-Cuckoo, **Doubleday's Hummingbird**, **Citreoline Trogon**, **Russet-crowned Motmot**, Rufous-naped, **Happy**, and Banded **wrens**, **Red-breasted Chat**, and **Orange-breasted Bunting**. It is also worth looking off the beach at Playa Zipolite for seabirds and around the small rocky islet, or stack, at the west end of the beach for Red-billed Tropicbirds.

Access/Birding Sites

Puerto Angel is a small fishing town on the Pacific Coast about 260 km (5–6 hours) by road almost due south of Oaxaca City. It is also about 60 km east of the larger resort of Puerto Escondido, served by a small airport with regular flights from Oaxaca City, and some 205 km (2.5–3.5 hours) west-southwest of Tehuantepec via the paved coastal highway (Route 200). There are a few hotels and restaurants in town, and one can also stay in *palapas* at Playa Zipolite.

To reach Zipolite, go down to the harbor in Puerto Angel and from there follow the most coastal road west; this road is now paved, and there may be a sign for Zipolite. The road passes through 4 km of variably cutover thorn forest before coming to Playa Zipolite, a popular spot with surfers and young people from North America. Sundry footpaths off the road and behind the beach lead inland into the thorn forest; those along (seasonally dry) stream beds—for example, at the far (west) side of Zipolite by the nursery school—tend to be the most productive. Early morning is very much the best time, as it gets hot here by 9 or 10 A.M., when the beach, surf, and a cold drink may command your attention.

To view the stack for seabirds, it is possible to follow a narrow path to a clifftop vantage point at the west end of the beach, although a telescope is needed to see the birds clearly. It is also possible to hire a small *lancha* (open fishing launch) in Puerto Angel to take you out to circle the stack and/or to go farther offshore in search of other species; **Townsend's Shearwater** has been found (January, April, September, October) 2–15 km off Puerto Angel (Binford 1989). In addition to birds, you may see Humpback Whales and Manta Rays. As always, make sure you agree upon a price before setting out.

Puerto Angel/Playa Zipolite Bird List

105 sp. R: rare; V: vagrant (not to be expected). †See p. 5–7 for notes on English names.

Audubon's Shearwater
Least Storm-Petrel
Black Storm-Petrel
Red-billed Tropicbird
Masked Booby
Brown Booby
Brown Pelican
Magnificent Frigatebird
Cattle Egret
Yellow-crowned Night-Heron
Black Vulture
Turkey Vulture
Osprey
Sharp-shinned Hawk
Common Black Hawk
Great Black Hawk
Grey Hawk
Roadside Hawk
Short-tailed Hawk
Zone-tailed Hawk
Red-tailed Hawk
American Kestrel
Peregrine Falcon
Spotted Sandpiper
Wandering Tattler
Red Phalarope
Laughing Gull
Gull-billed Tern
Caspian Tern
Elegant Tern
Common Tern
Sooty Tern
Black Tern
Black Skimmer
White-winged Dove
Inca Dove
Common Ground-Dove
Ruddy Ground-Dove
White-tipped Dove
Orange-fronted Parakeet
White-fronted Parrot
Squirrel Cuckoo
Lesser Ground-Cuckoo
Groove-billed Ani
Colima Pygmy-Owl†
Ferruginous Pygmy-Owl
White-collared Swift
Doubleday's Hummingbird†
Cinnamon Hummingbird
Ruby-throated Hummingbird
Citreoline Trogon
Russet-crowned Motmot
Golden-cheeked Woodpecker
Lineated Woodpecker
Pale-billed Woodpecker
Ivory-billed Woodcreeper
N. Beardless Tyrannulet
Willow Flycatcher
Least Flycatcher
Western Flycatcher†
Dusky-capped Flycatcher
Nutting's Flycatcher
Brown-crested Flycatcher
Great Kiskadee
Boat-billed Flycatcher
Tropical Kingbird
Thick-billed Kingbird
Masked Tityra
Rose-throated Becard
Grey-breasted Martin
N. Rough-winged Swallow
White-throated Magpie-Jay
Rufous-naped Wren
Banded Wren
Happy Wren
Rufous-backed Thrush†
Blue-grey Gnatcatcher
White-lored Gnatcatcher
Cedar Waxwing
Plumbeous Vireo†

Warbling Vireo
Bell's Vireo
Orange-crowned Warbler
Lucy's Warbler (R)
Yellow Warbler
Cape May Warbler (V)
Black-throated Grey Warbler
Black-and-white Warbler
MacGillivray's Warbler
Wilson's Warbler
Red-breasted Chat
Western Tanager
Northern Cardinal
Blue Bunting
Orange-breasted Bunting
Olive Sparrow
Blue-black Grassquit
Great-tailed Grackle
Orchard Oriole
Hooded Oriole (R)
Streak-backed Oriole
Spot-breasted Oriole
Altamira Oriole
Bullock's Oriole
Yellow-winged Cacique

Site 11.11: Puerto Escondido

(April 1988)

Introduction/Key Species

Given a choice, if you have to drive from Oaxaca City to the coast, even to Puerto Escondido, the fastest route (pending a drastic improvement or paving of Route 135) is Route 175 (the Puerto Angel road). However, the tortuous and often washboarded Route 135 does pass through some good humid evergreen forest, where **White-throated Jay** can, with luck, be found, and **Great Swallow-tailed Swift** occurs over the first high ridge you cross coming from the coast.

Note: Parts of this road do not have the best reputation for safety and might best be avoided, although the lower, coastal-slope stretches should be safe, are easily birded in a morning from Puerto Escondido, and can be good for **Cinnamon-sided Hummingbird** plus an assortment of Pacific Slope species.

Access/Birding Sites/Bird List

Puerto Escondido is a popular yet fairly small beach resort reached most easily by plane from Oaxaca City. If you're driving from Oaxaca City, via Puerto Angel, allow at least six hours (with no or minimal stops) for the journey. There are several hotels and restaurants in town.

To reach some birding areas, head inland (north) on Route 135 (toward Sola y Vega and Oaxaca), marking zero at the edge of Puerto Escondido. At Km 32 you cross a streambed with some taller trees. If the *Inga* flowers are in bloom here (usually March to May), **Cinnamon-sided Hummingbird** can be fairly common. Also between this point and the coast, stopping in thorn forest and the valley with humid oak/semi-deciduous woodland at Km 31, birds include **West Mexican Chachalaca**, Orange-fronted Parakeet, **Colima Pygmy-Owl**, **Golden-crowned Emerald**, **Doubleday's** and Cinnamon **hummingbirds**, Plain-capped Starthroat, **Citreoline Trogon**, **Russet-crowned Motmot**, **Golden-cheeked Woodpecker**, Nutting's Flycatcher, Tropical and Thick-billed kingbirds, White-throated Magpie-Jay, Rufous-naped and Banded wrens, White-lored Gnatcatcher, **Rufous-backed Thrush**, Olive Sparrow, Streak-backed and Altamira orioles, and **Yellow-winged Cacique**.

Site 11.12: Oaxaca City to Tehuantepec

(March 1982, February, November 1985, April 1988, December 1991, checked in part December 1996)

Introduction/Key Species

This is a drive of about 250 km along Route 190 from the arid temperate interior of the Oaxaca Valley to the arid Pacific Slope lowlands at the western side of the Isthmus of Tehuantepec. The avifauna along the drive changes accordingly—for example, from **Grey-breasted Woodpeckers** and **Boucard's Wrens** to Golden-fronted Woodpeckers and Rufous-naped Wrens. It is a good route along which to see **Green-fronted Hummingbird**.

Access/Birding Sites/Bird List

The Km Posts start at zero in Oaxaca City and end at around 250 in Tehuantepec. The first 50 km or so go through the extensively cultivated southeast arm of the Oaxaca Valley, past the turnings for Teotitlán del Valle (Site 11.2), Yagul (Site 11.3), and Mitla (nice ruins but of no birding interest) and through the town of Matatlán, where the main industry is the production of mezcal ("raw tequila").

Roadside birds along this stretch include Red-tailed and White-tailed hawks, Crested Caracara, American Kestrel, White-winged and Mourning doves, Vermilion Flycatcher, Cassin's Kingbird, Curve-billed Thrasher, Loggerhead Shrike, **White-throated Towhee**, and House Finch. In winter the open fields often have flocks of Clay-colored and Lark sparrows.

From Matatlán you can see radio towers (*microondas* towers) on a hill ahead to the south. A few kilometers southeast of Matatlán (and 47.3 km from the Route 175 intersection on the east edge of Oaxaca City) a cobbled road to the right leads up to the microwave towers, providing access to oak scrub. The turn should be signed **"Microondas Nueve Puntas"** but is somewhat concealed coming from Matatlán and on a convex slope, with oncoming traffic hidden; *take care making the turn*. Species along the road up to the towers include **Lesser Roadrunner**, **Pileated Flycatcher**, Western Scrub Jay, **Ocellated Thrasher**, **Grey Silky**, **Slaty Vireo,** and **Oaxaca Sparrow** (the last two at the eastern edge of their range).

The microwave junction is more or less at the pass where the highway begins its descent to the Pacific Slope. Stops anywhere from here onward, particularly where the highway crosses gulleys or small valleys at bends, can be good. At Km Posts 75 to 77 (1050–1250 m elevation), just west of Totolapan, birds include Ferruginous Pygmy-Owl, **Green-fronted Hummingbird**, Plain-capped Starthroat, Elegant Trogon, **Grey-breasted** and Golden-fronted **woodpeckers**, Rufous-naped Wren, White-lored Gnatcatcher, and Streak-backed Oriole. From here on the highway winds up and down over low ridges. The avifauna becomes more tropical and has mostly lost birds from the interior, although **Bridled Sparrow** occurs locally in pockets of habitat to around Km Post 152 (1050 m). At Km Posts 123/125 (around Las Animas, 700 m) birds in the thorn forest and along the river include Orange-fronted Parakeet, Pacific Screech-Owl, **Dusky** and **Green-fronted hummingbirds**, Golden-fronted Woodpecker, White-throated Magpie-Jay, **Rufous-backed Thrush**, **Orange-breasted Bunting**, **Sumichrast's Sparrow**, and **Yellow-winged Cacique**. By Km Post 227 (180 m) one can find **Doubleday's Hummingbird**, and the avifauna from here on is like that of Tehuantepec (Site 11.13).

Green-fronted Hummingbirds move seasonally and are often common (e.g., December, January, April) between Km Posts 72 and 187; they respond readily to imitations of Ferruginous Pygmy-Owl whistles. Birds here (of the subspecies *rowleyi*) generally have a distinct buffy wash to their sides, suggesting **Cinnamon-sided Hummingbird**, which has also been found in this area (around Km Post 80) but which appears to be rare here.

Site 11.13: Tehuantepec

(February 1983, November 1985, January 1987, April 1988, December 1991, October 1993, checked March 1996)

Introduction/Key Species

The thorn forest a short distance west of Tehuantepec along Route 190 is excellent for Lesser Ground-Cuckoo (seeing it, not just hearing it!), **Doubleday's Hummingbird** (December to May, at least), and **Sumichrast's Sparrow**. The sparrows are usually very vocal and hard to miss in the early morning, but from mid morning on, when it gets hot and usually windy, they can be very difficult to find.

Access/Birding Sites

Tehuantepec is a fairly large town on the southwest side of the isthmus of the same name. It is some 250 km (4.5–5 hours) southeast of Oaxaca City and about 205 km (3–4 hours) east of Puerto Angel (Site 11.10) and offers a variety of accommodations and restaurants. The birding sites below can be reached easily via second-class buses from Tehuantepec to Oaxaca.

To find thorn-forest habitat, head out of Tehuantepec on Route 190, toward Oaxaca City. Continue past the Salina Cruz junction (and mark zero at this point) and past a prison on the left. Between Km 7 and 9 you can pull off the highway and park on the left (south), at the entrance to one of the cattle-ranching tracks into the thorn forest (e.g., at Km Post 244). **Sumichrast's Sparrows** can be seen beside the highway here and favor overgrown edges; they respond to pishing and Ferruginous Pygmy-Owl imitations. To see the ground-cuckoo you will probably need to walk along a track into the thorn forest (the sparrows occur here also, at brushy openings). Due to grazing, the understory is fairly open, so you have a good chance of seeing a ground-cuckoo. Imitating a pygmy-owl is likely to bring in a band of mobbing birds, headed by White-lored Gnatcatchers and often including **Doubleday's Hummingbird**.

Tehuantepec Bird List

60 sp. Hummingbirds seasonal (Beautiful occurs in April, at least). †See p. 5–7 for notes on English names.

Cattle Egret
Black Vulture
Turkey Vulture
Grey Hawk
Roadside Hawk
Short-tailed Hawk
White-tailed Hawk
Zone-tailed Hawk
Red-tailed Hawk
Crested Caracara
American Kestrel
West Mexican Chachalaca
Northern Bobwhite
White-winged Dove
Mourning Dove
Inca Dove
Common Ground-Dove
Orange-fronted Parakeet
Lilac-crowned Parrot
Lesser Ground-Cuckoo
Lesser Roadrunner
Groove-billed Ani
Ferruginous Pygmy-Owl
Lesser Nighthawk
Doubleday's Hummingbird†
Plain-capped Starthroat
Beautiful Hummingbird
Ruby-throated Hummingbird
Citreoline Trogon
Golden-fronted Woodpecker
N. Beardless Tyrannulet
Least Flycatcher
Ash-throated Flycatcher
Nutting's Flycatcher
Brown-crested Flycatcher
Scissor-tailed Flycatcher
Tropical Kingbird
Western Kingbird
Barn Swallow
White-throated Magpie-Jay
Northern Raven†
Rufous-naped Wren
Banded Wren
Blue-grey Gnatcatcher
White-lored Gnatcatcher
Northern Mockingbird
Cedar Waxwing
Nashville Warbler
Yellow Warbler
Black-and-white Warbler
MacGillivray's Warbler
Orange-breasted Bunting
Olive Sparrow
Stripe-headed Sparrow
Sumichrast's Sparrow†
Bronzed Cowbird
Brown-headed Cowbird
Orchard Oriole
Streak-backed Oriole
Altamira Oriole

Site 11.14: La Ventosa Lagoon

(April 1988, December 1991, checked March 1998)

Introduction/Key Species

This coastal lagoon, on the eastern edge of Salina Cruz, is good for a variety of wading birds, shorebirds (including Hudsonian Godwit in spring), gulls, and terns. Like much of the Isthmus, though, it can be very windy here in the afternoons.

Access/Birding Sites

From Route 190 on the west edge of Tehuantepec, set to zero as you turn south on the highway to Salina Cruz. Stay on the main highway, past the right turn to Route 200 at Km 15, and you hit the coast (with shipyards and the like) at a T-junction at Km 18.2. Turn left here and in 0.5 km, just after a small naval base, set to zero as you turn left on a minor road (not necessarily in the best of shape but passable with care in a regular car) that may be signed to Bahía La Ventosa. This road (Avenida Bahía La Ventosa) winds over some low hills, and at 2.1 km there is a road to the right, which may be signed to Boca del Río or to La Ventosa, a small fishing village at the mouth of the large shallow lagoon visible ahead and to the right. It is 3.3 km from this turn to the edge of La Ventosa, and 3.9 km to the end of the road at the coast. If you turn here, the lagoon will be to your left (east) and can be reached via dirt tracks through cutover thorn forest and the yards of scattered small dwellings. The mouth of the lagoon can be viewed from La Ventosa itself, but here it can be crowded with people, so there may be fewer birds. It is unlikely that there isn't some form of public transport from Salina Cruz to La Ventosa (ask locally).

Alternatively, if you stay straight rather than take the right turn to La Ventosa, you soon cross a small bridge (the lagoon inflow) to Boca del Río and a fairly large chemical

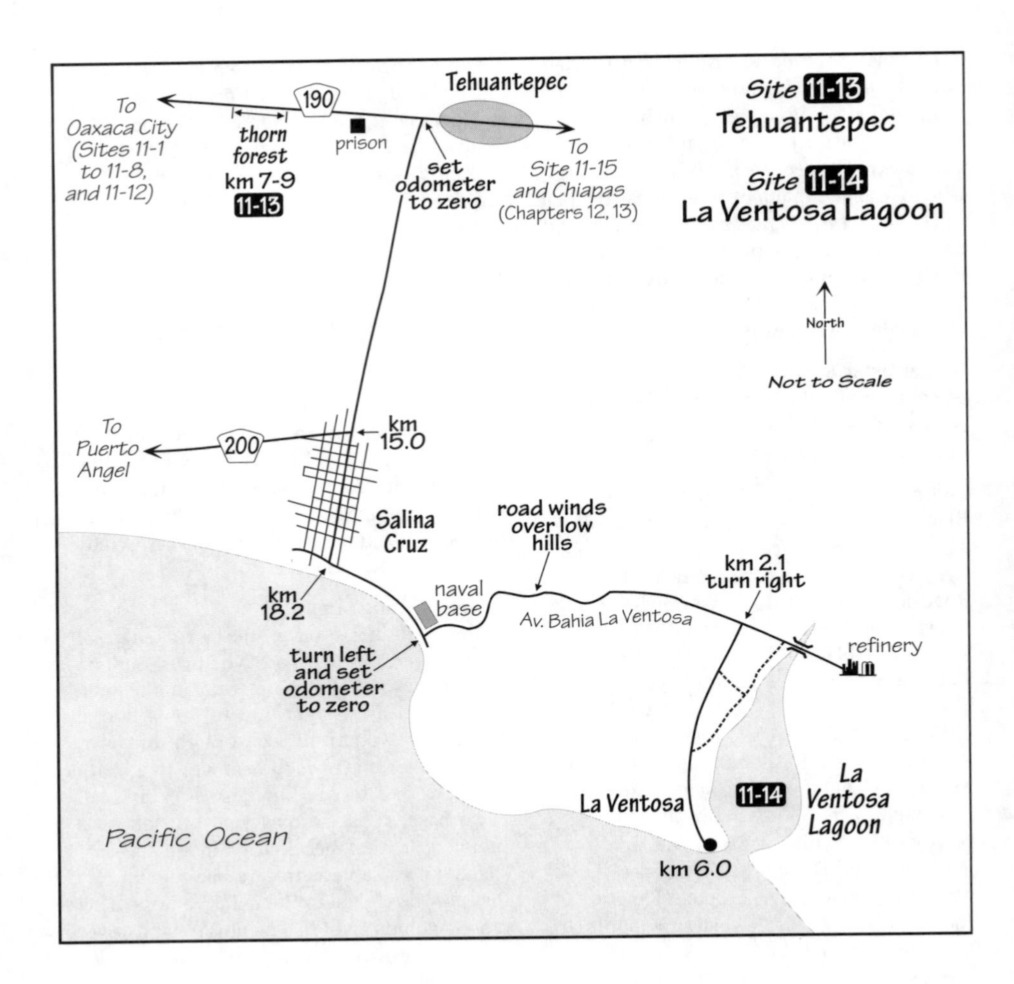

refinery. Just before this bridge there is a dirt track to the right along the edge of the lagoon, eventually coming out at the La Ventosa side road. Beyond the bridge to the right (south) are some seasonally dry marshes.

Birds found at La Ventosa include American White Pelican, Reddish Egret, White Ibis, Roseate Spoonbill, Wood Stork, Black-bellied Whistling-Duck, Peregrine Falcon, nineteen species of shorebirds (including American Golden Plover, Hudsonian Godwit, Pectoral and Stilt sandpipers, Wilson's Phalarope), Laughing, Franklin's, Heermann's (rare), Western (vagrant), and **Yellow-footed** (vagrant) **gulls**, and Gull-billed, Caspian, Royal, Sandwich, Common, Least, and Black terns. The adjacent thorn forest and scrub have Lesser Ground-Cuckoo, White-throated Magpie-Jay, and Stripe-headed and **Sumichrast's sparrows**.

Site 11.15: Tapanatepec Foothills

(March 1982, February 1983, January, February 1984, February, November 1985, January 1987, April 1988, January 1992, checked March 1996)

Introduction/Key Species

The thorn forest and gallery woodland of the foothills inland from Tapanatepec along Route 190 have an interesting variety of birds, including two sought-after species: Long-tailed Manakin and **Rosita's Bunting**. Other species here include King Vulture, **West Mexican Chachalaca**, **Green-fronted Hummingbird**, and **Citreoline Trogon**.

Access/Birding Sites

This site is about 140 km east of Tehuantepec and about 150 km from Tuxtla Gutierrez, Chiapas, so you'd probably be going out of your way to be here at dawn. Fortunately the main birds can be found at any time of day, and this makes a good stop to get out and stretch if you're driving between Oaxaca and Chiapas. Second-class buses running between Tuxtla and Tapanatepec could drop you in the right area.

From its junction with Route 200 just east of Tapanatepec, Route 190 (to Tuxtla Gutierrez, Chiapas) heads inland through foothills where burned hillsides and mango plantations alternate with patches of thorn forest. Marking the junction with Route 200 as zero, at Km 10 one can pull off the highway safely and bird the roadside brush for **Orange-breasted** and **Rosita's buntings**. Imitating Ferruginous Pygmy-Owl whistles will often stir up a mobbing band of birds, including the buntings. Between Km 14 and 15 a stream bordered by gallery woodland approaches the east side of highway. It is possible to pull off safely in this stretch and look for access along narrow paths down to the stream. Unfortunately, this area has undergone extensive clearing and burning in recent years, such as for papaya plantations, which may improve access to the stream but has decreased the extent of woodland. The main reason to get into the gallery woodland is the snappy Long-tailed Manakin, best found by listening for its distinctive whistles, which the males seem to make year-round and throughout most of the day. A few kilometers farther up the highway you cross into the state of Chiapas.

Flocks of **Green/Pacific Parakeets** can be seen in this area, but, as views tend not to be great and as these two forms seem essentially indistinguishable in the field, it is unclear which is involved—and both may occur! Green Parakeets occur a short distance inland, in the interior valley of Chiapas, while Pacific Parakeets are seasonally common in the coastal lowlands around Arriaga and Tonalá, in adjacent Chiapas.

Tapanatepec Foothills Bird List

65 sp. †See pp. 5–7 for notes on English names.

Black Vulture
Turkey Vulture
King Vulture
Sharp-shinned Hawk
Grey Hawk
Short-tailed Hawk
Swainson's Hawk
White-tailed Hawk
Crested Caracara
American Kestrel
West Mexican Chachalaca
White-winged Dove
Inca Dove
Common Ground-Dove
White-tipped Dove
Orange-fronted Parakeet
Green/Pacific Parakeet
White-fronted Parrot
Squirrel Cuckoo
Lesser Ground-Cuckoo
Groove-billed Ani
Ferruginous Pygmy-Owl
Lesser Nighthawk
Canivet's Emerald
Green-fronted Hummingbird
Plain-capped Starthroat
Citreoline Trogon
Russet-crowned Motmot
Golden-fronted Woodpecker
N. Beardless Tyrannulet
Yellow-olive Flycatcher
Yellow-bellied Flycatcher
Least Flycatcher
Brown-crested Flycatcher
Great Kiskadee
Sulphur-bellied Flycatcher
Scissor-tailed Flycatcher
Tropical Kingbird
Western Kingbird
Masked Tityra
Long-tailed Manakin
N. Rough-winged Swallow
Barn Swallow
White-throated Magpie-Jay
Banded Wren
Blue-grey Gnatcatcher
White-lored Gnatcatcher
Clay-colored Thrush†
Blue-headed Vireo†
Warbling Vireo
Yellow-green Vireo
Yellow Warbler
Magnolia Warbler
Black-throated Green Warbler
Black-and-white Warbler
Scrub Euphonia
Western Tanager
Blue Bunting
Rosita's Bunting†
Indigo Bunting
Orange-breasted Bunting
Painted Bunting
Stripe-headed Sparrow
Streak-backed Oriole
Yellow-winged Cacique

CHAPTER **12**

CENTRAL AND SOUTHERN CHIAPAS

Giant Wren

Chiapas, Mexico's southernmost state, has an avifauna similar to that of neighboring Guatemala, with the addition of some Mexican endemics in the west and southwest. Together with the adjacent states of Oaxaca (Chapter 11) and Veracruz (Chapter 10), Chiapas has the highest bird diversity in Mexico, with over 650 species. These include many sought-after species in central and southern Chiapas, such as White-breasted Hawk, White-bellied Chachalaca, Highland and Horned guans, Maroon-chested Ground-Dove, Pacific and Barred parakeets, Bearded Screech-Owl, Fulvous Owl, Unspotted Saw-whet Owl, Blue-tailed, Green-fronted, and Wine-throated hummingbirds, Resplendent Quetzal, Blue-throated Motmot, Belted Flycatcher, Black-throated Jay, Giant Wren, Slate-colored Solitaire, Pink-headed Warbler, Cabanis' Tanager, Rosita's Bunting,

White-naped Brushfinch, Prevost's and White-eared ground-sparrows, Bar-winged Oriole, and Black-capped Siskin.

From the point of view of a visiting birder, Chiapas can be divided into the north (Palenque and other sites; Chapter 13) and the central and south (this chapter). These can, of course, be combined into one trip, although at least three weeks would be needed to do justice to such an itinerary. Alternatively, Palenque and northern Chiapas can be combined with the Yucatan Peninsula (Chapter 14) for a productive two-week trip, and central and southern Chiapas could be combined with Oaxaca (Chapter 11) for at least a two-week trip.

Points of access to central and southern Chiapas are Tuxtla Gutiérrez (the state's capital) and Tapachula in the coastal lowlands near the Guatemala border. There are daily flights to both places from Mexico City (note that the Tuxtla airport is quite a ways west of town, in the middle of nowhere), and car rental is possible at the airports; your choice of rental companies, and consequently your chance of a good deal, is distinctly better in Tuxtla, which is where most birders arrive by air, assuming they're not driving from Oaxaca or Palenque. There are numerous hotels and restaurants in both Tuxtla and Tapachula.

Suggested Itineraries

If you arrive into Tuxtla Gutiérrez, the nearby El Sumidero (Site 12.1) is worth at least a morning, after which you could drive down to Puerto Arista (Sites 12.2, 12.3) for some Pacific Coast specialities and then either backtrack to Tuxtla (and on to San Cristóbal, Site 12.8) or drive a loop via Mapastepec (Site 12.4), Unión Juarez (Site 12.5), and Montebello (Site 12.7) to San Cristóbal, and back to Tuxtla. It's also easy to add a few days in the Palenque area (Chapter 13) to any itinerary, via the Ocosingo Highway out of San Cristóbal. Starting in Tapachula, Unión Juarez would be a good first stop, after which any of the above sites can be combined into a loop itinerary depending on your time and interests.

El Triunfo requires more planning than your average site, but you may be able to arrange a visit for later in your trip when you first arrive in Tuxtla, then spend the intervening period at other sites.

Note that the unfortunate recent civil unrest in parts of Chiapas has made some areas effectively off-limits to birders concerned about reaching an old age. While areas in western Chiapas (including Sites 12.1, 12.2, 12.3, 12.4, and 12.9) seems to pose no problems to visitors (as of January 1998), other areas in the northern and eastern highlands might be best avoided until the situation settles down. As always, and as with almost any area, check on the local situation in advance of your trip.

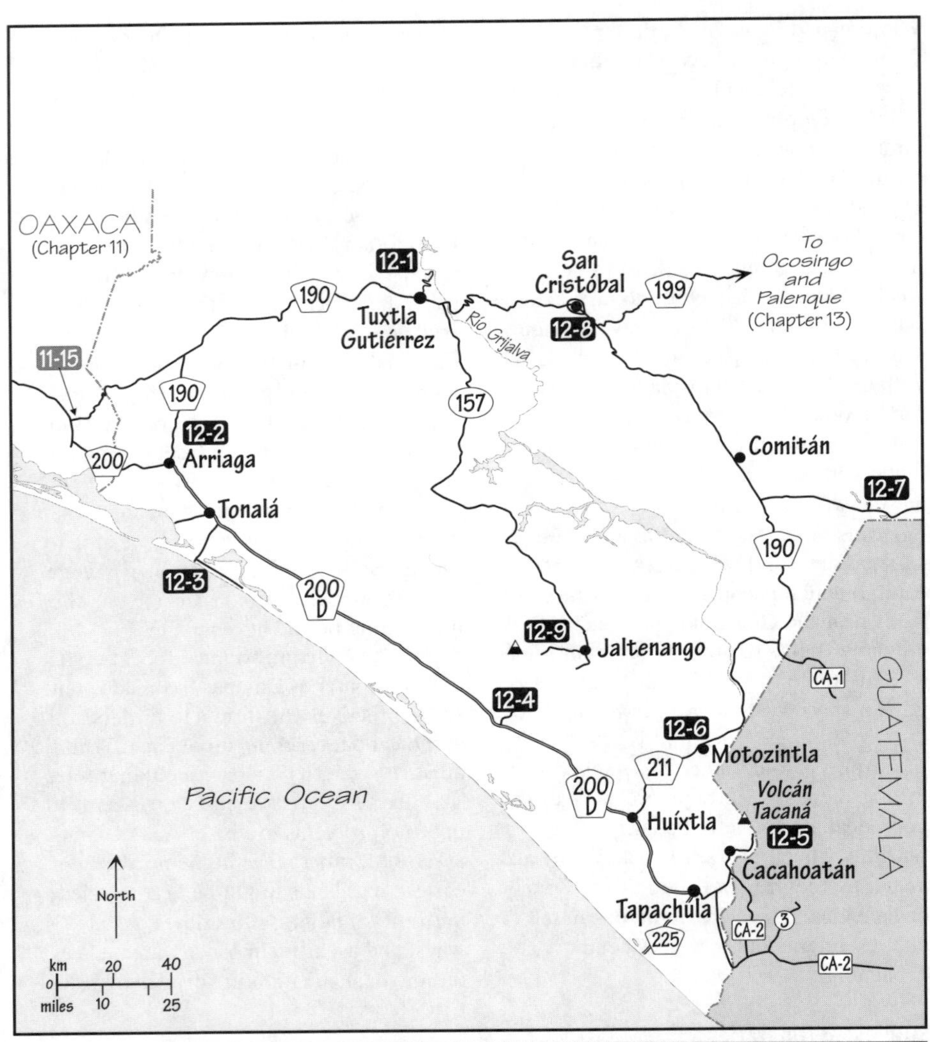

Chapter 12: CENTRAL and SOUTHERN CHIAPAS

- 12-1 El Sumidero
- 12-2 Arriaga Foothills
- 12-3 Puerto Arista / Boca del Cielo
- 12-4 Mapastepec Microwave Valley
- 12-5 Unión Juarez
- 12-6 The Motozintla Road
- 12-7 Lagos de Montebello
- 12-8 San Cristóbal de las Casas
- 12-9 El Triunfo

CENTRAL AND SOUTHERN CHIAPAS SITE LIST

Site 12.1 El Sumidero
Site 12.2 Arriaga Foothills
Site 12.3 Puerto Arista/Boca del Cielo
Site 12.4 Mapastepec Microwave Valley
Site 12.5 Unión Juarez
Site 12.6 The Motozintla Road
Site 12.7 Lagos de Montebello
Site 12.8 San Cristóbal de las Casas
Site 12.9 El Triunfo

Site 12.1: El Sumidero

(February 1982, 1983, 1984, 1985, November 1985, May 1986, January, February 1987, April, May, July 1988, checked March 1996)

Introduction/Key Species

El Sumidero is a deeply carved gorge by which the Río Grijalva, having drained the Central Valley of Chiapas, cuts its exit north to the Gulf of Mexico. This site is a short distance northeast of Tuxtla Gutiérrez (elevation 550 m), whence it is reached by a paved road that winds through an interesting mix of tropical and subtropical oak and thorn-scrub woodland to a restaurant and spectacular overlook some 23 km from Tuxtla. Birds along this road include Maroon-chested Ground-Dove, Pheasant Cuckoo, **Lesser Roadrunner**, **Great Swallow-tailed Swift**, **Green-fronted Hummingbird**, **Slender Sheartail**, **Belted and Flammulated flycatchers**, **Grey-collared Becard**, **Ridgway's Rough-winged Swallow**, **Blue-and-white Mockingbird**, **Fan-tailed Warbler**, **Red-breasted Chat**, Yellow Grosbeak, Blue Seedeater, and **Bar-winged Oriole**.

The increased tourist attraction value of Sumidero has led to the best birding areas' being designated a national park; ironically, this means that you're no longer allowed to enter before dawn or stay after dusk (although it appears that clearing of habitat and quarrying continue beyond the entrance gate).

Access

The road to Sumidero heads out of the northeast suburbs of Tuxtla Gutiérrez (Colonias Albania and Reforma). From downtown Tuxtla, head east from the *zócalo* on the main through street, Avenida 14 de Septiembre, and (about 1 km from the *zócalo*) turn left onto 11A Oriente Norte on the near (west) side of Parque Morelos. Follow this road north, bending right and then back left through Parque Francisco I. Madero, a large green area (which includes the botanical garden and regional anthropology museum). Try to stay on 11A Oriente Norte on the far side of Parque Madero (where there should be signs for Sumidero); this becomes the Sumidero road as it leaves Tuxtla, about 4 km from Parque Madero.

If you're using public transport, take a bus or *colectivo* taxi to Colonia (Col.) Reforma, which is about 4 km out along the Sumidero Road. From there you can try to hitch a ride into the park. Alternatively, a taxi can be hired for the day, which should be reasonably economical if split three or four ways.

Birding Sites

The entrance gateway to the park is at about Km 5 on the Sumidero road, after you have climbed through the outskirts of Tuxtla. Here you are still in arid scrub, with species such as White-throated Magpie-Jay and White-lored Gnatcatcher. From the gateway the road winds up through thorn forest (**Flammulated Flycatcher** occurs near the bend around Km 11) and some open grassy areas (e.g., around Km 14; Northern Bobwhite, Grey-crowned Yellowthroat, Botteri's Sparrow). The first *mirador*, La Ceiba, is on the right near Km 14 (?), and **Great Swallow-tailed Swift** has been seen from here (and could occur anywhere overhead) in late afternoons. From here on the habitat is mostly scrubby woodland, with oaks higher up. There are three other *miradores*, all to the right, before the road ends at a cafe and the main *mirador* (Los Chiapas) at Km 22 (elevation 1250 m). **Red-breasted Chat** and **Bar-winged Oriole** (the latter mainly during April to October) occur along the road from at least Km 7 onward, while most of the other target species, including Maroon-chested Ground-Dove (July/August at least) and **Belted Flycatcher**, occur from Km 16 (near the junction to the *mirador* named La Coyota) to the end of the road. Traffic is generally light, and birding just about anywhere along the road can be excellent, particularly when birds are singing strongly (e.g., April to June). Keep an eye out for seeding bamboo (Blue Seedeater) and flowering trees (orioles), and be aware that many orioles and hummingbirds are highly seasonal in their occurrence at Sumidero.

The weather at Sumidero is notably variable: the generic bright sunny days (as seen in Tuxtla) tend to be quiet for birds after early morning, but often the canyon lip is enshrouded in thick, cool cloud and, if it's not too dense, this cloud can keep birding good all day.

Boat trips up the canyon can be arranged from either the *embarcadero* (dock) on the northwest side of the Route 190 bridge over the river, 7.5 km east of downtown Tuxtla, or at the *embarcadero* in Chiapa de Corzo, 15 km east of Tuxtla on Route 190. Boat trips offer spectacular views of the canyon but not too much in the way of birds, although Great Curassows have been seen along the forested shores.

El Sumidero Bird List

(many hummingbirds are highly seasonal, and other species are local/elevational migrants): 153 sp. †See pp. 5–7 for notes on English names.

Thicket Tinamou
Brown Pelican
Neotropic Cormorant
Black Vulture
Turkey Vulture
White-tailed Kite
Sharp-shinned Hawk
Cooper's Hawk
Short-tailed Hawk
Red-tailed Hawk
American Kestrel
Bat Falcon
Peregrine Falcon
Plain Chachalaca
Singing Quail
Northern Bobwhite
Red-billed Pigeon
White-winged Dove
Mourning Dove
Inca Dove
Common Ground-Dove
Maroon-chested Ground-Dove
White-tipped Dove
Green Parakeet
Mangrove Cuckoo
Squirrel Cuckoo
Pheasant Cuckoo
Lesser Ground-Cuckoo
Lesser Roadrunner
Groove-billed Ani
Vermiculated Screech-Owl
Ferruginous Pygmy-Owl
Mottled Owl
Lesser Nighthawk
Pauraque
Buff-collared Nightjar
Northern Potoo
Vaux's Swift
Great Swallow-tailed Swift
Canivet's Emerald
White-bellied Emerald
Azure-crowned Hummingbird
Berylline Hummingbird
Buff-bellied Hummingbird
Green-fronted Hummingbird
Plain-capped Starthroat
Slender Sheartail
Ruby-throated Hummingbird
Violaceous Trogon
Collared Trogon
Blue-crowned Motmot
Russet-crowned Motmot
Emerald Toucanet
Golden-fronted Woodpecker
Golden-olive Woodpecker
Lineated Woodpecker
Olivaceous Woodcreeper
Ivory-bellied Woodcreeper
Barred Antshrike
N. Beardless Tyrannulet
Greenish Elaenia
Yellow-olive Flycatcher
Belted Flycatcher
Western Pewee†
Willow/Alder Flycatcher
Least Flycatcher
Dusky-capped Flycatcher
Nutting's Flycatcher
Flammulated Flycatcher
Boat-billed Flycatcher
Social Flycatcher
Sulphur-bellied Flycatcher
Couch's Kingbird
Western Kingbird
Eastern Kingbird
Masked Tityra
Grey-collared Becard
Rose-throated Becard
Ridgway's Rough-w. Swallow†
Bank Swallow
Cliff Swallow
Cave Swallow
Barn Swallow
White-throated Magpie-Jay
Green Jay
Northern Raven†
Band-backed Wren
Canyon Wren
Banded Wren
Plain Wren
Southern House Wren†
Blue-grey Gnatcatcher
White-lored Gnatcatcher
Orange-billed N.-Thrush
Swainson's Thrush
Clay-colored Thrush†
White-throated Thrush†
Grey Catbird
Blue-and-white Mockingbird
Cedar Waxwing
White-eyed Vireo
Plumbeous Vireo†
Yellow-throated Vireo
Warbling Vireo
Yellow-green Vireo
Rufous-browed Peppershrike
Blue-winged Warbler
Golden-winged Warbler
Tennessee Warbler
Nashville Warbler
Northern Parula

Magnolia Warbler
Myrtle Warbler†
Black-throated Green Warbler
Blackburnian Warbler
Black-and-white Warbler
Worm-eating Warbler
Ovenbird
Mourning Warbler
MacGillivray's Warbler
Grey-crowned Yellowthroat
Wilson's Warbler
Fan-tailed Warbler
Rufous-capped Warbler
Red-breasted Chat
Yellow-throated Euphonia
Blue-hooded Euphonia
Yellow-winged Tanager
Summer Tanager
Western Tanager
Red-throated Ant-Tanager
Black-headed Saltator
Yellow Grosbeak
Rose-breasted Grosbeak
Blue Grosbeak
Blue Bunting
Indigo Bunting
Varied Bunting
Blue-black Grassquit
Blue Seedeater
Olive Sparrow
Botteri's Sparrow
Rusty Sparrow
Melodious Blackbird
Great-tailed Grackle
Bronzed Cowbird
Black-vented Oriole
Bar-winged Oriole
Yellow-backed Oriole
Streak-backed Oriole
Altamira Oriole
Baltimore Oriole
Yellow-billed Cacique

Site 12.2: Arriaga Foothills

(November 1985, May 1988, November 1993)

Introduction/Key Species

The thorn forest and gallery woodland of the foothills inland from Arriaga, along Route 195, are similar to those inland from Tapanatepec, Oaxaca (Site 11.15). Species of interest include **Green-fronted Hummingbird**, Long-tailed Manakin, and **Rosita's Bunting**.

Access/Birding Sites/Bird List

Good birding areas can be reached easily in 15–30 minutes from Arriaga, where there are several hotels and restaurants, or when traveling between here and Tuxtla Gutiérrez. Obviously, mornings are best, but the main species can be found at any time of day. If you're using public transport, any second-class buses plying this route can let you off at any point you see good habitat, but the same birds may be easier to see above Tapanatepec (Site 11.15).

From the junction with Route 200 (the coastal highway) on the eastern edge of Arriaga, Route 195 heads inland through foothills to join Route 190, the highway to Tuxtla Gutiérrez, Chiapas. The lower slopes are somewhat cutover, but as the road begins to climb, with numerous switchback bends, it goes through good areas of thorn forest, with gallery woodland along some narrow valleys on the sharp bends. It is possible to pull off at numerous bends; those overlooking the scrubby and wooded valleys are particularly good vantage points.

Birds are much the same as those listed for the nearby Tapanatepec Foothills, Oaxaca (Site 11.15). Thus, imitating a Ferruginous Pygmy-Owl's whistles is likely to bring in a band of mobbing birds that often includes **Canivet's Emerald**, **Green-fronted Hummingbird** (seasonal), and **Rosita's** and **Orange-breasted buntings**. At some of the lusher-looking valleys, with tall fig trees, Long-tailed Manakin can be seen from the highway, although you have a better chance if you walk up a drainage a short distance into the woodland. Flocks of **Green and/or Pacific Parakeets** occur in these foothills (October/November, at least)—good luck trying to figure out which you see!

Site 12.3: Puerto Arista/Boca del Cielo

(March, November 1985, May 1986, January 1987, May 1988, January, October 1993, checked March 1996)

Introduction/Key Species

The coastal village of Puerto Arista is a popular destination for beach-seekers and also offers good birding nearby. Habitats include coastal lagoons, weedy pastures, orchards, second-growth thickets, and a nearby estuary (Boca del Cielo). **Giant Wrens** are fairly common in the area, and other birds of note include **White-bellied Chachalaca**, Pacific Screech-Owl, and Spot-breasted Oriole.

Access

To reach Puerto Arista, turn off Route 200 (the coastal highway) at the well-signed junction a few kilometers southeast of Tonalá

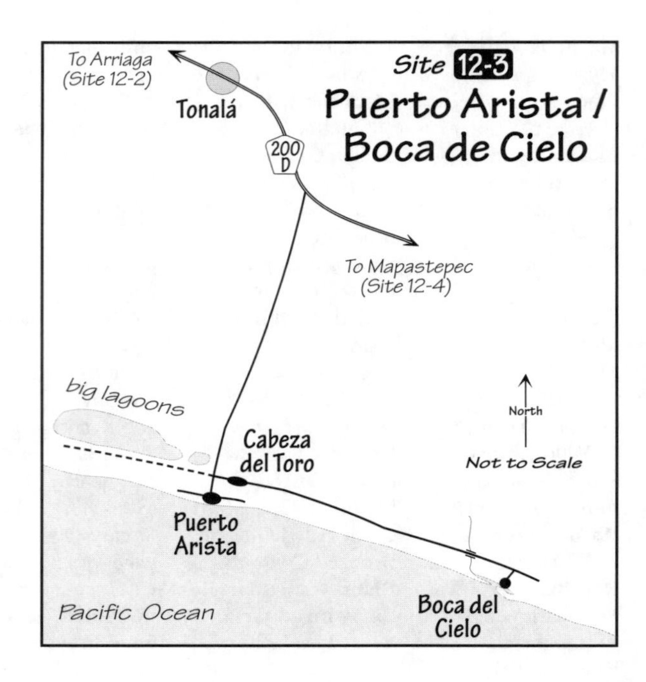

(where **Pacific Parakeets** can be common in orchards on the edge of town). It is about 17 km to Puerto Arista, through pasture land and mango orchards. There are several (mostly basic) hotels and a number of restaurants in Puerto Arista, or this area can be birded easily from a base in Tonalá (where, however, the infrastructure may be no better) or in Arriaga (nicer hotels, but it's less convenient, some 40 km from Puerto Arista).

Public buses run regularly between Arriaga/Tonalá and Puerto Arista; there may be some public transport out to Boca del Cielo (ask locally), but it also has been easy to walk and hitch along the road to the estuary from Cabeza del Toro (see below).

Birding Sites

About a kilometer before Puerto Arista is a paved road to the left (east) which should be signed to Boca del Cielo and/or Cabeza del Toro. Opposite this junction is a dirt track that leads (west) for 10 km alongside shallow, mangrove-fringed lagoons to your right (north), the first of which you can see from the highway. Some small fishponds at the east end of the lagoon, near the highway, are also worth checking. Waterbirds can be numerous here, including Reddish Egret, White Ibis, Roseate Spoonbill, Wilson's Phalarope, and Gull-billed Tern. **Giant Wren** can be found along the fencerows and hedges at the Cabeza del Toro junction, while **White-bellied Chachalaca** and Pacific Screech-Owl occur between the junction and Puerto Arista.

Puerto Arista itself and the beach offer little in the way of birds, although in spring (e.g., early May) it can be a good place to watch passing flocks of jaegers, gulls, and terns, including Pomarine Jaeger, Franklin's and Sabine's gulls, and Common/Arctic Terns.

If you take the Boca del Cielo road, it quickly enters the strung-out town of Cabeza del Toro; stay straight on the main road and after a kilometer or so you will emerge into more pasture, humid second-growth thickets, and orchards. Birding along the road can be productive, and **Giant Wrens** are fairly common. At about Km 16 from the Puerto Arista junction, not long after crossing a small bridge over a mangrove-lined channel, an obvious dirt road heads right (south) a short distance to the small fishing village of Boca del Cielo, visible across the fields. The tidal sand flats out to the west of the village usually have a good variety of shorebirds, gulls, and terns. Small restaurants offer fresh seafood and much-needed cold drinks.

Note: The Black Hawks in the mangroves along the coast here, for example, at Boca del Cielo, are "Mangrove Black Hawks." Common Black Hawk may also occur, but as the two forms are all-but-indistinguishable (and questionably distinct as species), whether you can identify one safely is moot! Also, Melodious Blackbirds colonized the Pacific Slope of Chiapas in the 1980s, but this was inadvertently overlooked by Howell and Webb (1995).

Puerto Arista/Boca del Cielo Bird List

177 sp. R: rare (not to be expected). †See pp. 5–7 for notes on English names.

Least Grebe
Brown Booby
American White Pelican
Brown Pelican
Neotropic Cormorant
Anhinga
Magnificent Frigatebird
Bare-throated Tiger-Heron
Great Blue Heron
Great Egret
Snowy Egret
Little Blue Heron
Tricolored Heron
Reddish Egret
Cattle Egret
Green Heron
White Ibis
White-faced Ibis
Roseate Spoonbill
Wood Stork
Northern Pintail
Blue-winged Teal
Northern Shoveler
American Wigeon
Lesser Scaup
Black Vulture
Turkey Vulture
Lesser Yellow-hd. Vulture
Osprey
White-tailed Kite
Northern Harrier
Sharp-shinned Hawk
Cooper's Hawk
Mangrove/Common Black Hawk
Harris' Hawk
Roadside Hawk
Short-tailed Hawk
Zone-tailed Hawk
Red-tailed Hawk
Crested Caracara
Laughing Falcon
American Kestrel
Merlin
White-bellied Chachalaca
Northern Bobwhite
Black-bellied Plover
American Golden Plover
Collared Plover
Snowy Plover
Wilson's Plover
Semipalmated Plover
Killdeer
Black-necked Stilt
American Avocet
Northern Jacana
Greater Yellowlegs
Lesser Yellowlegs
Willet
Spotted Sandpiper
Wandering Tattler
Whimbrel
Long-billed Curlew (R)
Marbled Godwit
Ruddy Turnstone
Sanderling
Semipalmated Sandpiper
Western Sandpiper
Least Sandpiper
Baird's Sandpiper
Stilt Sandpiper
Long-billed Dowitcher
Short-billed Dowitcher
Common Snipe
Wilson's Phalarope
Pomarine Jaeger
Parasitic Jaeger
Laughing Gull
Franklin's Gull
Ring-billed Gull
Herring Gull
Sabine's Gull
Gull-billed Tern
Caspian Tern
Royal Tern
Elegant Tern
Sandwich Tern
Common Tern
Black Tern
Least Tern
Black Skimmer
Red-billed Pigeon
White-winged Dove
Mourning Dove
Inca Dove
Common Ground-Dove
Ruddy Ground-Dove
Pacific Parakeet
Orange-fronted Parakeet
Orange-chinned Parakeet
White-fronted Parrot
Lesser Ground-Cuckoo
Groove-billed Ani
Pacific Screech-Owl
Ferruginous Pygmy-Owl
Lesser Nighthawk
Pauraque
Black Swift
Cinnamon Hummingbird
Ruby-throated Hummingbird
Citreoline Trogon
Russet-crowned Motmot
Ringed Kingfisher
Belted Kingfisher
Golden-fronted Woodpecker
Lineated Woodpecker
Barred Antshrike
Common Tody-Flycatcher
Willow Flycatcher
Least Flycatcher
Brown-crested Flycatcher
Great Kiskadee
Boat-billed Flycatcher
Social Flycatcher
Scissor-tailed Flycatcher
Tropical Kingbird
Western Kingbird
Rose-throated Becard
Grey-breasted Martin
Tree Swallow
Mangrove Swallow
N. Rough-winged Swallow
Bank Swallow
Cliff Swallow
Barn Swallow

White-throated Magpie-Jay
Giant Wren
Plain Wren
Blue-grey Gnatcatcher
Clay-colored Thrush†
Bell's Vireo
Nashville Warbler
Yellow Warbler
Magnolia Warbler
Black-and-white Warbler
Mourning Warbler
Northern Waterthrush
Common Yellowthroat
Grey-crowned Yellowthroat
Yellow-breasted Chat
Yellow-winged Tanager
Summer Tanager
Western Tanager
Greyish Saltator
Rose-breasted Grosbeak
Blue Grosbeak
Indigo Bunting
Painted Bunting
Dickcissel
Blue-black Grassquit
White-collared Seedeater
Ruddy-breasted Seedeater
Stripe-headed Sparrow
Lark Sparrow (R)
Lincoln's Sparrow
Melodious Blackbird
Great-tailed Grackle
Bronzed Cowbird
Orchard Oriole
Ochre Oriole†
Streak-backed Oriole
Spot-breasted Oriole
Altamira Oriole
Baltimore Oriole
Yellow-winged Cacique

Site 12.4: Mapastepec Microwave Valley

(February 1983, March 1985, May 1986, May 1988, checked March 1996)

Introduction/Key Species

Most of the Pacific lowlands of Chiapas have been cleared successfully of forest, such that Scarlet Macaws, once common here, are a dim memory. Route 200 (and the recently constructed *cuota* highway, Route 200D) run along the inland edge of the coastal plain at the base of foothills of the Sierra Madre de Chiapas. In places there are forest patches and some good birding near the highway. Higher in the Sierra there are extensive areas of evergreen forest, but worthwhile access to them requires an organized hike and overnight camping (see El Triunfo, Site 12.9). The diminishing forest patches, second-growth thickets, and brushy pastures near the microwave towers a short distance southeast of Mapastepec still offer good birding, including **White-bellied Chachalaca**, Yellow-naped Parrot, Lesser Swallow-tailed Swift, Long-billed Starthroat, Turquoise-browed Motmot, **Rufous-breasted Spinetail**, **Giant Wren**, Chestnut-capped Warbler, and **Prevost's Ground-Sparrow**.

Access/Birding Sites

The town of Mapastepec is just south of Route 200. Taking the junction into Mapastepec off the old highway as zero, continue southeast on Route 200 for 9 km and you come to a bridge (Puente Sececapa) over a small river. A hill with radio masts overlooks this river, on the north side of the highway just west of the bridge. If you're using public transport, take a second-class bus from Mapastepec southeast (toward Huixtla) along the old highway (Route 200) and watch for the microwave tower hill, which indicates where to get off the bus. If you're heading southeast on the new *cuota* highway, you have to go beyond the microwave tower hill to the first exit and and then double back to the bridge, which is inconspicuous.

Birding along the cobbled road up to the towers has been good, although much of the hillside is now cleared; still, about 500 m up the microwave road is a good vantage point from which to overlook the coastal plain and valley in late afternoon for Yellow-naped Parrots, which feed out on the plain and come to roost in trees in the foothills (e.g., 105 Yellow-naped, plus 160 White-fronted Parrots, the evening of 5 March 1985). The microwave tower hill also can be a good place from which to watch waves of Turkey Vultures and Swainson's and Broad-winged hawks migrating northwest along the foothills in spring.

A well-worn trail leads up the east bank of the river below the microwave tower hill, with room to park safely off the highway. This can be followed through second-growth woodland, thickets, and overgrown pastures for several kilometers, which will involve crisscrossing the river if you continue far enough inland. A good variety of birds can be seen along this trail, including **White-bellied Chachalaca**, Long-billed Starthroat, Turquoise-browed Motmot, **Rufous-breasted Spinetail**, **Giant** and Rufous-naped (rufous-backed form) **wrens**, Long-billed Gnatwren, Chestnut-capped Warbler (uncommon, at the northwest edge of its range), **Prevost's Ground-Sparrow**, and Spot-breasted Oriole.

Mapastepec Microwave Valley Bird List
145 sp. †See pp. 5–7 for notes on English names.

Thicket Tinamou
Brown Pelican
Great Egret
Snowy Egret
Little Blue Heron
Cattle Egret
Green Heron
Black-crowned Night-Heron
Black Vulture
Turkey Vulture
Grey Hawk
Broad-winged Hawk
Short-tailed Hawk
Swainson's Hawk
Red-tailed Hawk
Crested Caracara
American Kestrel
White-bellied Chachalaca
Spotted Sandpiper
Red-billed Pigeon
Inca Dove
Ruddy Ground-Dove
White-tipped Dove
Orange-fronted Parakeet
Orange-chinned Parakeet
White-fronted Parrot
Yellow-naped Parrot
Squirrel Cuckoo
Groove-billed Ani
Lesser Ground-Cuckoo
Ferruginous Pygmy-Owl
Mottled Owl
Lesser Nighthawk
Pauraque
Black Swift
Chestnut-collared Swift
White-collared Swift
Vaux's Swift
Lesser Swallow-tailed Swift
Green-breasted Mango
Cinnamon Hummingbird
Long-billed Starthroat
Ruby-throated Hummingbird
Violaceous Trogon
Ringed Kingfisher
Amazon Kingfisher
Green Kingfisher
Blue-crowned Motmot
Turquoise-browed Motmot
Collared Aracari
Golden-fronted Woodpecker
Golden-olive Woodpecker
Smoky-brown Woodpecker
Lineated Woodpecker
Ivory-billed Woodcreeper
Streak-headed Woodcreeper
Rufous-breasted Spinetail
Barred Antshrike
N. Beardless Tyrannulet
Yellow-bellied Elaenia
Greenish Elaenia
Ochre-bellied Flycatcher
Northern Bentbill
Common Tody-Flycatcher
Yellow-olive Flycatcher
Tropical Pewee
Yellow-bellied Flycatcher
Alder Flycatcher
Willow Flycatcher
Least Flycatcher
Bright-rumped Attila
Dusky-capped Flycatcher
Brown-crested Flycatcher
Great Crested Flycatcher
Great Kiskadee
Boat-billed Flycatcher
Social Flycatcher
Sulphur-bellied Flycatcher
Piratic Flycatcher
Scissor-tailed Flycatcher
Tropical Kingbird
Western Kingbird
Eastern Kingbird
Rose-throated Becard
Masked Tityra
Grey-breasted Martin
N. Rough-winged Swallow
Bank Swallow
Cliff Swallow
Barn Swallow
White-throated Magpie-Jay
Green Jay
Rufous-naped Wren
Giant Wren
Banded Wren
Plain Wren
Spot-breasted Wren
Long-billed Gnatwren
Blue-grey Gnatcatcher
Orange-billed N.-Thrush
Swainson's Thrush
Clay-colored Thrush†
Bell's Vireo
Yellow-throated Vireo
Warbling Vireo
Yellow-green Vireo
Lesser Greenlet
Blue-winged Warbler
Tennessee Warbler
Yellow Warbler
Magnolia Warbler
Black-and-white Warbler
American Redstart
Ovenbird
Louisiana Waterthrush
Kentucky Warbler
Mourning Warbler
Grey-crowned Yellowthroat
Chestnut-capped Warbler†
Yellow-breasted Chat
Red-legged Honeycreeper
Scrub Euphonia
Yellow-throated Euphonia
Blue-grey Tanager
Yellow-winged Tanager
Western Tanager
Red-throated Ant-Tanager
Greyish Saltator
Black-headed Saltator
Rose-breasted Grosbeak
Indigo Bunting
Painted Bunting
Dickcissel
Prevost's Ground-Sparrow
Blue-black Grassquit
White-collared Seedeater
Melodious Blackbird
Great-tailed Grackle
Bronzed Cowbird
Orchard Oriole
Spot-breasted Oriole
Altamira Oriole
Baltimore Oriole
Yellow-billed Cacique

Site 12.5: Unión Juarez

(December 1985)

Introduction/Key Species

This small town (elevation 1500 m?) offers access by a footpath to remnant patches of cloud forest on the southwest slopes of Volcán Tacaná (**Highland Guan**, **Green-throated Mountain-gem**, Mountain Thrush, **Pink-headed Warbler**, **Black-capped Siskin**) as well as good birding in the coffee plantations near town (**Blue-tailed Hummingbird**, Rufous-and-white Wren, **Prevost's Ground-Sparrow**). It is best to be in reasonable physical shape to make the hike from Unión Juarez to the higher slopes of Tacaná, the peak of which reaches about 4050 m. I walked to Tregales in a day, birding along the way, and it would be possible to do this as a long day hike from and returning to Unión Juarez, but you would have little time for birding. However, good birds can be found in the forest patches and brushy second growth well before Tregales, making for a less hurried day's birding.

Access/Birding Sites

From Tapachula head east toward the Guatemala border and the northern crossing (which may be signed Frontera/Talisman), also following signs for Cacahoatán. You come to a junction near the border: the choice is left for Cacahoatán, right for Guatemala. Turn left toward and through Cacahoatán (staying on the main road) and up through coffee plantations to Unión Juarez, about 17 km north of Cacahoatán. There is a small hostel in Unión Juarez (**Black-capped Swallows** occur over town), used as a base by hikers who want to ascend Tacaná. (Around Christmas this is a very popular hike, and hundreds of people may be encountered along the trail, making it easy to ask for directions.) Alternatively, the coffee plantations between Cacahoatán and Unión Juarez can be done as a day trip from Tapachula. Buses run regularly from Tapachula to Unión Juarez, from which the birding areas are reached on foot.

The trail up Tacaná to the small settlement of Tregales is not well marked in places, and a local guide may be useful. At the same time, even taking a "wrong" trail can result in good birding. The street along the right-hand (east) side of the *zócalo* in Unión Juarez continues north as a cobbled road, through gardens and coffee plots (**Blue-tailed Hummingbird**, **Sparkling-tailed Woodstar**, **Rufous-capped Warbler**, **Cinnamon-bellied Flowerpiercer**) before coming to a good patch of humid evergreen forest where the trail narrows and winds steeply up through the forest (**Green-throated Mountain-gem**, Mountain Thrush, **Pink-headed Warbler**, **Blue-crowned Chlorophonia**).

Emerging from the forest, the trail has high banks on either side and comes shortly to an obelisk where the trail is wider and overlooks the valley (the far side of which is Guatemala) down to the east. Within 1 km (?) look for an inconspicuous, narrow trail off to the left, through second-growth brush and cornfields; this is the trail to Tregales, which passes by a forested valley down to the left before coming to more cornfields, patches of alders (**Black-capped Siskin**), and a row of two or three houses (Tregales) on the left-hand side of the path. Hikers can stay here (it's pretty basic—a floor) on their climb to the top of Tacaná. Above Tregales is a patch of forest in the valley to the right of the trail (**Highland Guan**, **Garnet-throated Hummingbird**, **Pink-headed Warbler**), and to the left and above is mostly open bunch grass. The elevation here (well over 3000 m) is apt to make walking, let alone climbing, a slow process, and there appears to be no ornithological reason to go much above Tregales. Other species that could be expected on this hike include Buffy-crowned Wood-Partridge, White-breasted Hawk, Unspotted Saw-whet Owl, Rufous Sabrewing, Yellow-throated Brushfinch, and perhaps even Slaty Finch.

Birding is easier below Unión Juarez: I birded on foot for a morning from Unión Juarez about 5 km along the road back toward Cacahoatán. The habitat is largely coffee plantations with shade trees (**Blue-tailed Hummingbird**, **Sparkling-tailed Woodstar**, Paltry Tyrannulet, Rufous-and-white Wren, **Prevost's Ground-Sparrow**). Other species to watch for here include Salvin's [Fork-tailed] Emerald, Chestnut-capped Warbler, and White-eared Ground-Sparrow.

Unión Juarez Bird List

119 sp. [A]above Unión Juarez to Tregales (82 sp.); [B]below Unión Juarez (69 sp.); no code: both sites. †See pp. 5–7 for notes on English names.

Cattle Egret[B]
Black Vulture[B]
Turkey Vulture[B]
Hook-billed Kite[A]
Sharp-shinned Hawk[A]
Grey Hawk[B]
Broad-winged Hawk[A]
Red-tailed Hawk[A]
Highland Guan[A]
Spotted Wood-Quail[A]
Red-billed Pigeon[B]
Band-tailed Pigeon[A]
White-tipped Dove
Green/Pacific Parakeet
Squirrel Cuckoo[A]
Mountain Pygmy-Owl†[A]
Mexican Whip-poor-will†[A]
White-collared Swift
Vaux's Swift
White-throated Swift[A]
Lesser Swallow-tailed Swift[B]
Violet Sabrewing[B]
Green Violet-ear[A]
White-eared Hummingbird[A]
White-bellied Emerald[B]
Azure-crowned Hummingbird[A]
Blue-tailed Hummingbird
Cinnamon Hummingbird[B]
Green-throated Mountain-gem[A]
Amethyst-throated Hummingbird[A]
Garnet-throated Hummingbird[A]
Magnificent Hummingbird[A]
Sparkling-tailed Woodstar†
Ruby-throated Hummingbird
Mountain Trogon[A]
Golden-fronted Woodpecker
Hairy Woodpecker[A]
Smoky-brown Woodpecker[B]
Golden-olive Woodpecker
Lineated Woodpecker
Guatemalan Flicker†[A]
Paltry Tyrannulet
N. Beardless Tyrannulet[B]
Greenish Elaenia[B]
Yellow-olive Flycatcher[B]
Common Tufted Flycatcher†[A]
Greater Pewee[A]
Least Flycatcher[B]
Hammond's Flycatcher[A]
Black Phoebe[B]
Dusky-capped Flycatcher[B]
Social Flycatcher[B]
Grey-breasted Martin[B]
Violet-green Swallow[A]
Black-capped Swallow
White-throated Magpie-Jay[B]
Green Jay[A]
Band-backed Wren[A]
Spot-breasted Wren[B]
Rufous-and-white Wren[B]
Plain Wren[B]
Southern House Wren†
Blue-grey Gnatcatcher[B]
Brown-backed Solitaire
Ruddy-capped N.-Thrush
Swainson's Thrush[B]
Hermit Thrush[A]
Mountain Thrush†[A]
Rufous-collared Thrush†[A]
Blue-and-white Mockingbird[A]
Grey Silky†[A]
Blue-headed Vireo†
Hutton's Vireo[A]
Warbling Vireo
Tennessee Warbler
Nashville Warbler
Crescent-chested Warbler[A]
Magnolia Warbler[B]
Audubon's Warbler†[A]
Townsend's Warbler
Hermit Warbler[A]
Black-throated Green Warbler[B]
Grace's Warbler[A]
Black-and-white Warbler
Worm-eating Warbler
Ovenbird[B]
Louisiana Waterthrush[B]
MacGillivray's Warbler[A]
Common Yellowthroat[B]
Grey-crowned Yellowthroat[B]
Wilson's Warbler
Pink-headed Warbler[A]
Slate-throated Whitestart†[A]
Rufous-capped Warbler
Golden-browed Warbler[A]
Yellow-breasted Chat[B]
Red-legged Honeycreeper
Blue-crowned Chlorophonia
Blue-grey Tanager[B]
Yellow-winged Tanager
Summer Tanager[B]
Western Tanager
White-winged Tanager
Black-headed Saltator
Rose-breasted Grosbeak
Indigo Bunting
Painted Bunting[B]
Chestnut-capped Brushfinch[A]
Prevost's Ground-Sparrow[B]
Spotted Towhee[A]
White-collared Seedeater[B]
Cinnamon-bellied Flowerpiercer[A]
Rufous-collared Sparrow
Guatemalan Junco†[A]
Great-tailed Grackle
Baltimore Oriole
Black-capped Siskin[A]
Black-headed Siskin[A]
Lesser Goldfinch[A]

Site 12.6: The Motozintla Road

(February 1983, March 1985, January 1992)

Introduction/Key Species

Route 211 runs from the Pacific Slope lowlands at Huixtla over the Sierra Madre de Chiapas to Motozintla on the dry interior slopes, and thence along the eastern edge of the Central Valley of Chiapas to the junction with Route 190 near the Guatemala border at Ciudad Cuauhtemoc. A couple of basic hotels in Motozintla (which is easy to reach by bus) allow morning birding in the nearby arid scrub and pine-oak woodland (where birds include **White-breasted Hawk**, Unspotted Saw-whet Owl, **Black-capped Swallow**, **Blue-and-white Mockingbird**, and Yellow-throated Brushfinch).

Birding Sites/Bird List

From the junction with Route 200 in Huixtla (Km zero), the Motozintla road climbs up through cutover humid foothills, with coffee plantations and forest patches, into tropical semi-deciduous and then pine-oak and humid evergreen forest patches, which are almost always away from the highway, off on slopes across deforested valleys. Birds overall are similar to those listed for Unión Juarez (Site 12.5), taking into account habitat and elevation.

Areas lower down, nearer Huixtla, have Green-breasted Mango (winter) and Cinnamon Hummingbird, and should be checked for Salvin's [Fork-tailed] Emerald. Higher up, stops at wooded/coffee plantation valleys (e.g., at Km Posts 20 and 36) can produce **Blue-tailed Hummingbird**, both Chestnut-capped and **Rufous-capped warblers**, and White-eared Ground-Sparrow.

Continuing up, the highway crests at about 1800 m (the continental divide) and then passes through a short stretch of pine-oak woodland at Km Posts 50/51.5 (around 1700 m), just above the town of Motozintla, which is at about Km Post 55. Birds in the pine-oak here include **White-breasted** and Red-tailed **hawks**, Whiskered Screech-Owl, Great Horned Owl, Unspotted Saw-whet Owl, Mexican Whip-poor-will, **White-eared**, **Azure-crowned**, and **Amethyst-throated hummingbirds**, **Mountain Trogon**, Acorn and Hairy woodpeckers, **Guatemalan** [Northern] **Flicker**, **Black-capped Swallow**, Steller's Jay, Bushtit, Band-backed Wren, Eastern Bluebird, **Brown-backed Solitaire**, Orange-billed Nightingale-Thrush, **Rufous-collared Thrush**, **Blue-and-white Mockingbird**, Painted Whitestart, **Rufous-capped Warbler**, Hepatic Tanager, Yellow-throated Brushfinch, and Yellow-backed Oriole. Wintering migrants include Ruby-throated Hummingbird, Nashville, Audubon's [Yellow-rumped], Townsend's, MacGillivray's, and Wilson's warblers, and Western Tanager.

Motozintla is at the southeastern corner of the dry interior valley of Chiapas. Birds in the arid scrub just north of town, around Km Post 58 (1250 m), include **Green Parakeet**, **Lesser Roadrunner**, Plain-capped Starthroat, **Russet-crowned Motmot**, Banded Wren, White-lored Gnatcatcher, Tropical Mockingbird, **Rufous-capped Warbler**, and **Black-vented** and Streak-backed **orioles**.

Continuing north, the road winds along a broad valley; the river to your left flows northwest to join the Río Grijalva, which ends eventually at the Gulf of Mexico near Villahermosa! After a stretch of dry oak scrub (**Green-fronted Hummingbird** occurs seasonally around Km Post 84, in December/January, at least) and farmland, the road drops into a humid valley with coffee plantations (Squirrel Cuckoo, Ferruginous Pygmy-Owl, **Azure-crowned Hummingbird**, Yellow-olive Flycatcher, Clay-colored Thrush; wintering Yellow-bellied and Great Crested flycatchers, Tennessee Warbler, etc.). The junction with Ciudad Cuauhtemoc is about 70 km north of Motozintla.

From the Cuauhtemoc junction, Route 190 continues through thorn-forest patches, along a river with large trees, across the intensely cultivated floor of the Central Valley (about 800 m elevation), up through some thorn forest and oaks, and then drops to the dry, cultivated plains around Comitán (see Lagos de Montebello, Site 12.7).

Site 12.7: Lagos de Montebello

(February 1982, 1983, 1984, 1985, March, November 1985, May 1986, February 1987, October 1993)

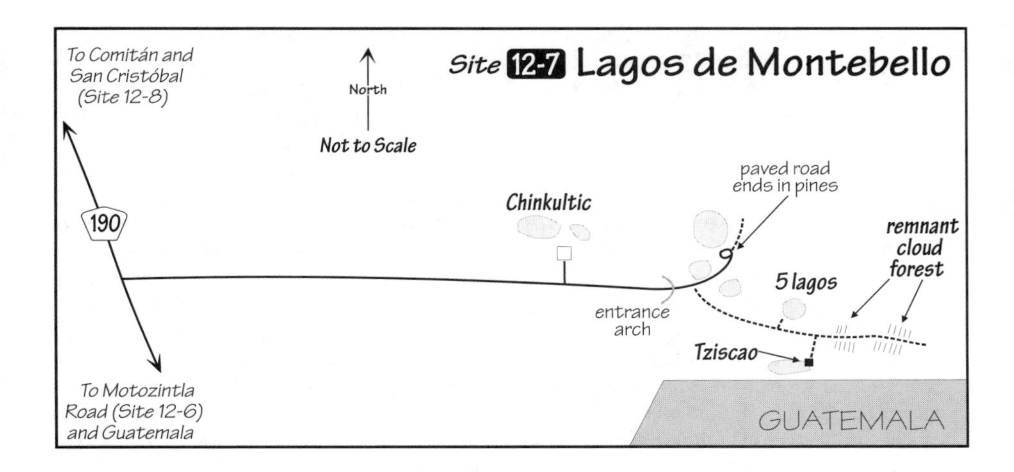

Introduction/Key Species

This site (1200–1500 m elevation), a national park on the Guatemala border in eastern Chiapas, is famous for its scenic lakes set amid attractive pine-evergreen forest. Of more interest to birders are some remnant patches of cloud forest, where there is a chance to see Resplendent Quetzal. Other species at Montebello include **Highland Guan**, **White-faced Quail-Dove**, Barred Parakeet, **Emerald-chinned**, Stripe-tailed, and **Wine-throated hummingbirds**, Black-crested Coquette, **Green-throated Mountain-gem**, **Sparkling-tailed Woodstar**, Ruddy Foliage-gleaner, Tawny-throated Leaftosser, Strong-billed Woodcreeper, **Mexican Antthrush**, **Black-throated** and **Unicolored jays**, **Slate-colored Solitaire**, Black-headed, and Spotted nightingale-thrushes, Shining Honeycreeper, **Blue-crowned Chlorophonia**, **White-naped Brushfinch**, and **Prevost's Ground-Sparrow**. Like all cloud forests, it can seem devoid of birds on sunny days after the first hour of light, and on the frequent cool and misty or rainy days, birding can be good but frustratingly challenging. January to April is perhaps the best season to visit, while autumn and early winter (e.g., October) seem to be the quietest times for birds.

Note: At the Route 190/Montebello junction and/or just north or south of Comitán, as well as at any points on the Montebello road, you may encounter military and immigration checkpoints. These guys are just doing their job: smile and be polite, show your passport if asked, and you should be on your way with a wave. At the same time, before you consider going to Montebello, be aware of the current political situation, in case a birding visit there might not be the best idea.

Access/Birding Sites

Montebello is about 50 km east-southeast of Comitán de Dominguez, where there are several hotels and restaurants (allow 1.5 hours to get to cloud-forest habitat from Comitán). Comitán is about 90 km (1–1.5 hours) east-southeast of San Cristóbal (Site 12.8). Buses from Comitán to "Montebello" go (or have gone) only to the end of the paved road (see below), but other buses (to destinations unknown) do transit the dirt road through the cloud forest (ask in Comitán, e.g., about buses to Tziscao; see below). It is also possible to walk (good birding along the road) and hitch rides from the archway (see below) to Tziscao.

From Comitán head south on Route 190 toward the Guatemala border. After about 16 km watch for a major junction to the left, immediately before La Trinitaria, which should be signed to Montebello or something similar. Turn left here, almost due east, on an undulating, straight, paved road running through cornfields and open pine woods. Eastern Bluebirds and Tropical Mockingbirds are often on the roadside wires, and this is a good stretch for White-tailed Hawk. Cave Swallows (summer) nest under some of the road bridges, and Botteri's Sparrows occur in the fields.

At around 30 km (?) look for a dirt road to the left, running north through the fields for about a kilometer to some overgrown-looking "hills": these are the Maya ruins of Chinkultic. There is a gate (and often small children hanging about to open it) at the start of this road, and there may even be a sign indicating the ruins, which recently have achieved the status of levying an official entrance fee. Birding in and around the ruins can be productive (**Slender Sheartail**, Barred Antshrike, **Prevost's Ground-Sparrow**), and a distinctive local race of Northern Bobwhite occurs in the fields here. Birding Chinkultic can take at least a couple of hours, and it often seems very birdy after a quiet day in the cloud forest.

Back at the main road, continue 5–10 km east (about 37 km? in total from the Route 190 junction) to a concrete archway over the road; this is the entrance to the park, where someone may ask your nationality and wave you on. Often, however, it has been unmanned, and as of October 1993 there was no entrance fee. The paved road continues straight from here and ends in about 2.5 km (?) at a parking lot overlooking a large lake. The pines around here are good for Strong-billed Woodcreeper (often with mixed-species flocks), and brushy areas have **White-naped Brushfinch**. A dirt road continues from this point into the woods and some *grutas* (caves); **Mountain Trogon** and **Unicolored Jay** can be seen along this track.

Immediately through the archway a graded dirt road branches off to the right. This is the road to more lakes and to the cloud forest. (Eventually, after skirting around the Mexico/Guatemala border, this "road" comes out near Palenque, in the Atlantic Slope lowlands.) However, driving past the birding areas described is not recommended, even if the road is passable (and see the cautionary note in the introductory paragraph).

This dirt road (set to zero at the archway junction) runs through humid pine-oak forest at first, then alternating secondary deciduous forest and patches of cloud forest, with side roads to various lakes (early morning on the quiet side road to 5 Lagos has produced good views of **Singing Quail**, Spotted Wood-Quail, and **White-faced Quail-Dove** on the roadside). At Km 9.7 there is a right-hand junction to the village of Tziscao, about a kilometer from the road and on the shore of a large lake (Limpkin, **Ruddy Crake**, Virginia Rail), the far side of which is Guatemalan territory. There is a small hostel in Tziscao by the lake. The hostel is potentially convenient for the cloud forest but, unless things have changed in recent years, a chronic case of delayed maintenance syndrome may make you question whether you want to stay there. If you think about staying, taking your own food has been a good idea in past years. The humid second-growth hedges and thickets in Tziscao have **Slender Sheartail**, **Blue-and-white Mockingbird**, and **Prevost's Ground-Sparrow**.

At 0.8–2.8 km east of the Tziscao turnoff (10.5–12.5 km from the archway junction) are two or three roadside patches of cloud forest, where **Highland Guan** and Resplendent Quetzal can be seen from the road. Keep an ear/eye out for Barred Parakeet and **Hooded Grosbeak** flying overhead, and look for fruiting trees (thrushes, **Blue-crowned Chlorophonia**) and flowering trees (hummingbirds, Shining Honeycreeper). Numerous footpaths provide access into the forest (the best way to see Yellowish Flycatcher and Spotted Nightingale-Thrush, among many others) but, depressingly, usually end in small to large clearings. However, the edges of overgrown clearings can have good flower banks with Black-crested Coquette, **Sparkling-tailed Woodstar**, and **Wine-throated Hummingbird** among other species.

It is possible to drive on a few kilometers past these cloud-forest patches, although much of the forest has been cleared, and birds more typical of lower elevations are soon found (e.g., Red-billed Pigeon, Long-tailed Hermit, Keel-billed Toucan, Lesser Greenlet, Bananaquit, Black-faced Grosbeak; not included below in the bird list); Nightingale Wren occurs in forest patches by a (usually dry) streambed with large boulders (at about Km 15/17?). Beyond this there are villages, refugee camps, and very little forest.

Note: Awareness of tourism around Montebello has increased in recent years, and with it may have come an increase in facilities. Sadly, this awareness has not extended to protecting the cloud forest.

Lagos de Montebello Bird List
(including Chinkultic): 219 sp. †See pp. 5–7 for notes on English names.

Thicket Tinamou
Slaty-breasted Tinamou
Least Grebe
Pied-billed Grebe
Neotropic Cormorant
Least Bittern
Great Egret
Little Blue Heron
Tricolored Heron
Cattle Egret
Green Heron
Black-bellied Whistling-Duck
Blue-winged Teal
Ruddy Duck
Black Vulture
Turkey Vulture
Hook-billed Kite
White-tailed Kite
Swallow-tailed Kite
Northern Harrier
White-breasted Hawk†
Great Black Hawk
Broad-winged Hawk
Short-tailed Hawk
White-tailed Hawk
Red-tailed Hawk
Barred Forest-Falcon
American Kestrel
Bat Falcon
Plain Chachalaca
Highland Guan
Spotted Wood-Quail
Singing Quail
Northern Bobwhite
Limpkin
Ruddy Crake
Virginia Rail
American Coot
Killdeer
Solitary Sandpiper
Spotted Sandpiper
Common Snipe
Band-tailed Pigeon
White-winged Dove
Mourning Dove
Inca Dove
Common Ground-Dove
White-tipped Dove
White-faced Quail-Dove
Barred Parakeet
White-crowned Parrot
Squirrel Cuckoo
Groove-billed Ani
Crested Owl
Mottled Owl
Lesser Nighthawk
Mexican Whip-poor-will†
White-collared Swift
Vaux's Swift
White-throated Swift
Violet Sabrewing
Green Violet-ear
Emerald-chinned Hummingbird
Black-crested Coquette
White-eared Hummingbird
Azure-crowned Hummingbird
Stripe-tailed Hummingbird
Magnificent Hummingbird
Green-throated Mountain-gem
Amethyst-throated Hummingbird
Garnet-throated Hummingbird
Sparkling-tailed Woodstar†
Ruby-throated Hummingbird
Wine-throated Hummingbird
Mountain Trogon
Collared Trogon
Resplendent Quetzal
Green Kingfisher
Acorn Woodpecker
Yellow-bellied Sapsucker
Hairy Woodpecker
Smoky-brown Woodpecker
Golden-olive Woodpecker
Lineated Woodpecker
Pale-billed Woodpecker
Guatemalan Flicker†
Emerald Toucanet
Scaly-throated Fol.-gleaner†
Ruddy Foliage-gleaner
Tawny-throated Leaftosser
Olivaceous Woodcreeper
Strong-billed Woodcreeper
Spotted Woodcreeper
Spot-crowned Woodcreeper
Barred Antshrike
Mexican Antthrush†
Paltry Tyrannulet
Greenish Elaenia
Ochre-bellied Flycatcher
Northern Bentbill
Eye-ringed Flatbill
Stub-tailed Spadebill
Olive-sided Flycatcher
Greater Pewee
Western Pewee†
Eastern Pewee†
Least Flycatcher
Hammond's Flycatcher
Yellowish Flycatcher
Pine Flycatcher
Bright-rumped Attila
Dusky-capped Flycatcher
Sulphur-bellied Flycatcher
Social Flycatcher
Eastern Kingbird
Rose-throated Becard
Masked Tityra
Lovely Cotinga
Tree Swallow
Rough-winged Swallow sp.†
Black-capped Swallow
Cave Swallow
Green Jay
Azure-hooded Jay
Black-throated Jay
Unicolored Jay
Northern Raven†
Band-backed Wren
Spot-breasted Wren
Plain Wren
Southern House Wren†
Grey-breasted Wood-Wren
Blue-grey Gnatcatcher
Eastern Bluebird
Brown-backed Solitaire
Slate-colored Solitaire
Orange-billed N.-Thrush
Black-headed N.-Thrush
Spotted N.-Thrush
Wood Thrush
Swainson's Thrush
Black Thrush†

Clay-colored Thrush†
White-throated Thrush†
Rufous-collared Thrush†
Grey Catbird
Tropical Mockingbird
Blue-and-white Mockingbird
Cedar Waxwing
Grey Silky†
Blue-headed Vireo†
Brown-capped Vireo
Philadelphia Vireo
Rufous-browed Peppershrike
Golden-winged Warbler
Blue-winged Warbler
Tennessee Warbler
Nashville Warbler
Crescent-chested Warbler
Tropical Parula
Magnolia Warbler
Hermit Warbler
Townsend's Warbler
Black-throated Green Warbler
Golden-cheeked Warbler
Blackburnian Warbler
Grace's Warbler
Myrtle Warbler†
Audubon's Warbler†
Black-and-white Warbler
Worm-eating Warbler
Ovenbird
Louisiana Waterthrush
MacGillivray's Warbler
Common Yellowthroat
Hooded Warbler
Wilson's Warbler
Slate-throated Whitestart†
Golden-crowned Warbler
Rufous-capped Warbler
Golden-browed Warbler
Yellow-breasted Chat
Green Honeycreeper
Shining Honeycreeper
Blue-crowned Chlorophonia
Yellow-winged Tanager
Red-crowned Ant-Tanager
Hepatic Tanager
Summer Tanager
Western Tanager
Flame-colored Tanager
White-winged Tanager
Common Bush-Tanager
Greyish Saltator
Rose-breasted Grosbeak
Blue Grosbeak
Indigo Bunting
White-naped Brushfinch
Chestnut-capped Brushfinch
Prevost's Ground-Sparrow
Blue-black Grassquit
White-collared Seedeater
Cinnamon-bellied Flowerpiercer
Rusty Sparrow
Botteri's Sparrow
Chipping Sparrow
Lincoln's Sparrow
Rufous-collared Sparrow
Eastern Meadowlark
Red-winged Blackbird
Melodious Blackbird
Great-tailed Grackle
Bronzed Cowbird
Yellow-backed Oriole
Baltimore Oriole
Yellow-billed Cacique
Red Crossbill
Black-headed Siskin
Lesser Goldfinch
Hooded Grosbeak

Site 12.8: San Cristóbal de las Casas

(February 1982, 1983, 1984, 1985, November, December 1985, May 1986, January, February 1987, January 1992, October 1993)

Introduction/Key Species

San Cristóbal de Las Casas (2100 m elevation), or simply San Cristóbal, sits in a mostly cleared valley surrounded by mountains covered with pine-oak forest. The colonial architecture, cool mountain climate, and abundance of local crafts have combined to make San Cristóbal a popular tourist destination. Market days in particular are a colorful affair, when Indians from numerous surrounding villages gather in town and the close cultural ties with Guatemala are readily apparent.

Note that photography of the indigenous people is potentially dangerous (at best you could have your camera taken and opened), so always respect the local views and ask first, or simply don't take a camera. When birding in the area, be aware of this situation, and make sure that your binoculars are not mistaken for a camera. I have always found the local people to be friendly, if initially suspicious of outsiders, and have not had any problems, but note that recent civil unrest may have altered this situation (see the cautionary note in the chapter introduction).

Several good birding sites are within 30–45 minutes of San Cristóbal, and birds possible here include **Ocellated Quail**, **Bearded Screech-Owl**, Unspotted Saw-whet Owl, **Garnet-throated Hummingbird**, **Mountain Trogon**, **Blue-throated Motmot**, **Black-capped Swallow**, **Black-throated Jay**, **Rufous-browed Wren**, **Rufous-collared Thrush**, **Blue-and-white Mockingbird**, Golden-cheeked (migrant) and **Pink-headed warblers**, and **White-naped Brushfinch**.

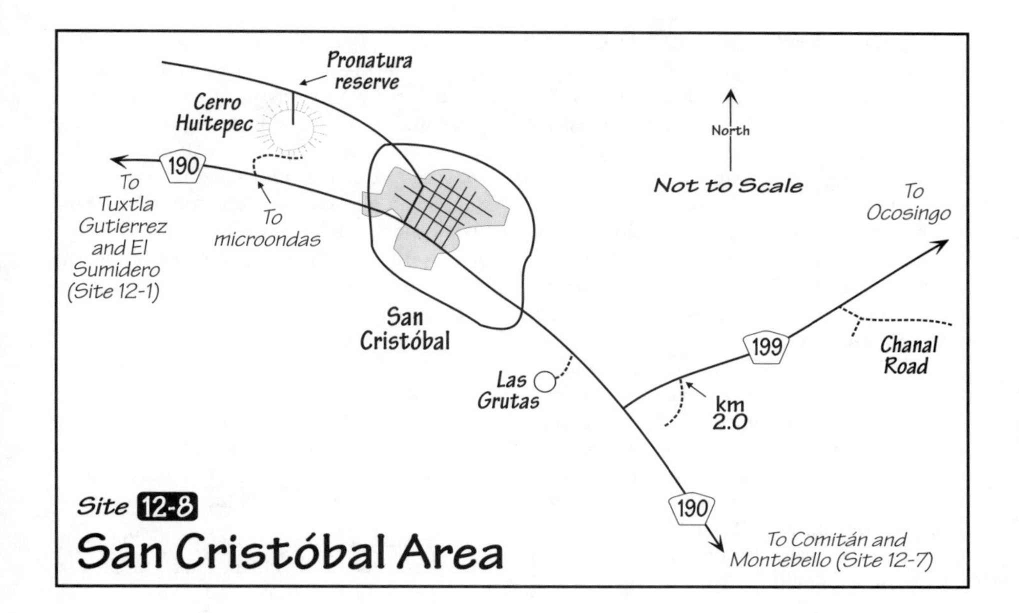

Access

San Cristóbal is 83 km (allow 1.5–2 hours because of winding roads and slow traffic) east of Tuxtla Gutiérrez via Route 190. There is also a *periferico*, so if you're in transit (e.g., from Tuxtla straight to Montebello), you can avoid going through the edge of San Cristóbal (although this is unlikely to save you much time). A wide range of accommodations and restaurants, to cover all budgets, is available, and the town is well served by first-class and second-class buses.

San Cristóbal is also about 190 km south of Palenque (Site 13.1) via Route 199, the Ocosingo Highway. Longtime visitors to Mexico may remember this as an all-day drive, or a grueling twelve-hour bus ride, but the road is now in much better shape (all of it is paved, give or take an occasional landslide). A non-stop drive to San Cristóbal from Palenque can take 4–5 hours, but you're likely to stop and bird along the way. Also be aware that, given recent political unrest, the Ocosingo Highway can be closed by roadblocks and/or dotted liberally with military checkpoints.

Birding Sites

Gardens in and around town have **Rufous-collared Thrush** and Rufous-collared Sparrow; Cave Swallows nest around the *zócalo*, **Black-capped Swallows** often can be seen over town, and Lesser Nighthawks (winter) often hunt over the *zócalo* in the evenings. During spring migration, at least, the *zócalo* can act as a magnet for flycatchers and warblers which get grounded by overnight rains: for instance, in May 1986, Yellow-bellied, Acadian, Least, and Hammond's flycatchers and Mourning and MacGillivray's warblers could be watched side by side as children screamed by! Calzada Mexico (the main road through the south side of town) passes through some wet grassy areas as well as beside a small dammed pond (south of the road) just west of town (Pied-billed Grebe, Virginia Rail).

One of the best-known birding sites near town is **Cerro Huitepec**, a largely forested hill overlooking San Cristóbal from the northwest. There are two points of access to the hill: one is the 135-hectare (350-acre) nature reserve, run under the auspices of PRONATURA-Chiapas; the other is via the microwave tower road.

To reach the **Cerro Huitepec reserve**, take the minor, paved road toward the village of Chamula (a popular tourist destination), which heads out northwest of San Cristóbal, off the *periferico*. A few kilometers (3.5?)

along the Chamula road look for a sign indicating the reserve on your left (south), which is a steep, forested, small peak (*cerro*). Second-class buses to Chamula can drop you off at the gate. A short entrance road goes to a few buildings, where an entrance fee should be paid and the trail system is explained. Some of the trails are fairly steep, and they can be muddy, but none is really long or difficult. The trails loop through humid pine-oak forest, where species include **Singing Quail**, **Bearded Screech-Owl**, Unspotted Saw-whet Owl (which has nested here in nest boxes), **Amethyst-throated** and **Garnet-throated hummingbirds**, **Blue-throated Motmot**, **Black-throated Jay**, **Rufous-browed Wren**, **Blue-and-white Mockingbird**, **Pink-headed Warbler**, and **White-naped Brushfinch**. Mixed-species warbler flocks often have Golden-cheeked Warbler (August to March).

The south side of Cerro Huitepec, outside the reserve, can be reached via the Cerro Huitepec microwave tower road, off Route 190 a few kilometers west of San Cristóbal. As you climb out of the valley of San Cristóbal, a few kilometers from the *periferico* intersection look for a gray-brick water storage tank on the right (north) side of the road; this tank looks rather like a small, flat-topped pyramid. At this point a cobbled road heads steeply uphill and may be signed to Microondas Cerro Huitepec. Follow this road steely uphill, through a village, and a few kilometers from the highway there is a stretch of humid pine-oak forest, just before you come to the edge of a large clearing and a second small village. It is possible to continue to the top of the microwave tower hill, through another patch of forest on steep slopes, but the road surface can be loose and should be approached with caution. Birds here are basically the same as for the adjacent Cerro Huitepec reserve.

There are also a couple of sites worth checking to the east of San Cristóbal. **Las Grutas** (The Caves) are a tourist attraction (now a municipal park) off Route 190 shortly before the junction with Route 199, the Ocosingo Road, which leads over to the Atlantic Slope lowlands at Palenque. Look for a sign labeled *"Grutas"* pointing to a dirt road off to the right (south) of the highway several kilometers (8–9?) east of downtown (and ? km before the Ocosingo Highway). This road goes a short distance to a parking area and some small buildings where an entrance fee may be charged; guides may be available to show you the caves. Numerous foot trails wend along the slopes of the ridge to the northwest, and mixed-species flocks in the more open pine woods here often include Strong-billed Woodcreeper, **Unicolored Jay**, and, in autumn-winter, Golden-cheeked Warbler. A wide trail heads steeply up the ridge, behind the *grutas*, and gets into humid pine-oak forest with species much like those at Cerro Huitepec. Second-class buses heading to towns southeast of San Cristóbal can drop you off at the entrance to Las Grutas.

The side road to Chanal is another spot that has been productive. This graded dirt road heads east off Route 199, 11 km from its junction with Route 190, near the ridge top, and can be reached easily by second-class buses heading to Ocosingo. It is about 30 km to Chanal through small settlements, cleared and grazed areas, cornfields, and patches of humid pine-oak forest. The road to Chanal starts at the highway and more or less immediately goes into a small village: when you come to a T-junction, less than 1 km from the highway, turn left; this is the Chanal road (right goes into the village). Species here are similar to those at Cerro Huitepec (including **Black-throated Jay**), plus **Ocellated Quail**, which has been seen at Km 23 and Km 29: cognizantly sharing time and space with this elusive species seems to be a matter of great luck.

Another site with most of the suite of "San Cristóbal birds" is at **Km 2 on the Ocosingo Road**, that is, about 2 km from the junction with Route 190. At this point look for a narrow dirt road to the right, which cuts sharply back to the right and can be missed (it is often best found by looking for a pile of garbage). There is room to pull off the highway and park safely. Birding can be good along the "road" for 1–2 km and along side trails up into the woods (**Singing Quail**, Whiskered and **Bearded screech-owls**, Unspotted Saw-whet Owl, **Blue-throated Motmot**, **Pink-headed Warbler**, etc.).

San Cristóbal Area Bird List

119 sp. R: rare; V: vagrant (not to be expected). †See pp. 5–7 for notes on English names.

Pied-billed Grebe
Great Blue Heron
Great Egret
Little Blue Heron
Cattle Egret
Green Heron
Black Vulture
Turkey Vulture
Red-tailed Hawk
American Kestrel
Singing Quail
Common Moorhen
American Coot
Band-tailed Pigeon
Mourning Dove
Inca Dove
Whiskered Screech-Owl
Bearded Screech-Owl
Mountain Pygmy-Owl†
Unspotted Saw-whet Owl
Lesser Nighthawk
Mexican Whip-poor-will†
White-collared Swift
White-throated Swift
Green Violet-ear
White-eared Hummingbird
Amethyst-throated Hummingbird
Garnet-throated Hummingbird
Magnificent Hummingbird
Mountain Trogon
Blue-throated Motmot
Acorn Woodpecker
Yellow-bellied Sapsucker
Hairy Woodpecker
Guatemalan Flicker†
Strong-billed Woodcreeper
Spot-crowned Woodcreeper
Common Tufted Flycatcher†
Greater Pewee
Yellow-bellied Flycatcher
Acadian Flycatcher
Least Flycatcher
Hammond's Flycatcher
Pine Flycatcher
Buff-breasted Flycatcher
Say's Phoebe (R)
Martin sp.
Tree Swallow
Violet-green Swallow
Black-capped Swallow
Bank Swallow
Cliff Swallow
Cave Swallow
Steller's Jay
Unicolored Jay
Bushtit
Brown Creeper
Band-backed Wren
Southern House Wren†
Rufous-browed Wren
Grey-breasted Wood-Wren
Eastern Bluebird
Brown-backed Solitaire
Ruddy-capped N.-Thrush
Hermit Thrush
Rufous-collared Thrush†
Tropical Mockingbird
Blue-and-white Mockingbird
Cedar Waxwing
Grey Silky†
Solitary Vireo ssp.
Hutton's Vireo
Nashville Warbler
Crescent-chested Warbler
Magnolia Warbler
Black-throated Blue Warbler (V)
Audubon's Warbler†
Townsend's Warbler
Hermit Warbler
Black-throated Green Warbler
Golden-cheeked Warbler
Blackburnian Warbler
Black-and-white Warbler
Ovenbird
Northern Waterthrush
Mourning Warbler
MacGillivray's Warbler
Common Yellowthroat
Wilson's Warbler
Canada Warbler
Red-faced Warbler
Pink-headed Warbler
Painted Whitestart†
Slate-throated Whitestart†
Rufous-capped Warbler
Golden-browed Warbler
Olive Warbler
Blue-hooded Euphonia
Hepatic Tanager
Common Bush-Tanager
Rose-breasted Grosbeak
White-naped Brushfinch
Spotted Towhee
Cinnamon-bellied Flowerpiercer
Rusty Sparrow
Chipping Sparrow
Lincoln's Sparrow
Rufous-collared Sparrow
Chiapas Junco†
Eastern Meadowlark
Red-winged Blackbird
Great-tailed Grackle
Bronzed Cowbird
Yellow-backed Oriole
Baltimore Oriole
Bullock's Oriole
House Finch
Red Crossbill
Black-headed Siskin
Lesser Goldfinch

Site 12.9: El Triunfo

(May 1988, April 1992; checked April 1996)

Introduction/Key Species

This site, in the Sierra Madre de Chiapas, is famous for the endangered **Horned Guan**, along with a host of other sought-after species, including **Highland Guan**, Barred Parakeet, **Fulvous Owl**, **Emerald-chinned** and **Wine-throated hummingbirds**, Resplendent Quetzal, **Blue-throated Motmot**,

Tawny-throated Leaftosser, Scaled Antpitta, **Black-throated Jay**, Spotted Nightingale-Thrush, **Black** and Mountain **thrushes**, **Cabanis' Tanager**, **Blue-crowned Chlorophonia**, White-eared Ground-Sparrow, and Yellow and **Hooded grosbeaks**.

El Triunfo is a wardened reserve administered by the Instituto de Historia Natural (IHN), with offices in Tuxtla Gutiérrez and Jaltenango. Permission is needed to visit the reserve, and a flat fee of $100.00 U.S. per person per visit (of any length) is payable to the IHN for camping or staying at the dormitory in the reserve.

It is possible to reach El Triunfo from the Pacific Slope of Chiapas, but this generally involves an overnight hike from near sea level to 2000 m elevation. Access from the interior, via Jaltenango and Finca Prusia, is far easier, although this still involves a hike of about 8 km (3–5 hours without stopping much to bird) and an overall climb of about 1000 m. Camping is allowed in the reserve, and there is a basic dormitory that works on a first-come, first-served basis unless you have made reservations to stay there. The reserve employs two or three wardens who know the birds well, and where to find them. If you decide to hire a guide, ask for Rafael Solis Galvez or his cousins Ismael and Enelfo Galvez Galvez. The best season to visit El Triunfo is March-April (when Horned Guans are most vocal). February and May are also good. At other seasons, bird song is reduced, weather tends to be rainier, and roads and trails can be in bad shape.

The IHN address is: Instituto de Historia Natural (El Triunfo), Calzada de los Hombres de la Revolución s/n, Apartado Postal #6, Tuxtla Gutiérrez, Chiapas, 29000 Phone: (011-52) 961-23663, 961-23754. Fax: 961-29943.

Access

The best way to visit El Triunfo is to contact the IHN at their offices in Tuxtla Gutiérrez (address above), ask permission, pay the $100 fee, and, if you wish, enlist optional extra services. The IHN can provide you with transport from Tuxtla or Jaltenango to the start of the El Triunfo trail, as well as food, guide, cook, and mules for transport. Prices in 1996 were about $15/day for food, $10/day each for cook and guide, and about $10/day for mules. Unless you've been to El Triunfo before and have a good memory, you'll need guidance to find the start of the trail, and mules are very useful at least for the walk in (when you tend to be laden with supplies), allowing you more freedom for birding along the trail. If you want to do this "alone," you'll need a high-clearance vehicle and all your camping gear and food.

Getting to the El Triunfo Trail: From Tuxtla Gutiérrez, head east on Route 190 (toward San Cristóbal). About 7.5 km from the center of Tuxtla you come to the Grijalva River bridge; just before this bridge, look for a paved road to the right (also signed Route 190, confusingly), which should have a sign for Angostura and/or La Concordia. Turn right here and mark zero at this junction. At Km 17 (a Pemex station and the village of America Libre), the road to Angostura and Venustiano Carranza goes off to the left; stay right here on what appears to be the main road, which then winds along through scorched, cutover hills, passing the small town of Revolución Mexicana (and the road to Villa Albino Corso and Villa Flores) off to the right at about Km 80. At Km 113 you come to the village of Independencia, where you should take the right-hand road fork, likely signed to Jaltenango (the left fork goes to La Concordia). Continue about 12.5 km, through Benito Juarez, to a left-hand fork for Jaltenango (right goes to La Tigrilla), and from here it is 22 km to Jaltenango. The total distance from Tuxtla Gutiérrez to Jaltenango is about 160 km, and the drive takes 2.5–4 hours, depending on road conditions and allowing for a couple of brief birding and stretching stops. In 1996 the road had many potholes and numerous unpaved but graded dirt stretches but presented no problems to a normal car. Birds that can be seen along the drive to Jaltenango include Northern Bobwhite, **Green Parakeet**, White-crowned Parrot, Striped Cuckoo, Lesser Ground-Cuckoo, **Lesser Roadrunner**, Ferruginous Pygmy-Owl, **Russet-crowned Motmot**, Fork-tailed Flycatcher, White-throated Magpie-Jay, Banded and Plain wrens, Rufous-browed Peppershrike, Rusty Sparrow, and **Yellow-winged Cacique**.

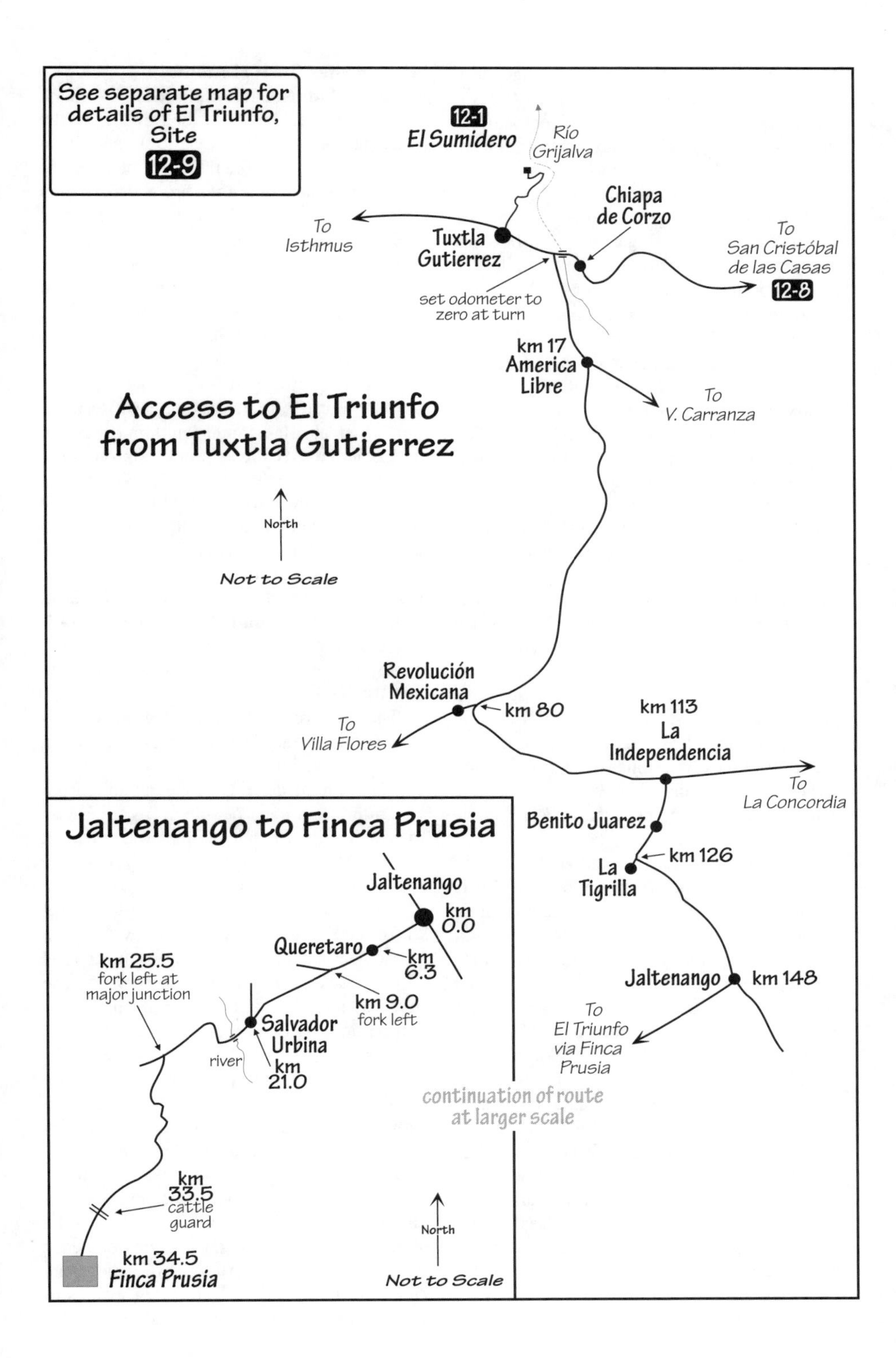
See separate map for details of El Triunfo, Site
12-9
12-1
El Sumidero
Río Grijalva
Chiapa de Corzo
To Isthmus
Tuxtla Gutierrez
To San Cristóbal de las Casas
12-8
set odometer to zero at turn
km 17
America Libre
To V. Carranza
Access to El Triunfo from Tuxtla Gutierrez
North
Not to Scale
Revolución Mexicana
km 80
To Villa Flores
km 113
La Independencia
To La Concordia
Benito Juarez
km 126
La Tigrilla
Jaltenango
km 148
To El Triunfo via Finca Prusia
continuation of route at larger scale
Jaltenango to Finca Prusia
Jaltenango
km 0.0
Queretaro
km 6.3
km 9.0
fork left
km 25.5
fork left at major junction
Salvador Urbina
river
km 21.0
km 33.5
cattle guard
km 34.5
Finca Prusia
North
Not to Scale

Note that Jaltenango (or Jaltenango de la Paz) has had its name changed many times in the last ten or so years; the other common name for it is Angel Albino Corso (not to be confused with Villa Albino Corso, reached from Revolución Mexicana).

Pending drastic improvements, the road from Jaltenango to Finca Prusia is best undertaken in a high-clearance vehicle, although it has been done in a VW Rabbit. Finding the road out of Jaltenango to Finca Prusia isn't intuitive, and you will need to ask. Most locals know of the El Triunfo reserve and Finca Prusia, but if you draw blank looks, ask for the road to Querétaro, the first village en route to Prusia. From Jaltenango to Prusia is about 32 km and takes 1.5–2 hours in a pickup truck. Basically you follow the main "road," bearing left at any forks. If you have made it this far (rather than left your vehicle in Jaltenango and arranged for IHN logistical support), you will need to find somewhere to leave your vehicle in Finca Prusia or Santa Rita, a small village 3 km beyond Prusia. Recent social unrest has left relations between Santa Rita and the El Triunfo reserve rather strained, and it may be best to leave your vehicle at Prusia, although this adds about 3 km to your walk. With luck, relations will improve in the near future—check when you go. If not, at least birding along the extra 3 km can be good!

In order to find the El Triunfo trail, you will need to get guidance from someone at Prusia or Santa Rita, as the maze of trails running through the coffee plantations has good potential for getting you well and truly lost. Once you're on the El Triunfo trail, it is about 8 km to the park headquarters and campground, set in a clearing surrounded by cloud forest. Initially the trail leads through coffee *finca* (plantation) and tropical broadleaf forest (Ornate Hawk-Eagle, Pheasant Cuckoo, Tody Motmot, **Chestnut-sided** and Green **shrike-vireos**, **Cabanis' Tanager**, White-eared Ground-Sparrow) and then enters cloud forest (**Blue-throated Motmot**, Ruddy Foliage-gleaner, Tawny-throated Leaftosser, Spotted Nightingale-Thrush, **Blue-crowned Chlorophonia**).

Birding Sites (from El Triunfo camp, including Cañada Honda)

Various trails lead into the forest from the clearing (at 1900 m elevation), and local sketch maps are available from the wardens at the park headquarters. From Prusia, you enter the El Triunfo clearing from the north, and the other main trails lead out of the clearing northwest (to Palo Gordo) and south (to Cañada Honda). Basically, any trail is good for birds: possibly the best for Horned Guan are the Palo Gordo and Prusia trails, but the wardens will probably have up-to-date information. Birding in and around the clearing is good: fruiting trees can have Resplendent Quetzal, **Black** and Mountain **thrushes**, and **Hooded Grosbeak**; thickets around the campsite have **Wine-throated Hummingbird**, **Blue-and-white Mockingbird**, and Yellow-throated Brushfinch; and keep an eye/ear out for Barred Parakeets overhead. Species seen regularly along the forest trails near the clearing include **Highland Guan**, **Singing Quail**, **White-faced Quail-Dove**, **Fulvous Owl**, Ruddy Foliage-gleaner, Tawny-throated Leaftosser, Scaled Antpitta, and **Black-throated Jay**.

The most reliable site for **Cabanis' Tanager** is on the Pacific Slope at Cañada Honda (1450 m elevation), in tropical evergreen forest. This can be done as a long day hike from El Triunfo, but most people make at least an overnight trip of it and camp by the stream at Cañada Honda. This is the trail you would take if coming to El Triunfo from the Pacific Slope lowlands; it's well-worn by local foot traffic and is pretty easy to follow without fear of getting lost (any apparent forks tend to join back shortly to the main trail). The hike down to Cañada Honda from El Triunfo can be done in two hours. The return trip is more strenuous and, depending on your fitness, can take 2.5–4 hours. From El Triunfo, after climbing up and out of the valley, you drop through pine-cypress woods and then into tropical forest. Cañada Honda is the second obvious stream crossing below the pine-cypress woods, and there is a well-marked campsite on the right, just after you cross the stream. **Cabanis' Tanager** is often quite easy to find right here and from the trail along the first 250 m beyond the campsite. Species to look for around and below Cañada Honda in-

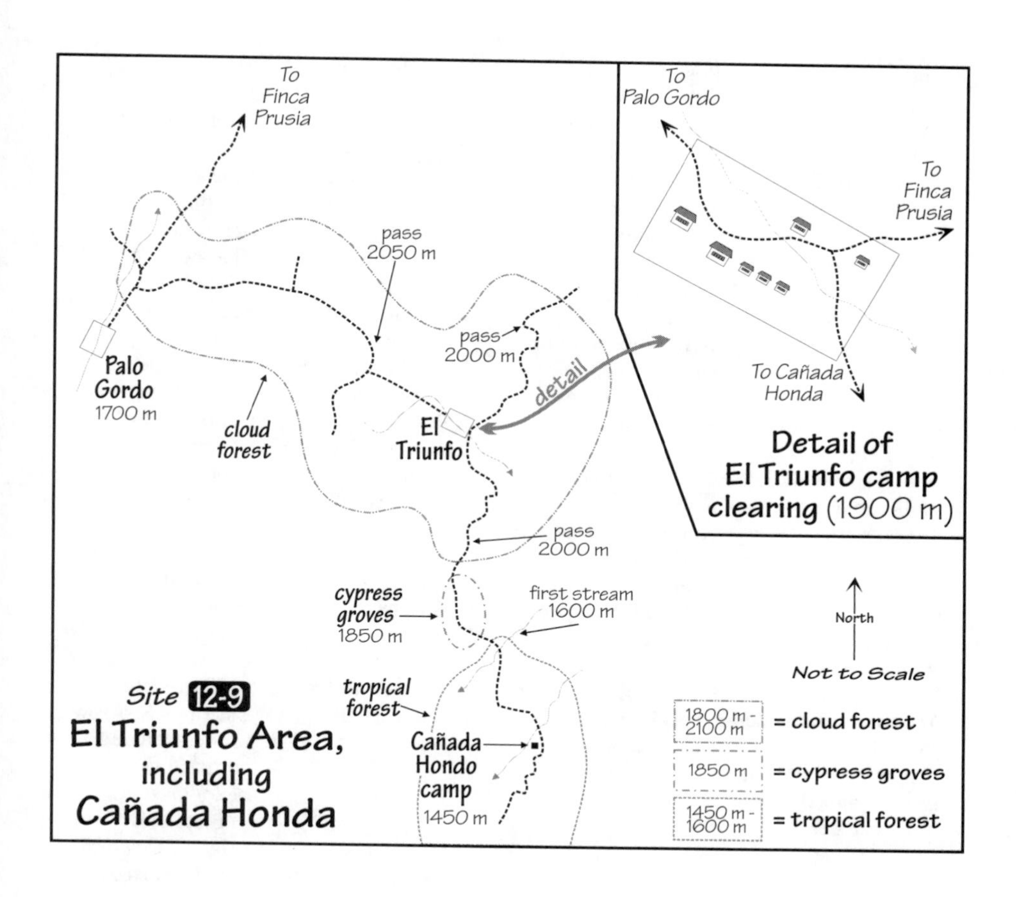

clude **Rufous Sabrewing** (en route from El Triunfo), **Emerald-chinned Hummingbird**, Long-tailed Manakin, Rufous-and-white Wren, and all of the species listed for the broadleaf forest above Prusia.

Finca Prusia to El Triunfo Bird List (including Cañada Honda)

132 sp. †See pp. 5–7 for notes on English names.

Thicket Tinamou
Black Vulture
Turkey Vulture
White-breasted Hawk†
Red-tailed Hawk
Ornate Hawk-Eagle
Barred Forest-Falcon
Highland Guan
Horned Guan
Singing Quail
Band-tailed Pigeon
White-winged Dove
Inca Dove
White-tipped Dove
White-faced Quail-Dove
Green/Pacific Parakeet
Barred Parakeet
Pheasant Cuckoo
Groove-billed Ani
Mottled Owl
Fulvous Owl
Mexican Whip-poor-will†
Chestnut-collared Swift
White-collared Swift
Vaux's Swift
Violet Sabrewing
Rufous Sabrewing
Emerald-chinned Hummingbird
White-eared Hummingbird
Berylline Hummingbird
Green-throated Mountain-gem
Amethyst-throated Hummingbird
Wine-throated Hummingbird
Violaceous Trogon
Mountain Trogon
Collared Trogon
Resplendent Quetzal
Tody Motmot
Blue-throated Motmot
Blue-crowned Motmot
Emerald Toucanet
Acorn Woodpecker
Hairy Woodpecker
Smoky-brown Woodpecker
Golden-olive Woodpecker
Scaly-throated Fol.-gleaner†

Ruddy Foliage-gleaner
Tawny-throated Leaftosser
Olivaceous Woodcreeper
Spotted Woodcreeper
Spot-crowned Woodcreeper
Scaled Antpitta
Paltry Tyrannulet
Greenish Elaenia
Eye-ringed Flatbill
Yellow-olive Flycatcher
Stub-tailed Spadebill
Common Tufted Flycatcher†
Olive-sided Flycatcher
Western Pewee†
Least Flycatcher
Yellowish Flycatcher
Black Phoebe
Bright-rumped Attila
Dusky-capped Flycatcher
Social Flycatcher
Boat-billed Flycatcher
Sulphur-bellied Flycatcher
Rose-throated Becard
Eastern Kingbird
Long-tailed Manakin
Black-capped Swallow
Bank Swallow
Cliff Swallow
Green Jay
Black-throated Jay
Unicolored Jay
Spot-breasted Wren
Rufous-and-white Wren
Southern House Wren†
Rufous-browed Wren
Grey-breasted Wood-Wren
Brown-backed Solitaire
Orange-billed N.-Thrush
Ruddy-capped N.-Thrush
Spotted N.-Thrush
Swainson's Thrush
Wood Thrush
Black Thrush†
Mountain Thrush†
Clay-colored Thrush†
White-throated Thrush†
Grey Catbird
Blue-and-white Mockingbird
Cedar Waxwing
Grey Silky†
Brown-capped Vireo
Yellow-green Vireo
Chestnut-sided Shrike-Vireo
Green Shrike-Vireo
Crescent-chested Warbler
Townsend's Warbler
Black-throated Green Warbler
Blackburnian Warbler
Black-and-white Warbler
MacGillivray's Warbler
Wilson's Warbler
Slate-throated Whitestart†
Golden-crowned Warbler
Golden-browed Warbler
Cabanis' Tanager†
Yellow-winged Tanager
Western Tanager
Flame-colored Tanager
White-winged Tanager
Blue-crowned Chlorophonia
Scrub Euphonia
Blue-hooded Euphonia
Red-legged Honeycreeper
Red-crowned Ant-Tanager
Common Bush-Tanager
Yellow Grosbeak
Yellow-throated Brushfinch
Chestnut-capped Brushfinch
White-eared Ground-Sparrow
Cinnamon-bellied Flowerpiercer
Lincoln's Sparrow
Rufous-collared Sparrow
Melodious Blackbird
Bronzed Cowbird
Yellow-backed Oriole
Hooded Grosbeak

CHAPTER 13 NORTHERN CHIAPAS

Black-throated Shrike-Tanager

The lowlands of northern Chiapas are home to an array of birds whose names come to the minds of birders when they hear the word "Neotropics"—tinamous, hawk-eagles, sungrebes, curassows, macaws, trogons, motmots, puffbirds, jacamars, toucans, woodcreepers, leaftossers, antbirds, cotingas, manakins, honeycreepers, tanagers, oropendolas. Conversely, relatively few endemics occur in this region.

Avian diversity in the rain forests of Chiapas is high by North American or Mexican standards but pales by comparison to southern Central America and South America. The good side of this situation is that, while the birding in Mexico is exciting, it isn't overwhelming, so you have a chance to figure out what you're seeing and to get a feel for a host of new families. Thus, time spent in southern Mexico can lay a good foundation for travels farther south, when there will be many more woodcreepers, antbirds, and flycatchers.

Fortunately for the visiting birder, there are still some areas of easily accessible rain forest around the famous Maya ruins of Palenque, as well as fairly extensive tracts of rain forest farther east, in the Selva Lacandona (often referred to simply as the Lacandon), such as around Bonampak and Yaxchilán, two more Maya sites. Note that trips to the Lacandon require camping, and for a first-time visitor, Palenque has plenty of birds.

It would be easy to make a two-week trip to this area; the best time is probably from February to April, in terms of both bird activity and road conditions into the Lacandon. One week (or longer) based in Palenque, at any time of year, also makes a good trip, and many birders return again and again to this rich area. Alternatively, Palenque could be combined with the Yucatan Peninsula (Chapter 14) for a very productive two-week trip, or with sites in Central and Southern Chiapas (Chapter 12) for at least a two-week trip. Given the unfortunate state of civil unrest in parts of Chiapas, however, it would be wise to find out the local situation before you consider adding the highlands around San Cristóbal (Chapter 12) to a birding itinerary.

If you're not driving to this area, the main point of access for Palenque and northern Chiapas is Villahermosa, some 150 km (1.5–2 hours) west of Palenque and the state capital of Tabasco. Villahermosa is served daily by several flights from Mexico. City and car rental is available at the airport with the usual suite of companies. An afternoon/evening birding option west of Villahermosa is Las Choapas, Veracruz (Site 10.8), a site for Spot-tailed Nightjar.

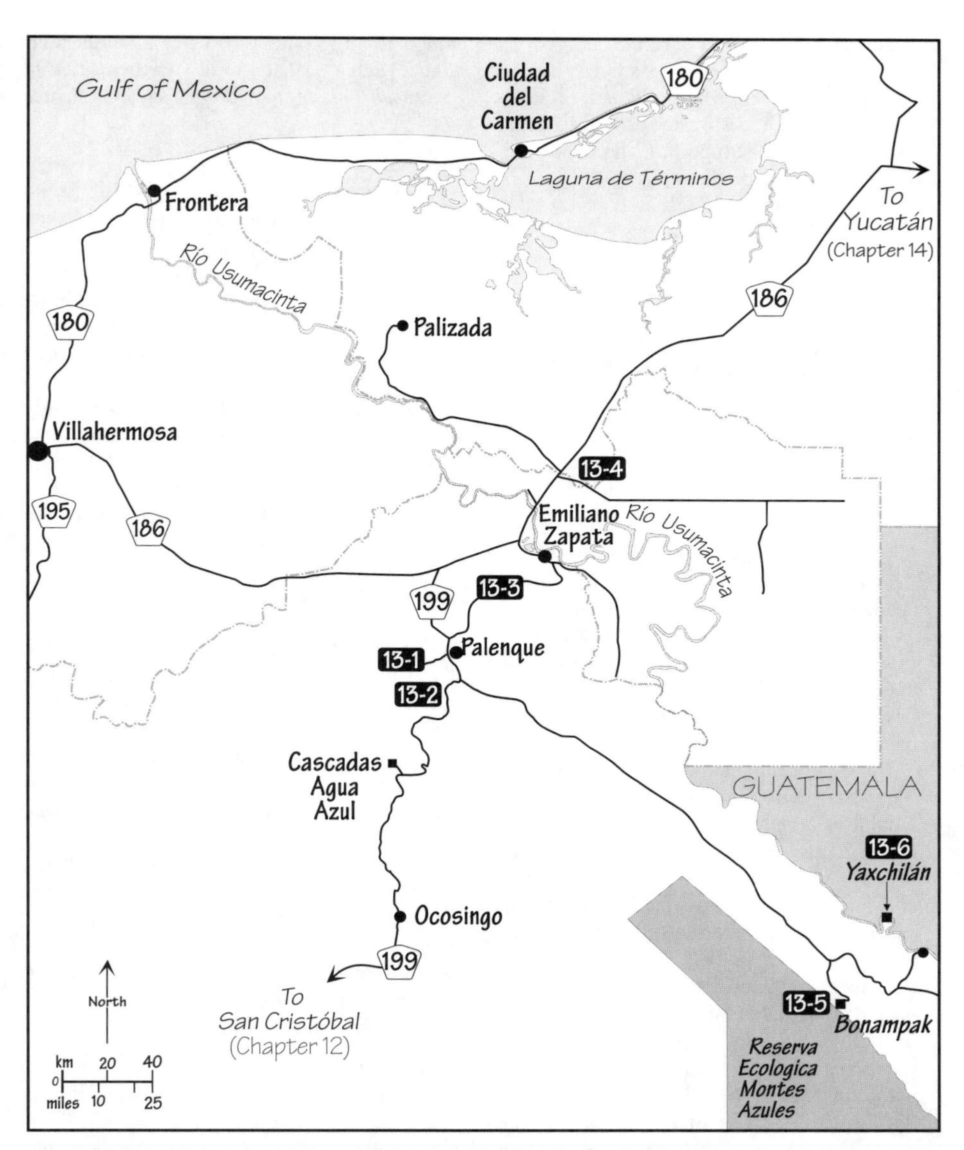

Chapter 13: NORTHERN CHIAPAS

- 13-1 Palenque, Chiapas
- 13-2 The San Manuel Road, Chiapas
- 13-3 The La Libertad Road, Chiapas
- 13-4 The Usumacinta Marshes, Chiapas / Campeche
- 13-5 Bonampak, Chiapas
- 13-6 Yaxchilán, Chiapas

NORTHERN CHIAPAS SITE LIST

Site 13.1 Palenque, Chiapas
Site 13.2 The San Manuel Road, Chiapas
Site 13.3 The La Libertad Road, Chiapas
Site 13.4 The Usumacinta Marshes, Chiapas/Campeche
Site 13.5 Bonampak, Chiapas
Site 13.6 Yaxchilán, Chiapas

Site 13.1: Palenque, Chiapas

(February 1982, 1983, January, February, September 1984, March, November 1985, May 1986, January, February, March 1987, July 1988, February 1989, April 1992, checked December 1996)

Introduction/Key Species

Palenque is one of the most popular birding sites in Mexico. The rain forest around the spectacular Maya ruins here offers a chance to see many species typical of Mexico's Atlantic Slope lowlands. Sadly, although it looks as if the forest is extensive back into the hills behind the ruins, this is now an isolated remnant of rain forest; birding remains good, however.

Species of interest here include Great and Slaty-breasted tinamous, Hook-billed and Double-toothed kites, White Hawk, Black Hawk-Eagle, Spotted Wood-Quail, Grey-headed and Grey-chested doves, Crested and Black-and-white owls, Central American Pygmy-Owl, **Wedge-tailed Sabrewing**, White-necked Jacobin, four trogons, three toucans, Tody Motmot, White-necked Puffbird, Rufous-tailed Jacamar, Great Antshrike, Scaled Antpitta, Royal and Sulphur-rumped flycatchers, White-collared and Red-capped manakins, Long-billed Gnatwren, Green Shrike-Vireo, **Black-throated Shrike-Tanager**, Golden-hooded, Scarlet-rumped, and Crimson-collared tanagers, Green Honeycreeper, and Orange-billed and **Green-backed sparrows**.

Many rain-forest species at Palenque are at the edge of their range and are commoner and more easily seen at Bonampak and Yaxchilán (Sites 13.5, 13.6). To make the most of a trip to either or both of these exciting sites, spending time at Palenque to become familiar with the commoner species is recommended. Nearby forest patches on the Ocosingo Road (Route 199) offer a similar avifauna (Site 13.2), and to the northeast lies an extensive area of marshes and savannas (Sites 13.3, 13.4).

Access

The town of Palenque is just off Route 199, about 26 km by road south of Route 186, the Villahermosa to Yucatan highway. Coming from Villahermosa, turn right (south) off Route 186 at the well-signed junction (actually a crossroads) for Palenque, Ocosingo, and so on. As you enter Palenque from the north, there is a traffic circle with a large white statue of a Maya head (commonly called "La Cabeza Maya"). At this point one can bear left into Palenque town or turn right and continue on Route 199 (the Ocosingo Road), which goes over the mountains to San Cristóbal de las Casas (Site 12.8). Palenque has a wide range of hotels, campgrounds, and restaurants, including several along the road to the ruins.

To get to the ruins from Palenque, take the Ocosingo Road from the Maya Head and in 300 m take the right-hand fork (which should be signed to the ruins). In about 8 km the road ends at the parking lot adjacent to Palenque ruins. One can walk from town to the ruins, or vice versa, birding in the fields, hedges, and second growth. This can be quite birdy, but there is little in the way of shade, and doing it once will probably be enough for most people. Public transport, in the form of *colectivo combis* (VW minibuses), runs regularly between town and the ruins, starting around 6 or 7 A.M.; this enables you to get there early, if not quite in time for dawn.

Birding Sites

The ruins are in the first low ridge of hills that rises from the wide coastal plain. Birding is often good in the early morning (before the armies of tour buses arrive) from the road for the last kilometer or so as it winds up through forest to the parking lot (trogons, White-necked Puffbird, Rufous-tailed Jacamar, toucans, Great Antshrike, flycatchers, tanagers, orioles, oropendolas). Right around the edge of the parking lot can also be good in the early morning (Grey-headed Dove, Streak-headed and other woodcreepers). At least two narrow trails lead into the forest along this last kilometer of road: one to the

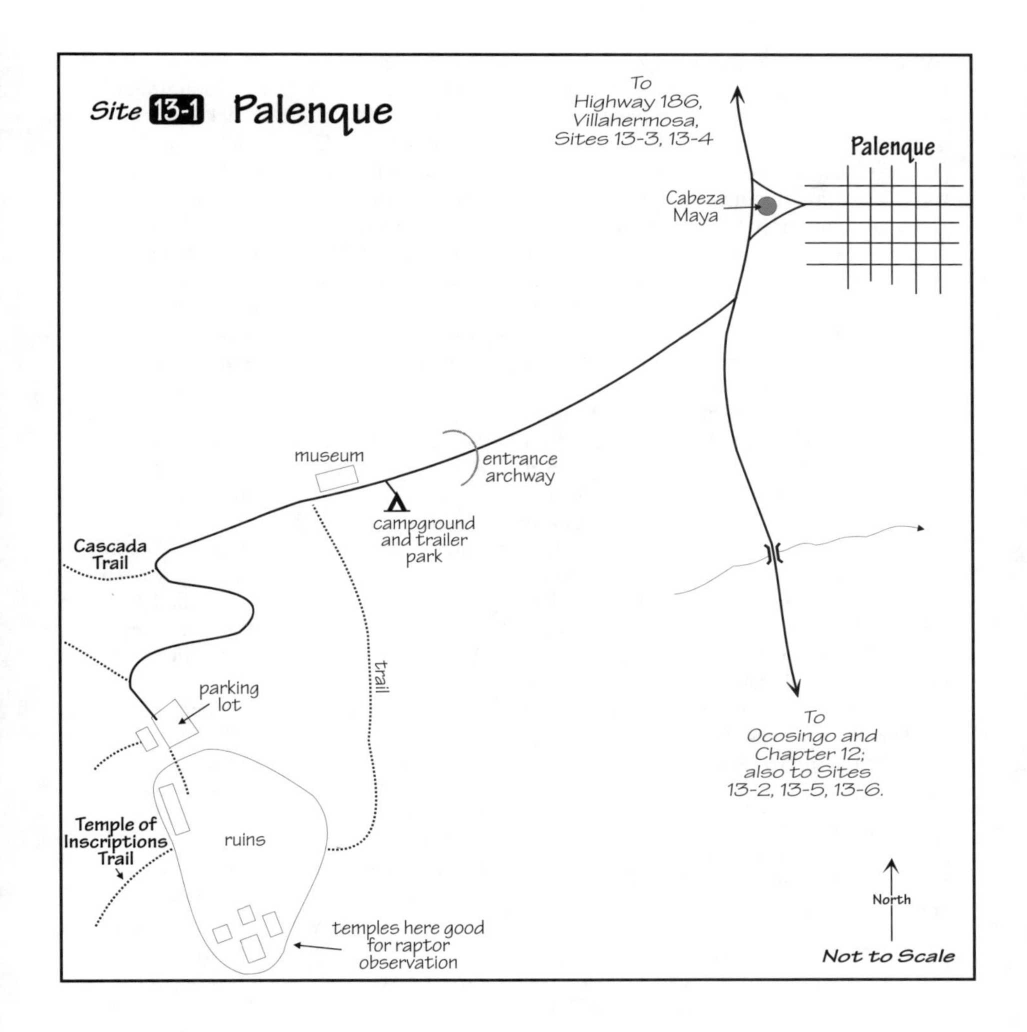

right at the first sharp left-hand bend, known as the *cascada* (waterfall) trail (Grey-chested Dove, Scaled Antpitta, Orange-billed Sparrow), and another to the right just before the parking lot and administrative buildings, at the last road bend (Pygmy Kingfisher, White-collared Manakin).

The ruins are open to the public from 8 A.M. to 5 P.M. and cost 16 pesos/person (December 1996; free on Sundays and holidays); tripods are not allowed into the ruins (at least not without obtaining prior permission). Birding along the forest edge from the clearing of the ruins can be good in early morning, or throughout the day if there are fruiting and/or flowering trees, and climbing some of the temples and other structures gives good views into the forest canopy. As you enter the ruins, the first large structure on the right is the Temple of the Inscriptions, and just past it on the right is a trail that leads steeply up into the forest, at first following the stream course. This is famous among birders as the Temple of Inscriptions Trail, and it is quite easy to spend all day on this trail, which provides good access to the forest interior (tinamous, Spotted Wood-Quail, trogons, Tody Motmot, Scaled Antpitta, Royal Flycatcher, and mixed-species flocks). The trail eventually leads to a Maya village, and local foot traffic can be quite heavy at times. If you go far enough, you will come to barbed wire fences and a depressing view over extensive pastures, showing that the Palenque forest is an isolated patch (hence, Crested Guan and Great Curassow have long been extirpated). Other trails lead off into the forest at various points, including one near the small museum, at the "back left" corner from the entrance, which leads along the stream and eventually comes out at the road below the ruins near the new museum (formerly the Hotel Las Ruinas).

The temples also offer good points from which to watch for raptors. Any one of the two or three main structures on the hillsides at the "back right" of the ruins (as seen from the entrance) offers an excellent view as well as some shade. Starting at 8 A.M., it can be good for raptors, and the best window of time tends to be 9:30 to 11:30 A.M. Regularly seen species include Hook-billed, Double-toothed, and Plumbeous (summer) kites, White, Common Black, and Great Black hawks, Black Hawk-Eagle, and Bat Falcon. During spring and autumn, one can also see spectacular kettles of Broad-winged Hawks and smaller numbers of other migrating raptors, including Swallow-tailed and Mississippi kites.

Palenque Bird List

(ruins and the road from town): 253 sp. R: rare (not to be expected). †See pp. 5–7 for notes on English names.

Great Tinamou
Slaty-breasted Tinamou
Cattle Egret
Green Heron
Yellow-crowned Night-Heron
Black Vulture
Turkey Vulture
King Vulture (R)
Hook-billed Kite
White-tailed Kite
Double-toothed Kite
Plumbeous Kite
Sharp-shinned Hawk
Cooper's Hawk
White Hawk
Common Black Hawk
Great Black Hawk
Grey Hawk
Roadside Hawk
Broad-winged Hawk
Short-tailed Hawk
Red-tailed Hawk
Black Hawk-Eagle
Laughing Falcon
American Kestrel
Bat Falcon
Peregrine Falcon
Barred Forest-Falcon
Collared Forest-Falcon
Plain Chachalaca
Spotted Wood-Quail
Northern Bobwhite
Ruddy Crake
Killdeer
Northern Jacana
Pale-vented Pigeon
Red-billed Pigeon
Short-billed Pigeon
Ruddy Ground-Dove
Blue Ground-Dove
White-tipped Dove
Grey-headed Dove†
Grey-chested Dove
Ruddy Quail-Dove
Aztec Parakeet†
Brown-hooded Parrot
White-crowned Parrot
White-fronted Parrot
Red-lored Parrot
Yellow-billed Cuckoo
Squirrel Cuckoo
Striped Cuckoo
Groove-billed Ani
Barn Owl
Crested Owl
Mottled Owl
Black-and-white Owl
Central American Pygmy-Owl†
Ferruginous Pygmy-Owl
Lesser Nighthawk
Pauraque
White-collared Swift
Vaux's Swift
Lesser Swallow-tailed Swift
Long-tailed Hermit
Little Hermit
Wedge-tailed Sabrewing
Violet Sabrewing
White-necked Jacobin
Green-breasted Mango
White-bellied Emerald
Rufous-tailed Hummingbird
Stripe-tailed Hummingbird
Long-billed Starthroat
Ruby-throated Hummingbird
Black-headed Trogon
Violaceous Trogon
Collared Trogon
Slaty-tailed Trogon
Tody Motmot
Blue-crowned Motmot
Ringed Kingfisher
Belted Kingfisher
Amazon Kingfisher
Green Kingfisher
Pygmy Kingfisher†
White-necked Puffbird
Rufous-tailed Jacamar
Emerald Toucanet
Collared Aracari
Keel-billed Toucan
Black-cheeked Woodpecker
Golden-fronted Woodpecker
Yellow-bellied Sapsucker
Smoky-brown Woodpecker
Golden-olive Woodpecker
Chestnut-colored Woodpecker
Lineated Woodpecker
Pale-billed Woodpecker
Rufous-breasted Spinetail
Buff-throated Fol.-gleaner
Plain Xenops
Tawny-winged Woodcreeper
Olivaceous Woodcreeper
Wedge-billed Woodcreeper
Barred Woodcreeper
Ivory-billed Woodcreeper
Streak-headed Woodcreeper

Great Antshrike
Barred Antshrike
Dot-winged Antwren
Dusky Antbird
Mexican Antthrush†
Scaled Antpitta
Yellow-bellied Tyrannulet
Greenish Elaenia
Yellow-bellied Elaenia
Ochre-bellied Flycatcher
Sepia-capped Flycatcher
Northern Bentbill
Slate-headed Tody-Flycatcher
Common Tody-Flycatcher
Eye-ringed Flatbill
Yellow-olive Flycatcher
Stub-tailed Spadebill
Royal Flycatcher
Sulphur-rumped Flycatcher
Eastern Pewee†
Tropical Pewee
Yellow-bellied Flycatcher
Acadian Flycatcher
Alder Flycatcher
Least Flycatcher
Vermilion Flycatcher
Bright-rumped Attila
Rufous Mourner
Dusky-capped Flycatcher
Great Crested Flycatcher
Brown-crested Flycatcher
Great Kiskadee
Boat-billed Flycatcher
Social Flycatcher
Streaked Flycatcher
Sulphur-bellied Flycatcher
Piratic Flycatcher
Tropical Kingbird
Couch's Kingbird
Eastern Kingbird
Thrushlike Mourner†
Grey-collared Becard
Rose-throated Becard
Masked Tityra
Black-crowned Tityra
Rufous Piha
Lovely Cotinga
White-collared Manakin
Red-capped Manakin
Grey-breasted Martin
N. Rough-winged Swallow
Brown Jay
Band-backed Wren
Spot-breasted Wren
Southern House Wren†
White-bellied Wren
White-breasted Wood-Wren
Long-billed Gnatwren
Blue-grey Gnatcatcher
Tropical Gnatcatcher
Veery
Swainson's Thrush
Wood Thrush
Clay-colored Thrush†
Grey Catbird
White-eyed Vireo
Blue-headed Vireo†
Yellow-throated Vireo
Philadelphia Vireo
Red-eyed Vireo
Yellow-green Vireo
Tawny-crowned Greenlet
Lesser Greenlet
Green Shrike-Vireo
Blue-winged Warbler
Tennessee Warbler
Yellow Warbler
Chestnut-sided Warbler
Magnolia Warbler
Myrtle Warbler†
Black-throated Green Warbler
Blackburnian Warbler
Yellow-throated Warbler
Black-and-white Warbler
American Redstart
Worm-eating Warbler
Ovenbird
Northern Waterthrush
Louisiana Waterthrush
Kentucky Warbler
Mourning Warbler
Common Yellowthroat
Grey-crowned Yellowthroat
Hooded Warbler
Wilson's Warbler
Canada Warbler
Golden-crowned Warbler
Yellow-breasted Chat
Bananaquit
Golden-hooded Tanager†
Green Honeycreeper
Red-legged Honeycreeper
Scrub Euphonia
Yellow-throated Euphonia
Olive-backed Euphonia
Blue-grey Tanager
Yellow-winged Tanager
Grey-headed Tanager
Black-throated Shrike-Tanager
Red-crowned Ant-Tanager
Red-throated Ant-Tanager
Summer Tanager
Western Tanager
Crimson-collared Tanager
Scarlet-rumped Tanager
Greyish Saltator
Buff-throated Saltator
Black-headed Saltator
Black-faced Grosbeak
Rose-breasted Grosbeak
Blue-black Grosbeak
Blue Grosbeak
Indigo Bunting
Painted Bunting
Dickcissel
Orange-billed Sparrow
Green-backed Sparrow
Blue-black Grassquit
White-collared Seedeater
Variable Seedeater
Yellow-faced Grassquit
Rusty Sparrow
Lincoln's Sparrow
Eastern Meadowlark
Melodious Blackbird
Great-tailed Grackle
Bronzed Cowbird
Giant Cowbird
Black-cowled Oriole
Orchard Oriole
Yellow-tailed Oriole
Altamira Oriole
Baltimore Oriole
Yellow-billed Cacique
Chestnut-headed Oropendola
Montezuma Oropendola

Site 13.2: The San Manuel Road, Chiapas

(February, March 1987, July 1988, February 1989, checked December 1996)

Introduction/Key Species

This site, along the Ocosingo Road, offers a good overlook into the forest canopy as well as some forest edge and second-growth birding. Species overall are similar to Palenque, but San Manuel seems to be a better site for Lovely Cotinga, and it's a good place from which to watch for raptors in mid to late morning.

Access/Birding Sites/Bird List

From Palenque (mark zero at the Maya Head) take the Ocosingo Road (Route 199) up into the foothills for 12.6 km, at which point there is a dirt side road down to the right (west) which may be signed to San Manuel. Along the highway about 300 m past this turn is a pulloff on the left, and just past this on the right is a low concrete wall and an overlook across a valley with emergent trees. This is just before Km Post 40. It should be possible to pull off here, but if not there is plenty of room to park at the San Manuel junction and walk to the overlook. Check the treetops from this vantage point for Lovely Cotinga. Birding from the highway here can be good, traffic permitting, and the first kilometer or so of the road down to San Manuel also offers good birding. At least during February to April, flowering trees (especially *Inga*) can be good for hummingbirds, including White-necked Jacobin and Black-crested Coquette. Other birds here include Little Tinamou, White Hawk, Black Hawk-Eagle, Barred Forest-Falcon, Brown-hooded Parrot, Emerald Toucanet, Paltry Tyrannulet, White-collared Manakin, Green Shrike-Vireo, Red-legged and Green honeycreepers, and many other forest and edge species listed for Palenque.

This site can be reached easily using second-class buses traveling between Palenque and Ocosingo.

Site 13.3: The La Libertad Road, Chiapas

(November 1985, February 1987, February 1989, April 1992, checked December 1996)

Introduction/Key Species

This paved (if potholed) road through fields and savannas just northeast of Palenque is a good area for Aplomado Falcon, Double-striped Thick-knee, Plain-breasted Ground-Dove, Acorn Woodpecker, Fork-tailed Flycatcher, Grassland Yellow-Finch, and Botteri's and Grasshopper sparrows. There are also some marshes just north of La Libertad, along the road to Emiliano Zapata, where one can find Boat-billed Heron, Black-collared Hawk, Sungrebe, and Grey-necked Wood-Rail; many other species listed for the Usumacinta Marshes (Site 13.4) can also be seen. The La Libertad Road makes a good afternoon trip from Palenque, when birding in the forest and around the ruins tends to be quiet.

Access/Birding Sites/Bird List

Head out of Palenque on Route 199, north toward Route 186. After about 4 km you pass through the village of Pakalna along the road, with a railway crossing at the far (north) side. Shortly north of the railway crossing, look for a paved road to the right, which should be signed to La Libertad. It is about 40 km to La Libertad. Driving slowly along the road and looking/listening for birds is the best approach. Large, short-grass fields with scattered bushes are good to scan carefully for sleeping Double-striped Thick-knees, crouched down and overlooked easily. The tall dead trees are good for raptors, including Laughing and Aplomado falcons. Most of the key species can be seen within the first 5–10 km, but if you haven't seen Aplomado Falcons by then it is worth going on until, with luck, you spot one. Nearer La Libertad there are wetter areas and more extensive thickets. When you come to a T-junction, at La Libertad, turn left toward Emiliano Zapata. This road goes through an area of marshes (Bare-throated Tiger-Heron, Snail Kite, Grey-necked Wood-Rail), although these can be largely dry in summer (part of this stretch is in the state of Tabasco, but figuring out the state border here is a challenge). The ponds and riverbanks with overhanging vegetation are worth checking for Sungrebe, and at dusk (nearly dark) there is often a flight of Boat-billed Herons over the road at the second small bridge shortly north of the La Libertad

T-junction. From here it is possible to continue on to Emiliano Zapata and thence out to Route 186 and the Usumacinta Marshes (Site 13.4): after the La Libertad marshes fork left on a potholed road for a few kilometers before you hit a new, divided highway, where you turn left to reach Route 186 in a few more kilometers.

This area can be covered using public transport (second-class buses to La Libertad), but the nature of the habitat and birding makes it much easier to cover using your own vehicle.

Site 13.4: The Usumacinta Marshes, Chiapas/Campeche

(February 1982, January, February, 1984, March, November/December 1985, January, February, March 1987, July 1988, February 1989, April 1992, December 1996)

Introduction/Key Species

The name "Usumacinta Marshes" is given by birders to the area of savannas and marshes out to the east of Palenque; the consistently best areas are along and off Route 186 just west of and for several kilometers east of the toll bridge over the Usumacinta River.

Birds of interest here include Pinnated Bittern, Glossy Ibis, Jabiru (often relatively common during July to November), Masked Duck, Black-collared Hawk, Aplomado Falcon, Double-striped Thick-knee, Plain-breasted Ground-Dove, **Yellow-headed Parrot**, Fork-tailed Flycatcher, and Grassland Yellow-Finch. This area tends to be good at any time of day (although best in the morning) and, like the La Libertad Road (Site 13.3), makes a good afternoon trip from Palenque, when forest birding tends to be quiet.

Access/Birding Sites

From Palenque, head north on Route 199 (keeping an eye out for Aplomado Falcon, Double-striped Thick-knee, and Fork-tailed Flycatcher) to Route 186 and turn right (east, toward the Yucatan). It's about 21 km to the turnoff on the right for Emiliano Zapata (and La Libertad, Site 13.3), marked by a large Pemex station. Between Palenque and Emiliano Zapata is a side road north to El Cuyo, where **Yellow-headed Parrot** has been seen. Beyond (east of) the Emiliano Zapata junction you start to drive through seasonally extensive marshes (the best stretch for Black-collared Hawk), and after several kilometers you cross the Usumacinta River at a large toll bridge (Bat Falcons often sit on the cables and towers of the bridge). After the bridge you pass through a short stretch of the state of Tabasco and then enter the state of Campeche. Shortly after the toll bridge is a dirt road to the left (northwest) to Playa Larga, then (about 10 km from the bridge?) a paved road to the right (southeast) to Balancan and, in another 2 km, a paved road to the left (northwest) to Palizada. Driving along any of these roads can be good for birds (Pinnated Bittern, Jabiru, Aplomado Falcon, Double-striped Thick-knee, Plain-breasted Ground-Dove, Fork-tailed Flycatcher, Grassland Yellow-Finch). Birding can be excellent along Route 186 out to the vicinity of Km Post 175 (20 km past the Palizada junction).

Birds move about depending on water levels. If water is generally scarce, the rushy, roadside irrigation channels and waterways act like magnets for wading birds: Pinnated Bitterns are often common and easy to see, and Masked Ducks can be found with luck and careful scanning. When fields are flooded (especially July to December), bird concentrations can be spectacular: hundreds of herons, egrets, ibis, and jacanas, thousands of whistling-duck and Blue-winged Teal, and up to twenty Pinnated Bitterns and twenty-five Jabirus in a single field! If any fields are being burned or harvested, large aggregations of raptors (mainly Lesser Yellow-headed Vultures, White-tailed Hawks, and Crested Caracaras) can often be seen.

Like the La Libertad Road, this area can be birded by using public transport, but the large extent of habitat and nature of the birding make it not the best area for this option. One approach is to get off at the Emiliano Zapata junction, by the Pemex station and walk the highway east toward, or to, the Usumacinta bridge.

Note: While birding from Route 186 is possible, traffic is often fast and heavy. Taking any of the numerous side roads is safer and more pleasant; out near Km 170 there are dirt roads and tracks beside the highway and also along levees into the fields (remember to ask permission to drive on the levees, assuming there's anyone around to ask).

The Usumacinta Marshes Bird List
(birding the wooded patches more heavily would undoubtedly produce additional species): 196 sp. R: rare; V: vagrant (not to be expected). †See pp. 5–7 for notes on English names. [a]See Martin (1997).

Thicket Tinamou
Least Grebe
Pied-billed Grebe
American White Pelican
Neotropic Cormorant
Anhinga
Pinnated Bittern
American Bittern
Least Bittern
Bare-throated Tiger-Heron
Great Blue Heron
Great Egret
Snowy Egret
Little Blue Heron
Tricolored Heron
Cattle Egret
Green Heron
Black-crowned Night-Heron
Yellow-crowned Night-Heron
Boat-billed Heron
White Ibis
Glossy Ibis
White-faced Ibis (R)
Roseate Spoonbill
Jabiru
Wood Stork
Fulvous Whistling-Duck
Black-bellied Whistling-Duck
Muscovy Duck
Green-winged Teal
Northern Pintail
Blue-winged Teal
Cinnamon Teal (R)
Northern Shoveler
American Wigeon
Lesser Scaup
Masked Duck
Black Vulture
Turkey Vulture
Lesser Yellow-hd. Vulture
Osprey
White-tailed Kite
Snail Kite
Mississippi Kite
Northern Harrier
Crane Hawk
Common Black Hawk
Great Black Hawk
Black-collared Hawk
Grey Hawk
Roadside Hawk
White-tailed Hawk
Red-tailed Hawk
Crested Caracara
Laughing Falcon
American Kestrel
Merlin
Aplomado Falcon
Bat Falcon
Peregrine Falcon
Plain Chachalaca
Northern Bobwhite
Ruddy Crake
Grey-necked Wood-Rail
Sora
Purple Gallinule
Common Moorhen
American Coot
Sungrebe
Limpkin
Double-striped Thick-knee
Southern Lapwing (V[a])
Black-bellied Plover
Collared Plover
Wilson's Plover (R)
Semipalmated Plover
Killdeer
Black-necked Stilt
American Avocet
Northern Jacana
Greater Yellowlegs
Lesser Yellowlegs
Solitary Sandpiper
Spotted Sandpiper
Semipalmated Sandpiper
Western Sandpiper
Least Sandpiper
Pectoral Sandpiper
Stilt Sandpiper
Long-billed Dowitcher
Common Snipe
Laughing Gull
Franklin's Gull
Gull-billed Tern
Caspian Tern
Black Skimmer
Pale-vented Pigeon
White-winged Dove
Mourning Dove
Plain-breasted Ground-Dove
Ruddy Ground-Dove
White-tipped Dove
Aztec Parakeet†
White-fronted Parrot
Red-lored Parrot
Yellow-headed Parrot (R)
Squirrel Cuckoo
Striped Cuckoo
Groove-billed Ani
Barn Owl
Great Horned Owl
Lesser Nighthawk
Pauraque
Northern Potoo
White-collared Swift
Green-breasted Mango
Azure-crowned Hummingbird
Rufous-tailed Hummingbird
Buff-bellied Hummingbird
Ruby-throated Hummingbird
Ringed Kingfisher
Belted Kingfisher
Amazon Kingfisher
Green Kingfisher
Acorn Woodpecker
Golden-fronted Woodpecker
Ladder-backed Woodpecker
Lineated Woodpecker
Rufous-breasted Spinetail
Barred Antshrike
N. Beardless Tyrannulet
Yellow-bellied Elaenia
Common Tody-Flycatcher
Least Flycatcher
Vermilion Flycatcher
Dusky-capped Flycatcher
Brown-crested Flycatcher
Great Kiskadee
Social Flycatcher
Scissor-tailed Flycatcher
Fork-tailed Flycatcher
Tropical Kingbird
Couch's Kingbird
Grey-breasted Martin
Tree Swallow
Mangrove Swallow
N. Rough-winged Swallow
Barn Swallow
Brown Jay
Yucatan Jay
Spot-breasted Wren
Sedge Wren
Blue-grey Gnatcatcher

Clay-colored Thrush†
Grey Catbird
Tropical Mockingbird
White-eyed Vireo
Northern Parula
Yellow Warbler
Magnolia Warbler
Myrtle Warbler†
Yellow-throated Warbler
Black-and-white Warbler
American Redstart
Northern Waterthrush
Common Yellowthroat
Grey-crowned Yellowthroat
Wilson's Warbler
Yellow-breasted Chat
Blue-grey Tanager
Summer Tanager
Greyish Saltator
Rose-breasted Grosbeak
Blue Grosbeak
Indigo Bunting
Painted Bunting
Dickcissel
Blue-black Grassquit
White-collared Seedeater
Thick-billed Seedfinch
Grassland Yellow-Finch
Botteri's Sparrow
Grasshopper Sparrow
Savannah Sparrow
Red-winged Blackbird
Eastern Meadowlark
Melodious Blackbird
Great-tailed Grackle
Bronzed Cowbird
Black-cowled Oriole
Orchard Oriole
Hooded Oriole
Yellow-tailed Oriole
Altamira Oriole
Baltimore Oriole
Yellow-billed Cacique

Site 13.5: Bonampak, Chiapas

(March 1985, April 1992)

Introduction/Key Species

This is another Maya ruin site, a little off the beaten track but well worth visiting for birds. You need to take camping gear, food, and water purification tablets. Because the regulations about visiting and camping seem prone to vary, you may be allowed to stay no more than one night. Some tour operators offer day trips from Palenque, but these are too rushed to be any good for birding. There is also a runway, and a small plane or two often comes daily with tourists from Palenque, where local tour operators can arrange this transport option.

Bonampak lies in the Montes Azules Biosphere Reserve in the famous Lacandon rain forest. If you're going to see a Harpy Eagle in Mexico, Bonampak and Yaxchilán (Site 13.6) are as good as anywhere to look, and these sites also have an excellent variety of other species, including Scarlet Macaw, Mealy Parrot, Great Potoo, Short-tailed Nighthawk, Scaly-breasted Hummingbird, Purple-crowned Fairy, Tody Motmot, White-whiskered Puffbird, Scaly-throated Leaftosser, Russet Antshrike, Plain Antvireo, Ruddy-tailed Flycatcher, and Speckled Mourner; there also are unconfirmed reports of White-winged Becard and White-vented Euphonia. The relatively large tracts of forest, combined with reduced hunting pressures, also mean you have a fair chance of seeing tinamous, Crested Guan, Great Curassow, Spotted Wood-Quail, and mammals. Scarlet Macaws may be seen seasonally at Bonampak but are more reliably seen at Yaxchilán. The best season to visit is January to May (i.e., the drier season).

Access/Birding Sites

From the Maya Head at Palenque, head south on Route 199 (the Ocosingo highway) for 9.2 km until a paved road off to the left, which may be signed to Chancalá. Mark zero at this junction. Pavement gives way to graded dirt, and at Km 33.8 (about an hour from Palenque) there is a crossroads and a few wooden buildings (Chancalá) where cold drinks can be purchased. Stay straight on the main road, and at Km 95.8 (2.5–3 hours from Palenque), after a few villages and mostly cleared farmland, there is a *caseta* announcing the Bosques Azules Biosphere Reserve. Shortly after this is a stretch of forest along the road. At Km 120 (3–4 hours from Palenque) there is a dirt road off to the right. At the junction there may be a couple of buildings and another sign declaring the Montes Azules Biosphere Reserve. This side road goes to the village of Lacanjá, but before that, after about 4 km (i.e., immediately after/at the edge of the small village of Crucero Bethel), look for an unsigned grassed-over track (little more than two ruts) off to the left (southeast). This is the track to Bonampak and is a 9 km walk, with good birding all the way (Purple-crowned Fairy, Rufous-tailed Jacamar, Russet Antshrike, Plain Antvireo, Royal Flycatcher, mixed-species flocks). If you take your own vehicle, it is possible to leave it at Crucero Bethel and pay someone to watch it for you. The locals

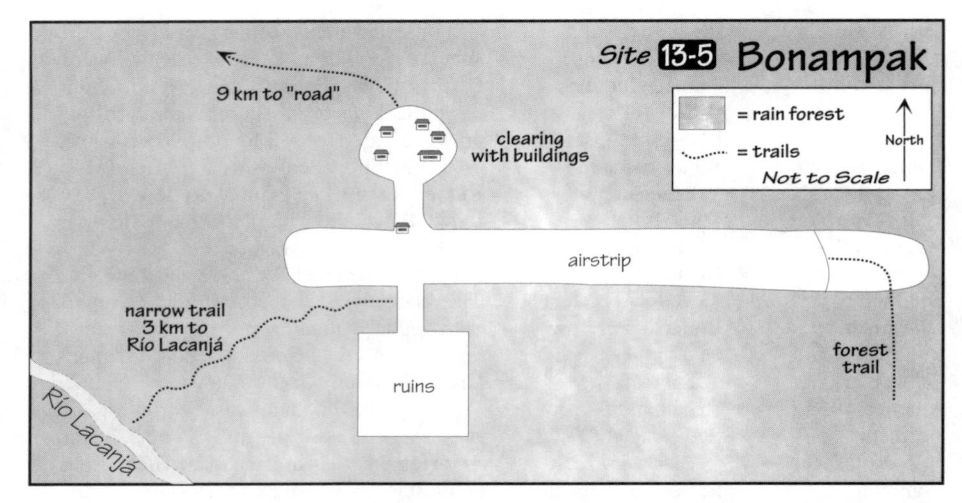

also can show you the start of the Bonampak trail. The first kilometer or so could be driven in a Jeep, but then it narrows to a footpath. The walk can take two hours to most of the day, depending on how much you stop to look at birds.

When you finally reach the site, you enter a clearing with a few buildings and should seek out the caretaker, who lives in one of the houses on the right. Camping has been allowed in this clearing, and there are *palapas* under which you may be able to set up camp. About 1.5 km before reaching Bonampak, you cross a stream (the source of local drinking water). Beyond camp is the runway, and across that a broad, cleared walk leads to the ruins. The caretaker will probably require that you have a "guide" with you when in the ruins, but after they see that you really are there for birds, they may realize it is easier to let you wander on your own.

Anywhere around the clearings and the edge of the runway is good birding (White-whiskered Puffbird, Great Antshrike, Slate-headed Tody-Flycatcher, Cinnamon Becard), especially early and late in the day, and Great Potoo can be found here at night. The main temple makes a good observation point for raptors in mid to late morning. Just after crossing the runway toward the ruins, look for a path off to the right into the forest. This is an excellent trail for birds (Great Curassow, Tody Motmot, Scaly-throated Leaftosser, Thrushlike and Speckled mourners, Rufous Piha, White-collared Manakin, Ruddy-tailed Flycatcher, **Grey-throated Chat**) and ends in about 3 km at the Río Lacanjá. The small clearing by the riverbank makes a good campsite, if you can get permission to stay there, and the river is very refreshing in the heat of the day. Short-tailed Nighthawks can be seen over the river-edge canopy at dusk. Another good trail into the forest is at the far (south) end of the runway and goes on for several kilometers.

Using public transport, there are buses at least as far as the Chancalá junction, after which trucks (often crammed) serve as "buses," or you may be able to hitch; local drivers should know the Bonampak junction and also the turnoff for Frontera Corozal and Yaxchilán (Site 13.6).

Bonampak Bird List

221 sp. Because the avifaunas are broadly similar, the lists for Bonampak and Yaxchilán are combined (see pp. 280–82). Differences mostly reflect season of visits (more spring migrants at Bonampak) and habitat (e.g., more open, river-edge second growth at Yaxchilán).

Site 13.6: Yaxchilán, Chiapas

(March 1985, January 1989)

Introduction/Key Species

This is another Maya ruin site set amid excellent rain forest. The setting, on the south bank of the Usumacinta River (the north

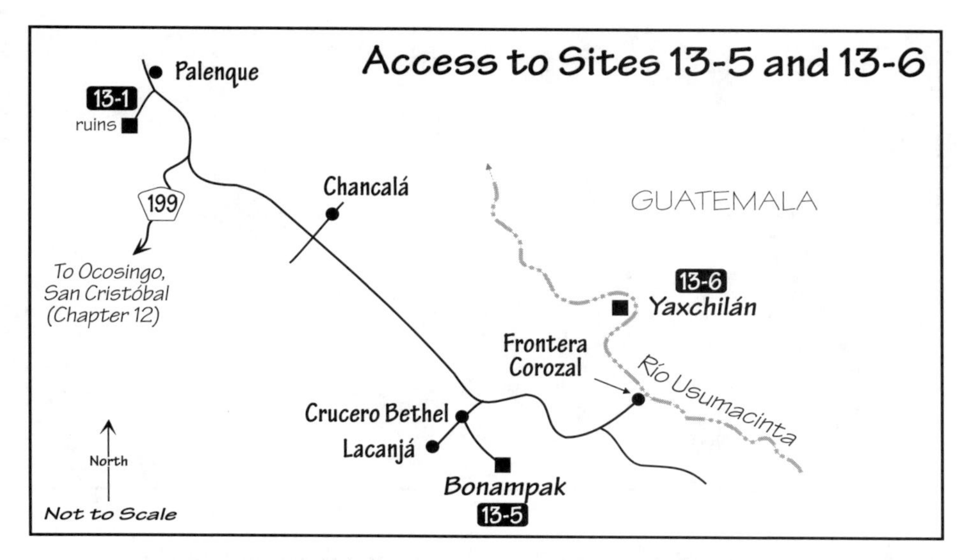

bank is Guatemala), means that birds here are slightly different from those at Bonampak (Site 13.5), with more second growth and river-edge species. This is the most readily accessible spot in Mexico for Scarlet Macaw, but the species still isn't guaranteed here.

As at Bonampak, you need to take camping gear, food, and water purification tablets. It has been easier to stay here for several nights than at Bonampak, but this may vary. As with Bonampak, tour operators offer day trips (too rushed for birding) from Palenque, and there is a runway where small planes occasionally come in with tourists from Palenque. This also is a popular stop for river-rafting trips along the Usumacinta, so you may end up sharing the camping space (where there are a couple of *palapas*) with groups of rafters. As with Bonampak, the best season to visit is January to May (i.e., the drier season).

Access/Birding Sites

Follow the directions for Bonampak, but rather than take the turning to Crucero Bethel and Lacanjá, stay straight on the main road for another 25 km (12 km?) to a fairly major junction off to the left (4–4.5 hours from Palenque). There are some buildings (basic food and drink stands) at this junction, which may be signed to the left for Frontera Corozal (formerly Frontera Echevería). If in doubt, there should be someone to ask for the Corozal road. It is about 13 km (7 km?) to Corozal, on the banks of the Usumacinta River (5–6 hours from Palenque), and the road can be in bad shape, muddy with deep ruts, but is generally passable in a *combi* or pickup truck. However, before you make the trip it is worth asking one of the tour operators in Palenque about road conditions.

Go on through the small town of Corozal to the river, where there is a small military base, and there's a chance a boatman will find you before you find him. If not, make it known that you are interested in a boat to Yaxchilán (downriver about 30–45 minutes, 45 minutes to an hour back against the current). The boats are open *lanchas* with outboard motors, and it doesn't hurt to ask about life vests. You should negotiate a price for the trip, although rates may be semi-fixed; and make sure the boatman knows when you want to be picked up again (pay half in advance). It's a good idea to have a tarp or plastic sheeting to protect your bags and supplies from water (including rain), and I would pack anything valuable in plastic (Ziplock) bags. Birding from the boat can be good. Look on the sand and gravel bars for Collared Plovers, and watch for King Vultures and Scarlet Macaws.

On arriving at Yaxchilán, there is a short steep walk up the riverbank to a grassy clearing with *palapas* where you can camp. Just beyond this and to your right is the runway,

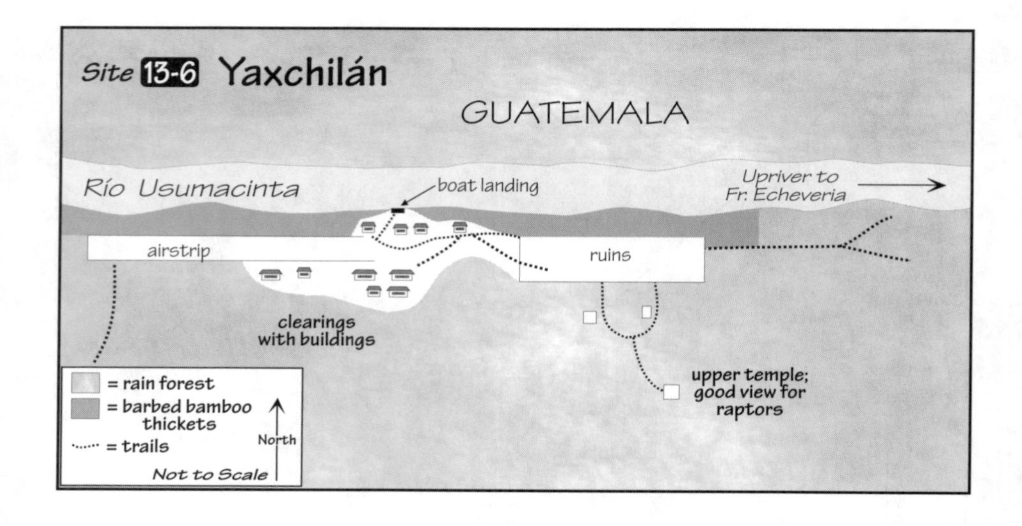

about halfway down which, on the left, are buildings where the caretakers may be found if there is nobody at the camp area. The ruins are off to your left as you climb the riverbank; and, as at Bonampak, the caretaker may insist that you need a guide but, if you watch birds and not ruins, you may well be left to your own devices, with a caution about not getting lost. Birding along the runway is good early and late in the day (Scarlet Macaw, Scaly-breasted Hummingbird, Rufous-tailed Jacamar, Great Antshrike); the clearings around the ruins can be good at any time (Scaly-breasted Hummingbird, White-whiskered Puffbird, Cinnamon Becard), especially if you find fruiting and flowering trees (White-necked Jacobin, Purple-crowned Fairy); and the upper temple is a good lookout for raptors. There are trails into the forest at the far end of the runway and at the far end of the ruin area (Great Tinamou, Thrushlike Mourner), but these seem not as long or easy to follow as the trails at Bonampak. Short-tailed Nighthawk can be seen over the canopy along the river and runway at dusk.

Yaxchilán Bird List

217 sp. Because the avifaunas are broadly similar, the lists for Bonampak and Yaxchilán are combined. Differences often reflect season (more spring migrants at Bonampak) and habitat (e.g., more open, river-edge second growth at Yaxchilán).

Bird Lists for Bonampak (Site 13.5) and Yaxchilán (Site 13.6)

[B]Bonampak; [Y]Yaxchilán (including boat ride from Corozal); no code: both sites. 271 sp. †See pp. 5–7 for notes on English names.

Great Tinamou
Little Tinamou
Slaty-breasted Tinamou
Neotropic Cormorant[Y]
Anhinga[Y]
Great Blue Heron[Y]
Snowy Egret[Y]
Little Blue Heron[Y]
Tricolored Heron[Y]
Cattle Egret
Green Heron[Y]
Yellow-crowned Night-Heron[Y]
Muscovy Duck[B]
Blue-winged Teal[Y]
Black Vulture
Turkey Vulture
King Vulture
Osprey
Grey-headed Kite[B]
Hook-billed Kite[B]
Swallow-tailed Kite[Y]
White-tailed Kite[Y]
Snail Kite[Y]
Double-toothed Kite
Mississippi Kite[B]
Plumbeous Kite[Y]
Sharp-shinned Hawk
White Hawk
Common Black Hawk[B]
Great Black Hawk
Roadside Hawk
Broad-winged Hawk
Short-tailed Hawk[Y]
Swainson's Hawk[B]
Black Hawk-Eagle[B]
Ornate Hawk-Eagle[B]
Laughing Falcon
Bat Falcon
Barred Forest-Falcon
Collared Forest-Falcon
Plain Chachalaca
Crested Guan[Y]
Great Curassow
Spotted Wood-Quail
Ruddy Crake
Collared Plover[Y]
Killdeer[Y]
Spotted Sandpiper[Y]
Laughing Gull[Y]

Pale-vented Pigeon[Y]
Scaled Pigeon
Short-billed Pigeon
Ruddy Ground-Dove
Blue Ground-Dove
White-tipped Dove
Grey-headed Dove†
Grey-chested Dove
Ruddy Quail-Dove
Scarlet Macaw[Y]
Aztec Parakeet†
Brown-hooded Parrot
White-crowned Parrot
White-fronted Parrot
Red-lored Parrot
Mealy Parrot
Squirrel Cuckoo
Striped Cuckoo[Y]
Groove-billed Ani
Vermiculated Screech-Owl[B]
Mottled Owl
Black-and-white Owl
Central American Pygmy-Owl†[B]
Great Potoo[B]
Short-tailed Nighthawk
Common Nighthawk[B]
Pauraque
White-collared Swift
Vaux's Swift
Lesser Swallow-tailed Swift
Long-tailed Hermit
Little Hermit
Wedge-tailed Sabrewing[B]
Scaly-breasted Hummingbird[Y]
Violet Sabrewing
White-necked Jacobin
Green-breasted Mango[Y]
Black-crested Coquette[B]
White-bellied Emerald
Rufous-tailed Hummingbird
Purple-crowned Fairy
Long-billed Starthroat
Ruby-throated Hummingbird[Y]
Black-headed Trogon
Violaceous Trogon
Collared Trogon
Slaty-tailed Trogon
Tody Motmot
Blue-crowned Motmot
Ringed Kingfisher
Amazon Kingfisher
Green Kingfisher
Pygmy Kingfisher†
White-necked Puffbird[B]
White-whisked Puffbird
Rufous-tailed Jacamar
Collared Aracari
Keel-billed Toucan
Black-chinned Woodpecker
Golden-fronted Woodpecker[Y]
Smoky-brown Woodpecker
Golden-olive Woodpecker
Chestnut-colored Woodpecker
Lineated Woodpecker[Y]
Pale-billed Woodpecker
Rufous-breasted Spinetail[Y]
Buff-throated Fol.-gleaner[B]
Plain Xenops
Scaly-throated Leaftosser[B]
Tawny-winged Woodcreeper[B]
Ruddy Woodcreeper[B]
Olivaceous Woodcreeper
Wedge-billed Woodcreeper
Barred Woodcreeper
Ivory-billed Woodcreeper
Streak-headed Woodcreeper
Great Antshrike
Barred Antshrike
Russet Antshrike[B]
Plain Antvireo[B]
Dot-winged Antwren
Dusky Antbird
Mexican Antthrush†
Yellow-bellied Tyrannulet
Greenish Elaenia
Yellow-bellied Elaenia[Y]
Ochre-bellied Flycatcher
Sepia-capped Flycatcher
Northern Bentbill[B]
Slate-headed Tody-Flycatcher[B]
Common Tody-Flycatcher[Y]
Yellow-olive Flycatcher
Ruddy-tailed Flycatcher[B]
Stub-tailed Spadebill
Royal Flycatcher
Sulphur-rumped Flycatcher
Olive-sided Flycatcher[B]
Eastern Pewee†[B]
Tropical Pewee
Yellow-bellied Flycatcher
Acadian Flycatcher[B]
Least Flycatcher
Bright-rumped Attila
Rufous Mourner
Dusky-capped Flycatcher
Great Crested Flycatcher
Brown-crested Flycatcher
Great Kiskadee[Y]
Boat-billed Flycatcher
Social Flycatcher
Streaked Flycatcher[B]
Sulphur-bellied Flycatcher[B]
Piratic Flycatcher
Couch's Kingbird
Eastern Kingbird[B]
Thrushlike Mourner†
Speckled Mourner[B]
Cinnamon Becard
Rose-throated Becard
Masked Tityra
Black-crowned Tityra[B]
Rufous Piha
Lovely Cotinga
White-collared Manakin
Red-capped Manakin
Grey-breasted Martin[B]
Mangrove Swallow[Y]
Rough-winged Swallow sp.†[Y]
Cliff Swallow[B]
Barn Swallow[Y]
Green Jay[Y]
Brown Jay
Band-backed Wren
Spot-breasted Wren
Southern House Wren†
White-bellied Wren[B]
White-breasted Wood-Wren
Long-billed Gnatwren
Tropical Gnatcatcher
Swainson's Thrush
Wood Thrush
Clay-colored Thrush†
Grey Catbird
Cedar Waxwing[Y]
White-eyed Vireo

Yellow-throated Vireo[Y]
Philadelphia Vireo
Red-eyed Vireo[B]
Yellow-green Vireo[Y]
Tawny-crowned Greenlet
Lesser Greenlet
Green Shrike-Vireo
Blue-winged Warbler[Y]
Golden-winged Warbler
Tennessee Warbler
Northern Parula[Y]
Yellow Warbler
Chestnut-sided Warbler
Magnolia Warbler
Black-throated Green Warbler
Blackburnian Warbler[B]
Cerulean Warbler[B]
Black-and-white Warbler
American Redstart
Worm-eating Warbler
Ovenbird
Northern Waterthrush
Louisiana Waterthrush
Kentucky Warbler
Mourning Warbler[B]
Common Yellowthroat
Hooded Warbler
Wilson's Warbler
Canada Warbler[B]
Golden-crowned Warbler
Grey-throated Chat[B]
Yellow-breasted Chat
Bananaquit
Golden-hooded Tanager†
Green Honeycreeper[B]
Red-legged Honeycreeper
Scrub Euphonia
Yellow-throated Euphonia
Olive-backed Euphonia
Blue-grey Tanager
Yellow-winged Tanager
Grey-headed Tanager
Black-throated Shrike-Tanager
Red-crowned Ant-Tanager
Red-throated Ant-Tanager
Summer Tanager
Scarlet Tanager[B]
Crimson-collared Tanager
Scarlet-rumped Tanager
Greyish Saltator[Y]
Buff-throated Saltator
Black-headed Saltator
Black-faced Grosbeak
Rose-breasted Grosbeak[B]
Blue-black Grosbeak
Blue Grosbeak
Indigo Bunting
Painted Bunting
Dickcissel[B]
Orange-billed Sparrow
Green-backed Sparrow
Blue-black Grassquit[B]
White-collared Seedeater
Variable Seedeater
Lincoln's Sparrow[B]
Red-winged Blackbird[Y]
Melodious Blackbird
Great-tailed Grackle
Bronzed Cowbird[B]
Giant Cowbird[Y]
Black-cowled Oriole
Orchard Oriole[Y]
Yellow-tailed Oriole
Baltimore Oriole
Yellow-billed Cacique
Chestnut-headed Oropendola[Y]
Montezuma Oropendola

CHAPTER 14 THE YUCATAN PENINSULA

Turquoise-browed Motmot

The Yucatan Peninsula (often referred to simply as the Yucatan) is a low-lying limestone shelf that protrudes north into the Caribbean Sea. The peninsula includes the adjacent parts of Guatemala and Belize and is home to many regional endemics (four restricted to Isla Cozumel) as well as a host of more widespread tropical species and an excellent variety of waterbirds, both residents and migrants.

Beyond birds, the Yucatan is famous for its Maya ruins (notably the impressive sites of Uxmal and Chichén Itzá), regional cuisine, and snorkeling along the Caribbean coast, including Cozumel. For these reasons the Yucatan has a well-developed infrastructure for tourism and makes a great place for a one-week or two-week birding trip which can be combined fairly easily with a more general-interest family vacation. For the more target-oriented birder, the Yucatan can be combined with Palenque (Site 13.1) for a good two-week trip, starting and ending in Cancún.

The two main ports of entry to the Yucatan are Mérida and Cancún, both of which, particularly the latter, are served daily by many flights from several North American cities. One can also fly direct from the U.S.A. to Cozumel. Car rental is available from numerous companies; relatively inexpensive rates (by Mexican standards) are possible in Cancún. Almost all of these sites can also be visited easily via public transport.

If you are driving, note that bicycles (and tricycles, often employed as taxis) are ubiquitous and common in the Yucatan and often have right of way (or act as if they do) in towns. Also, at night be particularly alert for bicycles without lights.

Suggested Itineraries

The sites are set out to be visited from either Mérida or Cancún, and directions are given accordingly. If you fly into Cancún, an itinerary might be: Jardín Botanico Dr. Alfredo Barrera M. (Site 14.1), then on to Cobá (14.2) and/or Felipe Carillo Puerto (14.3), then to Chichén Itzá (14.5) and Río Lagartos (14.5) and on to Mérida, whence one can visit Celestún (14.8), Progreso (14.6), Dzibilchaltún (14.7), and Uxmal (14.9). Alternatively, if you are coming into Mérida, this itinerary can be done more or less in reverse. Isla Cozumel (Site 14.10) is best visited at one end of your trip; particularly if you like swimming and snorkeling, it makes a good place to unwind before flying back home.

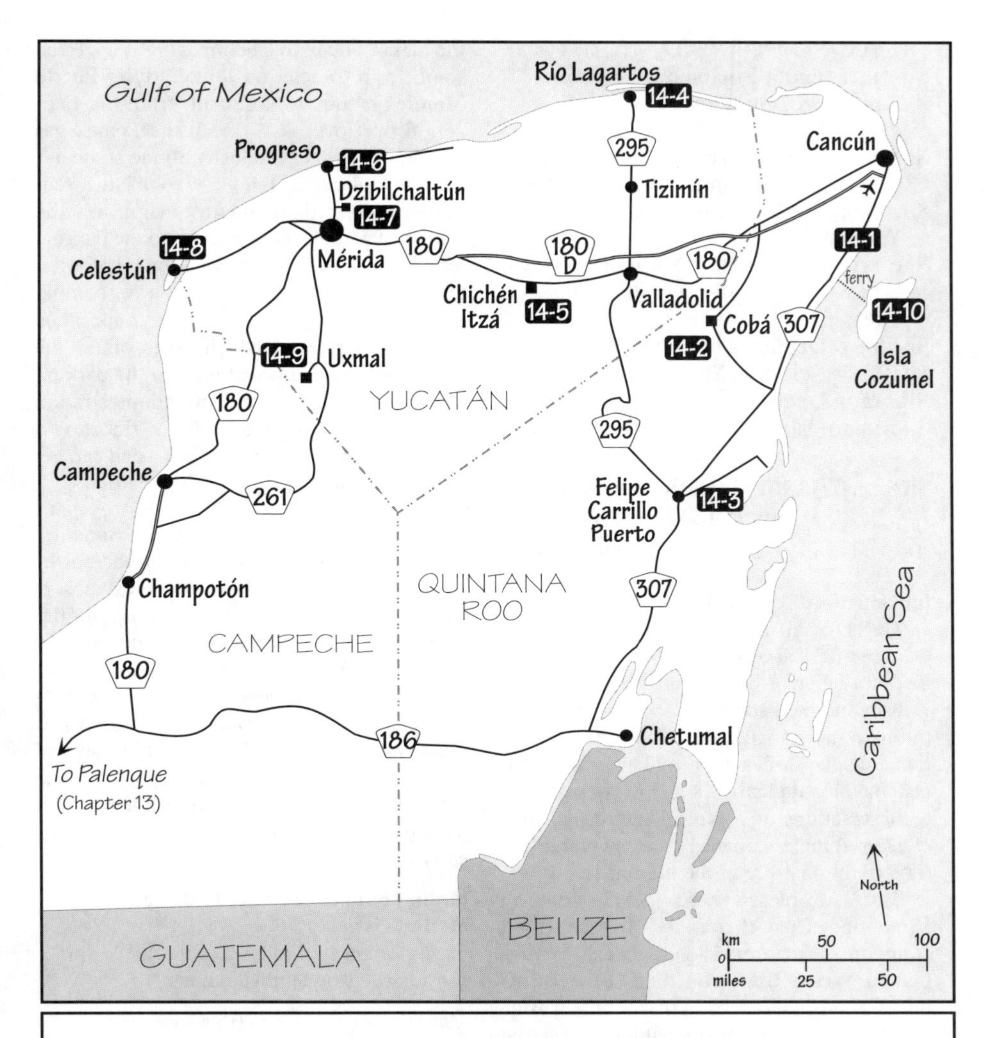

Chapter 14: The YUCATAN PENINSULA

14-1	Jardín Botanico Dr. Alfredo Barrera M., Quintana Roo
14-2	Cobá, Quintana Roo
14-3	Felipe Carrillo Puerto (Vigia Chico Road), Quintana Roo
14-4	Río Lagartos / Las Coloradas, Yucatán
14-5	Chichén Itzá, Yucatán
14-6	Progreso / Dzilam de Bravo, Yucatán
14-7	Dzibilchaltún, Yucatán
14-8	Celestún, Yucatán
14-9	Uxmal, Yucatán
14-10	Isla Cozumel, Quintana Roo

THE YUCATAN PENINSULA SITE LIST

Site 14.1 Jardín Botanico Dr. Alfredo Barrera M., Quintana Roo
Site 14.2 Cobá, Quintana Roo
Site 14.3 Felipe Carrillo Puerto (Vigia Chico Road), Quintana Roo
Site 14.4 Río Lagartos/Las Coloradas, Yucatán
Site 14.5 Chichén Itzá, Yucatán
Site 14.6 Progreso/Dzilam de Bravo, Yucatán
Site 14.7 Dzibilchaltún, Yucatán
Site 14.8 Celestún, Yucatán
Site 14.9 Uxmal, Yucatán
Site 14.10 Isla Cozumel, Quintana Roo

Site 14.1: Jardín Botanico Dr. Alfredo Barrera M., Quintana Roo

(December 1996, checked December 1997)

Introduction/Key Species

The Dr. Alfredo Barrera M. Botanical Garden offers access to a 60-hectare patch of medium-height tropical forest via 3 km of well-maintained trails. In addition to a path through the native forest there are educational displays of cacti, succulents, ornamentals, medicinal plants, a small Maya ruin, and reconstructions of a typical house and of a *chiclero* (chicle gatherer) camp, which illustrate traditional life in the region.

Birds of interest here include Yucatan Vireo and **Rose-throated Tanager** (both common), **Yucatan Woodpecker**, **White-bellied Wren**, **Black Catbird**, **Blue Bunting**, **Green-backed Sparrow**, and **Orange Oriole**. The nature of the habitat makes the garden an excellent place to actually see Spot-breasted Wren, as well as understory migrants such as Wood Thrush, Kentucky Warbler, and Ovenbird. Other species of interest that should be looked for here include Caribbean Dove and Caribbean Elaenia. The printed brochure and trail guide says "it is very common to see . . . Great Curassow," and park personnel confirmed that curassows occur, so you may be lucky. I saw Agoutis, Spider Monkeys, and a Grey Fox, suggesting that the area is genuinely protected.

Access/Birding Sites

This site is only about a 20- to 30-minute drive south from Cancún, beside Route 307, the main Cancún to Chetumal highway. Head south from Cancún on Route 307 to Puerto Morelos (about 21 km south from the Cancún airport overpass). From the Pemex gas station in Puerto Morelos continue south 1.1 km and look on the left (east) for a large, red-bordered, whitish concrete sign for the botanical garden. The entrance to the parking lot is right beside this sign, although not too conspicuous, especially if you're coming from the south; it is also right opposite a limestone dirt road which takes off to the west from the highway. Pull into the parking lot; at its south end are the administration buildings, where one buys a ticket (20 pesos/person in December 1996) and can get a map showing the layout of the garden.

The garden is open from 9 A.M. to 5 P.M., Monday to Saturday, but closed on Sundays. The trails are well marked, and in the northeast corner there is a small but sturdy observation platform about 8–10 m up in the canopy, reached via a small and not-so-sturdy ladder which, if you're not keen on heights, might best be avoided. The tower offers a nice view out over the mangroves to the coast (frigatebirds, vultures) but may not add much in the way of bird species. Note that biting insects (mainly mosquitoes) can be locally numerous in the forest, at least early and late in the day.

Jardín Botanico Dr. Alfredo Barrera M. Bird List

73 sp. R: rare (not to be expected). †See pp. 5–7 for notes on English names.

Magnificent Frigatebird
Black Vulture
Turkey Vulture
Lesser Yellow-hd. Vulture
Ruddy Ground-Dove
White-tipped Dove
Squirrel Cuckoo
Ferruginous Pygmy-Owl
Vaux's Swift
Buff-bellied Hummingbird
Cinnamon Hummingbird
Black-headed Trogon
Blue-crowned Motmot
Golden-fronted Woodpecker
Yucatan Woodpecker†
Yellow-bellied Sapsucker
Tawny-winged Woodcreeper
Barred Woodcreeper
Ivory-bellied Woodcreeper
N. Beardless Tyrannulet
Greenish Elaenia
Northern Bentbill

Yellow-olive Flycatcher
Tropical Pewee
Least Flycatcher
Bright-rumped Attila
Dusky-capped Flycatcher
Great Kiskadee
Boat-billed Flycatcher
Social Flycatcher
Tropical Kingbird
Masked Tityra
Green Jay
Brown Jay
Spot-breasted Wren
White-browed Wren†
White-bellied Wren
Tropical Gnatcatcher
Wood Thrush
Clay-colored Thrush†
Black Catbird
Tropical Mockingbird
White-eyed Vireo
Yellow-throated Vireo
Yucatan Vireo
Lesser Greenlet
Rufous-browed Peppershrike
Blue-winged Warbler
Northern Parula
Magnolia Warbler
Black-throated Green Warbler
Cape May Warbler (R?)
Black-and-white Warbler
American Redstart
Worm-eating Warbler
Ovenbird
Northern Waterthrush
Kentucky Warbler
Common Yellowthroat
Hooded Warbler
Yellow-throated Euphonia
Grey-headed Tanager
Red-crowned Ant-Tanager
Red-throated Ant-Tanager
Rose-throated Tanager
Blue Bunting
Green-backed Sparrow
Melodious Blackbird
Great-tailed Grackle
Hooded Oriole
Orange Oriole
Altamira Oriole
Yellow-backed Oriole

Site 14.2: Cobá, Quintana Roo

(February 1984, December 1985, January 1987, February 1989, October 1993, September 1995, December 1996, checked February 1998)

Introduction/Key Species

Cobá is a Maya ruin site distinguished by its adjacent permanent lake, which gained popularity in the 1980s as a site for Spotted Rail. Changing water levels have made seeing the rails more difficult in recent years (assuming they're still around), but the forest around the ruins offers good birding, being a poorer version of what one can find at Felipe Carrillo Puerto (Site 14.3). Other species of interest at Cobá include Ruddy Crake, **Yucatan Poorwill**, **Yucatan Nightjar**, Northern Potoo, and **Ridgway's Rough-winged Swallow**.

Access/Birding Sites

Cobá (served regularly by buses from both north and south) is about 170 km (2–2.5 hours) southwest of Cancún. Head south out of Cancún on Route 307 to Tulum, which is about 130 km (1.5–2 hours) by good paved highway. As you come into Tulum you pass on the left (east) turnoffs for the northern *Zona Hotelera* and the entry to the new and huge parking lot for the ruins and shopping complex, and then come to a crossroads, left (southeast) to the southern *Zona Hotelera* and right (northwest) signed to Cobá. Turn right here onto a two-lane paved highway and stay on it for about 44 km (20–35 minutes), until a right-hand bend with a paved road to the left signed for Cobá and the Villas Arqueologicas (Club Med) hotel. Turn here and in 2.3 km you enter the small village of Cobá and at 2.7 km come to the lakeshore, marked by a sharp left in the road. A dirt road to the right leads about 250 m to the Villas Arqueologicas hotel. Cobá is visited mainly as a day trip: there are numerous restaurants, a recently opened economical hotel near where the bus stops in the village, and the expensive Villas Arqueologicas hotel by the lake.

Turning sharply left at the lake you follow the reedy shore, and in 400 m the parking lot and entrance for the ruins are to your left (south). The lake can be birded from the car, or on foot, for about 1 km to the opposite (west) side, where the road ends in a gravel area where you can park easily. From here one can continue along the lake edge on foot, via a narrow path.

Spotted Rails can be seen anywhere around the lake; just before and at the gravel parking spot on the far side have been good in the past, even at midday with screaming kids and barking dogs around! Especially early and late in the day, swallows often come in to drink at the lake, including **Ridgway's Rough-wings** which sometimes perch on the wires at the southeast corner of the lake. The forest around the ruins is quite good for birding (although Site 14.3 is bet-

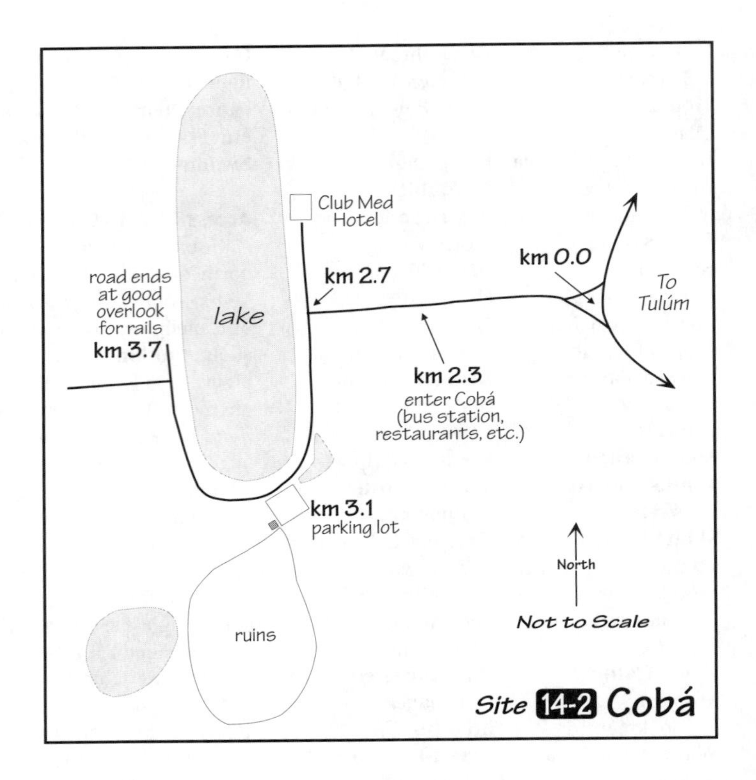

ter), as is the second-growth and woodland edge near the lakeshore.

For night birding, the short stretch of road into Cobá from the highway, and the main highway for 5 km both north and south of the Cobá entry junction, can be good; beware that traffic can travel very fast on the main highway at night. **Yucatan Poorwills** are common, while **Yucatan Nightjars** and Northern Potoos are fairly common, but calling regimes vary greatly with season, temperature, and state of the moon. Warm moonlit nights are always best, and except perhaps in the breeding season, calling tends to be concentrated in a short window around dusk and, to a lesser extent, just before dawn.

Cobá Bird List

(forest bird list notably incomplete): 143 sp. R: rare (not to be expected). †See pp. 5–7 for notes on English names.

Thicket Tinamou
Least Grebe
Pied-billed Grebe
Neotropic Cormorant
Anhinga
Great Blue Heron
Great Egret
Little Blue Heron
Tricolored Heron
Cattle Egret
Green Heron
Yellow-crowned Night-Heron
American Wigeon
Ring-necked Duck
Lesser Scaup
Black Vulture
Turkey Vulture
Osprey
Hook-billed Kite
Grey Hawk
Short-tailed Hawk
Laughing Falcon
Collared Forest-Falcon
Plain Chachalaca
Singing Quail
Ruddy Crake
Spotted Rail
Sora
Purple Gallinule
Limpkin
Killdeer
Northern Jacana
Spotted Sandpiper
Laughing Gull
Royal Tern
Ruddy Ground-Dove
White-tipped Dove
Aztec Parakeet†
White-fronted Parrot
Squirrel Cuckoo
Groove-billed Ani
Vermiculated Screech-Owl
Ferruginous Pygmy-Owl
Mottled Owl
Pauraque
Yucatan Poorwill
Yucatan Nightjar
Northern Potoo
Vaux's Swift
Wedge-tailed Sabrewing
Canivet's Emerald
White-bellied Emerald

Cinnamon Hummingbird
Buff-bellied Hummingbird
Black-headed Trogon
Blue-crowned Motmot
Turquoise-browed Motmot
Belted Kingfisher
Collared Aracari
Golden-fronted Woodpecker
Yucatan Woodpecker†
Olivaceous Woodcreeper
Ivory-billed Woodcreeper
N. Beardless Tyrannulet
Greenish Elaenia
Yellow-bellied Elaenia
Northern Bentbill
Yellow-olive Flycatcher
Eastern Pewee†
Tropical Pewee
Yellow-bellied Flycatcher
Least Flycatcher
Bright-rumped Attila
Yucatan Flycatcher
Dusky-capped Flycatcher
Great Crested Flycatcher
Great Kiskadee
Boat-billed Flycatcher
Social Flycatcher
Tropical Kingbird
Couch's Kingbird
Eastern Kingbird
Masked Tityra
Grey-breasted Martin
Mangrove Swallow
Ridgway's Rough-w. Swallow†
Bank Swallow
Cliff Swallow
Barn Swallow
Yucatan Jay
Brown Jay
Green Jay
Spot-breasted Wren
White-browed Wren†
Blue-grey Gnatcatcher
Wood Thrush
Clay-colored Thrush†
Grey Catbird
Black Catbird
Tropical Mockingbird
White-eyed Vireo
Rufous-browed Peppershrike
Blue-winged Warbler
Tennessee Warbler
Northern Parula
Yellow Warbler
Magnolia Warbler
Black-throated Blue Warbler (R)
Myrtle Warbler†
Black-throated Green Warbler
Palm Warbler
Yellow-throated Warbler
Bay-breasted Warbler
Black-and-white Warbler
American Redstart
Ovenbird
Northern Waterthrush
Common Yellowthroat
Grey-crowned Yellowthroat
Hooded Warbler
Scrub Euphonia
Yellow-throated Euphonia
Red-throated Ant-Tanager
Yellow-winged Tanager
Rose-throated Tanager
Summer Tanager
Greyish Saltator
Black-headed Saltator
Northern Cardinal
Blue Grosbeak
Indigo Bunting
Painted Bunting
Green-backed Sparrow
White-collared Seedeater
Melodious Blackbird
Great-tailed Grackle
Bronzed Cowbird
Orchard Oriole
Hooded Oriole
Black-cowled Oriole
Yellow-tailed Oriole
Altamira Oriole
Yellow-billed Cacique

Site 14.3: Felipe Carillo Puerto (Vigia Chico Road), Quintana Roo

(February, October, November 1984, December 1985, May 1986, January/February 1987, July 1988, February 1989, December 1996)

Introduction/Key Species

The small town of Felipe Carrillo Puerto (FCP), in the Maya heart of Quintana Roo, is surrounded by excellent areas of tropical semi-deciduous forest which hold a high diversity of interesting birds. The forests north of town adjoin the Sian Ka'an Biosphere Reserve and are subject to limited agricultural and hunting pressure, such that one can still see (with luck) Great Curassow and **Ocellated Turkey**.

Other birds of interest include Thicket Tinamou, King Vulture, Black and Ornate hawk-eagles, **Singing Quail**, **Yucatan Bobwhite**, Scaled Pigeon, Blue Ground-Dove, Caribbean Dove, **Yucatan Parrot**, **Yucatan Poorwill**, **Yucatan Nightjar**, **Wedge-tailed Sabrewing**, **Yucatan Woodpecker**, **Rufous-breasted Spinetail**, Tawny-winged, Ruddy, and Barred woodcreepers, **Mexican Antthrush**, Stub-tailed Spadebill, **Yucatan Flycatcher**, Thrushlike Mourner, **Grey-collared Becard**, **Ridgway's Rough-winged Swallow**, **White-bellied Wren**, Long-billed Gnatwren, Tropical Gnatcatcher, **Black Catbird**, Swainson's Warbler (winter), **Grey-throated Chat**, **Rose-throated Tanager**, **Blue Bunting**, **Green-backed Sparrow**, and **Orange Oriole**.

Access/Birding Sites

FCP (served by regular first-class and second-class bus service) is about 225 km

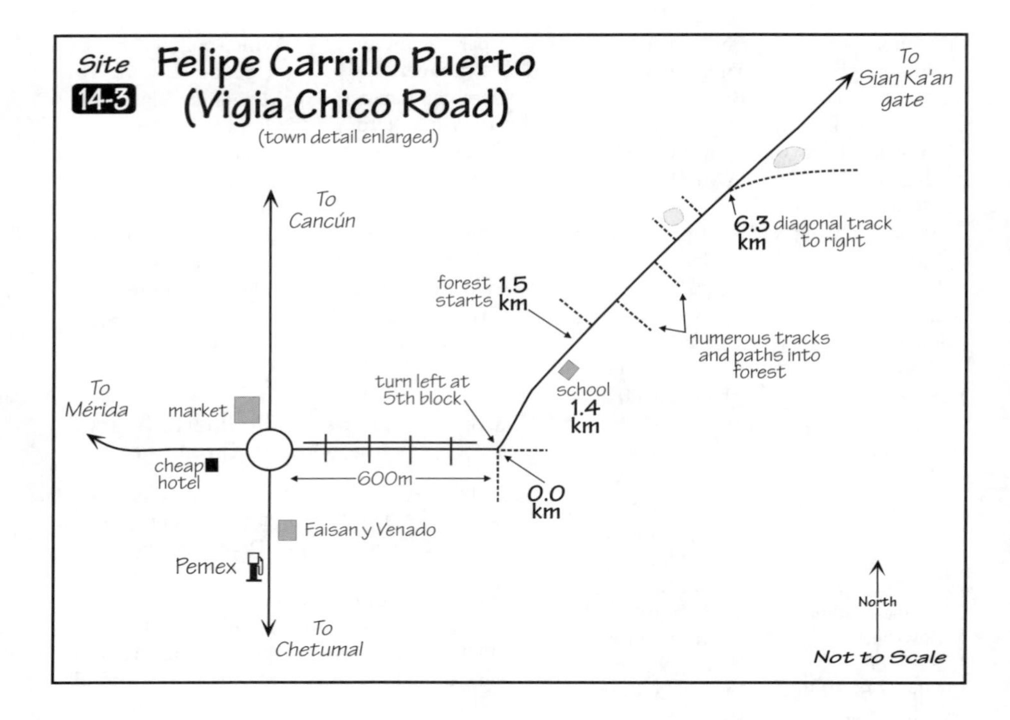

(2.5–3.5 hours) south of Cancún on Route 307. It is also about 320 km (3.5–5 hours) from Mérida, by way of Valladolíd and Tepich. On the north side of FCP is a *glorieta* with exits north to Cancún, west to Mérida (and Valladolíd via Tepich), south to Chetumal, and east (not always prominently signed) to Vigia Chico. To reach the Vigia Chico Road without a car, you have to walk the short distance from FCP.

There are a few basic hotels (birders have stayed at the Faisan y Venado) and restaurants in FCP, allowing you to stay near the forest. If nicer accommodations are needed, FCP can be done as a day trip from Tulum, 100 km to the north.

To reach good forest, take the Vigia Chico exit from the *glorieta*, go five blocks (600 m) east along a divided street to where the pavement ends, and turn left (northeast) at the fifth street, which becomes the Vigia Chico Road. Mark zero at this turning. The hard-dirt, white limestone road goes on for many kilometers, eventually reaching the coast in the Sian Ka'an Biosphere Reserve. Initially the road runs through the rural edge of town (often very birdy), with overgrown weedy fields (**Yucatan Bobwhite**), second growth, and orchards and, at Km 1.5, just past a school on the right (east), enters forest. Birding anywhere along the Vigia Chico Road can be good out to Km 15 or beyond. Trees shade much of the road, making it possible to bird throughout the day, and traffic is light, usually restricted to a truck or two and a few bicycles over the course of a morning.

As well as birding along the main road, numerous tracks and footpaths lead off into the forest, although many lead quickly to agricultural plots (*milpas*). At Km 5.0 there is an obvious track off to the left, about 150 m along which is a small *aguada* (pond) to the right (Least Grebe, **Ruddy Crake**). At Km 6.3 a track goes off diagonally ahead, to the right, and on through excellent forest for several kilometers (at least), if you want to get off the main road.

The *milpas*, both active and overgrown, are worth checking, especially if there are tall dead trees left standing (good for raptors, pigeons, parrots, toucans, woodpeckers, and, in summer, nesting Streaked and Sulphur-bellied flycatchers). Early morning at the edge of a *milpa*, as the sun hits the forest edge, can be very productive, and you'll probably see more birds than in the forest. Because the agriculture here works on a rotating system, cornfields of one year can be second-growth thickets a year or two later, so finding habitat is a matter of trial and error.

Birding on foot, or driving and stopping to walk every few hundred meters, is the best way to bird the Vigia Chico Road in the early to mid morning. Walking quietly and listening for rustlings in the leaf litter has proved a good way to find **Singing Quail**, and imitating a Ferruginous Pygmy-Owl (don't overdo it) usually stirs up a mobbing band of sundry species which can include **Yucatan Flycatcher**, wrens, Tropical Gnatcatcher, **Grey-throated Chat**, **Rose-throated Tanager**, and even Ruddy Woodcreeper or Thrushlike Mourner! When things get hot and activity dies down it can be worth driving slowly along the road, looking and listening for army-ant swarms (often with attendant Tawny-winged, Ruddy, and Barred woodcreepers) or finding a spot with a good vista of sky and looking/listening for hawk-eagles (10:30–11:45 A.M.is a good time) and other raptors.

Beyond about Km 15 the forest gets lower and scrubbier, and at Km 28 you come to the Santa Teresa Gate to **Sian Ka'an Biosphere Reserve,** which is effectively closed to casual birders (you need to apply well in advance for a permit, which involves too much red tape to worry about unless things change, and probably you would not be allowed to bird without a guide or security guard). Going into Sian Ka'an adds seasonally flooded savanna (Lesser Yellow-headed Vulture, White-tailed Hawk, Limpkin, Pale-vented Pigeon), coastal lagoons (herons, shorebirds), and mangroves (Bare-throated Tiger-Heron, Common Black Hawk, White-crowned Pigeon) to the habitats, and your chances of seeing Great Curassow and **Ocellated Turkey** are better than outside the reserve. There is another gate to Sian Ka'an at Chumpón, 50 km north of the FCP *glorieta* off Route 307: coming from FCP, look for an inconspicuous narrow road to the right (east), perhaps with a sign for Sian Ka'an. In summer, **Orange Orioles** nest colonially along the kilometer stretch between the highway and the Chumpón Gate and so can be seen without going into the reserve.

Felipe Carrillo Puerto Bird List
(Vigia Chico Road out to Km 15): 193 sp. R: rare (not to be expected). †See pp. 5–7 for notes on English names.

Thicket Tinamou
Least Grebe
Neotropic Cormorant
Blue-winged Teal
Black Vulture
Turkey Vulture
King Vulture
Hook-billed Kite
White-tailed Kite
Double-toothed Kite
Crane Hawk
Great Black Hawk
Grey Hawk
Roadside Hawk
Short-tailed Hawk
Black Hawk-Eagle
Ornate Hawk-Eagle
Laughing Falcon
Collared Forest-Falcon
Bat Falcon
Plain Chachalaca
Ocellated Turkey (R)
Singing Quail
Yucatan Bobwhite†
Ruddy Crake
Grey-necked Wood-Rail
Common Moorhen
Solitary Sandpiper
Common Snipe
Scaled Pigeon
Red-billed Pigeon
White-winged Dove
Ruddy Ground-Dove
Blue Ground-Dove
White-tipped Dove
Caribbean Dove
Aztec Parakeet†
White-crowned Parrot
White-fronted Parrot
Yucatan Parrot†
Squirrel Cuckoo
Groove-billed Ani
Vermiculated Screech-Owl
Ferruginous Pygmy-Owl
Mottled Owl
Pauraque
Yucatan Poorwill
Yucatan Nightjar
Vaux's Swift
Wedge-tailed Sabrewing
Green-breasted Mango
Canivet's Emerald
White-bellied Emerald
Cinnamon Hummingbird
Buff-bellied Hummingbird
Black-headed Trogon
Violaceous Trogon
Collared Trogon
Blue-crowned Motmot
Turquoise-browed Motmot
Belted Kingfisher
White-necked Puffbird
Collared Aracari
Keel-billed Toucan
Yucatan Woodpecker†
Golden-fronted Woodpecker
Ladder-backed Woodpecker
Smoky-brown Woodpecker

Golden-olive Woodpecker
Chestnut-colored Woodpecker
Lineated Woodpecker
Pale-billed Woodpecker
Rufous-breasted Spinetail
Plain Xenops
Tawny-winged Woodcreeper
Ruddy Woodcreeper
Olivaceous Woodcreeper
Barred Woodcreeper
Ivory-billed Woodcreeper
Barred Antshrike
Mexican Antthrush†
N. Beardless Tyrannulet
Greenish Elaenia
Caribbean Elaenia
Yellow-bellied Elaenia
Ochre-bellied Flycatcher
Northern Bentbill
Eye-ringed Flatbill
Yellow-olive Flycatcher
Stub-tailed Spadebill
Royal Flycatcher
Sulphur-rumped Flycatcher
Tropical Pewee
Yellow-bellied Flycatcher
Least Flycatcher
Bright-rumped Attila
Yucatan Flycatcher
Dusky-capped Flycatcher
Great Crested Flycatcher
Brown-crested Flycatcher
Great Kiskadee
Boat-billed Flycatcher
Social Flycatcher
Streaked Flycatcher
Sulphur-bellied Flycatcher
Piratic Flycatcher
Tropical Kingbird
Couch's Kingbird
Thrushlike Mourner†
Grey-collared Becard
Rose-throated Becard
Masked Tityra
Black-crowned Tityra
Red-capped Manakin
Purple Martin
Grey-breasted Martin
Mangrove Swallow
Ridgway's Rough-w. Swallow†
N. Rough-winged Swallow
Cave Swallow
Barn Swallow
Yucatan Jay
Brown Jay
Green Jay
Spot-breasted Wren
White-browed Wren†
White-bellied Wren
Southern House Wren†
Long-billed Gnatwren
Blue-grey Gnat-catcher
Tropical Gnatcatcher
Wood Thrush
Clay-colored Thrush†
Grey Catbird
Black Catbird
Tropical Mockingbird
White-eyed Vireo
Mangrove Vireo
Yellow-throated Vireo
Philadelphia Vireo
Red-eyed Vireo
Yellow-green Vireo
Tawny-crowned Greenlet
Lesser Greenlet
Rufous-browed Peppershrike
Blue-winged Warbler
Northern Parula
Yellow Warbler
Chestnut-sided Warbler
Magnolia Warbler
Black-throated Green Warbler
Black-and-white Warbler
American Redstart
Swainson's Warbler
Ovenbird
Northern Waterthrush
Kentucky Warbler
Common Yellowthroat
Grey-crowned Yellowthroat
Hooded Warbler
Golden-crowned Warbler
Yellow-breasted Chat
Grey-throated Chat
Red-legged Honeycreeper
Scrub Euphonia
Yellow-throated Euphonia
Red-crowned Ant-Tanager
Red-throated Ant-Tanager
Yellow-winged Tanager
Rose-throated Tanager
Summer Tanager
Greyish Saltator
Black-headed Saltator
Northern Cardinal
Rose-breasted Grosbeak
Blue Bunting
Olive Sparrow
Green-backed Sparrow
Blue-black Grassquit
White-collared Seedeater
Yellow-faced Grassquit
Red-winged Blackbird
Melodious Blackbird
Great-tailed Grackle
Bronzed Cowbird
Black-cowled Oriole
Orchard Oriole
Hooded Oriole
Yellow-backed Oriole
Yellow-tailed Oriole
Orange Oriole
Altamira Oriole
Yellow-billed Cacique

Site 14.4: Río Lagartos/Las Coloradas, Yucatán

(January, February 1982, February, September 1984, December 1985, January 1987, July 1988, February 1989, October 1993, February 1994, September 1995, December 1996)

Introduction/Key Species

This site, on the north coast of the Yucatan Peninsula, lies within the Parque Natural Ría Lagartos (note the spelling; the town itself is R*ío* Lagartos). It combines desert-like coastal scrub, thorn forest, a mangrove-fringed estu-

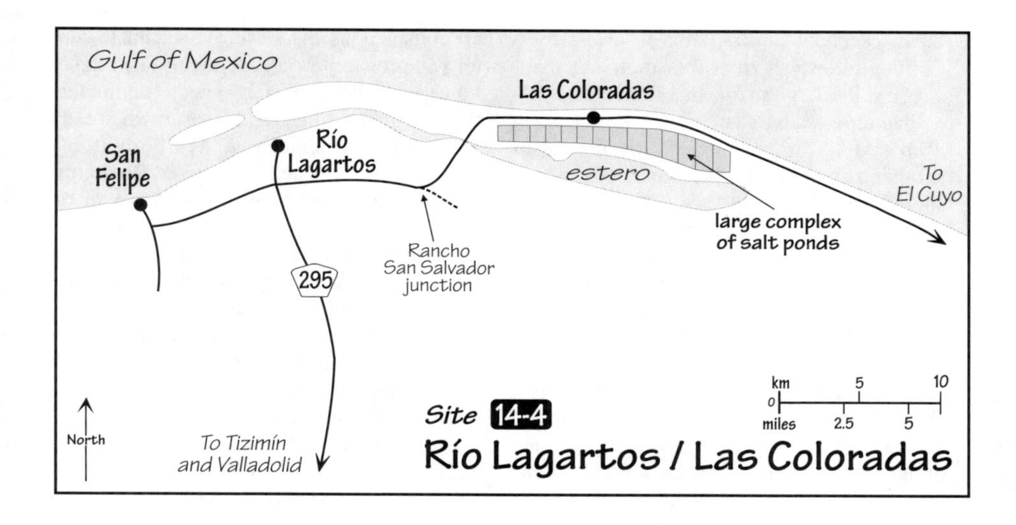

ary-lagoon complex, and a large area of salt evaporation ponds. Boatmen often hail you as you enter town, offering trips to see the flamingos, for which the area is famous. Besides flamingos, one can see an excellent variety of other waterbirds and shorebirds, particularly in winter, as well as species characteristic of the coastal scrub and thorn forest. Specialities here include Kelp and Lesser Black-backed (winter) gulls, **Yucatan Bobwhite**, Zenaida Dove, **Lesser Roadrunner**, **Mexican Sheartail**, and **Yucatan Wren**.

Access

Río Lagartos (with regular bus service from Valladolíd) is about equidistant from Cancún and Mérida; it's a 3–4 hour drive from either if you use the new Mérida/Cancún *cuota* highway. Take the *cuota* road to the exit signed for Valladolíd and Tizimín and head north, on a wide paved road that runs through thorn forest, cornfields, and (mostly overgrown) henequen plantations. Watch for Crested Caracara, Vermilion Flycatcher, **Yucatan Jay**, and **Ridgway's Rough-winged Swallow** on the drive.

Beware the Tizimín one-way system: as you enter the south side of Tizimín, about 1 km after an electricity substation on your left (west), you have to turn sharp left, marked by little more than an arrow pointing ahead but with a diagonal slash through it. Assuming you make this turn (many have missed it, especially at night), turn sharp right in a block, at a Uniroyal Tire Shop, and stay straight for about 1 km to the *zócalo*, which you intersect at its near right (southeast) corner. Continue straight and at the far side keep straight on, which is signed (look carefully) to Río Lagartos. Initially you are on a minor-looking street that, in about 2.5 km, leaves Tizimín and continues north as a narrow, two-lane paved road.

At 52 km from the Tizimín *zócalo* you go through a crossroads, signed left to San Felipe, right to Las Coloradas; Río Lagartos is straight on, about 3 km from this crossroads. The Hotel Nefertiti on the Río Lagartos waterfront closed down recently, and the choice for accommodation is now (December 1996) between some new *cabañas* on the southeast side of Río Lagartos or the new Hotel San Felipe de Jesus (basic) in the small fishing town of San Felipe, some 15 km by road west of Río Lagartos. There also are a few hotels in Tizimín, about an hour to the south.

Unleaded gas has been unavailable at Río Lagartos so it is a good idea to fill your tank in Tizimín; the main gas station is 800 m from the *zócalo*, on the road that leads west from the corner opposite where you enter from the south. Getting there and back to the *zócalo* involves common sense, following the one-way system.

Birding Sites

Río Lagartos, a fishing village protected from the Gulf of Mexico by a beach barrier, offers excellent birding from its waterfront. Coming into town from the south, the main street splits 2.8 km north of the Las Coloradas crossroads; bear right, pass the *mercado* (market) to your left in 200 m, and in another 200 m you hit the waterfront, opposite a small tourist-information kiosk where boat trips can be arranged. For the best views, turn left here and in about 100 m park on the wide, waterfront verge by the clock tower (which can provide good shelter from the wind). There is usually some exposed sand and mud (varying with the tide) where one can find Reddish Egret (white morphs are fairly common), American Oystercatcher, Sandwich Tern, and, not infrequently, groups of flamingos. In winter, flocks of Black Skimmers occur, and Piping Plover has been seen. Early-morning and evening roost flights can number hundreds to thousands of cormorants, White Ibis, and egrets—the occasional *Plegadis* ibis are presumably Glossy, but views are rarely good enough to be certain. Zenaida Doves often fly by, commuting to/from the beach barrier, and **Mexican Sheartails** and Cinnamon Hummingbirds occur in the village gardens.

A boat trip up the estuary (arranged easily at the waterfront) can be fun, and flamingos are basically guaranteed. The boats are open-topped fiberglass *lanchas* and are quite safe, assuming your boatmen is awake. Try to find a boatman who knows about *las cucas*, the local, onomatopoeic name for Boat-billed Herons, which roost in the mangroves, and make sure you negotiate and agree upon the price of your trip before you set out. Bare-throated Tiger-Heron and Pygmy Kingfisher are other species to watch for in the mangroves. If your boat trip goes "upriver" and under a road bridge to the upper estuary, look on the sandbars for Kelp Gulls—two adults have been seen here regularly in recent years.

Las Coloradas, a small village about 25 km by road east of Río Lagartos, is adjacent to a vast area of salt evaporation ponds. To reach this area, go back south out of Río Lagartos, about 3 km, to the first crossroads and turn left (east) on a narrow, two-lane paved road, signed (with luck) to Las Coloradas. At the northeast corner of this junction is a quarry where many pairs of Turquoise-browed Motmots nest—in the summer they are conspicuous on roadside wires, but in winter you probably won't see them. The road to Las Coloradas runs for several kilometers through thorn forest and overgrown weedy fields, all of which can be very birdy in the early morning. Watch for **Lesser Roadrunners** sunning themselves up in bushes or on stone walls, while Crested Caracaras patrol the road for flattened fauna. Other birds along the first few kilometers (reached easily by walking or hitching from Río Lagartos) include White-tailed Hawk, **Yucatan Bobwhite**, **Yucatan Woodpecker**, Mangrove Vireo, Grey-crowned Yellowthroat, and Yellow-backed and **Orange orioles**.

At Km 8 the road bends fairly sharply left, with a dirt track continuing straight on, probably signed to Rancho San Salvador. You can pull in about 100 m on this track and park easily in an open area where a track cuts back to the highway. In the thorn scrub within 100 m of this bend, along the main road in both directions, you should find **Mexican Sheartail** (common) and **Yucatan Wren**. The shallow ponds along the road often have Black-necked Stilt and other shorebirds, while the reed-fringed, freshwater pools should be checked for Least Grebe, Limpkin, and Ruddy Crake.

Back on the main road, continue east and at Km 12.8 (from the highway) cross a bridge (which you would go under in a boat from town; look here for Bare-throated Tiger-Heron, Common Black Hawk, and beware of the *topes*!). The road changes to hard sand at Km 13.2, and bends left to run through agave scrub behind the beach. Check the flowering agaves for male **Mexican Sheartails** and the stalks for **Yucatan Woodpeckers**; Zenaida Doves are often common, and **Yucatan Bobwhites** can be encountered anywhere. At Km 19.7 there is usually a mountain of salt to your right, just before a (new) track bridge over the road, connecting to a small pier. At Km 20.3 the road bends sharply right and becomes divided at a small traffic island (with a few *Casuarina* trees) before meeting a main, airstrip-wide sand road; turn left (right goes into the saltworks), trying not to stare at the unnatural-looking pink ponds, and you soon enter Las Coloradas, a small village with a market area and stores (selling cold drinks!) at Km 21.6. One can walk or drive on the sand tracks behind the market to the beach, some 250 m distant, and check the pier area for Lesser Black-backed Gull (up to three have been seen here).

From Las Coloradas, the hard-sand road continues behind the beach for many kilometers (**Yucatan Bobwhite** is possible anywhere), and inland is a vast complex of salt evaporation ponds with a maze of dikes you can drive around on. There are only a few main dikes on which driving looks safe—if in doubt, head back, as it's not a fun place to get your car stuck! If you see saltworkers, always smile and be polite. The Las Coloradas salt ponds are great for shorebirds, gulls, and terns: hundreds of Lesser Yellowlegs and Western Sandpipers winter, with smaller numbers of Stilt and Semipalmated sandpipers; Snowy and Wilson's plovers are common residents; hundreds of Black Terns occur in migration; Gull-billed Terns appear to be resident; and Least Terns occur in summer. From October through March it is rare to spend a morning birding here without seeing a Peregrine Falcon or two, and often a Merlin.

One route around the dikes that has been good begins with a right-hand (south) turn, into the salt ponds, 1.9 km past (east of) the Las Coloradas store. This road is marked by a small brick hut and small concrete bridge (watch the clearance here) a short way along the dike; the first pond or two is often good for shorebirds. If you reset to zero as you turn into the ponds here, stay on the main track until an obvious main fork and bear left until a T-junction at Km 4.0; turn right here and you are soon at the inland side of the salt-pond maze, overlooking the upper estuary, which you could have reached in a boat from Río Lagartos; check the sandbars for Kelp Gulls. Here and in any of the ponds (water and salt levels vary) you should see hundreds if not thousands of flamingos. Follow the edge of the estuary and come to another T-junction at Km 6.4; turn left here and stay on the main track, which brings you back to the beach road at Km 8.5. From here, you can turn left and go about 5 km back to Las Coloradas, or turn right and explore further, although the best areas seem to be nearer Las Coloradas. Heading back to Río Lagartos, remember the sharp right turn about 1 km out of Las Coloradas, marked by the *Casuarina* trees at the traffic island—otherwise you'll end up in the saltworks!

San Felipe: If you'd turned right (west) at the crossroads coming south from Río Lagartos, the narrow, two-lane paved road runs 8.7 km to a T-junction, signed right to San Felipe, left to Panaba; the first 5 km or of this road run through habitat where **Yucatan Bobwhite**, **Mexican Sheartail**, and **Yucatan Wren** can be found. Turn right at the T-junction and go 1.7 km through San Felipe to the waterfront. The Hotel San Felipe de Jesus is 300 m to the left from where you hit the water. Birds off the waterfront here can be much like Río Lagartos, with pelicans, Reddish Egrets, Roseate Spoonbills, gulls, terns, skimmers, and so on.

Río Lagartos/Las Coloradas Bird List

195 sp. R: rare; V: vagrant (not to be expected). †See pp. 5–7 for notes on English names.

Least Grebe
Pied-billed Grebe
American White Pelican
Brown Pelican
Double-crested Cormorant
Neotropic Cormorant
Anhinga
Magnificent Frigatebird
Least Bittern
Bare-throated Tiger-Heron
Great Blue Heron
"Great White Heron"
Great Egret
Snowy Egret
Little Blue Heron
Tricolored Heron
Reddish Egret
Green Heron
Yellow-crowned Night-Heron
White Ibis
Glossy/White-faced Ibis (R)
Roseate Spoonbill
Wood Stork
American Flamingo†
Black-bellied Whistling-Duck
Blue-winged Teal
Northern Shoveler
American Wigeon
Lesser Scaup
Black Vulture
Turkey Vulture
Lesser Yellow-hd. Vulture
Osprey
Northern Harrier
Cooper's Hawk
Common Black Hawk
Great Black Hawk
Grey Hawk
Short-tailed Hawk
White-tailed Hawk
Red-tailed Hawk
Crested Caracara
Laughing Falcon
American Kestrel
Merlin
Bat Falcon
Peregrine Falcon
Plain Chachalaca
Yucatan Bobwhite†

Ruddy Crake
Clapper Rail
Sora
Common Moorhen
American Coot
Limpkin
Black-bellied Plover
Snowy Plover
Wilson's Plover
Semipalmated Plover
Piping Plover (R)
Killdeer
American Oystercatcher
Black-necked Stilt
American Avocet
Northern Jacana
Greater Yellowlegs
Lesser Yellowlegs
Solitary Sandpiper
Willet
Spotted Sandpiper
Whimbrel
Long-billed Curlew
Marbled Godwit
Ruddy Turnstone
Red Knot
Sanderling
Semipalmated Sandpiper
Western Sandpiper
Least Sandpiper
Pectoral Sandpiper
Dunlin
Stilt Sandpiper
Buff-breasted Sandpiper (R?)
Ruff (V)
Short-billed Dowitcher
Long-billed Dowitcher
Wilson's Phalarope
Red-necked Phalarope (R)
Laughing Gull
Franklin's Gull (R)
Ring-billed Gull
Herring Gull
Lesser Black-backed Gull (R)
Kelp Gull (V)
Gull-billed Tern
Caspian Tern
Royal Tern
Sandwich Tern
Common Tern
Forster's Tern
Least Tern
Black Tern
Black Skimmer
Red-billed Pigeon
White-winged Dove
Zenaida Dove
White-tipped Dove
Common Ground-Dove
Ruddy Ground-Dove
Aztec Parakeet†
Yucatan Parrot†
White-fronted Parrot
Yellow-billed Cuckoo
Mangrove Cuckoo
Squirrel Cuckoo
Lesser Roadrunner
Groove-billed Ani
Ferruginous Pygmy-Owl
Great Horned Owl
Lesser Nighthawk
Pauraque
Canivet's Emerald
Cinnamon Hummingbird
Buff-bellied Hummingbird
Mexican Sheartail
Ruby-throated Hummingbird
Turquoise-browed Motmot
Belted Kingfisher
Green Kingfisher
Yucatan Woodpecker†
Golden-fronted Woodpecker
Ladder-backed Woodpecker
Lineated Woodpecker
N. Beardless Tyrannulet
Least Flycatcher
Vermilion Flycatcher
Bright-rumped Attila
Yucatan Flycatcher
Great Kiskadee
Social Flycatcher
Tropical Kingbird
Couch's Kingbird
Eastern Kingbird
Purple Martin
Tree Swallow
Mangrove Swallow
Bank Swallow
Cliff Swallow
Barn Swallow
Green Jay
Yucatan Jay
Yucatan Wren
White-browed Wren†
White-bellied Wren
Blue-grey Gnatcatcher
White-lored Gnatcatcher
Grey Catbird
Tropical Mockingbird
White-eyed Vireo
Mangrove Vireo
Rufous-browed Peppershrike
Blue-winged Warbler
Tennessee Warbler
Northern Parula
Yellow Warbler
Mangrove Warbler†
Myrtle Warbler†
Yellow-throated Warbler
Palm Warbler
Black-and-white Warbler
American Redstart
Prothonotary Warbler
Ovenbird
Northern Waterthrush
Common Yellowthroat
Grey-crowned Yellowthroat
Scrub Euphonia
Northern Cardinal
Rose-breasted Grosbeak
Blue Grosbeak
Indigo Bunting
Painted Bunting
Blue-black Grassquit
White-collared Seedeater
Yellow-faced Grassquit
Lark Sparrow (V)
Savannah Sparrow
Red-winged Blackbird
Eastern Meadowlark
Melodious Blackbird
Great-tailed Grackle
Orchard Oriole
Hooded Oriole
Yellow-backed Oriole
Yellow-tailed Oriole
Orange Oriole
Altamira Oriole
Lesser Goldfinch

Site 14.5: Chichén Itzá, Yucatán

(February 1984, May 1986, January, February 1987, February 1989, December 1996)

Introduction/Key Species

This Maya site is famous for its spectacular ruins, including the majestic El Castillo pyramid, and offers the birder a variety of birds typical of dry thorn forest as well some species of more humid forest. Birds at

Chichén Itzá include Ferruginous Pygmy-Owl, **Wedge-tailed Sabrewing**, **Canivet's Emerald**, Blue-crowned and Turquoise-browed motmots, Collared Aracari, **Ridgway's Rough-winged** and Cave **swallows**, **Yucatan Jay**, Rufous-browed Peppershrike, **Blue Bunting**, Olive Sparrow, **Orange Oriole**, and a variety of migrant vireos and warblers.

Access/Birding Sites

Chichén Itzá is about 200 km (2–2.5 hours) west of Cancún and about 115 km (1.5–2 hours) east of Mérida and can be reached easily via the old highway (Route 180) or the new *cuota* highway (Route 180D). The ruins are immediately south of Route 180, just to the east of the village of Piste, and are well signed. (If you miss them, you're doomed!) From the *cuota* highway, mark zero at the well-signed exit for Piste and Chichén Itzá (at one of the only two toll booths on the highway), head south into Piste, turn left at the stop sign at Km 3.4 (just past the *zócalo*, where you hit Route 180), and at Km 5.0 stay straight (Route 180 bends left), following signs into the parking lot for the ruins at Km 5.9.

The ruins are open daily from 8 A.M. to 5 P.M. and cost 30 pesos/person (December 1996; free on Sundays and holidays), plus 5 pesos for parking. Tripods are not allowed into the ruins (for no good reason) unless you apply in advance for a permit (good luck!). There is a second entrance into the ruins from the hotel zone immediately to the east, but in December 1996 this was closed. Although the ruin site can get impressively crowded with tourists, there are enough minor trails into the thorn forest (e.g., between Xtoloc Cenote and El Caracol, the observatory) and quiet areas of open woodland behind the structures (e.g., behind/west of the ball court) that one can see birds throughout the heat of the day.

There are several (mostly expensive) hotels and numerous restaurants available around the ruins and in Piste, or Chichén Itzá can be done as a day trip from Mérida or from Valladolíd, some 40 km (30–45 minutes) to the east. For those staying at hotels at the east end of the ruins, the hotel gardens (especially the Hacienda) provide good birding early and late in the day.

Chichén Itzá Bird List

91 sp. †See pp. 5–7 for notes on English names.

Black Vulture
Turkey Vulture
Grey Hawk
Short-tailed Hawk
Zone-tailed Hawk
Plain Chachalaca
White-winged Dove
Ruddy Ground-Dove
White-tipped Dove
White-fronted Parrot
Squirrel Cuckoo
Groove-billed Ani
Ferruginous Pygmy-Owl
Pauraque
Vaux's Swift
Wedge-tailed Sabrewing
Green-breasted Mango
Canivet's Emerald
Cinnamon Hummingbird
Buff-bellied Hummingbird
Violaceous Trogon
Blue-crowned Motmot
Turquoise-browed Motmot
Collared Aracari
Golden-fronted Woodpecker
Yellow-bellied Sapsucker
Golden-olive Woodpecker
Lineated Woodpecker
Olivaceous Woodcreeper
Greenish Elaenia
Tropical Pewee
Least Flycatcher
Great Crested Flycatcher
Brown-crested Flycatcher
Great Kiskadee
Boat-billed Flycatcher
Social Flycatcher
Couch's Kingbird
Rose-throated Becard
Masked Tityra
Purple Martin
Ridgway's Rough-w. Swallow†
Cave Swallow
Green Jay
Yucatan Jay
Spot-breasted Wren
White-bellied Wren
Southern House Wren†
Blue-grey Gnatcatcher
Clay-colored Thrush†
Grey Catbird
Tropical Mockingbird
Cedar Waxwing
White-eyed Vireo
Mangrove Vireo
Yellow-throated Vireo
Philadelphia Vireo
Yellow-green Vireo
Rufous-browed Peppershrike
Northern Parula
Yellow Warbler
Magnolia Warbler
Yellow-throated Warbler
Black-throated Green Warbler
Black-and-white Warbler
American Redstart

Common Yellowthroat
Hooded Warbler
Red-legged Honeycreeper
Scrub Euphonia
Yellow-throated Euphonia
Yellow-winged Tanager
Summer Tanager
Greyish Saltator
Black-headed Saltator
Northern Cardinal
Rose-breasted Grosbeak
Blue Grosbeak
Indigo Bunting
Painted Bunting
Blue Bunting
Olive Sparrow
Melodious Blackbird
Great-tailed Grackle
Hooded Oriole
Orange Oriole
Altamira Oriole
Baltimore Oriole
Lesser Goldfinch

Site 14.6: Progreso/Dzilam de Bravo, Yucatán

(January 1982, February, October 1984, December 1985, February 1994, December 1996)

Introduction/Key Species

This stretch of north Yucatan coast, between the port town of Progreso and the fishing village of Dzilam de Bravo, about 80 km to the east, is excellent for waterbirds, including American Flamingos, as well as for **Mexican Sheartail** and **Yucatan Wren**. It can be birded easily as a half-day trip (or even in 2–3 hours for those with little time) from Mérida, or can be made into a full day for those intent on waterbirds. Because it can get windy along the coast, early morning tends to be best for landbirds, but both the sheartail and, to a lesser extent, the wren can be found readily at any time of day. Progreso can be combined easily into a day trip with a morning or afternoon visit to Dzibilchaltún ruins (Site 14.7).

Other species of interest include Reddish Egret, Yellow-crowned Night-Heron, Roseate Spoonbill, Wood Stork, **Yucatan Bobwhite**, Clapper Rail, Wilson's Plover, Gull-billed Tern, Zenaida Dove, Common Tody-Flycatcher, and White-lored Gnatcatcher.

Access/Birding Sites

Progreso is only 20–40 minutes north of Mérida via a well-signed, fast, divided four-lane highway. A suggested birding route is as follows: head north from Mérida on the Progreso Road, and mark zero as you pass under the Mérida *periferico*. At Km 18.5, exit right up onto an overpass, signed "Yucalpeten," curve back left (west) over the highway, and continue west, through mangroves, for 4.4 km from the overpass bridge (5.0 km from the actual exit) to where the road bends sharply right (north) and there is a large, hard-crushed sand pulloff on the right, marked by two derelict ship hulls and an old building. This pulloff is a good vantage point for American White Pelican, herons (including Reddish Egret, Yellow-crowned Night-Heron), Peregrine Falcon, terns, and Black Skimmer (winter). The adjacent mangroves are worth checking for Clapper Rail and Mangrove [Yellow] Warbler, and the far end of the pulloff usually has Wilson's Plovers (resident, and up to 100 birds in winter). If you want to check some shallow ponds for peeps (fall through spring), turn right back on the road and continue 1.3 km, across the causeway, to a right turn onto a paved road between two shallow ponds. Depending on water levels, either or both ponds can have hundreds of sandpipers. From this point (or, if you're not a peep fan, from the first pullout) retrace your steps and, following the signs for Progreso, go up over the Yucalpeten overpass and turn right, curving back down to the Mérida-Progreso highway, and head north again toward the coast, resetting to zero as you go north under the overpass. At Km 1.9 (21 km north of the Mérida *periferico*) there is a major crossroads which signifies the Progreso bypass; a Pemex station is on the northeast corner. One option here is to turn left and check the lagoon edges for shorebirds and terns, or you can turn right, toward the flamingo lagoons and coastal scrub, for sheartails and wrens.

If you opt to go left, there is a large lagoon on your left and a trashy canal separating you from the southern edge of Progreso town on your right. At Km 2.3 from the crossroads there is usually a sand spit into the lagoon on your left (south), just past the current (December 1996) end of the canal. This spit and the next kilometer or so of lagoon edge can be good for shorebirds and terns; Reddish Egret and Clapper Rail are also possible. From here, turn around and head back to the crossroads.

If you opt to go right for flamingos and all

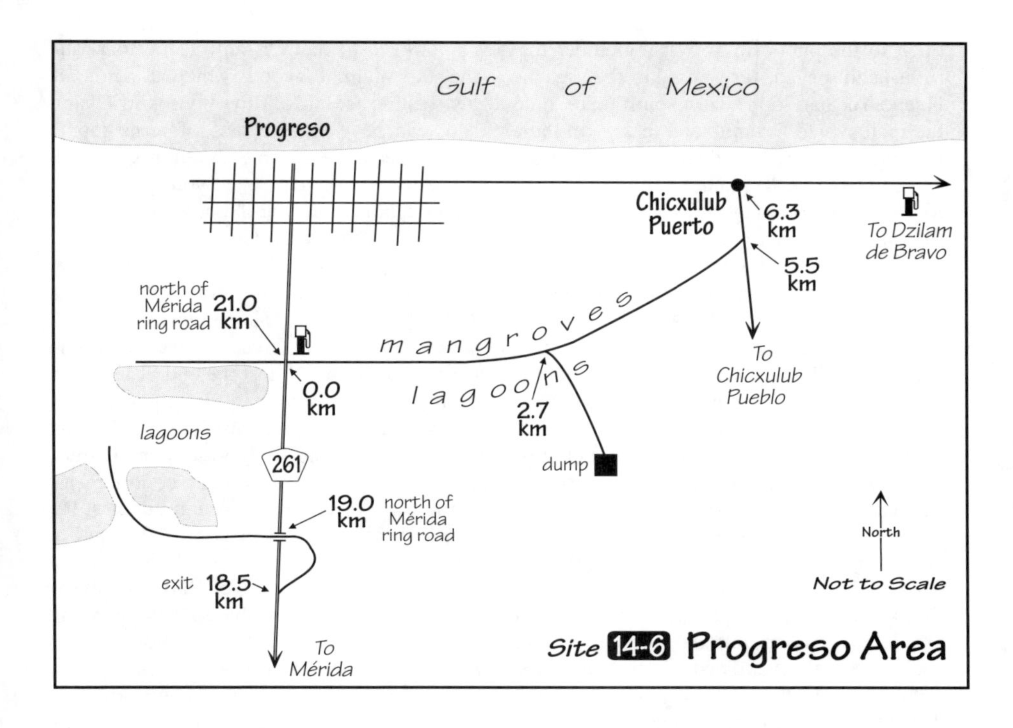

that good stuff, reset to zero as you make the turn. The road continues 5.5 km through mangrove lagoons (egrets, White Ibis, Roseate Spoonbill, Wood Stork, Black-necked Stilts) to a T-junction, signed right (south) to Chicxulub Pueblo, left (north) to Chicxulub Puerto. Turn left and in 750 m (Km 6.3) you hit the main coastal road and a stop sign, with the beach visible straight ahead; left is signed to Progreso, right to Uaymitun. Reset to zero again here and turn right (east).

The paved, two-lane coastal highway runs about 200 m behind the beach, past ever-increasing home and holiday developments between the road and the beach; any remnant patches of coastal scrub can have **Mexican Sheartail** (also in gardens with flowers) and **Yucatan Wren**, though more extensive areas of coastal scrub farther east are better. Anytime after about Km 6 look for groups of American Flamingos in the lagoons to your right (south), and at Km 8.9 (opposite a large sign for the development named Los Flamencos) is a short, inconspicuous, narrow track to the right where you can pull off safely. Usually there are flocks of flamingos here, often very close to the road, and **Mexican Sheartails** occur in the scrub. About 200–300 m back to your right (west) along the lagoon edge is a spit which often has good roosts of terns (including Gull-billed) and shorebirds.

If you continue east on the paved coastal highway, at Km 28.0 the road bends right (0.5 km after a narrow, paved right-hand turn signed to "Dzemul"), and a narrow paved road goes off straight. Stay straight on the latter road, a short "loop" to the Nuevo Yucatán development that rejoins the coastal highway in 1.7 km. Traffic is light along this loop, and there is good, undisturbed coastal scrub where **Mexican Sheartails** can be abundant (fighting for the agave flowers) and **Yucatan Wrens** occur. Other birds here include **Canivet's Emerald**, Cinnamon Hummingbird, White-lored Gnatcatcher, and Grey-crowned Yellowthroat, and **Yucatan Bobwhite** is possible.

If you're not too interested in waterbirds, this would be a good point from which to head back, although a site that may be worth checking, since you're here, is the small Maya ruin site of Xtampu (scheduled to be open to the public in 1997): remember that right-hand paved turn 0.5 km before the Nuevo Yucatán loop? Turn south there, onto the road signed Dzemul, and in 2.3 km there is a narrow paved road to the right (west), signed Cerro de Xtampu; turn here, and the road ends in 700 m at the small parking area for the ruins. There is some low woodland around the ruins, which are surrounded by an area of palmetto savanna. Little birding has been done here to date, although species found include Lesser Yellow-headed Vulture, Crane and Short-tailed hawks, Buff-bellied Hummingbird, Turquoise-browed Motmot, Tropical Pewee, and Green Jay.

If you head back from this point, note that as you get back into the edge of Progreso, the left-hand turn for Chicxulub Pueblo (and the Progreso bypass) is not conspicuous. At 1.3 km after you pass a Pemex station on the left (south), look for a left-hand turn signed Chicxulub Pueblo; take this street, which at first seems very minor, and in about 750 m you come to the familiar-looking road to the right, signed Progreso, which runs back through the mangrove lagoons to the main Mérida-Progreso road.

If you decide to go on, what you see is a matter of season, luck, and water level in the lagoons. In spring (May), large flocks of White-rumped Sandpipers can be found, and in winter Piping Plovers have occurred. Birds can be almost anywhere, although several spots in particular are worth checking. With zero marked where you turned onto the coast road back in Progreso, the end of the Nuevo Yucatán loop is Km 29.6. Shortly east of this point you come to an unused storm-harbor excavated on the left (north) and a large lagoon (Laguna Rosada) on the right (south); at Km 31.6, as the road bends left around the storm harbor, one can park beside the road and check the sand spit and islands in Laguna Rosada, usually an excellent spot for terns and shorebirds (American Avocet, Stilt and Semipalmated sandpipers, Gull-billed Tern). At Km 32.0 you can pull off to the left (north) on a sand track which allows good views of the breakwaters at the entrance to the storm harbor. The small town of Telchac Puerto (cold drinks) is shortly after this and, staying on the coast road, you now go through about 10 km of boring coconut plantations, to San Crisanto (Km 46.4) and the start again of lagoons which, at Km 52.8, just behind (south of) the houses in Chabihau, can have large roosts of terns (up to seven species) and some shorebirds (American Avocet). At Km 63.2 you go through Santa Clara, and the lagoons to the left (north) at Km 65–66 (signed Mina de Oro and marked by some large derelict buildings) are worth checking for shorebirds (Snowy, Wilson's, and Piping plovers, peeps).

At Km 76 you come to the *zócalo* in Dzilam de Bravo, almost the end of the road. From the *zócalo* you can turn right (south), on the paved road toward Temax, through mangrove lagoons for 500–800 m (egrets, ducks, shorebirds), or you can continue east to the harbor: if you stay straight on past the *zócalo*, in about 1 km you come to the harbor on your right (south) and a small tidal lagoon on the left (north) (Clapper Rail, shorebirds including a few Long-billed among many Short-billed Dowitchers). Also, 600 m from the *zócalo*, just before this lagoon, you can turn left a block to the coast and turn right on a hard-sand road for about 500 m to the harbor entrance. This stretch of waterfront often has a selection of gulls, terns, and shorebirds on the rocks along the shore.

From this point you can retrace your steps to Progreso and Mérida (note the caution above about finding the Progreso bypass when heading back) or, if you're feeling adventurous, cut overland on narrow paved roads south from Dzilam de Bravo and then east, via Temax, to Tizimín and up to Río Lagartos (Site 14.4).

Progreso/Dzilam de Bravo Bird List

135 sp. R: rare (not to be expected). †See pp. 5–7 for notes on English names.

Pied-billed Grebe
American White Pelican
Brown Pelican
Double-crested Cormorant
Neotropic Cormorant
Magnificent Frigatebird
Great Blue Heron
Great Egret
Snowy Egret
Little Blue Heron
Tricolored Heron
Reddish Egret
Cattle Egret
Green Heron
Yellow-crowned Night-Heron
White Ibis
Roseate Spoonbill
Wood Stork
American Flamingo†

Black-bellied Whistling-Duck
Northern Pintail
Blue-winged Teal
Northern Shoveler
Lesser Scaup
Red-breasted Merganser (R)
Black Vulture
Turkey Vulture
Osprey
Northern Harrier
Grey Hawk
Short-tailed Hawk
Zone-tailed Hawk
Crested Caracara
American Kestrel
Merlin
Peregrine Falcon
Plain Chachalaca
Yucatan Bobwhite†
Clapper Rail
American Coot
Black-bellied Plover
Wilson's Plover
Snowy Plover
Semipalmated Plover
Killdeer
Black-necked Stilt
American Avocet
Greater Yellowlegs
Lesser Yellowlegs
Willet
Spotted Sandpiper
Whimbrel
Long-billed Curlew
Marbled Godwit
Ruddy Turnstone
Red Knot
Sanderling
Semipalmated Sandpiper
Western Sandpiper
Least Sandpiper
Stilt Sandpiper
Short-billed Dowitcher
Long-billed Dowitcher
Wilson's Phalarope
Laughing Gull
Ring-billed Gull
Herring Gull
Gull-billed Tern
Caspian Tern
Royal Tern
Sandwich Tern
Common Tern
Forster's Tern
Least Tern
Black Tern
Black Skimmer
White-winged Dove
Zenaida Dove
Mourning Dove
Common Ground-Dove
Ruddy Ground-Dove
Aztec Parakeet†
Groove-billed Ani
Canivet's Emerald
Cinnamon Hummingbird
Buff-bellied Hummingbird
Mexican Sheartail
Turquoise-browed Motmot
Belted Kingfisher
Golden-fronted Woodpecker
Common Tody-Flycatcher
Tropical Pewee
Least Flycatcher
Vermilion Flycatcher
Dusky-capped Flycatcher
Tropical Kingbird
Purple Martin
Tree Swallow
Mangrove Swallow
Ridgway's Rough-w. Swallow†
Barn Swallow
Green Jay
Yucatan Wren
Southern House Wren†
Blue-grey Gnatcatcher
White-lored Gnatcatcher
Grey Catbird
Tropical Mockingbird
White-eyed Vireo
Mangrove Vireo
Rufous-browed Peppershrike
Tennessee Warbler
Northern Parula
Yellow Warbler
Mangrove Warbler†
Chestnut-sided Warbler
Magnolia Warbler
Myrtle Warbler†
Black-throated Green Warbler
Yellow-throated Warbler
Prairie Warbler (R)
Palm Warbler
Black-and-white Warbler
American Redstart
Northern Waterthrush
Common Yellowthroat
Grey-crowned Yellowthroat
Summer Tanager
Northern Cardinal
Indigo Bunting
Painted Bunting
Savannah Sparrow
Red-winged Blackbird
Great-tailed Grackle
Hooded Oriole
Baltimore Oriole

Site 14.7: Dzibilchaltún, Yucatán

(January 1982, October 1984, December 1996)

Introduction/Key Species

This very attractive Maya ruin, a short distance north of Mérida, offers access to thorn forest and overgrown fields, with their associated avifauna. Birds of interest include **Yucatan Bobwhite**, **Lesser Roadrunner**, Turquoise-browed Motmot, **Yucatan Woodpecker**, **Yucatan Flycatcher**, **Ridgway's Rough-winged** and Cave swallows, **Yucatan Jay**, **White-browed** [Carolina] and **White-bellied wrens**, White-lored Gnatcatcher, Rufous-browed Peppershrike, Botteri's Sparrow, and **Orange Oriole**.

Access/Birding Sites

From Mérida, head north on the well-signed, divided four-lane highway toward Progreso (Site 14.6). Marking zero as you pass under the overpass of the Mérida *periferico*, the signed right-hand turn to Dzibilchaltún is at Km 6.2. Mark zero at the turn and stay on the narrow, two-lane road through the village of Dzibilchaltún at Km 4.2; on the far side of the village, at Km 4.7, look for a right-hand turn, signed to the ruins. Turn right here and in 500 m you come to the parking lot for the ruins on the right, at the end of the road. Before the ruins open in the morning, this stretch of road can be worth birding (**Yucatan Bobwhite**, Turquoise-browed Motmot, Botteri's Sparrow, **Orange Oriole**).

In recent years Dzibilchaltún has changed greatly, from a sleepy little site to a popular destination with fancy new administration buildings and a major museum. It is still possible to swim in the refreshingly cool *cenote* (limestone sinkhole), but avoid weekends and holidays unless you like crowds. The ruins are open daily from 8 A.M. to 5 P.M. (the museum closes at 4 P.M.) and cost 22 pesos/person (December 1996; free on Sundays and holidays), plus 5 pesos for parking. A nicely produced book on the fauna and flora of Dzibilchaltún is available at the entrance buildings and costs 85 pesos (1996). The book includes a bird list, although this is incomplete and some species are listed in error (e.g., Zenaida Dove, Ringed Kingfisher, Yucatan Wren, Green-backed Sparrow) while others are of dubious occurrence (e.g., Ruddy Quail-Dove, Yucatan Poorwill, Yellow-tailed Oriole).

A gravel track leads about 150 m through thorn forest (good birding) from the parking lot to the administration buildings, whence another short gravel path leads to the ruins through more thorn forest. You enter the ruin complex with a view off to the left down a broad grassy "runway" to a temple, and the main structures and *cenote* are a short distance off to your right. The forest edge around the ruins (which provide good viewing platforms) is worth birding, and there are some narrow paths into the woods, such as at the far west end. If you arrive early, walk the edges of the grassy areas for **Yucatan Bobwhite**.

Dzibilchaltún Bird List

81 sp. †See pp. 5–7 for notes on English names.

Black Vulture
Turkey Vulture
Grey Hawk
American Kestrel
Plain Chachalaca
Yucatan Bobwhite†
Killdeer
Red-billed Pigeon
White-winged Dove
Common Ground-Dove
Ruddy Ground-Dove
White-tipped Dove
Aztec Parakeet†
Squirrel Cuckoo
Lesser Roadrunner
Groove-billed Ani
Ferruginous Pygmy-Owl
Vaux's Swift
Canivet's Emerald
Buff-bellied Hummingbird
Ruby-throated Hummingbird
Black-headed Trogon
Turquoise-browed Motmot
Blue-crowned Motmot
Belted Kingfisher
Yucatan Woodpecker†
Golden-fronted Woodpecker
Ladder-backed Woodpecker
Barred Antshrike
N. Beardless Tyrannulet
Least Flycatcher
Yucatan Flycatcher
Dusky-capped Flycatcher
Great Crested Flycatcher
Brown-crested Flycatcher
Great Kiskadee
Boat-billed Flycatcher
Social Flycatcher
Couch's Kingbird
Rose-throated Becard
Ridgway's Rough-w. Swallow†
Cave Swallow
Green Jay
Yucatan Jay
Spot-breasted Wren
White-browed Wren†
White-bellied Wren
Southern House Wren†
Blue-grey Gnatcatcher
White-lored Gnatcatcher
Clay-colored Thrush†
Tropical Mockingbird
White-eyed Vireo
Mangrove Vireo
Yellow-throated Vireo
Rufous-browed Peppershrike
Northern Parula
Blue-winged Warbler
Magnolia Warbler
Black-throated Green Warbler
Black-and-white Warbler
American Redstart
Ovenbird
Grey-crowned Yellowthroat
Hooded Warbler
Scrub Euphonia
Greyish Saltator
Northern Cardinal
Rose-breasted Grosbeak
Indigo Bunting
Painted Bunting
Olive Sparrow
Botteri's Sparrow
Blue-black Grassquit
Yellow-faced Grassquit
Melodious Blackbird
Great-tailed Grackle
Hooded Oriole
Orange Oriole
Altamira Oriole
Yellow-billed Cacique

Site 14.8: Celestún, Yucatán

(September/October 1984, January, February 1987, February 1989, 1994, December 1996)

Introduction/Key Species

This site in the Parque Natural Ría Celestún is famous for its flamingos (the largest numbers being present in winter) but is also good for a variety of waterbirds and a few speciality landbirds. Birding (or at least boat trips to see the flamingos) are important to the town's economy; consequently, awareness about protecting the environment has increased greatly in the past ten years or so.

Most of the birds here can be seen from land, but a boat trip up the Ría Celestún is recommended for getting close-up views of the flamingos and a variety of waterbirds typical of the mangroves. Species of interest that can be found in and around Celestún include Bare-throated Tiger-Heron, Boat-billed Heron, American Flamingo, Lesser Yellow-headed Vulture, **Yucatan Bobwhite**, Rufous-necked Wood-Rail, **Mexican Sheartail**, Pygmy Kingfisher, **Yucatan Woodpecker**, **Yucatan Wren**, and White-lored Gnatcatcher.

Access

The small coastal fishing town of Celestún (served by regular bus service) is about 95 km (1.5–2 hours) west of Mérida, whence there are two routes, one via Hunucma, the other via Uman. If you're not in a hurry and enjoy Maya villages, with their attendant *topes*, the former route may be for you. The quicker route, however, is via Uman: from Mérida, head south on Route 180, signed to Campeche, marking zero at the Mérida *periferico*. At Km 7.1, on the north side of Uman, the road splits into a one-way system, and you have to fork right. Follow the main route, past a Pemex station on the left at Km 7.3, to the *zócalo*, on your left (east) at Km 8.6. At the traffic lights on the southwest corner of the *zócalo* is a signed, hard right turn to Celestún (ahead are a huge church and a sign for Campeche Zona Arqueologica and Campeche Via Corta). Reset to zero as you turn right here, then bend sharply left in 600 m (there's not much choice!) and right again and you're out of Uman. Stay on the main road through Oxolon (Km 7), Samahil (Km 15.1), and Kinchil (Km 22; the road from Hunucma joins from the right at Km 23.2), whence it is about 50 km to Celestún.

At Km 58.5 look for a couple of reed-fringed, freshwater pools on your left (south), at the far (west) side of which are a couple of derelict concrete huts where you can pull off safely and park. Usually there's not too much here, but early and late in the day **Ridgway's Rough-winged** and Cave **swallows** often come in to drink. The bridge across the estuary into Celestún is at Km 72, and you drive across it into town (see Birding Sites).

When coming back via Uman, if you're returning to Mérida, at the *zócalo* junction go straight across and turn left at the far side onto the northbound (Mérida) one-way street.

There are one or two basic hotels in Celestún, and numerous restaurants; more and fancier hotels are apparently being built. There isn't (December 1996) unleaded gasoline available in Celestún, so it's best to fill up in Mérida or Uman if your tank is low.

Birding Sites

Several sites in and around Celestún are worth checking for birds. The first, as you approach the bridge over the estuary coming from Uman, is a small, trashy pullout on the left (south) immediately before you get onto the bridge. There is room to park here, and by walking quietly along the mangrove edge, you may find Rufous-necked Wood-Rail (the extent of suitable mud varies with the tide). Note that Grey-necked Wood-Rail occurs in the freshwater marshes just before (east of) the mangroves. You can either walk from here or drive onto the bridge and park (at least briefly), to scan the estuary for waterbirds (pelicans, herons including Reddish Egret, flamingos—usually long lines of pink are visible off in the distance to the north); early and late in the day the bridge is a good spot from which to watch roost flights of waterbirds, and flocks of flamingos sometimes fly overhead.

On the Celestún side of the bridge, the boat dock for flamingo trips is down to your left (south), and hard to miss, while any exposed mud along the mangrove edge to the north should be checked for Rufous-necked

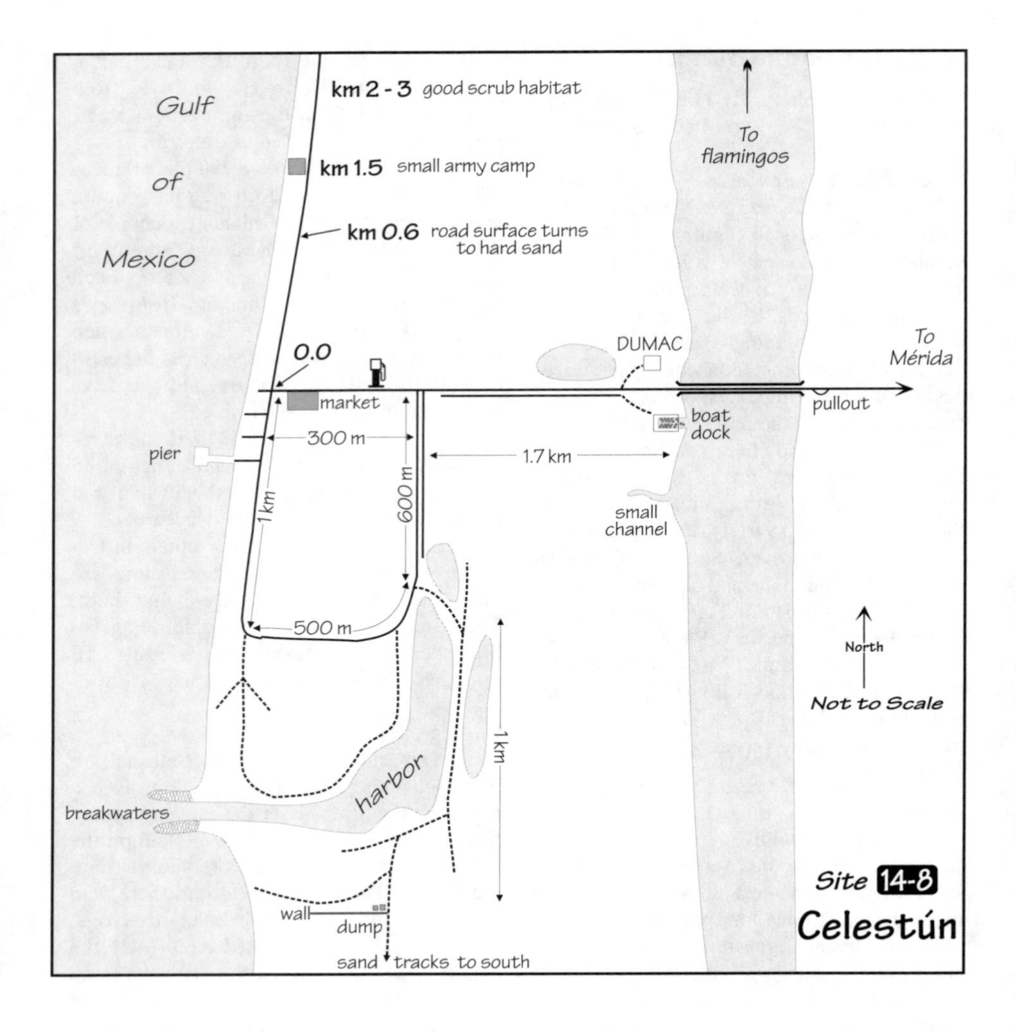

Wood-Rails. Due largely to increased traffic, the boats now have a more or less set route, to minimize disturbance, and with it a set price (though you may be able to bargain). As always, make sure you agree upon price and time before you set out. Most trips go upriver (north), to where the flamingos are, and also take in an island where frigatebirds and both species of cormorants nest. You may not see Boat-billed Herons on this route, however; try to find a boatman who knows the *cucas* (the local onomatopoeic name for this great bird) and can take you to some. One site you can mention is very close to the bridge—a small channel into the mangroves a few hundred meters south of the bridge and boat dock, on the Celestún side of the estuary: Bare-throated Tiger-Heron and Pygmy Kingfisher also can be seen here, but access depends upon water levels.

Marking the bridge as zero, the road runs into town (the ponds on your right, immediately after the bridge, are worth checking for herons and shorebirds), with a crossroads at Km 1.7 (the left-hand road, a divided street, was under construction in December 1996), a Pemex station on the right at Km 1.8, and the *mercado* and *zócalo* at Km 2.0; you hit the waterfront road at Km 2.1 and can go either left (south) or right (north).

If you turn left, in about 250 m turn right (easy to miss) at the third block and you'll be at the base of the Celestún pier, often a good roosting place for gulls, terns, and skimmers, which tend to be acclimated to people and thus allow excellent, close-range views. If it's windy, scanning out to sea can produce a jaeger or two, which sometimes come right over the pier to harass terns.

Continue south from the pier. The pavement ends in about 750 m, with a small naval camp on the right-hand corner and a dirt soccer field on the left. You can continue straight here for a few hundred meters on hard-sand tracks to scan the harbor mouth and beach or turn left to get around to the south side of the harbor and the dump. If you turn left, turn left again in 300 m and in another 200 m turn right onto a narrow, diked-up dirt road that cuts diagonally across the head of the harbor; in 200 m, at the far side, turn right again on a hard-sand dike along the east side of the harbor. This main track can be followed, bending right and then left, for another kilometer to the dump, on the far south side of town. The harbor and adjacent pools can be good for herons (Reddish Egret, Yellow-crowned Night Heron), shorebirds, gulls, terns, and skimmers.

A hundred meters before the dump, which is marked by a couple of small buildings and the inevitable vultures and distinctive odor, a sand track branches off to the right. This can be followed, between the dump and the harbor, for 400 m to the beach on the south side of the breakwaters. It is often possible to get close-range views of perched Lesser Yellow-headed Vultures along this track (as well as elsewhere at the dump), and the beach here sometimes has a roost of gulls and terns. If you continue south through the dump, hard-sand tracks go for many kilometers through beach scrub and mangroves, where **Yucatan Bobwhite**, **Mexican Sheartail**, **Yucatan Wren**, and White-lored Gnatcatcher can be found.

Back in town, turning north instead of south on hitting the waterfront road, the pavement turns to hard sand in 600 m, then passes a small army camp at Km 1.5. **Yucatan Bobwhite**, **Mexican Sheartail**, **Yucatan Wren**, and White-lored Gnatcatcher all occur in the semi-open coastal scrub between Km 2 and 3. This hard-sand road continues north behind the beach for many kilometers, with more of the same in terms of birds, and Zenaida Doves can be common, at least seasonally (September/October 1984).

Celestún Bird List

164 sp. R: rare; V: vagrant (not to be expected). †See pp. 5–7 for notes on English names.

Pied-billed Grebe
American White Pelican
Brown Pelican
Double-crested Cormorant
Neotropic Cormorant
Anhinga
Magnificent Frigatebird
Bare-throated Tiger-Heron
Great Blue Heron
"Great White Heron" (R)
Great Egret
Snowy Egret
Little Blue Heron
Tricolored Heron
Reddish Egret
Cattle Egret
Green Heron
Black-crowned Night-Heron
Yellow-crowned Night-Heron
Boat-billed Heron
White Ibis
Roseate Spoonbill
Wood Stork
American Flamingo†
Fulvous Whistling-Duck (R)
Northern Pintail
Blue-winged Teal
Northern Shoveler
American Wigeon
Ring-necked Duck
Lesser Scaup
Red-breasted Merganser (R)
Black Vulture
Turkey Vulture
Lesser Yellow-hd. Vulture
Osprey
Northern Harrier
Common Black Hawk
Great Black Hawk
Short-tailed Hawk
White-tailed Hawk
Zone-tailed Hawk
Crested Caracara
Laughing Falcon
Collared Forest-Falcon
Merlin
Bat Falcon
Peregrine Falcon
Plain Chachalaca
Yucatan Bobwhite†
Grey-necked Wood-Rail
Rufous-necked Wood-Rail
Common Moorhen
American Coot
Black-bellied Plover
Snowy Plover
Wilson's Plover
Semipalmated Plover
Killdeer
American Oystercatcher
Black-necked Stilt
Greater Yellowlegs
Lesser Yellowlegs
Willet
Spotted Sandpiper
Ruddy Turnstone
Sanderling
Semipalmated Sandpiper
Western Sandpiper
Least Sandpiper
Stilt Sandpiper
Short-billed Dowitcher
Wilson's Phalarope
Parasitic Jaeger

Laughing Gull
Ring-billed Gull
Herring Gull
Sabine's Gull (V)
Caspian Tern
Royal Tern
Sandwich Tern
Common Tern
Forster's Tern
Least Tern
Black Skimmer
White-winged Dove
Zenaida Dove
Mourning Dove
Common Ground-Dove
Ruddy Ground-Dove
White-tipped Dove
Aztec Parakeet†
Lesser Roadrunner
Groove-billed Ani
Great Horned Owl
Lesser Nighthawk
Pauraque
Yucatan Nightjar
Canivet's Emerald
Buff-bellied Hummingbird
Cinnamon Hummingbird
Mexican Sheartail
Belted Kingfisher
Green Kingfisher
Pygmy Kingfisher†
Yucatan Woodpecker†
Golden-fronted Woodpecker
Ladder-backed Woodpecker
Golden-olive Woodpecker
Lineated Woodpecker
Pale-billed Woodpecker
Ivory-billed Woodcreeper
N. Beardless Tyrannulet
Least Flycatcher
Vermilion Flycatcher
Dusky-capped Flycatcher
Social Flycatcher
Bright-rumped Attila
Tropical Kingbird
Couch's Kingbird
Purple Martin
Tree Swallow
Mangrove Swallow
Ridgway's Rough-w. Swallow†
Bank Swallow
Cave Swallow
Barn Swallow
Green Jay
Yucatan Wren
Blue-grey Gnatcatcher
White-lored Gnatcatcher
Grey Catbird
Black Catbird (R)
Tropical Mockingbird
American Pipit
White-eyed Vireo
Mangrove Vireo
Rufous-browed Peppershrike
Northern Parula
Tennessee Warbler
Yellow Warbler
Mangrove Warbler†
Myrtle Warbler†
Yellow-throated Warbler
Prairie Warbler (R)
Palm Warbler
Black-and-white Warbler
American Redstart
Prothonotary Warbler
Worm-eating Warbler
Ovenbird
Northern Waterthrush
Common Yellowthroat
Hooded Warbler
Northern Cardinal
Indigo Bunting
White-collared Seedeater
Yellow-faced Grassquit
Savannah Sparrow
Red-winged Blackbird
Melodious Blackbird
Great-tailed Grackle
Hooded Oriole
Yellow-backed Oriole
Altamira Oriole
Yellow-billed Cacique

Site 14.9: Uxmal, Yucatán

(January 1982, January, February 1987, February 1989, December 1996)

Introduction/Key Species

Uxmal is a famous site of impressive Maya ruins amid deciduous thorn forest. The woods around the ruins hold a variety of species typical of the peninsula, including **Yucatan Poorwill**, **Yucatan Nightjar**, Blue-crowned and Turquoise-browed motmots, **Yucatan Flycatcher**, **Ridgway's Rough-winged** and Cave **swallows**, Green and **Yucatan jays**, **White-browed** [Carolina] and **White-bellied wrens**, Rufous-browed Peppershrike, **Grey-throated Chat**, **Blue Bunting**, and **Orange Oriole**.

Access

Uxmal (served by regular bus service) is about 75 km (1.3–2 hours) south of Mérida, on Route 261. From Mérida, head south on Route 180, the main highway to Campeche, marking zero at the Mérida *periferico*. At Km 7.1 bear right into the one-way system through Uman, pass the *zócalo* (and the turn for Celestún, Site 14.8) at Km 8.6 and stay more or less straight, bearing right immediately past the *zócalo* in response to a potentially ambiguous sign for Campeche Zona Arqueologica and Campeche Via Corta. At Km 9.4 watch for a left-hand turn (straight is signed prominently for Campeche, and left to Chetumal, but the sign for Uxmal is less conspicuous and could be missed at night); turn left here and then right in 500 m, signed for Uxmal, among other places. Stay on this road through Muna (the *zócalo* is at Km 56.2), and then (a shock if you've been in the north and east Yucatan for a few days) the road winds sharply up over some low hills and down again, reaching Uxmal at about Km 73. Returning to Mérida, retrace your steps, following the signed one-way systems through Muna and Uman.

At Uxmal there are a few (mostly expensive) hotels near the ruins, and several restaurants. For more economical accommodation one could stay in Ticul, some 30 km to the east, or you can visit the ruins as a day trip from Mérida.

The ruins are open daily from 8 A.M. to 5 P.M. and cost 30 pesos/person (December 1996; free on Sundays and holidays), plus 5 pesos for parking. Tripods are not allowed unless a permit has been obtained in advance.

Birding Sites

Coming from Muna, before reaching the ruins of Uxmal you pass the Rancho Uxmal and Misión Uxmal on your right (west). About 1.5 km south of the latter, the highway bends sharply left, with the Hacienda Uxmal on your left (1.7 km) and the entry to the Villas Arqueologicas Hotel and the ruins themselves on your right (1.8 km). At the bend there is a wide gravel verge, and at the near (north) side of this ample parking area look for a dirt track into the thorn forest to the right (west); the track's entrance is 1.4 km south of the Misión Uxmal. Opposite the track, across the road, are the gardens of the Hacienda Uxmal and a weedy second-growth area, with papaya plants, which can be good for flocks of jays, saltators, and orioles. The track itself leads into good forest habitat and forks about 250 m from the highway. One can go either way, and between the arms of the fork, well hidden by the vegetation, are some seasonal shallow ponds where Masked Duck, Snail Kite, and Limpkin have been seen. The thorn forest here can be birded profitably before the ruins open, and it hosts Collared Forest-Falcon (often heard at dawn), **Yucatan Flycatcher**, Green and **Yucatan jays**, **White-browed** [Carolina] and **White-bellied wrens**, **Grey-throated Chat**, and **Blue Bunting**.

From the parking lot of the ruins, note that a narrow paved road continues on to the south. Birding along here in the morning, before the parking lot is crowded, can be good (jays, saltators, orioles, etc.). In the ruins themselves, any edge and open woodland away from noisy tourist groups can be good; Turquoise-browed Motmots are common and acclimated to being approached closely (the good side of noisy tourists!). Hundreds of Cave Swallows nest in the Magician's Pyramid, and probably in other structures, and flocks can often be seen leaving the rooms around the central courtyard between 8 and 9:30 A.M.

For night birding, work the main road and any side tracks north to the vicinity of the Rancho Uxmal, where **Yucatan Poorwill** and **Yucatan Nightjar** can be found.

Uxmal Bird List

101 sp. †See pp. 5–7 for notes on English names.

Pied-billed Grebe
Black Vulture
Turkey Vulture
Sharp-shinned Hawk
Crane Hawk
Grey Hawk
Roadside Hawk
Short-tailed Hawk
Collared Forest-Falcon
Merlin
Plain Chachalaca
Purple Gallinule
Common Moorhen
Red-billed Pigeon
White-winged Dove
Common Ground-Dove
Ruddy Ground-Dove
White-tipped Dove
Aztec Parakeet†
Squirrel Cuckoo
Groove-billed Ani
Ferruginous Pygmy-Owl
Great Horned Owl
Lesser Nighthawk
Pauraque
Yucatan Poorwill
Yucatan Nightjar
Vaux's Swift
Green-breasted Mango
Canivet's Emerald
Cinnamon Hummingbird
Buff-bellied Hummingbird
Blue-crowned Motmot
Turquoise-browed Motmot
Golden-fronted Woodpecker
Golden-olive Woodpecker
Lineated Woodpecker
Ivory-billed Woodcreeper
N. Beardless Tyrannulet
Tropical Pewee
Yellow-olive Flycatcher
Least Flycatcher
Yucatan Flycatcher
Dusky-capped Flycatcher
Brown-crested Flycatcher
Social Flycatcher
Boat-billed Flycatcher
Bright-rumped Attila
Great Kiskadee
Tropical Kingbird
Couch's Kingbird
Masked Tityra
Purple Martin
Ridgway's Rough-w. Swallow†
Cave Swallow
Green Jay
Spot-breasted Wren

White-browed Wren†
White-bellied Wren
Southern House Wren†
Blue-grey Gnatcatcher
Clay-colored Thrush†
Grey Catbird
Tropical Mockingbird
White-eyed Vireo
Mangrove Vireo
Yellow-throated Vireo
Rufous-browed Peppershrike
Blue-winged Warbler
Northern Parula
Magnolia Warbler
Black-throated Green Warbler
Yellow-throated Warbler
Black-and-white Warbler
American Redstart
Ovenbird
Common Yellowthroat
Hooded Warbler
Grey-throated Chat
Scrub Euphonia
Greyish Saltator
Black-headed Saltator
Northern Cardinal
Rose-breasted Grosbeak
Blue Grosbeak
Indigo Bunting
Painted Bunting
Blue Bunting
Olive Sparrow
Blue-black Grassquit
Yellow-faced Grassquit
Melodious Blackbird
Great-tailed Grackle
Black-cowled Oriole
Hooded Oriole
Orange Oriole
Altamira Oriole
Yellow-billed Cacique
Lesser Goldfinch

Site 14.10: Isla Cozumel, Quintana Roo

(February 1982, 1984, May 1986, February 1987, July 1988, February 1989, October 1993, December 1996)

Introduction/Key Species

Isla Cozumel (Cozumel Island), often referred to simply as Cozumel, is a low limestone island about 45 km long and 15 km wide. It lies 20 km off the Caribbean coast of Quintana Roo, some 60 km south of Cancún, and, despite its proximity to the mainland, has a distinctive avifauna: of the approximately 50 breeding species, four are endemic, about fourteen are represented by endemic subspecies, and a few others are of Caribbean origin and occur only rarely on the mainland. Conversely, many species common on the adjacent mainland do not occur on Cozumel. In addition, several species of migrants occur on Cozumel much more commonly than they do on the Mexican mainland.

Birds of interest on Cozumel include **Ruddy Crake**, Bridled Tern (summer), White-crowned Pigeon, Caribbean Dove, **Yucatan Parrot**, Mangrove Cuckoo, Smooth-billed Ani, **Yucatan Nightjar**, **Cozumel Emerald**, **Yucatan Woodpecker**, Caribbean Elaenia, **Yucatan Flycatcher**, **Cozumel Wren**, **Black Catbird**, **Cozumel Thrasher**, **Cozumel** and Yucatan **vireos**, Cape May, Black-throated Blue, Prairie, and Swainson's warblers (all winter), Stripe-headed and **Rose-throated tanagers**, and distinctive endemic forms of Roadside Hawk, Blue-grey Gnatcatcher, Rufous-browed Peppershrike, Yellow Warbler, Bananaquit, and Yellow-faced Grassquit.

Access

As a tropical Caribbean island, with the attendant reefs and beaches, Cozumel is a major tourist destination, so getting there is not a problem. There are daily flights direct from several U.S. cities and from Mexico City, as well as regular short (twenty-minute) flights during the day (8 A.M. to 8 P.M.; December 1996) from Cancún. An inexpensive passenger ferry (thirty minutes to an hour, depending upon sea state) runs twelve times a day (from 4 or 5 A.M. to 8 P.M.; December 1996) between Playa del Carmen and Cozumel, and a vehicle ferry runs between Puerto Morelos and Cozumel. It is possible, and far easier, to leave your car at a locked and guarded lot in Playa del Carmen for up to several days, rather than going through the hassle of trying to get it over to Cozumel (and back) on the vehicle ferry.

There is no shortage of places to stay and eat, although the budget traveler may be pushed to find cheap accommodation. Rental of vehicles ranging from bicycles to cars is readily available, but note that Cozumel is subject to irregular but not infrequent heavy tropical downpours of rain at any time of day or year. Usually these aren't long-lived, but a few seconds is all you need to get you and your equipment drenched, so you might think about this when renting transport, or when taking your expensive tape-recording equipment out on an unprotected moped.

Life on Cozumel centers around the town

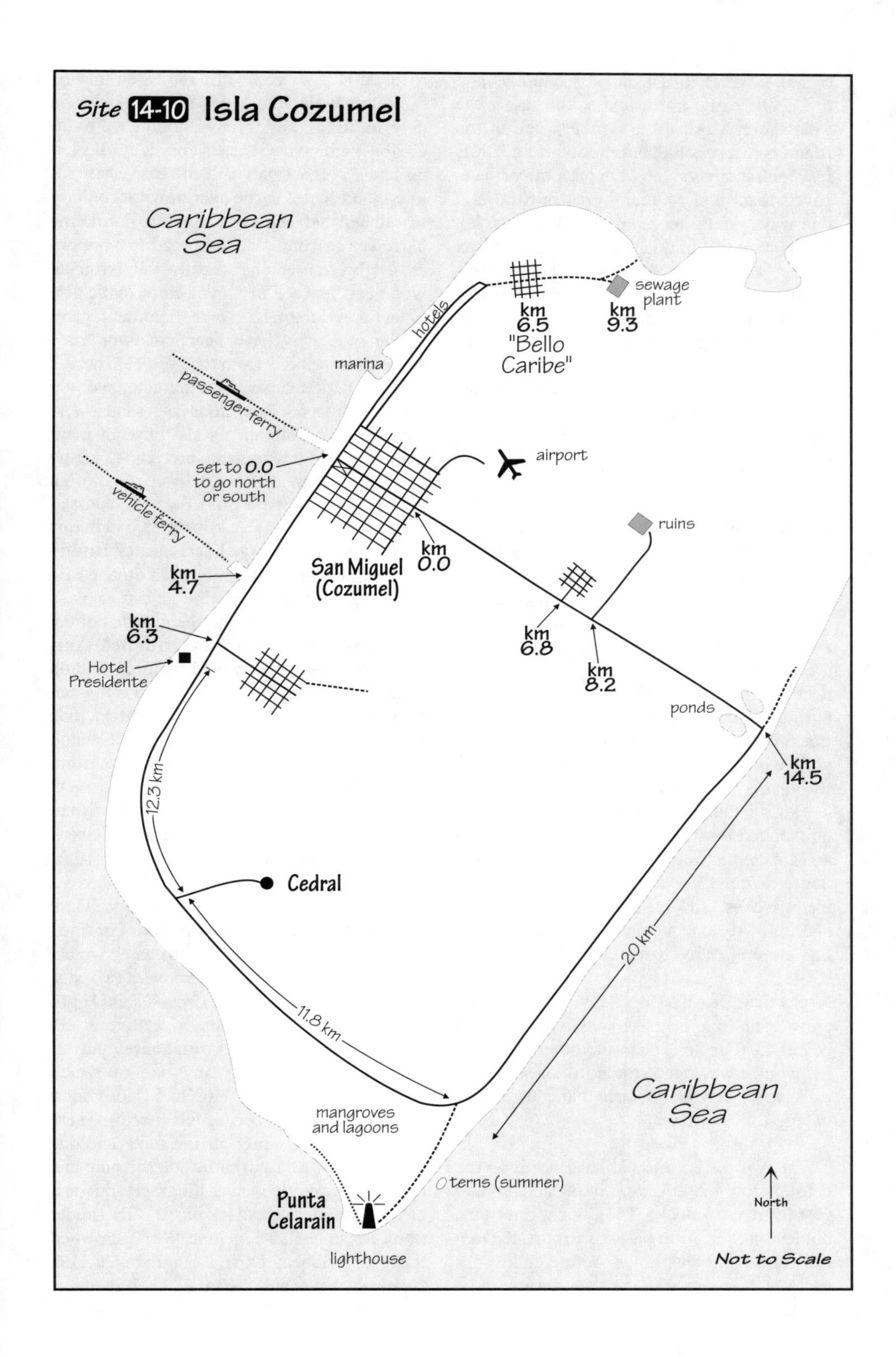
Site 14-10 Isla Cozumel
Caribbean Sea
hotels
km 6.5
"Bello Caribe"
km 9.3
sewage plant
marina
passenger ferry
airport
set to 0.0 to go north or south
vehicle ferry
ruins
km 0.0
San Miguel (Cozumel)
km 4.7
km 6.3
Hotel Presidente
km 6.8
km 8.2
ponds
km 14.5
12.3 km
Cedral
20 km
11.8 km
Caribbean Sea
mangroves and lagoons
terns (summer)
Punta Celarain
lighthouse
North
Not to Scale

of San Miguel (often called Cozumel) and the hotels to its north and south, along the west shore. Most of the northern half of the island is inaccessible, but the southern half is reached readily via a 65 km loop of two-lane paved road. San Miguel works mostly on a one-way system (maps are readily available), and it is worth noting that the main Pemex station, at the intersection of Avenidas Benito Juarez and 30 Sur, is best approached by coming up Juarez, despite the temptation (and misleading signs) to turn into it off 30 Sur, even with a moped. Technically, you would drive a few feet the wrong way down Juarez, and the traffic cops who linger there may well pull you over (while turning a blind eye to the residents to whom this maneuver is second nature!).

Birding Sites

Most of the species of interest occur in the dense thicket woodland and edge that covers much of Cozumel, and a morning in this habitat anywhere is likely to produce most or even all of the specialities. Even the overgrown lots around the edge of town, such as near the airport, can be very birdy, especially during migration and winter, and fruiting trees here can hold both tanagers when you've been looking without success in quiet woodland elsewhere!

Having said this, a few areas can be recommended for their general species diversity and for being away from traffic and barking dogs, as much as that is possible. Like so many places, however, tourist development continues to expand on Cozumel, so areas may be subject to change.

Areas near Town. Depending on where you are staying, one of these sites is likely to be nearer to your hotel than the others but, unless you're way out to the north or south, all three will likely be no more than a fifteen- to twenty-minute drive.

The North. From town, head north on the coastal road, marking zero at the *zócalo* (and passenger ferry dock). The divided four-lane boulevard runs north past the airport, the marina, and numerous hotels before ending (as such) at Km 5.9, whence you can continue north on a bumpy but passable hard-sand road. At Km 6.5 you will be in an unfinished housing development (possibly marked by a deteriorating sign announcing "Bello Caribe") with paved streets and sidewalks but no houses. The small grid of streets here allows good access to the overgrown secondary woodland and edge. Continuing north, the narrower and bumpier sand road is more enclosed by denser and slightly taller thicket woodland, and at Km 9.3 you come to the gate of a fancy-looking sewage treatment plant (which you may have been smelling for a while if the wind is from the north!). There is room to park a car outside the gate, and you can continue on foot along the sand track, which runs for about 2.5 km through mangroves, scrubby woodland, and some marshy palmetto savanna to a small sandy (and often buggy) beach. There's probably not much reason to walk more than a kilometer, and in this stretch keep an eye (and ear) out for **Ruddy Crakes**, which often can be seen quite easily if you're quiet and patient. One can also ask permission to look around the grounds of the sewage plant, where Smooth-billed Anis occur (this species can be difficult to find on Cozumel). Other species found between Bello Caribe and the sewage plant include White-crowned Pigeon, Mangrove Cuckoo (in scrub, not just mangroves), **Cozumel Emerald**, **Yucatan Woodpecker**, Caribbean Elaenia, **Cozumel Wren**, **Black Catbird**, **Cozumel** and **Yucatan vireos**, and, in winter, Prairie and Swainson's warblers.

The East (Cross-Island Highway). Mark zero at the Pemex station and continue straight east on Avenida Benito Juarez which, after a *glorieta* in 800 m, becomes a two-lane paved road at Km 1.5: the Cross-Island Highway. The road runs through relatively tall, humid forest, punctuated with houses and numerous narrow tracks, any one of which could be good birding. At Km 6.8 look for a gated, broad track to your left (north). There is room to park safely off the road, and one can step over a low wall beside the gate and bird the forest edge along the track and various crossroads without having to worry about traffic. Species here include White-crowned Pigeon, Caribbean Dove, Mangrove Cuckoo, **Cozumel Emerald**, **Yucatan Woodpecker**, **Cozumel Wren**, **Black Catbird**, **Cozumel** and **Yucatan vireos**, Swainson's Warbler (winter), Stripe-headed Tanager, and so on.

The South. Head south from San Miguel along the coast road, marking zero again at the *zócalo*. At Km 4.7 you pass the vehicle ferry terminal on your right (west); at Km 6.3 look for a narrow paved road to the left, which may be signed "Palmar Ranch, Horseback Riding." This turn is about 250 m before the Hotel Presidente on your right (west), so if you reach that you've gone too far. Turn left onto the side road and, in 500 m, pass the "ranch" with horses on your left. At 1.8 km from the highway, the road forks and becomes another seemingly failed housing development with paved streets and sidewalks and a few derelict buildings. You can go right or left at the fork; both lead into the small grid of paved streets. This grid area has most of the island specialities: White-crowned Pigeon, Caribbean Dove, Yucatan Parrot, Mangrove Cuckoo, **Cozumel Emerald**, **Yucatan Woodpecker**, Caribbean Elaenia, **Cozumel Wren**, **Black Catbird**, **Cozumel** and **Yucatan vireos**, Swainson's Warbler (winter), Stripe-headed and **Rose-throated tanagers**, and so on.

You'll have noticed that **Cozumel Thrasher** is conspicuously absent from these lists. This distinctive endemic seems to have been affected adversely by a hurricane which devastated Cozumel in autumn 1988. All the other island species have recovered (although the tanagers still seem less numerous than before the hurricane), but the thrasher, which was common and widespread before, now appears to be distinctly uncommon, even rare. A few birds have been reported in recent years, however, and you could come across it almost anywhere.

Farther Afield on Cozumel. For those with more time on Cozumel it can be worth checking a few other areas, and you'll probably find some good spots yourself. If you make the round-island circuit by road, the following areas may be worth checking.

Assuming you start by heading east out of town, on the Cross-Island Highway (see under "The East," above), at Km 8.2 from the Pemex station look for a left turn signed to San Gervasio Maya Ruins. There is a gate at the highway junction, where you will be waved through (the ruins are open 8 A.M. to 4 P.M. daily), and it's about a 3 km drive via a narrow paved road, through some nice habitat, to the parking lot for the ruins; entry to the ruins cost 17 pesos/person in December 1996. This site can get crowded, so if you plan any birding there, go early, as soon as it's open. Birds are much the same as at the sites mentioned above, although you may have greater visibility into the shady understory for species like Swainson's and Kentucky warblers (winter).

Staying on the Cross-Island Highway, you hit the windward, east side of the island at Km 14.5, with a nice view of waves rolling in onto white-sand beaches. Just before the coast, on either side of the highway, are a couple of mangrove- and reed-fringed seasonal pools which, if not dry, are worth checking. Species seen here include Least Bittern and Masked Duck.

Turning right (south), the highway runs behind the beach, through low scrub that usually doesn't have too much in the way of birds, until Km 34.8, where it bends right to cut back across the island. At this bend, a hard-sand road goes off straight ahead and leads to the **Punta Celarain** lighthouse, which you can see in the distance. The lighthouse road is bumpy but passable with care, although if it's rained recently you might want to check some puddle depths before driving through them. Bird-wise, there's not much reason to visit the lighthouse itself (4.2 km from the highway), although the mangrove lagoons beyond, reached via a continuation of the sand road (closed to vehicle traffic in December 1996) can be good for wading birds (Reddish Egret, White Ibis, Roseate Spoonbill). En route to the lighthouse, at 2.4 km from the highway, look for a pulloff on the left (east), possibly marked by a small *palapa*. The spit of sand here, and the small rocky islet just off the point (both hidden from the road), are worth checking in spring through autumn for roosting terns (Roseate and Least, Brown Noddy), and a few pairs of Bridled Terns nest on the islet.

Back on the main road, at Km 46.6 (about 11.8 km from the Punta Celarain turnoff), there is a paved right turn (east) signed Cedral (and probably Cedral Maya Ruins). The weedy fields and second growth around the village of **Cedral** offer a slightly different variety of species from the woodland and mangroves, and this is a good spot for Smooth-billed Ani (often together with Groove-billeds). Turn right toward Cedral; at 2.8 km from the highway the road bends sharply right. Look here for a long, straight, white-limestone road to the left. You can turn here and continue a few hundred meters, almost to the end, where you turn right, go another few hundred meters to a stop sign (with a huge, overgrown weedy field ahead; December 1996), and turn right again which comes back to the parking area for the ruins, about 250 m on from where you turned off the pavement to make this circuit. A block before you get back to the pavement you can turn left for a kilometer or so along the edge of the huge weedy field which, together with the surrounding animal pens and fields, can be good for anis (both species). The whole circuit can be very birdy, especially during winter when numerous migrants are present. Birds here include Prairie and Black-throated Blue warblers (winter), Grey-crowned Yellowthroat, Blue Grosbeak (winter), Indigo and Painted buntings (winter), Blue-black Grassquit, White-collared Seedeater, and Lesser Goldfinch (escapes?).

Back at the highway around the island, turn right (north) and it's about 19 km to downtown San Miguel. Alternatively, the Cedral road is 12.3 km south of the Hotel Presidente.

Isla Cozumel Bird List

168 sp. A more complete list for the island, with status for each species, is given in Howell and Webb (1995; Appendix D); note that Rose-throated Becard, Warbling Vireo, and Lesser Goldfinch (escapes?) have been recorded since Howell and Webb (1995), Grey-crowned Yellowthroat has been confirmed, and Mourning Warbler and Blue-black Grassquit were overlooked by Howell and Webb (1995). R: rare; V: vagrant (not to be expected) †See pp. 5–7 for notes on English names.

Least Grebe
Pied-billed Grebe
Brown Pelican
Cormorant sp.
Anhinga
Magnificent Frigatebird
Least Bittern
Great Blue Heron
"Great White Heron"
Great Egret
Snowy Egret
Little Blue Heron
Tricolored Heron
Reddish Egret
Cattle Egret
Green Heron
Yellow-crowned Night-Heron
White Ibis
Glossy Ibis (R)
Roseate Spoonbill
Blue-winged Teal
Northern Shoveler
Ring-necked Duck
Black Vulture
Turkey Vulture
Osprey
Hook-billed Kite (R)
Swallow-tailed Kite
Roadside Hawk
Short-tailed Hawk
American Kestrel
Merlin
Bat Falcon (R)
Peregrine Falcon
Ruddy Crake
Sora
Common Moorhen
American Coot
Black-bellied Plover
Wilson's Plover
Semipalmated Plover
Killdeer
Black-necked Stilt
Northern Jacana
Greater Yellowlegs
Solitary Sandpiper
Willet
Spotted Sandpiper
Ruddy Turnstone
Red Knot
Sanderling
Least Sandpiper
Common Snipe
Laughing Gull
Royal Tern
Sandwich Tern
Roseate Tern (R?)
Common Tern
Least Tern
Bridled Tern
Brown Noddy
Black Skimmer
White-crowned Pigeon
White-winged Dove
Mourning Dove
Common Ground-Dove
Ruddy Ground-Dove
Caribbean Dove
Yucatan Parrot†
Yellow-billed Cuckoo
Mangrove Cuckoo
Smooth-billed Ani
Groove-billed Ani
Stygian Owl (R?)
Lesser Nighthawk
Common Nighthawk
Pauraque
Yucatan Nightjar
Chimney Swift
Vaux's Swift
Green-breasted Mango
Cozumel Emerald
Belted Kingfisher
Yucatan Woodpecker†
Golden-fronted Woodpecker
Yellow-bellied Sapsucker
N. Beardless Tyrannulet
Greenish Elaenia
Caribbean Elaenia
Eastern Pewee†
Least Flycatcher
Bright-rumped Attila
Yucatan Flycatcher
Dusky-capped Flycatcher
Brown-crested Flycatcher
Scissor-tailed Flycatcher
Tropical Kingbird
Eastern Kingbird
Rose-throated Becard (R?)
Purple Martin
Grey-breasted Martin
Tree Swallow

Mangrove Swallow
N. Rough-winged Swallow
Bank Swallow
Cliff Swallow
Barn Swallow
Cozumel Wren†
Blue-grey Gnatcatcher
Grey-cheeked Thrush
Swainson's Thrush
Wood Thrush
American Robin (V)
Grey Catbird
Black Catbird
Tropical Mockingbird
Cozumel Thrasher
White-eyed Vireo
Cozumel Vireo
Warbling Vireo (R)
Red-eyed Vireo
Yucatan Vireo
Rufous-browed Peppershrike
Blue-winged Warbler
Tennessee Warbler
Northern Parula
Yellow Warbler
Golden Warbler†
Chestnut-sided Warbler
Magnolia Warbler
Cape May Warbler
Black-throated Blue Warbler
Myrtle Warbler†
Black-throated Green Warbler
Yellow-throated Warbler
Prairie Warbler
Palm Warbler
Bay-breasted Warbler
Black-and-white Warbler
American Redstart
Worm-eating Warbler
Swainson's Warbler
Ovenbird
Northern Waterthrush
Mourning Warbler
Kentucky Warbler
Common Yellowthroat
Grey-crowned Yellowthroat
Hooded Warbler
Yellow-breasted Chat
Cozumel Bananaquit†
Red-legged Honeycreeper (R)
Stripe-headed Tanager
Rose-throated Tanager
Summer Tanager
Scarlet Tanager
Northern Cardinal
Rose-breasted Grosbeak
Blue Grosbeak
Indigo Bunting
Painted Bunting
Blue-black Grassquit
White-collared Seedeater
Yellow-faced Grassquit
Great-tailed Grackle
Bronzed Cowbird
Orchard Oriole
Hooded Oriole
Baltimore Oriole
Lesser Goldfinch

LITERATURE CITED

Alden, P. 1969. *Finding the Birds in Western Mexico*. Tucson: University of Arizona Press.

American Ornithologists' Union (AOU). 1983. *Checklist of North American Birds*, 6th ed. Washington, D.C.: AOU.

———. 1985–1997. Thirty-fifth to forty-first supplements to AOU *Checklist of North American Birds*, 6th ed. *Auk* 102:680–686; 104:591–595; 106:532–538; 108:750–754; 110:675–682; 112:819–830; 114:542–552.

Andrle, R. F. 1966. North American migrants in the Sierra de Tuxtla (*sic*) of southern Veracruz, Mexico. *Condor* 68:177–184.

———. 1967. Birds of the Sierra de Tuxtla (*sic*) in Veracruz, Mexico. *Wilson Bulletin* 79:163–187.

Arvin, J. C. 1990. *A Checklist of the Birds of the Gomez Farias Region, Southwestern Tamaulipas*. Austin, Tex.: Self-published.

Baptista, L. F., and S. L. L. Gaunt. 1997. Bioacoustics as a tool in conservation studies. Pp. 212–242 in J. R. Clemmons and R. Buchholz, eds., *Behavioural Approaches to Conservation in the Wild*. Cambridge: Cambridge University Press.

Behrstock, R. A., and T. L. Eubanks. 1997. Additions to the avifauna of Nuevo León, Mexico, with notes on new breeding records and infrequently seen species. *Cotinga* 7:27–30.

Behrstock, R. A., C. W. Sexton, G. W. Lasley, T. L. Eubanks, and J. P. Gee. 1997. First nesting records of Worthen's Sparrow *Spizella wortheni* for Nuevo León, Mexico, with a habitat characterisation of the nest site and notes on ecology, voice, additional recent sightings and leg coloration. *Cotinga* 8:27–33.

Binford, L. C. 1989. A distributional survey of the birds of the Mexican state of Oaxaca. *Ornithological Monographs* No. 43. Washington, D.C.: AOU.

Buckingham, R. 1997. Reader's Letters. *Wildlife Sound* 8(1):5–6.

Coates-Estrada, R., and A. Estrada. 1985. *Lista de las Aves de la Estación de Biología Los Tuxtlas*. México, D.F.: Instituto de Biología de la UNAM.

Delaney, D. 1987. Finding the Flammulated Flycatcher (*Deltarhynchus flammulatus*) in Colima. *MBA Bulletin Board* 1(2):3.

Edwards, E. P., and R. E. Tashian. 1959. Avifauna of the Catemaco Basin of southern Veracruz, Mexico. *Condor* 61:325–327.

Erickson, R. A., and T. E. Wurster. 1998. Confirmation of nesting in Mexico for four bird species from the Sierra San Pedro Mártir, Baja California. *Wilson Bulletin* 110:118–120.

Erickson, R. A., A. D. Barron, and T. E. Wurster. 1994. Northern Saw-whet Owl in the Sierra San Pedro Mártir: First Baja California record. *Western Birds* 25:66–68.

Gomez de Silva G., H., and S. Aguilar R. 1994. The Bearded Wood-Partridge in central Veracruz and suggestions for finding and conserving the species. *Euphonia* 3:8–12.

Gomez de Silva G., H., and A. M. Sada. 1997. Nava's Wren *Hylorchilus navai* in Oaxaca, Mexico. *Cotinga* 7:20.

Holt, H. 1997. *A Birder's Guide to Colorado*. Colorado Springs, Colo.: ABA.

Howell, S. N. G. 1992. The Short-crested Coquette: Mexico's least-known endemic. *Birding* 24:86–91.

———. 1996. *A Checklist of the Birds of Mexico*. Berkeley, Calif.: Golden Gate Audubon Society.

Howell, S. N. G., and S. Webb. 1987. Birding at Tecolutla, Veracruz. *MBA Bulletin Board* 1(3):4.

———. 1990. A site for Buff-collared Nightjar (*Caprimulgus ridgwayi*) and Mexican Sheartail (*Calothorax [=Doricha] eliza*) in Veracruz. *Aves Mexicanas* 2(2):1–2.

———. 1992. A little-known cloud forest in Hidalgo, Mexico. *Euphonia* 1:7–11.

———. 1995. *A Guide to the Birds of Mexico and Northern Central America*. Oxford: Oxford University Press.

Howell, S. N. G., P. Snetsinger, and R. G. Wilson. 1997. A sight record of the White-fronted Swift *Cypseloides storeri* in Michoacán, Mexico. *Cotinga* 7:23–26.

Ireland, D. 1997. Reader's Letters. *Wildlife Sound* 8(1):6–7.

Martin, J. P. 1997. The first Southern Lapwing *Vanellus chilensis* in Mexico. *Cotinga* 8:52–53.

Novick, R., and L. S. Wu. 1994. *Where to Find Birds in San Blas, Nayarit*, 3rd ed. San Blas, Nay., Mexico: Self-published.

Russell, S. M., and G. Monson. In press (1998). *The Birds of Sonora*. Tucson: University of Arizona Press.

Schaldach, W. J., Jr. 1963. The avifauna of Colima and adjacent Jalisco, Mexico. *Proceedings of the Western Foundation of Vertebrate Zoology* 1:1–100.

Schaldach, W. J., Jr., and B. P. Escalante-Pliego. 1997. Lista de Aves. In E. González S., R. Dirzo, and R. C. Vogt, eds., *Historia Natural de Los Tuxtlas*. México, D.F.: Instituto de Biología de la UNAM.

Schaldach, W. J., Jr., and A. R. Phillips. 1961. The Eared Poor-Will. *Auk* 78:567–572.

Sibley, C. G., and B. L. Monroe, Jr. 1990. *Distribution and Taxonomy of Birds of the World*. New Haven, Conn.: Yale University Press.

Wetmore, A. 1943. The birds of southern Veracruz, Mexico. *Proceedings of the U.S. National Museum* 93:215–340.

Wilson, R. G., and H. Ceballos-L. 1993. *The Birds of Mexico City*, 2nd ed. Burlington, Ontario, Canada: BBC Printing and Graphics.

Zimmerman, D. A. 1957. Spotted-tailed Nightjar nesting in Veracruz, Mexico. *Condor* 59:124–127.

APPENDIX A

CHECKLIST OF SPECIES INCLUDED

The following table lists by chapter all species included in the site accounts of this guide (see Appendix B for details of regional specialities). Most species can be located in the Bird Lists of a chapter; note, however, that some may be mentioned only within the accounts for Birding Sites. Species endemic to Mexico and northern Central America are **boldfaced**.

In the listings below, abbreviated genera indicate that a species has the same genus as the preceding entry. Lowercase names indicate taxa treated (provisionally) as subspecies, following Howell (1996). If the "parent" species is listed first, as in "MALLARD *A. platyrhynchos,*" then it is not repeated for the subspecies following, as in "Mexican duck *A. [p.] diazi.*"

Key

- • Occurs regularly (common to rare); most such species can be found readily in the relevant habitat and season; some require time and/or luck
- ▲ Occurs rarely and/or irregularly, not found reliably (often refers to rare migrants/vagrants)

	1: Baja	2: NW Mexico	3: Monterrey	4: NE Mexico	5: Mazatlán	6: San Blas	7: Colima	8: Cen. Mexico	9: Guerrero	10: Veracruz	11: Oaxaca	12: Cen. Chiapas	13: N. Chiapas	14: Yucatán
GREAT TINAMOU *Tinamus major*										•	•		•	
LITTLE TINAMOU *Crypturellus soui*										•	•		•	
THICKET TINAMOU *C. cinnamomeus*				•					•	•	•	•	•	•
SLATY-BREASTED TINAMOU *C. boucardi*										•		•	•	
RED-THROATED LOON *Gavia stellata*	•													
ARCTIC LOON *G. arctica*	▲													
PACIFIC LOON *G. pacifica*	•													
LEAST GREBE *Tachybaptus dominicus*	•		•		•	•	•		•	•	•	•	•	•
PIED-BILLED GREBE *Podilymbus podiceps*	•	•	•		•	•	•	•	•	•	•	•	•	•
HORNED GREBE *Podiceps auritus*	•													

	1: Baja	2: NW Mexico	3: Monterrey	4: NE Mexico	5: Mazatlán	6: San Blas	7: Colima	8: Cen. Mexico	9: Guerrero	10: Veracruz	11: Oaxaca	12: Cen. Chiapas	13: N. Chiapas	14: Yucatán
EARED GREBE *P. nigricollis*	●	●	●	●		●	●			●	●			
WESTERN GREBE *Aechmophorus occidentalis*	●	●	▲											
CLARK'S GREBE *A. [o.] clarki*	●	▲					●							
NORTHERN FULMAR *Fulmarus glacialis*	●													
COOK'S PETREL *Pterodroma cookii*	▲													
PINK-FOOTED SHEARWATER *Puffinus creatopus*	●													
WEDGE-TAILED SHEARWATER *P. pacificus*	▲													
SOOTY SHEARWATER *P. griseus*	●	▲												
BLACK-VENTED SHEARWATER *P. opisthomelas*	●				●	●	▲							
TOWNSEND'S SHEARWATER *P. auricularis*	▲										▲			
AUDUBON'S SHEARWATER *P. lherminieri*									●		●			
WILSON'S STORM-PETREL *Oceanites oceanicus*	▲													
LEACH'S STORM-PETREL *Oceanodroma leucorhoa*	●													
GALAPAGOS STORM-PETREL *O. tethys*	●													
ASHY STORM-PETREL *O. homochroa*	▲													
BLACK STORM-PETREL *O. melania*	●	●				●	●		●		●			
LEAST STORM-PETREL *O. microsoma*	●	●				●	●		●		●			
RED-BILLED TROPICBIRD *Phaethon aethereus*						●	●		●		●			
MASKED BOOBY *Sula dactylatra*	▲						▲				▲			
BLUE-FOOTED BOOBY *S. nebouxii*	●	●			●	●	▲							
BROWN BOOBY *S. leucogaster*	●	●			●	●	●		●		●	●		
AM. WHITE PELICAN *Pelecanus erythrorhynchos*	●	●		●		●	●			●	●	●	●	●
BROWN PELICAN *P. occidentalis*	●	●		●	●	●	●		●	●	●	●		●
DOUBLE-CRESTED CORMORANT *Phalacrocorax auritus*	●	●	●	●						●				●
NEOTROPIC CORMORANT *P. brasilianus*	●	●	●	●	●	●	●		●	●	●	●	●	●
BRANDT'S CORMORANT *P. penicillatus*	●	▲												
PELAGIC CORMORANT *P. pelagicus*	●													
ANHINGA *Anhinga anhinga*				●	●	●	●			●		●	●	●

	1: Baja	2: NW Mexico	3: Monterrey	4: NE Mexico	5: Mazatlán	6: San Blas	7: Colima	8: Cen. Mexico	9: Guerrero	10: Veracruz	11: Oaxaca	12: Cen. Chiapas	13: N. Chiapas	14: Yucatán
MAGNIFICENT FRIGATEBIRD *Fregata magnificens*	●	●			●	●	●		●	●	●	●		●
PINNATED BITTERN *Botaurus pinnatus*										●			●	
AMERICAN BITTERN *B. lentiginosus*							●	●		●			●	
LEAST BITTERN *Ixobrychus exilis*	●					●	●	●		●		●	●	●
BARE-THROATED TIGER-HERON *Tigrisoma mexicanum*				●		●				●		●	●	●
GREAT BLUE HERON *Ardea herodias*	●	●	●	●	●	●	●	●		●	●	●	●	●
GREAT EGRET *Egretta alba*	●	●	●	●	●	●	●	●	●	●	●	●	●	●
SNOWY EGRET *E. thula*	●	●	●	●	●	●	●	●	●	●	●	●	●	●
LITTLE BLUE HERON *E. caerulea*	●			●	●	●	●	●		●	●	●	●	●
TRICOLORED HERON *E. tricolor*	●			●		●	●	●		●	●	●	●	●
REDDISH EGRET *E. rufescens*	●	●		●		●	●			●	●	●		●
CATTLE EGRET *Bubulcus ibis*	●	●	●	●	●	●	●	●	●	●	●	●	●	●
GREEN HERON *Butorides virescens*	●	●	●	●		●	●	●		●	●	●	●	●
AGAMI HERON *Agamia agami*										▲				
BLACK-CROWNED NIGHT-HERON *Nycticorax nycticorax*	●	●	●			●	●	●		●	●	●	●	●
YELLOW-CROWNED NIGHT-HERON *N. violaceus*	●			●		●	●			●	●		●	●
BOAT-BILLED HERON *Cochlearius cochlearius*						●	●			●			●	●
WHITE IBIS *Eudocimus albus*				●	●	●	●			●	●	●	●	●
GLOSSY IBIS *Plegadis falcinellus*							▲						●	▲
WHITE-FACED IBIS *P. chihi*	●	●		●		●	●	●		●	●	●	●	
ROSEATE SPOONBILL *Platalea ajaja*				●		●	●			●	●	●	●	●
JABIRU *Jabiru mycteria*													●	
WOOD STORK *Mycteria americana*				●	●	●	●			●	●	●	●	●
AMERICAN FLAMINGO *Phoenicopterus ruber*														●
FULVOUS WHISTLING-DUCK *Dendrocygna bicolor*						●	●			●			●	▲
BLACK-BELLIED WHISTLING-DUCK *D. autumnalis*	▲			●	●	●	●			●	●	●	●	●
TUNDRA SWAN *Cygnus columbianus*	▲													
WHITE-FRONTED GOOSE *Anser albifrons*	●		▲											
SNOW GOOSE *A. caerulescens*	●	●					●							

	1: Baja	2: NW Mexico	3: Monterrey	4: NE Mexico	5: Mazatlán	6: San Blas	7: Colima	8: Cen. Mexico	9: Guerrero	10: Veracruz	11: Oaxaca	12: Cen. Chiapas	13: N. Chiapas	14: Yucatán
ROSS' GOOSE *A. rossii*	▲	●												
BRANT *Branta bernicla*	●					▲								
CANADA GOOSE *B. canadensis*	●													
MUSCOVY DUCK *Cairina moschata*				●		●	●			●			●	
WOOD DUCK *Aix sponsa*							▲			▲				
GREEN-WINGED TEAL *Anas crecca*	●	●	●			●	●	●		●			●	
MALLARD *A. platyrhynchos*	●	●												
Mexican Duck *A. [p.] diazi*		●	●	●			●	●						
MOTTLED DUCK *A. fulvigula*				●										
NORTHERN PINTAIL *A. acuta*	●	●	●			●	●	●		●		●	●	●
BLUE-WINGED TEAL *A. discors*	●		●	●		●	●	●		●	●	●	●	●
CINNAMON TEAL *A. cyanoptera*	●	●	●	●		●	●	●		●				
NORTHERN SHOVELER *A. clypeata*	●	●	●	●		●	●	●		●		●	●	●
GADWALL *A. strepera*	●	●	●	●		●	●	●		●				
EURASIAN WIGEON *A. penelope*	●						▲							
AMERICAN WIGEON *A. americana*	●	●	●	●		●	●	●		●		●	●	●
CANVASBACK *Aythya valisineria*	●	●	●											
REDHEAD *A. americana*	●		●				●			●				
RING-NECKED DUCK *A. collaris*	●	●	●	●		●	●	●						●
GREATER SCAUP *A. marila*	▲													
LESSER SCAUP *A. affinis*	●		●	●		●	●	●		●	●	●	●	●
BLACK SCOTER *Melanitta nigra*	▲													
SURF SCOTER *M. perspicillata*	●		▲											
WHITE-WINGED SCOTER *M. fusca*	●													
COMMON GOLDENEYE *Bucephala clangula*	●	●												
BUFFLEHEAD *B. albeola*	●	●	●	●										
HOODED MERGANSER *Lophodytes cucullatus*	▲		▲				▲			▲				
COMMON MERGANSER *Mergus merganser*	▲													
RED-BREASTED MERGANSER *M. serrator*	●	●		●										▲
RUDDY DUCK *Oxyura jamaicensis*	●	●	●			●	●	●				●		
MASKED DUCK *O. dominica*													●	▲
BLACK VULTURE *Coragyps atratus*		●		●	●	●	●	●	●	●	●	●	●	●
TURKEY VULTURE *Cathartes aura*	●	●	●	●	●	●	●	●	●	●	●	●	●	●

	1: Baja	2: NW Mexico	3: Monterrey	4: NE Mexico	5: Mazatlán	6: San Blas	7: Colima	8: Cen. Mexico	9: Guerrero	10: Veracruz	11: Oaxaca	12: Cen. Chiapas	13: N. Chiapas	14: Yucatán
LESSER YELLOW-HEADED VULTURE *C. burrovianus*										•		•	•	•
KING VULTURE *Sarcoramphus papa*									•		•		•	•
OSPREY *Pandion haliaetus*	•	•	•	•	•	•	•	•	•	•	•	•	•	•
GREY-HEADED KITE *Leptodon cayanensis*										•			•	
HOOK-BILLED KITE *Chondrohierax uncinatus*					•	•	•		•	•		•	•	•
SWALLOW-TAILED KITE *Elanoides forficatus*												•	•	
WHITE-TAILED KITE *Elanus leucurus*	•			•	•	•	•			•	•	•	•	•
SNAIL KITE *Rostrhamus sociabilis*										•	•		•	▲
DOUBLE-TOOTHED KITE *Harpagus bidentatus*							•		•	•			•	•
MISSISSIPPI KITE *Ictinia mississippiensis*										•			•	
PLUMBEOUS KITE *I. plumbea*													•	
BALD EAGLE *Haliaeetus leucocephalus*	•	•												
NORTHERN HARRIER *Circus cyaneus*	•	•	•	•	•	•	•	•		•	•	•	•	•
SHARP-SHINNED HAWK *Accipiter striatus*	•	•	•	•	•	•	•	•	•	•	•	•	•	•
WHITE-BREASTED HAWK *A. chionogaster*												•		
BICOLORED HAWK *A. bicolor*										•				
COOPER'S HAWK *A. cooperi*	•	•	•	•	•	•	•	•	•	•	•	•	•	
NORTHERN GOSHAWK *A. gentilis*		•												
CRANE HAWK *Geranospiza caerulescens*						•	•			•	•		•	•
BLACK-COLLARED HAWK *Busarellus nigricollis*													•	
WHITE HAWK *Leucopternis albicollis*										•	•		•	
COMMON BLACK HAWK *Buteogallus anthracinus*				•		•	•		•	•	•	?	•	•
Mangrove Black Hawk *B. [a.] subtilis*												•		
GREAT BLACK-HAWK *B. urubitinga*				•		•			•	•	•	•	•	•
SOLITARY EAGLE *Harpyhaliaetus solitarius*								▲						
HARRIS' HAWK *Parabuteo unicinctus*	•		•	•	•	•	•			•		▲		
GREY HAWK *Buteo nitidus*		•		•	•	•	•	•	•	•	•	•	•	•
ROADSIDE HAWK *B. magnirostris*				•			•			•	•	•	•	•
RED-SHOULDERED HAWK *B. lineatus*	•		•	•										
BROAD-WINGED HAWK *B. platypterus*	▲			•		•	•		•	•	•	•	•	

	1: Baja	2: NW Mexico	3: Monterrey	4: NE Mexico	5: Mazatlán	6: San Blas	7: Colima	8: Cen. Mexico	9: Guerrero	10: Veracruz	11: Oaxaca	12: Cen. Chiapas	13: N. Chiapas	14: Yucatán
SHORT-TAILED HAWK *B. brachyurus*				●	●	●	●	●	●	●	●	●	●	●
SWAINSON'S HAWK *B. swainsoni*	▲	●	●	●			●			●	●	●	●	
WHITE-TAILED HAWK *B. albicaudatus*							●				●	●	●	●
ZONE-TAILED HAWK *B. albonotatus*	●			●	●	●	●		●	●	●	●		●
RED-TAILED HAWK *B. jamaicensis*	●	●	●	●	●	●	●	●	●	●	●	●	●	●
FERRUGINOUS HAWK *B. regalis*	●	●	●											
GOLDEN EAGLE *Aquila chrysaetos*	●	●	●											
BLACK HAWK-EAGLE *Spizaetus tyrannus*									●	●	●		●	●
ORNATE HAWK-EAGLE *S. ornatus*								●		●	●	●	●	●
CRESTED CARACARA *Caracara plancus*	●	●		●	●	●	●			●	●	●	●	●
LAUGHING FALCON *Herpetotheres cachinnans*						●	●		●	●	●	●	●	●
BARRED FOREST-FALCON *Micrastur ruficollis*								●	●	●	●	●	●	
COLLARED FOREST-FALCON *M. semitorquatus*				●	●	●	●		●	●	●		●	●
AMERICAN KESTREL *Falco sparverius*	●	●	●	●	●	●	●	●	●	●	●	●	●	●
MERLIN *F. columbarius*	●	●			●	●	●			●		●	●	●
APLOMADO FALCON *F. femoralis*				●						●			●	
BAT FALCON *F. rufigularis*				●					●	●	●	●	●	●
PEREGRINE FALCON *F. peregrinus*	●		●	●		●	●			●	●	●	●	●
PRAIRIE FALCON *F. mexicanus*	●	●	●											
PLAIN CHACHALACA *Ortalis vetula*			●	●						●	●	●	●	●
WEST MEXICAN CHACHALACA *O. poliocephala*							●	●	●		●			
RUFOUS-BELLIED CHACHALACA *O. wagleri*		●			●	●								
WHITE-BELLIED CHACHALACA *O. leucogastra*												●		
HIGHLAND GUAN *Penelopina nigra*												●		
CRESTED GUAN *Penelope purpurascens*					●	●	●	●	●	●	●		●	
HORNED GUAN *Oreophasis derbianus*												●		
GREAT CURASSOW *Crax rubra*													●	●
OCELLATED TURKEY *Meleagris ocellata*														●

	1: Baja	2: NW Mexico	3: Monterrey	4: NE Mexico	5: Mazatlán	6: San Blas	7: Colima	8: Cen. Mexico	9: Guerrero	10: Veracruz	11: Oaxaca	12: Cen. Chiapas	13: N. Chiapas	14: Yucatán
LONG-TAILED WOOD-PARTRIDGE *Dendrortyx macroura*							●	●	●		●			
BEARDED WOOD-PARTRIDGE *D. barbatus*								●		●				
BUFFY-CROWNED WOOD-PARTRIDGE *D. leucophrys*												▲		
SPOTTED WOOD-QUAIL *Odontophorus guttatus*										●	●	●	●	
SINGING QUAIL *Dactylortyx thoracicus*				●			●		●			●		●
OCELLATED QUAIL *Cyrtonyx [montezumae] ocellatus*												▲		
NORTHERN BOBWHITE *Colinus virginianus*				●						●		●	●	
YUCATAN BOBWHITE *C. nigrogularis*														●
BANDED QUAIL *Philortyx fasciatus*							●	●	●					
SCALED QUAIL *Callipepla squamata*			●											
GAMBEL'S QUAIL *C. gambelii*	●	●												
CALIFORNIA QUAIL *C. californica*	●													
MOUNTAIN QUAIL *C. picta*	●													
RUDDY CRAKE *Laterallus ruber*							●			●	●	●	●	●
BLACK RAIL *L. jamaicensis*	▲									▲				
CLAPPER RAIL *Rallus longirostris*	●					●								●
King Rail *R. [l.] elegans*							●	●		●				
VIRGINIA RAIL *R. limicola*	●					●	●	●		●		●		
SPOTTED RAIL *Pardirallus maculatus*						▲		●		●				●
GREY-NECKED WOOD-RAIL *Aramides cajanea*						●								●
RUFOUS NECKED WOOD-RAIL *A. axillaris*										●	●		●	●
SORA *Porzana carolina*	●			●		●	●	●		●			●	●
YELLOW-BREASTED CRAKE *P. flaviventer*										●				
PURPLE GALLINULE *Porphyrula martinica*						●	●			●			●	●
COMMON MOORHEN *Gallinula chloropus*	●	●		●	●	●	●	●		●		●	●	●
AMERICAN COOT *Fulica americana*	●	●	●	●	●	●	●	●		●	●	●	●	●
SUNGREBE *Heliornis fulica*										●	●		●	
LIMPKIN *Aramus guarauna*							●			●		●	●	●
SANDHILL CRANE *Grus canadensis*		●												

	1: Baja	2: NW Mexico	3: Monterrey	4: NE Mexico	5: Mazatlán	6: San Blas	7: Colima	8: Cen. Mexico	9: Guerrero	10: Veracruz	11: Oaxaca	12: Cen. Chiapas	13: N. Chiapas	14: Yucatán
DOUBLE-STRIPED THICK-KNEE *Burhinus bistriatus*										●			●	
SOUTHERN LAPWING *Vanellus chilensis*													▲	
BLACK-BELLIED PLOVER *Pluvialis squatarola*	●	●	▲	●		●	●			●		●	●	●
AMERICAN GOLDEN PLOVER *P. dominica*				●			●			●	●	●		
COLLARED PLOVER *Charadrius collaris*						●	●			●	●	●	●	
SNOWY PLOVER *C. alexandrinus*	●	●	●	●		●	●			●		●		●
WILSON'S PLOVER *C. wilsonia*	●	●		●		●	●			●		●	▲	●
SEMIPALMATED PLOVER *C. semipalmatus*	●	●		●		●	●			●		●	●	●
PIPING PLOVER *C. melodus*				●		▲								●
KILLDEER *C. vociferus*	●	●	●	●	●	●	●	●		●	●	●	●	●
MOUNTAIN PLOVER *C. montanus*	●													
AMERICAN OYSTERCATCHER *Haematopus palliatus*	●	●		●		●	●							●
BLACK OYSTERCATCHER *H. bachmani*	●													
BLACK-NECKED STILT *Himantopus mexicanus*	●	●	●	●		●	●			●	●	●	●	●
AMERICAN AVOCET *Recurvirostra americana*	●	●		●	●	●	●			●		●	●	●
NORTHERN JACANA *Jacana spinosa*						●	●			●	●	●	●	●
GREATER YELLOWLEGS *Tringa melanoleuca*	●	●	●	●	●	●	●	●		●	●	●	●	●
LESSER YELLOWLEGS *T. flavipes*	●	●		●		●	●	●		●	●	●	●	●
SOLITARY SANDPIPER *T. solitaria*	●			●		●	●	●		●	●	●	●	●
WILLET *Catoptrophorus semipalmatus*	●	●		●	●	●	●			●		●		●
WANDERING TATTLER *Heteroscelus incanus*	●	●			●	●	●				●	●		
SPOTTED SANDPIPER *Actitis macularia*	●	●	●	●	●	●	●	●		●	●	●	●	●
UPLAND SANDPIPER *Bartramia longicauda*			●	●				●		●				
WHIMBREL *Numenius phaeopus*	●	●		●		●				●		●		●
LONG-BILLED CURLEW *N. americanus*	●	●		●		●	●			●		▲		●
HUDSONIAN GODWIT *Limosa haemastica*											●			
MARBLED GODWIT *L. fedoa*	●	●				●	●					●		●
RUDDY TURNSTONE *Arenaria interpres*	●	●		●		●	●			●		●		●
BLACK TURNSTONE *A. melanocephala*	●	●				▲								
SURFBIRD *Aphriza virgata*	●	●				●	●							
RED KNOT *Calidris canutus*	●	●		●		●								●

	1: Baja	2: NW Mexico	3: Monterrey	4: NE Mexico	5: Mazatlán	6: San Blas	7: Colima	8: Cen. Mexico	9: Guerrero	10: Veracruz	11: Oaxaca	12: Cen. Chiapas	13: N. Chiapas	14: Yucatán
SANDERLING *C. alba*	●	●		●		●	●			●		●		●
SEMIPALMATED SANDPIPER *C. pusilla*	▲			●						●		●	●	●
WESTERN SANDPIPER *C. mauri*	●	●		●	●	●	●	●		●		●	●	●
LEAST SANDPIPER *C. minutilla*	●	●	●	●	●	●	●	●		●	●	●	●	●
WHITE-RUMPED SANDPIPER *C. fuscicollis*										●				●
BAIRD'S SANDPIPER *C. bairdii*	●		●	●				●		●		●		
PECTORAL SANDPIPER *C. melanotus*	●			●			●	●		●	●		●	●
DUNLIN *C. alpina*	●	●		●		●								●
STILT SANDPIPER *C. himantopus*	▲					●	●			●	●	●	●	●
BUFF-BREASTED SANDPIPER *Tryngites subruficollis*										●				●
RUFF *Philomachus pugnax*	▲													▲
SHORT-BILLED DOWITCHER *Limnodromus griseus*	●			●		●						●		●
LONG-BILLED DOWITCHER *L. scolopaceus*	●	●	●	●	●	●	●			●	●	●	●	●
COMMON SNIPE *Gallinago gallinago*	●	●	●	●		●	●	●		●	●	●	●	●
WILSON'S PHALAROPE *Steganopus tricolor*	●	●		●		●	●	●		●	●	●		●
RED-NECKED PHALAROPE *Phalaropus lobatus*	●	●				●	●							▲
RED PHALAROPE *P. fulicaria*	●					●	●					●		
POMARINE JAEGER *Stercorarius pomarinus*	●					●	●							
PARASITIC JAEGER *S. parasiticus*	●					●	●					●		
LONG-TAILED JAEGER *S. longicaudus*	●													
LAUGHING GULL *Larus atricilla*	●	●	●	●	●	●	●		●	●	●	●	●	●
FRANKLIN'S GULL *L. pipixcan*				●		●	●			●	●	●	●	▲
LITTLE GULL *L. minutus*										▲				
BONAPARTE'S GULL *L. philadelphia*	●	●	●	●	●	●								
HEERMANN'S GULL *L. heermanni*	●	●			●	●	●				▲			
MEW GULL *L. canus*	●													
RING-BILLED GULL *L. delawarensis*	●	●	●	●	●	●	●			●		●		●
CALIFORNIA GULL *L. californicus*	●	●												
HERRING GULL *L. argentatus*	●	●		●	●	●	●			●	▲	●		●
Thayer's [ICELAND] Gull *L. [glaucoides.] thayeri*	●													

	1: Baja	2: NW Mexico	3: Monterrey	4: NE Mexico	5: Mazatlán	6: San Blas	7: Colima	8: Cen. Mexico	9: Guerrero	10: Veracruz	11: Oaxaca	12: Cen. Chiapas	13: N. Chiapas	14: Yucatán
LESSER BLACK-BACKED GULL *L. fuscus*				●										●
KELP GULL *L. dominicanus*														▲
YELLOW-FOOTED GULL *L. livens*	●	●			▲						▲			
WESTERN GULL *L. occidentalis*	●	●				▲					▲			
GLAUCOUS-WINGED GULL *L. glaucescens*	●	▲												
GLAUCOUS GULL *L. hyperboreus*	▲	▲												
BLACK-LEGGED KITTIWAKE *L. tridactylus*						▲								
SABINE'S GULL *L. sabini*	●						●					●		▲
GULL-BILLED TERN *Sterna nilotica*	●	●		●		●	●			●	●	●	●	●
CASPIAN TERN *S. caspia*	●	●		●	●	●	●			●	●	●	●	●
ROYAL TERN *S. maxima*	●	●		●	●	●	●			●	●	●		●
ELEGANT TERN *S. elegans*	●	●				●	●				●	●		
SANDWICH TERN *S. sandvicensis*				●			▲			●	●	●		●
ROSEATE TERN *S. dougallii*														▲
COMMON TERN *S. hirundo*	●	●		●		●	●			●	●	●		●
ARCTIC TERN *S. paradisaea*	●													
FORSTER'S TERN *S. forsteri*	●	●		●	●	●	●			●				●
LEAST TERN *S. antillarum*	●	●		●		●	●			●	●	●		●
BRIDLED TERN *S. anaethetus*						●			●					●
SOOTY TERN *S. fuscata*						●					●			
BLACK TERN *Chlidonias niger*	▲	●			●	●	●		●	●	●	●		●
BROWN NODDY *Anous stolidus*						●			●					●
BLACK SKIMMER *Rynchops niger*	●			●		●	●			●	●	●	●	●
MARBLED MURRELET *Brachyramphus marmoratus*	▲													
XANTUS' MURRELET *Endomychura hypoleuca*	●													
CRAVERI'S MURRELET *E. craveri*	●	▲												
ANCIENT MURRELET *Synthliboramphus antiquus*	▲													
CASSIN'S AUKLET *Ptychoramphus aleucticus*	●													
RHINOCEROS AUKLET *Cerorhinca monocerata*	●													
PALE-VENTED PIGEON *Columba cayennensis*													●	●

	1: Baja	2: NW Mexico	3: Monterrey	4: NE Mexico	5: Mazatlán	6: San Blas	7: Colima	8: Cen. Mexico	9: Guerrero	10: Veracruz	11: Oaxaca	12: Cen. Chiapas	13: N. Chiapas	14: Yucatán
SCALED PIGEON *C. speciosa*										●	●		●	●
WHITE-CROWNED PIGEON *C. leucocephala*														●
RED-BILLED PIGEON *C. flavirostris*		●		●	●	●	●	●	●	●	●	●	●	●
BAND-TAILED PIGEON *C. fasciata*	●		●		●		●	●	●		●	●		
SHORT-BILLED PIGEON *C. nigrirostris*										●	●		●	
WHITE-WINGED DOVE *Zenaida asiatica*	●	●	●	●	●	●	●	●		●	●	●	●	●
ZENAIDA DOVE *Z. aurita*														●
MOURNING DOVE *Z. macroura*	●	●	●	●	●	●	●	●	●	●	●	●	●	●
INCA DOVE *Columbina inca*		●	●	●	●	●	●	●	●	●	●	●		
COMMON GROUND-DOVE *C. passerina*	●	●		●	●	●	●	●	●	●	●	●		●
PLAIN-BREASTED GROUND-DOVE *C. minuta*										●			●	
RUDDY GROUND-DOVE *C. talpacoti*	▲				▲	●	●		●	●	●	●	●	●
BLUE GROUND-DOVE *Claravis pretiosa*				●						●			●	●
MAROON-CHESTED GROUND-DOVE *C. mondetoura*										●		●		
WHITE-TIPPED DOVE *Leptotila verreauxi*		●	●	●	●	●	●	●	●	●	●	●	●	●
GREY-HEADED DOVE *L. plumbeiceps*										●	●		●	
CARIBBEAN DOVE *L. jamaicensis*														●
GREY-CHESTED DOVE *L. cassinii*													●	
WHITE-FACED QUAIL-DOVE *Geotrygon albifacies*								●	●	●	●	●		
TUXTLA [PURPLISH-BACKED] **QUAIL-DOVE *G.*** *[lawrencii]* ***carrikeri***										●				
RUDDY QUAIL-DOVE *G. montana*				●		●			●	●			●	
GREEN PARAKEET *Aratinga holochlora*				●				●		●	?	●		
Pacific Parakeet *A. [h.] strenua*											?	●		
AZTEC PARAKEET *A. astec*										●	●		●	●
ORANGE-FRONTED PARAKEET *A. canicularis*					●	●	●		●		●	●		
MILITARY MACAW *Ara militaris*		●		●	●	●	●							
SCARLET MACAW *A. macao*													●	
THICK-BILLED PARROT *Rhynchopsitta pachyrhyncha*					▲		●							
MAROON-FRONTED PARROT *R. terrisi*			●											
BARRED PARAKEET *Bolborhynchus lineola*											●	●		

	1: Baja	2: NW Mexico	3: Monterrey	4: NE Mexico	5: Mazatlán	6: San Blas	7: Colima	8: Cen. Mexico	9: Guerrero	10: Veracruz	11: Oaxaca	12: Cen. Chiapas	13: N. Chiapas	14: Yucatán
MEXICAN PARROTLET *Forpus cyanopygius*		•			•	•	•							
ORANGE-CHINNED PARAKEET *Brotogeris jugularis*												•		
BROWN-HOODED PARROT *Pionopsitta haematotis*										•	•		•	
WHITE-CROWNED PARROT *Pionus senilis*				•				•		•	•	•	•	•
WHITE-FRONTED PARROT *Amazona albifrons*		•			•	•					•	•	•	•
YUCATAN PARROT *A. xantholora*														•
RED-CROWNED PARROT *A. viridigenalis*				•										
LILAC-CROWNED PARROT *A. finschi*					•	•	•		•		•			
RED-LORED PARROT *A. autumnalis*				•						•	•		•	
MEALY PARROT *A. farinosa*										•			•	
YELLOW-HEADED PARROT *A. oratrix*				•									•	
YELLOW-NAPED PARROT *A. auropalliata*												•		
BLACK-BILLED CUCKOO *Coccyzus erythropthalmus*										•				
YELLOW-BILLED CUCKOO *C. americanus*		•				•				•			•	•
MANGROVE CUCKOO *C. minor*						•						•		•
SQUIRREL CUCKOO *Piaya cayana*				•	•	•	•	•	•	•	•	•	•	•
STRIPED CUCKOO *Tapera naevia*										•	•	•	•	
PHEASANT CUCKOO *Dromococcyx phasianellus*									•		•	•		
LESSER GROUND-CUCKOO *Morococcyx erythropygus*					•	•	•	•	•		•	•		
LESSER ROADRUNNER *Geococcyx velox*					•	•	•	•	•		•	•		•
GREATER ROADRUNNER *G. californianus*	•	•	•											
SMOOTH-BILLED ANI *Crotophaga ani*														•
GROOVE-BILLED ANI *C. sulcirostris*		•	•	•	•	•	•	•	•	•	•	•	•	•
BARN OWL *Tyto alba*	•					•	•			•			•	
FLAMMULATED OWL *Otus flammeolus*			•											
WESTERN SCREECH-OWL *O. kennicottii*	•													
BALSAS SCREECH-OWL *O. seductus*							•	•	•					
PACIFIC SCREECH-OWL *O. cooperi*											•	•		
WHISKERED SCREECH-OWL *O. trichopsis*			•		•		•	•				•		

	1: Baja	2: NW Mexico	3: Monterrey	4: NE Mexico	5: Mazatlán	6: San Blas	7: Colima	8: Cen. Mexico	9: Guerrero	10: Veracruz	11: Oaxaca	12: Cen. Chiapas	13: N. Chiapas	14: Yucatán
VERMICULATED SCREECH-OWL *O. guatemalae*						●						●	●	●
BEARDED SCREECH-OWL *O. barbarus*												●		
CRESTED OWL *Lophostrix cristata*												●	●	
SPECTACLED OWL *Pulsatrix perspicillata*										●				
GREAT HORNED OWL *Bubo virginianus*	●	●					●					●	●	●
MOUNTAIN PYGMY-OWL *Glaucidium gnoma*		●	●		●		●	●	●		●	●		
CAPE PYGMY-OWL *G. hoskinsii*	●													
COLIMA PYGMY-OWL *G. palmarum*		●			●	●	●	●			●			
TAMAULIPAS PYGMY-OWL *G. sanchezi*				●										
CENTRAL AMERICAN PYGMY-OWL *G. griseiceps*										●	●		●	
FERRUGINOUS PYGMY-OWL *G. brasilianum*		●		●	●	●	●	●	●	●	●	●	●	●
ELF OWL *Micrathene whitneyi*	●													
BURROWING OWL *Athene cunicularia*	●	●	●				▲		●					
MOTTLED OWL *Strix virgata*		●			●	●	●	●	●	●	●	●	●	●
BLACK-AND-WHITE OWL *S. nigrolineata*										●			●	
SPOTTED OWL *S. occidentalis*			●				●							
FULVOUS OWL *S. fulvescens*												●		
LONG-EARED OWL *Asio otus*														
STYGIAN OWL *A. stygius*					●		●							
STRIPED OWL *A. clamator*											●			
SHORT-EARED OWL *A. flammeus*	●									▲				
NORTHERN SAW-WHET OWL *Aegolius acadicus*	●		●		●									
UNSPOTTED SAW-WHET OWL *A. ridgwayi*												●		
SHORT-TAILED NIGHTHAWK *Lurocalis semitorquatus*													●	
LESSER NIGHTHAWK *Chordeiles acutipennis*	●	●		●	●	●	●	●	●	●	●	●	●	●
COMMON NIGHTHAWK *C. minor*		●								●			●	●
PAURAQUE *Nyctidromus albicollis*				●	●	●	●		●	●	●	●	●	●
COMMON POORWILL *Phalaenoptilus nuttallii*	●													
EARED POORWILL *Nyctiphrynus mcleodii*							●		●					
YUCATAN POORWILL *N. yucatanicus*														●

	1: Baja	2: NW Mexico	3: Monterrey	4: NE Mexico	5: Mazatlán	6: San Blas	7: Colima	8: Cen. Mexico	9: Guerrero	10: Veracruz	11: Oaxaca	12: Cen. Chiapas	13: N. Chiapas	14: Yucatán
CHUCK-WILL'S-WIDOW *Caprimulgus carolinensis*				•						•				
TAWNY-COLLARED NIGHTJAR *C. salvini*			•	•										
YUCATAN NIGHTJAR *C. badius*														•
BUFF-COLLARED NIGHTJAR *C. ridgwayi*		•			•	•	•	•		•		•		
MEXICAN WHIP-POOR-WILL *C. arizonae*	•	•	•		•		•				•	•		
SPOT-TAILED NIGHTJAR *C. maculicaudus*										•				
GREAT POTOO *Nyctibius grandis*													•	
NORTHERN POTOO *N. jamaicensis*						•	•			•			•	•
BLACK SWIFT *Cypseloides niger*		•			•	•		•	•			•		
White-fronted [WHITE-CHINNED.] **Swift *C.[cryptus.] storeri***								•						
CHESTNUT-COLLARED SWIFT *C. rutilus*					•	•	•	•	•	•	•	•		
WHITE-COLLARED SWIFT *Streptoprocne zonaris*				•			•	•	•	•	•	•	•	
WHITE-NAPED SWIFT *S. semicollaris*					•	•		•	•					
CHIMNEY SWIFT *Chaetura pelagica*				•						•				•
VAUX'S SWIFT *C. vauxi*				•	•	•	•	•	•	•	•	•	•	•
WHITE-THROATED SWIFT *Aeronautes saxatalis*	•	•	•		•		•	•			•	•		
LESSER SWALLOW-TAILED SWIFT *Panyptila cayennensis*										•		•	•	
GREAT SWALLOW-TAILED SWIFT *P. sanctihieronymi*							•	•	•		•	•		
LONG-TAILED HERMIT *Phaethornis superciliosus*										•	•		•	
MEXICAN HERMIT *P. mexicanus*						•	•		•		•			
LITTLE HERMIT *Pygmornis longuemareus*										•	•		•	
SCALY-BREASTED HUMMINGBIRD *Phaeochroa cuvierii*													•	
WEDGE-TAILED SABREWING *Campylopterus curvipennis*				•						•	•		•	•
LONG-TAILED SABREWING *C. excellens*										•				
RUFOUS SABREWING *C. rufus*												•		
VIOLET SABREWING *C. hemileucurus*									•	•	•	•	•	
WHITE-NECKED JACOBIN *Florisuga mellivora*													•	

	1: Baja	2: NW Mexico	3: Monterrey	4: NE Mexico	5: Mazatlán	6: San Blas	7: Colima	8: Cen. Mexico	9: Guerrero	10: Veracruz	11: Oaxaca	12: Cen. Chiapas	13: N. Chiapas	14: Yucatán
GREEN-BREASTED MANGO *Anthracothorax prevostii*										•	•	•	•	•
GREEN VIOLET-EAR *Colibri thalassinus*							•	•	•		•	•		
EMERALD-CHINNED HUMMINGBIRD *Abeillia abeillei*											•	•		
SHORT-CRESTED COQUETTE *Lophornis brachylopha*									•					
BLACK-CRESTED COQUETTE *L. helenae*										•	•	•	•	
CANIVET'S EMERALD *Chlorostilbon canivetii*										•	•	•		•
SALVIN'S EMERALD *C. salvini*												•		
GOLDEN-CROWNED EMERALD *C. auriceps*					•	•	•	•	•		•			
COZUMEL EMERALD *C. forficatus*														•
DUSKY HUMMINGBIRD *Cynanthus sordidus*								•	•		•			
BROAD-BILLED HUMMINGBIRD *C. latirostris*		•		•	•	•	•	•						
DOUBLEDAY'S HUMMINGBIRD *C. doubledayi*									•		•			
MEXICAN WOODNYMPH *Thalurania ridgwayi*						•	•							
WHITE-EARED HUMMINGBIRD *Basilinna leucotis*		•			•	•	•	•	•		•	•		
XANTUS' HUMMINGBIRD *B. xantusii*	•													
WHITE-BELLIED EMERALD *Amazilia candida*								•		•	•	•	•	•
AZURE-CROWNED HUMMINGBIRD *A. cyanocephala*				•						•	•	•	•	
BERYLLINE HUMMINGBIRD *A. beryllina*					•	•	•	•	•		•	•		
BLUE-TAILED HUMMINGBIRD *A. cyanura*												•		
RUFOUS-TAILED HUMMINGBIRD *A. tzacatl*										•	•		•	
BUFF-BELLIED HUMMINGBIRD *A. yucatanensis*				•						•		•	•	•
CINNAMON HUMMINGBIRD *A. rutila*					•	•	•		•		•	•		•
VIOLET-CROWNED HUMMINGBIRD *A. violiceps*		•			•	•	•	•	•		•			
GREEN-FRONTED HUMMINGBIRD *A. viridifrons*									•		•	•		

	1: Baja	2: NW Mexico	3: Monterrey	4: NE Mexico	5: Mazatlán	6: San Blas	7: Colima	8: Cen. Mexico	9: Guerrero	10: Veracruz	11: Oaxaca	12: Cen. Chiapas	13: N. Chiapas	14: Yucatán
CINNAMON-SIDED HUMMINGBIRD *A. wagneri*											●			
STRIPE-TAILED HUMMINGBIRD *Eupherusa eximia*										●		●	●	
WHITE-TAILED HUMMINGBIRD *E. poliocerca*									●					
BLUE-CAPPED HUMMINGBIRD *E. cyanophrys*											●			
GREEN-THROATED MOUNTAIN-GEM *Lampornis viridipallens*												●		
AMETHYST-THROATED HUMMINGBIRD *L. amethystinus*				●			●	●	●	●	●	●		
BLUE-THROATED HUMMINGBIRD *L. clemenciae*		●			●	●	●	●	●		●			
GARNET-THROATED HUMMINGBIRD *Lamprolaima rhami*									●		●	●		
MAGNIFICENT HUMMINGBIRD *Eugenes fulgens*		●	●		●	●	●	●	●	●	●	●		
PURPLE-CROWNED FAIRY *Heliothryx barroti*													●	
LONG-BILLED STARTHROAT *Heliomaster longirostris*									●	●		●	●	
PLAIN-CAPPED STARTHROAT *H. constantii*		●			●	●	●		●		●	●		
SPARKLING-TAILED WOODSTAR *Philodice dupontii*					●	●	●		●		●	●		
MEXICAN SHEARTAIL *Calothorax eliza*										●				●
SLENDER SHEARTAIL *C. enicura*												●		
LUCIFER HUMMINGBIRD *C. lucifer*			●				●	●						
BEAUTIFUL HUMMINGBIRD *C. pulcher*								?			●			
RUBY-THROATED HUMMINGBIRD *Archilochus colubris*	▲			●	●	●	●	●	●	●	●	●	●	●
BLACK-CHINNED HUMMINGBIRD *A. alexandri*	●		●		●	●	●							
ANNA'S HUMMINGBIRD *A. anna*	●													
COSTA'S HUMMINGBIRD *A. costae*	●	●			●	●	▲							
CALLIOPE HUMMINGBIRD *A. calliope*	●				●	●	●	●	●					
BROAD-TAILED HUMMINGBIRD *Selasphorus platycercus*		●	●				●	●			●			

	1: Baja	2: NW Mexico	3: Monterrey	4: NE Mexico	5: Mazatlán	6: San Blas	7: Colima	8: Cen. Mexico	9: Guerrero	10: Veracruz	11: Oaxaca	12: Cen. Chiapas	13: N. Chiapas	14: Yucatán
RUFOUS HUMMINGBIRD *S. rufus*	•	•			•	•	•		•	▲	•			
ALLEN'S HUMMINGBIRD *S. sasin*	•				?	•	•							
BUMBLEBEE HUMMINGBIRD *S. heloisa*					•	•	•	•	•	•	•			
WINE-THROATED HUMMINGBIRD *S. ellioti*												•		
BLACK-HEADED TROGON *Trogon melanocephalus*										•	•		•	•
CITREOLINE TROGON *T. citreolus*					•	•	•		•		•	•		
VIOLACEOUS TROGON *T. violaceus*										•	•	•	•	•
MOUNTAIN TROGON *T. mexicanus*		•		•	•		•	•	•		•	•		
ELEGANT TROGON *T. elegans*		•	•	•	•	•	•	•			•			
COLLARED TROGON *T. collaris*									•	•	•	•	•	•
SLATY-TAILED TROGON *T. massena*										•			•	
EARED QUETZAL *Euptilotus neoxenus*		•			•		•							
RESPLENDENT QUETZAL *Pharomachrus mocinno*												•		
TODY MOTMOT *Hylomanes momotula*												•	•	
BLUE-THROATED MOTMOT *Aspatha gularis*												•		
BLUE-CROWNED MOTMOT *Momotus momota*			•	•				•		•	•	•	•	•
RUSSET-CROWNED MOTMOT *M. mexicanus*					•	•	•	•	•		•	•		
TURQUOISE-BROWED MOTMOT *Eumomota superciliosa*												•		•
RINGED KINGFISHER *Ceryle torquata*			•	•		•	•			•	•	•	•	
BELTED KINGFISHER *C. alcyon*	•	•	•	•	•	•	•	•		•	•	•	•	•
AMAZON KINGFISHER *Chloroceryle amazona*				•						•	•	•	•	
GREEN KINGFISHER *C. americana*				•		•	•	•		•	•	•	•	•
PYGMY KINGFISHER *C. aenea*										•			•	•
WHITE-NECKED PUFFBIRD *Notharchus macrorhynchos*													•	•
WHITE-WHISKERED PUFFBIRD *Malacoptila panamensis*													•	
RUFOUS-TAILED JACAMAR *Galbula ruficauda*										•	•		•	
EMERALD TOUCANET *Aulacorhynchus prasinus*								•	•	•	•	•	•	
COLLARED ARACARI *Pteroglossus torquatus*										•	•	•	•	•
KEEL-BILLED TOUCAN *Ramphastos sulfuratus*										•	•		•	•

	1: Baja	2: NW Mexico	3: Monterrey	4: NE Mexico	5: Mazatlán	6: San Blas	7: Colima	8: Cen. Mexico	9: Guerrero	10: Veracruz	11: Oaxaca	12: Cen. Chiapas	13: N. Chiapas	14: Yucatán
LEWIS' WOODPECKER *Melanerpes lewis*	▲													
ACORN WOODPECKER *M. formicivorus*	●	●	●	●	●	●	●	●	●	●	●	●	●	
BLACK-CHEEKED WOODPECKER *Centurus pucherani*										●			●	
GOLDEN-CHEEKED WOODPECKER *C. chrysogenys*					●	●	●	●	●		●			
GREY-BREASTED WOODPECKER *C. hypopolius*								●			●			
YUCATAN WOODPECKER *C. pygmaeus*														●
GOLDEN-FRONTED WOODPECKER *C. aurifrons*			●	●			●			●	●	●	●	●
GILA WOODPECKER *C. uropygialis*	●	●			●	●								
YELLOW-BELLIED SAPSUCKER *Sphyrapicus varius*	▲		●	●		●	●	●		●	●	●	●	●
Red-naped Sapsucker *S. [v.] nuchalis*	●													
Red-breasted Sapsucker *S. [v.] ruber*	●													
WILLIAMSON'S SAPSUCKER *S. thyroideus*	●				●									
LADDER-BACKED WOODPECKER *Picoides scalaris*	●	●	●	●	●	●	●	●	●	●	●		●	●
NUTTALL'S WOODPECKER *P. nuttallii*	●													
HAIRY WOODPECKER *P. villosus*	●	●	●				●	●	●		●	●		
ARIZONA WOODPECKER *P. arizonae*					●	●	●							
STRICKLAND'S WOODPECKER *P. stricklandi*								●						
SMOKY-BROWN WOODPECKER *Veniliornis fumigatus*				●			●	●	●	●	●	●	●	●
GOLDEN-OLIVE WOODPECKER *Piculus rubiginosus*										●	●	●	●	●
BRONZE-WINGED WOODPECKER *P. aeruginosus*			●	●				●		●				
GREY-CROWNED WOODPECKER *P. auricularis*					●	●	●		●		●			
Red-shafted [NORTHERN] Flicker *Colaptes [auratus] cafer*	●	●	●		●		●	●	●		●			
Gilded [NORTHERN] Flicker *C. [a.] chrysoides*	●	●												
Guatemalan [NORTHERN] **Flicker *C. [a.] mexicanoides***												●		

	1: Baja	2: NW Mexico	3: Monterrey	4: NE Mexico	5: Mazatlán	6: San Blas	7: Colima	8: Cen. Mexico	9: Guerrero	10: Veracruz	11: Oaxaca	12: Cen. Chiapas	13: N. Chiapas	14: Yucatán
CHESTNUT-COLORED WOODPECKER *Celeus castaneus*										●			●	●
LINEATED WOODPECKER *Dryocopus lineatus*				●	●	●	●		●	●	●	●	●	●
PALE-BILLED WOODPECKER *Campephilus guatemalensis*				●	●	●	●	●	●	●	●	●	●	●
RUFOUS-BREASTED SPINETAIL *Synallaxis eryhtrothorax*										●	●	●	●	●
SCALY-THROATED (SPECTACLED) FOLIAGE-GLEANER *Anabacerthia variegaticeps*									●	●	●	●		
BUFF-THROATED FOLIAGE-GLEANER *Automolus ochrolaemus*										●	●		●	
RUDDY FOLIAGE-GLEANER *A. rubiginosus*								●	●	●	●	●		
PLAIN XENOPS *Xenops minutus*										●			●	●
TAWNY-THROATED LEAFTOSSER *Sclerurus mexicanus*								●			●	●		
SCALY-THROATED LEAFTOSSER *S. guatemalensis*													●	
TAWNY-WINGED WOODCREEPER *Dendrocincla anabatina*										●			●	●
RUDDY WOODCREEPER *D. homochroa*													●	●
OLIVACEOUS WOODCREEPER *Sittasomus griseicapillus*				●		●	●	●	●	●	●	●	●	●
WEDGE-BILLED WOODCREEPER *Glyphorynchus spirurus*										●			●	
STRONG-BILLED WOODCREEPER *Xiphocolaptes promeropirhynchus*								●			●	●		
BARRED WOODCREEPER *Dendrocolaptes certhia*									●	●	●		●	●
IVORY-BILLED WOODCREEPER *Xiphorhynchus flavigaster*				●	●	●	●		●	●	●	●	●	●
SPOTTED WOODCREEPER *X. erythropygius*								●	●		●	●		
WHITE-STRIPED WOODCREEPER *Lepidocolaptes leucogaster*					●	●	●	●						
STREAK-HEADED WOODCREEPER *L. souleyetii*									●	●		●	●	
SPOT-CROWNED WOODCREEPER *L. affinis*				●				●	●	●	●	●		
GREAT ANTSHRIKE *Taraba major*										●			●	
BARRED ANTSHRIKE *Thamnophilus doliatus*				●						●	●	●	●	●

	1: Baja	2: NW Mexico	3: Monterrey	4: NE Mexico	5: Mazatlán	6: San Blas	7: Colima	8: Cen. Mexico	9: Guerrero	10: Veracruz	11: Oaxaca	12: Cen. Chiapas	13: N. Chiapas	14: Yucatán
RUSSET ANTSHRIKE *Thamnistes anabatinus*													•	
PLAIN ANTVIREO *Dysithamnus mentalis*													•	
DOT-WINGED ANTWREN *Microrhopias quixensis*										•			•	
DUSKY ANTBIRD *Cercomacra tyrannina*										•	•		•	
MEXICAN ANTTHRUSH *Formicarius moniliger*										•		•	•	•
SCALED ANTPITTA *Grallaria guatimalensis*								•	•			•	•	
PALTRY TYRANNULET *Zimmerius vilissimus*												•	•	
YELLOW-BELLIED TYRANNULET *Ornithion semiflavum*										•			•	
N. BEARDLESS TYRANNULET *Camptostoma imberbe*		•		•	•	•	•	•	•	•	•	•	•	•
GREENISH ELAENIA *Myiopagis viridicata*				•		•	•	•	•	•	•	•	•	•
CARIBBEAN ELAENIA *Elaenia martinica*														•
YELLOW-BELLIED ELAENIA *E. flavogaster*										•	•	•	•	•
OCHRE-BELLIED FLYCATCHER *Mionectes oleaginus*										•	•	•	•	•
SEPIA-CAPPED FLYCATCHER *Leptopogon amaurocephalus*										•	•		•	
NORTHERN BENTBILL *Oncostoma cinereigulare*										•	•	•	•	•
SLATE-HEADED TODY-FLYCATCHER *Todirostrum sylvia*										•	•		•	
COMMON TODY-FLYCATCHER *T. cinereum*										•	•	•	•	•
EYE-RINGED FLATBILL *Rhynchocyclus brevirostris*									•	•	•	•	•	•
YELLOW-OLIVE FLYCATCHER *Tolmomyias sulphurescens*										•	•	•	•	•
STUB-TAILED SPADEBILL *Platyrinchus cancrominus*										•		•	•	•
ROYAL FLYCATCHER *Onychorhynchus coronatus*										•			•	•
RUDDY-TAILED FLYCATCHER *Terenotriccus erythrurus*													•	
SULPHUR-RUMPED FLYCATCHER *Myiobius sulphureipygius*										•			•	•
BELTED FLYCATCHER *Xenotriccus callizonus*												•		
PILEATED FLYCATCHER *X. mexicanus*								•			•			

	1: Baja	2: NW Mexico	3: Monterrey	4: NE Mexico	5: Mazatlán	6: San Blas	7: Colima	8: Cen. Mexico	9: Guerrero	10: Veracruz	11: Oaxaca	12: Cen. Chiapas	13: N. Chiapas	14: Yucatán
COMMON TUFTED FLYCATCHER *Mitrephanes phaeocercus*					●	●	●	●	●	●	●	●		
OLIVE-SIDED FLYCATCHER *Contopus borealis*	●						●	●	●	●	●	●	●	
GREATER PEWEE *C. pertinax*		●	●	●	●	●	●	●	●	●	●	●		
WESTERN PEWEE *C. sordidulus*	●	●						●	●	●	●	●		
EASTERN PEWEE *C. virens*										●		●	●	●
TROPICAL PEWEE *C. cinereus*										●		●	●	●
YELLOW-BELLIED FLYCATCHER *Empidonax flaviventris*				●		▲				●	●	●	●	●
ACADIAN FLYCATCHER *E. virescens*										●		●	●	
ALDER FLYCATCHER *E. alnorum*										●		●	●	
WILLOW FLYCATCHER *E. traillii*	●	●				●	●		●	●	●	●		
WHITE-THROATED FLYCATCHER *E. albigularis*						●	●	●						
LEAST FLYCATCHER *E. minimus*				●	●	●	●	●	●	●	●	●	●	●
HAMMOND'S FLYCATCHER *E. hammondii*	●		●	●	●	●	●	●	●		●	●		
DUSKY FLYCATCHER *E. oberholseri*	●			●	●	●	●	●			●			
GREY FLYCATCHER *E. wrightii*	●		●				●				●			
PINE FLYCATCHER *E. affinis*			●		●		●	●			●	●		
WESTERN FLYCATCHER *E. difficilis*	●	●	●		●	●	●	●	●	●	●			
YELLOWISH FLYCATCHER *E. flavescens*												●		
BUFF-BREASTED FLYCATCHER *E. fulvifrons*		●	●	●	●	●	●	●	●			●		
BLACK PHOEBE *Sayornis nigricans*	●	●	●	●				●		●	●	●		
EASTERN PHOEBE *S. phoebe*	▲		●	●			▲	●		▲				
SAY'S PHOEBE *S. saya*	●	●	●	●				●			●	▲		
VERMILION FLYCATCHER *Pyrocephalus rubinus*	●	●	●	●	●	●	●	●		●	●		●	●
BRIGHT-RUMPED ATTILA *Attila spadiceus*					●	●	●	●	●	●	●	●	●	●
RUFOUS MOURNER *Rhytipterna holerythra*													●	
YUCATAN FLYCATCHER *Myiarchus yucatanensis*														●
DUSKY-CAPPED FLYCATCHER *M. tuberculifer*		●		●	●	●	●	●	●	●	●	●	●	●
ASH-THROATED FLYCATCHER *M. cinerascens*	●	●			●	●	●	●			●			
NUTTING'S FLYCATCHER *M. nuttingi*		●			●	●	●	●	●		●	●		

	1: Baja	2: NW Mexico	3: Monterrey	4: NE Mexico	5: Mazatlán	6: San Blas	7: Colima	8: Cen. Mexico	9: Guerrero	10: Veracruz	11: Oaxaca	12: Cen. Chiapas	13: N. Chiapas	14: Yucatán
GREAT CRESTED FLYCATCHER *M. crinitus*				•						•	•	•	•	•
BROWN-CRESTED FLYCATCHER *M. tyrannulus*		•		•	•	•	•	•	•	•	•	•	•	•
FLAMMULATED FLYCATCHER *Deltarhynchus flammulatus*					•		•					•		
GREAT KISKADEE *Pitangus sulphuratus*				•	•	•	•	•	•	•	•	•	•	•
BOAT-BILLED FLYCATCHER *Megarynchus pitangua*				•		•	•	•	•	•	•	•	•	•
SOCIAL FLYCATCHER *Myiozetetes similis*				•	•	•	•	•	•	•	•	•	•	•
STREAKED FLYCATCHER *Myiodynastes maculatus*														•
SULPHUR-BELLIED FLYCATCHER *M. luteiventris*		•		•	•	•		•	•	•	•	•	•	•
PIRATIC FLYCATCHER *Legatus leucophaius*										•		•	•	•
TROPICAL KINGBIRD *Tyrannus melancholicus*	▲	•		•	•	•	•	•	•	•	•	•	•	•
COUCH'S KINGBIRD *T. couchii*				•						•	•	•	•	•
CASSIN'S KINGBIRD *T. vociferans*	•	•	•	•	•	•	•	•	•		•			
THICK-BILLED KINGBIRD *T. crassirostris*	▲	•			•	•	•	•	•		•			
WESTERN KINGBIRD *T. verticalis*	•	•			•	•	•		•	▲	•	•		
EASTERN KINGBIRD *T. tyrannus*				•						•		•	•	•
SCISSOR-TAILED FLYCATCHER *T. forficatus*			•	•						•	•	•	•	▲
FORK-TAILED FLYCATCHER *T. savana*										•		•	•	
THRUSHLIKE MOURNER *Schiffornis turdinus*													•	•
SPECKLED MOURNER *Laniocera rufescens*													•	
CINNAMON BECARD *Pachyramphus cinnamomeus*													•	
GREY-COLLARED BECARD *P. major*				•	•	•	•	•		•		•	•	•
ROSE-THROATED BECARD *P. aglaiae*		•		•		•	•	•	•	•	•	•	•	•
MASKED TITYRA *Tityra semifasciata*				•	•	•	•	•	•	•	•	•	•	•
BLACK-CROWNED TITYRA *T. inquisitor*										•			•	•
RUFOUS PIHA *Lipaugus unirufus*										•			•	
LOVELY COTINGA *Cotinga amabilis*										•	•	•	•	
WHITE-COLLARED MANAKIN *Manacus candei*											•		•	
LONG-TAILED MANAKIN *Chiroxiphia linearis*											•	•		

	1: Baja	2: NW Mexico	3: Monterrey	4: NE Mexico	5: Mazatlán	6: San Blas	7: Colima	8: Cen. Mexico	9: Guerrero	10: Veracruz	11: Oaxaca	12: Cen. Chiapas	13: N. Chiapas	14: Yucatán
RED-CAPPED MANAKIN *Pipra mentalis*										•			•	•
HORNED LARK *Eremophila alpestris*	•	•	•	•				•						
PURPLE MARTIN *Progne subis*	•					•				•				•
SINALOA MARTIN *P. sinaloae*							•							
GREY-BREASTED MARTIN *P. chalybea*						•	•			•	•	•	•	•
TREE SWALLOW *Tachycineta bicolor*	•	•	•	•		•	•	•		•	•	•	•	•
MANGROVE SWALLOW *T. albilinea*					•	•	•			•	•	•	•	•
VIOLET-GREEN SWALLOW *T. thalassina*	•	•	•		•	•	•	•			•	•		
BLACK-CAPPED SWALLOW *Notiochelidon pileata*												•		
N. ROUGH-WINGED SWALLOW *Stelgidopteryx serripennis*	•	•	•	•	•	•	•	•	•	•	•	•	•	•
RIDGWAY'S ROUGH-WINGED SWALLOW *S. ridgwayi*										•	•	•		•
BANK SWALLOW *Riparia riparia*	•					•	•			•		•		•
CLIFF SWALLOW *Hirundo pyrrhonota*	•	•				•	•	•		•	•	•	•	•
CAVE SWALLOW *H. fulva*			•									•		•
BARN SWALLOW *H. rustica*	•	•	•	•		•	•	•		•	•	•	•	•
STELLER'S JAY *Cyanocitta stelleri*		•	•		•			•	•		•	•		
BLACK-THROATED MAGPIE-JAY *Calocitta colliei*		•			•	•								
WHITE-THROATED MAGPIE-JAY *C. formosa*							•	•	•		•	•		
TUFTED JAY *Cyanocorax dickeyi*					•									
GREEN JAY *C. yncas*			•	•		•	•		•	•	•	•	•	•
BROWN JAY *C. morio*			•	•						•	•		•	•
PURPLISH-BACKED JAY *C. beecheii*		•			•	•								
SAN BLAS JAY *C. sanblasianus*						•	•							
YUCATAN JAY *C. yucatanicus*													•	•
AZURE-HOODED JAY *Cyanolyca cucullata*								•			•	•		
BLACK-THROATED JAY *C. pumilo*												•		
DWARF JAY *C. nana*											•			
WHITE-THROATED JAY *C. mirabilis*									•		•			
WESTERN SCRUB JAY *Aphelocoma californica*	•		•					•	•		•			
GREY-BREASTED JAY *A. ultramarina*		•	•	•	•		•	•						

	1: Baja	2: NW Mexico	3: Monterrey	4: NE Mexico	5: Mazatlán	6: San Blas	7: Colima	8: Cen. Mexico	9: Guerrero	10: Veracruz	11: Oaxaca	12: Cen. Chiapas	13: N. Chiapas	14: Yucatán
UNICOLORED JAY *A. unicolor*							•		•		•	•		
PINYON JAY *Gymnorhinus cyanocephalus*	•													
CLARK'S NUTCRACKER *Nucifraga columbiana*	▲		•											
AMERICAN CROW *Corvus brachyrhynchos*	•													
TAMAULIPAS CROW *C. imparatus*				•						•				
SINALOA CROW *C. sinaloae*		•			•	•								
CHIHUAHUAN RAVEN *C. cryptoleucus*		•	•	•										
NORTHERN RAVEN *C. corax*	•	•	•	•	•	•	•	•			•	•		
MEXICAN CHICKADEE *Parus sclateri*		•	•		•		•	•	•		•			
MOUNTAIN CHICKADEE *P. gambeli*	•													
BRIDLED TITMOUSE *P. wollweberi*		•	•	•	•		•	•	•		•			
California [PLAIN] Titmouse *P. [i.] inornatus*	•													
Grey [PLAIN] Titmouse *P. [i.] ridgwayi*		•												
BLACK-CRESTED TITMOUSE *P. atricristatus*			•	•				•		•				
VERDIN *Auriparus flaviceps*	•	•	•					•			•	•		
BUSHTIT *Psaltriparus minimus*	•	•	•				•	•	•		•	•		
RED-BREASTED NUTHATCH *Sitta canadensis*	▲													
WHITE-BREASTED NUTHATCH *S. carolinensis*	•	•	•		•		•	•						
PYGMY NUTHATCH *S. pygmaea*	•		•				•	•						
BROWN CREEPER *Certhia americana*		•	•		•		•	•	•		•			
BAND-BACKED WREN *Campylorhynchus zonatus*										•	•	•	•	
GREY-BARRED WREN *C. megalopterus*							•	•			•			
GIANT WREN *C. chiapensis*												•		
RUFOUS-NAPED WREN *C. rufinucha*							•			•	•	•		
SPOTTED WREN *C. gularis*				•	•	•	•	•						
BOUCARD'S WREN *C. jocosus*								•	•		•			
YUCATAN WREN *C. yucatanicus*														•
CACTUS WREN *C. brunneicapillus*	•	•	•				•							
ROCK WREN *Salpinctes obsoletus*	•	•	•								•			
CANYON WREN *Catherpes mexicanus*	•	•	•	•	•		•	•	•		•			

	1: Baja	2: NW Mexico	3: Monterrey	4: NE Mexico	5: Mazatlán	6: San Blas	7: Colima	8: Cen. Mexico	9: Guerrero	10: Veracruz	11: Oaxaca	12: Cen. Chiapas	13: N. Chiapas	14: Yucatán
SUMICHRAST'S WREN *Hylorchilus sumichrasti*										•	•			
NAVA'S WREN *H. navai*										•				
SPOT-BREASTED WREN *Thryothorus maculipectus*			•	•						•	•	•	•	•
HAPPY WREN *T. felix*					•	•	•	•	•		•			
RUFOUS-AND-WHITE WREN *T. rufalbus*												•		
SINALOA WREN *T. sinaloa*		•			•	•	•		•					
BANDED WREN *T. pleurostictus*								•	•		•	•		
CAROLINA WREN *T. ludovicianus*			•	•										
White-browed Wren *T. [l.] albinucha*														•
PLAIN WREN *T. modestus*												•		
WHITE-BELLIED WREN *Uropsila leucogastra*							•			•	•		•	•
BEWICK'S WREN *Thryomanes bewickii*	•	•	•					•			•			
Northern House [HOUSE] Wren *Troglodytes [a.] aedon*	•	•	•	•	•	•	•	•	•	•	•			
Brown-throated [HOUSE] Wren *T. [a.] brunneicollis*		•	•		•		•	•	•		•			
Southern House [HOUSE] Wren *T. [a.] musculus*										•	•	•	•	•
COZUMEL WREN *T. beani*														•
RUFOUS-BROWED WREN *T. rufociliatus*												•		
SEDGE WREN *Cistothorus platensis*								•		•			•	
MARSH WREN *C. palustris*	•	•	•	•		•	•	•		•	▲			
WHITE-BREASTED WOOD-WREN *Henicorhina leucosticta*										•	•		•	
GREY-BREASTED WOOD-WREN *H. leucophrys*				•			•	•	•	•	•	•		
NIGHTINGALE WREN *Microcerculus philomela*												•		
AMERICAN DIPPER *Cinclus mexicanus*		•						•	•					
GOLDEN-CROWNED KINGLET *Regulus satrapa*	▲						•	•						
RUBY-CROWNED KINGLET *R. calendula*	•	•	•	•	•	•	•	•	•	•	•			
LONG-BILLED GNATWREN *Ramphocaenus melanurus*										•	•	•	•	•
BLUE-GREY GNATCATCHER *Polioptila caerulea*	•	•	•	•	•	•	•	•	•	•	•	•	•	•

	1: Baja	2: NW Mexico	3: Monterrey	4: NE Mexico	5: Mazatlán	6: San Blas	7: Colima	8: Cen. Mexico	9: Guerrero	10: Veracruz	11: Oaxaca	12: Cen. Chiapas	13: N. Chiapas	14: Yucatán
BLACK-TAILED GNATCATCHER *P. melanura*	●	●	●											
CALIFORNIA GNATCATCHER *P. californica*	●													
BLACK-CAPPED GNATCATCHER *P. nigriceps*		●			●		●							
WHITE-LORED GNATCATCHER *P. albiloris*									●		●	●		●
TROPICAL GNATCATCHER *P. plumbea*										●			●	●
EASTERN BLUEBIRD *Sialia sialis*				●	●	●	●	●	●			●		
WESTERN BLUEBIRD *S. mexicana*	●	●	●		●			●						
MOUNTAIN BLUEBIRD *S. currucoides*	●	●	●											
TOWNSEND'S SOLITAIRE *Myadestes townsendi*	▲	●	▲			●								
BROWN-BACKED SOLITAIRE *M. occidentalis*			●	●	●	●	●	●	●	●	●	●		
SLATE-COLORED SOLITAIRE *M. unicolor*								●		●	●	●		
ORANGE-BILLED NIGHTINGALE-THRUSH *Catharus aurantiirostris*				●	●	●	●	●	●	●	●	●		
RUSSET NIGHTINGALE-THRUSH *C. occidentalis*			●		●	●	●	●	●		●			
RUDDY-CAPPED NIGHTINGALE-THRUSH *C. frantzii*							●	●	●	●	●	●		
BLACK-HEADED NIGHTINGALE-THRUSH *C. mexicanus*				●				●		●	●	●		
SPOTTED NIGHTINGALE-THRUSH *C. dryas*												●		
VEERY *C. fuscescens*										●			●	
GREY-CHEEKED THRUSH *C. minimus*														●
SWAINSON'S THRUSH *C. ustulatus*	●			●	●	●	●		●	●	●	●	●	●
HERMIT THRUSH *C. guttatus*	●	●	●	●	●		●	●			●	●		
WOOD THRUSH *C. mustelinus*										●	●	●	●	●
BLACK THRUSH *Turdus infuscatus*								●	●		●	●		
MOUNTAIN THRUSH *T. plebejus*												●		
CLAY-COLORED THRUSH *T. grayi*			●	●						●	●	●	●	●
WHITE-THROATED THRUSH *T. assimilis*				●	●	●	●	●	●	●	●	●		
RUFOUS-BACKED THRUSH *T. rufopalliatus*		●			●	●	●	●	●		●			
AMERICAN ROBIN *T. migratorius*	●	●	●		●		●	●	●		●			▲
San Lucas Robin *T. [m.] confinis*	●													

	1: Baja	2: NW Mexico	3: Monterrey	4: NE Mexico	5: Mazatlán	6: San Blas	7: Colima	8: Cen. Mexico	9: Guerrero	10: Veracruz	11: Oaxaca	12: Cen. Chiapas	13: N. Chiapas	14: Yucatán
AZTEC THRUSH *Zoothera pinicola*			●		●		●		●		●			
WRENTIT *Chamea fasciata*	●													
GREY CATBIRD *Dumetella carolinensis*				●				▲		●	●	●	●	●
BLACK CATBIRD *D. glabrirostris*														●
BLUE MOCKINGBIRD *Melanotis caerulescens*				●	●	●	●	●	●	●	●			
BLUE-AND-WHITE MOCKINGBIRD *M. hypoleucus*												●		
NORTHERN MOCKINGBIRD *Mimus polyglottos*	●	●	●	●	●	●	●	●		●	●			
TROPICAL MOCKINGBIRD *M. gilvus*											▲	●	●	●
SAGE THRASHER *Oreoscoptes montanus*	●													
LONG-BILLED THRASHER *Toxostoma longirostre*			●	●										
COZUMEL THRASHER *T. guttatum*														●
GREY THRASHER *T. cinereum*	●													
BENDIRE'S THRASHER *T. bendirei*		●												
CURVE-BILLED THRASHER *T. curvirostre*		●	●		●			●			●			
OCELLATED THRASHER *T. ocellatum*								●			●			
CALIFORNIA THRASHER *T. redivivum*	●													
CRISSAL THRASHER *T. crissale*			●											
LE CONTE'S THRASHER *T. lecontei*	●	●												
YELLOW WAGTAIL *Motacilla flava*	▲													
RED-THROATED PIPIT *Anthus cervinus*	▲						▲							
AMERICAN PIPIT *A. rubescens*	●	●	●			●	●	●		●	●			
SPRAGUE'S PIPIT *A. spragueii*		●	●				●			●				
CEDAR WAXWING *Bombycilla cedrorum*	●	●	●	●			●	●	●	●	●	●	●	●
GREY SILKY *Ptilogonys cinereus*			●	●	●	●	●	●	●	●	●	●		
PHAINOPEPLA *Phainopepla nitens*	●	●	●											
LOGGERHEAD SHRIKE *Lanius ludovicianus*	●	●	●	●	●	●	●	●		●	●		▲	
EUROPEAN STARLING *Sturnus vulgaris*	●	●						▲						
SLATY VIREO *Vireo brevipennis*							●	●			●			
WHITE-EYED VIREO *V. griseus*			●	●						●	●	●	●	●
MANGROVE VIREO *V. pallens*						●								●
COZUMEL VIREO *V. bairdi*														●

	1: Baja	2: NW Mexico	3: Monterrey	4: NE Mexico	5: Mazatlán	6: San Blas	7: Colima	8: Cen. Mexico	9: Guerrero	10: Veracruz	11: Oaxaca	12: Cen. Chiapas	13: N. Chiapas	14: Yucatán
BELL'S VIREO *V. bellii*	●	●	●	●	●	●	●		●	●	●	●		
BLACK-CAPPED VIREO *V. atricapillus*					●	●	●				●			
DWARF VIREO *V. nelsoni*							●				●			
GREY VIREO *V. vicinior*	●													
Blue-headed [SOLITARY] Vireo *V. [s.] solitarius*			●	●				●	●	●	●	●	●	
Cassin's [SOLITARY] Vireo *V. [s.] cassini*	●		●	●	●	●	●	●			●			
Plumbeous [SOLITARY] Vireo *V. [s.] plumbeus*	●	●	●		●	●	●	●	●		●	●		
YELLOW-THROATED VIREO *V. flavifrons*				●		▲				●	●	●	●	●
HUTTON'S VIREO *V. huttoni*	●	●	●		●	●	●	●	●		●	●		
GOLDEN VIREO *V. hypochryseus*					●	●	●	●	●		●			
WARBLING VIREO *V. gilvus*	●	●		●	●	●	●	●	●	●	●	●		▲
BROWN-CAPPED VIREO *V. leucophrys*								●			●	●		
PHILADELPHIA VIREO *V. philadelphicus*										●		●	●	●
YELLOW-GREEN VIREO *V. flavoviridis*				●		●		●	●	●	●	●	●	●
RED-EYED VIREO *V. olivaceus*				●						●			●	●
YUCATAN VIREO *V. magister*														●
TAWNY-CROWNED GREENLET *Hylophilus ochraceiceps*										●			●	●
LESSER GREENLET *H. decurtatus*										●	●	●	●	●
CHESTNUT-SIDED SHRIKE-VIREO *Vireolanius melitophrys*							●	●	●		●	●		
GREEN SHRIKE-VIREO *V. pulchellus*										●		●	●	
RUFOUS-BROWED PEPPERSHRIKE *Cyclarhis gujanensis*				●						●		●		●
BLUE-WINGED WARBLER *Vermivora pinus*				●						●	●	●	●	●
GOLDEN-WINGED WARBLER *V. chrysoptera*						▲				●		●	●	
TENNESSEE WARBLER *V. peregrina*	▲								●	●	●	●	●	●
ORANGE-CROWNED WARBLER *V. celata*	●	●	●	●	●	●	●	●	●	●	●			
COLIMA WARBLER *V. crissalis*			●				●	●	●		▲			
NASHVILLE WARBLER *V. ruficapilla*		●	●	●	●	●	●	●	●	●	●	●		
VIRGINIA'S WARBLER *V. virginiae*		●			●		●	●			●			
LUCY'S WARBLER *V. luciae*		●			●	●	●				▲			

	1: Baja	2: NW Mexico	3: Monterrey	4: NE Mexico	5: Mazatlán	6: San Blas	7: Colima	8: Cen. Mexico	9: Guerrero	10: Veracruz	11: Oaxaca	12: Cen. Chiapas	13: N. Chiapas	14: Yucatán
CRESCENT-CHESTED WARBLER ***V. superciliosa***			●	●	●	●	●	●	●		●	●		
NORTHERN PARULA *Parula americana*	▲		●	●						●		●	●	●
TROPICAL PARULA *P. pitiayumi*			●	●		●	●		●	●	●			
YELLOW WARBLER *Dendroica petechia*	●	●			●	●	●	●	●	●	●	●	●	●
Mangrove Warbler *D. [p.] erithachorides*		●				●								●
CHESTNUT-SIDED WARBLER *D. pensylvanica*						▲			▲	●			●	●
MAGNOLIA WARBLER *D. magnolia*						▲	▲			●	●	●	●	●
CAPE MAY WARBLER *D. tigrina*										▲	▲			●
BLACK-THROATED BLUE WARBLER *D. caerulescens*												▲		●
Myrtle [YELLOW-RUMPED] Warbler *D. [c.] coronata*	●	●	●	●			●			●	●	●	●	●
Audubon's [YELLOW-RUMPED] Warbler *D. [c.] auduboni*	●	●	●	●	●	●	●	●	●	●	●	●		
BLACK-THROATED GREY WARBLER *D. nigrescens*	●	●	●	●	●	●	●	●			●			
TOWNSEND'S WARBLER *D. townsendi*	●	●	●	●	●	●	●	●	●		●	●		
HERMIT WARBLER *D. occidentalis*	●	●	●	●	●	●	●	●	●		●	●		
BLACK-THROATED GREEN WARBLER *D. virens*			●	●		●	●	●	●	●	●	●	●	●
GOLDEN-CHEEKED WARBLER *D. chrysoparia*												●		
BLACKBURNIAN WARBLER *D. fusca*	▲									●		●	●	
YELLOW-THROATED WARBLER *D. dominica*				●						●			●	●
GRACE'S WARBLER *D. graciae*					●	●	●	●	●		●	●		
PRAIRIE WARBLER *D. discolor*	▲													●
PALM WARBLER *D. palmarum*	▲			▲			▲			▲				●
BAY-BREASTED WARBLER *D. castanea*										●				●
BLACKPOLL WARBLER *D. striata*	▲													
CERULEAN WARBLER *D. cerulea*										●			●	
BLACK-AND-WHITE WARBLER *Mniotilta varia*	●		●	●	●	●	●	●	●	●	●	●	●	●
AMERICAN REDSTART *Setophaga ruticilla*	●			●		●	●		●	●	●	●	●	●
PROTHONOTARY WARBLER *Protonotaria citrea*										●				●
WORM-EATING WARBLER *Helmitheros vermivorus*				●		▲				●	●	●	●	●

	1: Baja	2: NW Mexico	3: Monterrey	4: NE Mexico	5: Mazatlán	6: San Blas	7: Colima	8: Cen. Mexico	9: Guerrero	10: Veracruz	11: Oaxaca	12: Cen. Chiapas	13: N. Chiapas	14: Yucatán
SWAINSON'S WARBLER *Limnothlypis swainsonii*														•
OVENBIRD *Seiurus aurocapillus*				•	•	•	•	•	•	•	•	•	•	•
NORTHERN WATERTHRUSH *S. noveboracensis*					•	•	•	•		•	•	•	•	•
LOUISIANA WATERTHRUSH *S. motacilla*				•	•	•	•	•	•	•	•	•	•	
KENTUCKY WARBLER *Oporornis formosus*				•		▲				•	•	•	•	•
MOURNING WARBLER *O. philadelphia*										•		•	•	•
MACGILLIVRAY'S WARBLER *O. tolmiei*	•	•	•	•	•	•	•	•	•	•	•	•		
COMMON YELLOWTHROAT *Geothlypis trichas*	•	•	•	•	•	•	•	•	•	•	•	•	•	•
ALTAMIRA YELLOWTHROAT *G. flavovelata*				•						•				
BELDING'S YELLOWTHROAT *G. beldingi*	•													
HOODED YELLOWTHROAT *G. speciosa*			•					•			•			
BLACK-POLLED YELLOWTHROAT *G. nelsoni*								•						
GREY-CROWNED YELLOWTHROAT *Chamaethlypis poliocephala*				•	•	•	•	•		•	•	•	•	•
HOODED WARBLER *Wilsonia citrina*				•		▲				•	•	•	•	•
WILSON'S WARBLER *W. pusilla*	•	•	•	•	•	•	•	•	•	•	•	•	•	
CANADA WARBLER *W. canadensis*										•		•	•	
RED-FACED WARBLER *Cardellina rubrifrons*		•			•	•	•	•	•		•	•		
RED WARBLER *Ergaticus ruber*					•		•	•	•		•			
PINK-HEADED WARBLER *E. versicolor*												•		
PAINTED WHITESTART *Myioborus pictus*		•	•	•	•	•	•	•	•		•	•		
SLATE-THROATED WHITESTART *M. miniatus*		•	•		•	•	•	•	•	•	•	•		
FAN-TAILED WARBLER *Basileuterus lachrymosa*				•	•	•	•		•	•		•		
GOLDEN-CROWNED WARBLER *B. culicivorus*			•	•		•	•	•	•	•	•	•	•	•
RUFOUS-CAPPED WARBLER *B. rufifrons*		•	•	•	•	•	•	•	•	•	•	•		
CHESTNUT-CAPPED WARBLER *B. delattrii*												•		
GOLDEN-BROWED WARBLER *B. belli*					•		•	•	•	•	•	•		
YELLOW-BREASTED CHAT *Icteria virens*	•	•		•	•	•	•		•	•	•	•	•	•
RED-BREASTED CHAT *Granatellus venustus*					•	•	•		•		•	•		
GREY-THROATED CHAT *G. sallaei*													•	•
OLIVE WARBLER *Peucedramus taeniatus*		•	•		•		•	•	•		•	•		

	1: Baja	2: NW Mexico	3: Monterrey	4: NE Mexico	5: Mazatlán	6: San Blas	7: Colima	8: Cen. Mexico	9: Guerrero	10: Veracruz	11: Oaxaca	12: Cen. Chiapas	13: N. Chiapas	14: Yucatán
BANANAQUIT *Coereba flaveola*										•	•		•	
Cozumel Bananaquit *C. [f.] caboti*														•
CABANIS' TANAGER *Tangara cabanisi*												•		
GOLDEN-HOODED TANAGER *T. larvata*											•		•	
GREEN HONEYCREEPER *Chlorophanes spiza*											•		•	
SHINING HONEYCREEPER *Cyanerpes lucidus*												•		
RED-LEGGED HONEYCREEPER *C. cyaneus*									•	•	•	•	•	•
BLUE-CROWNED CHLOROPHONIA *Chlorophonia occipitalis*										•	•	•		
SCRUB EUPHONIA *Euphonia affinis*				•						•	•	•	•	•
Godman's Euphonia *E. [a.] godmani*					•	•	•		•					
YELLOW-THROATED EUPHONIA *E. hirundinacea*				•				•		•	•	•	•	•
BLUE-HOODED EUPHONIA *E. elegantissima*			•	•	•		•	•	•	•	•	•		
OLIVE-BACKED EUPHONIA *E. gouldi*										•	•		•	
BLUE-GREY TANAGER *Thraupis episcopus*				▲						•	•	•	•	
YELLOW-WINGED TANAGER *T. abbas*				•						•	•	•	•	•
STRIPE-HEADED TANAGER *Spindalis zena*														•
GREY-HEADED TANAGER *Eucometis penicillata*										•			•	•
BLACK-THROATED SHRIKE-TANAGER *Lanio aurantius*										•			•	
RED-CROWNED ANT-TANAGER *Habia rubica*						•	•		•	•	•	•	•	•
RED-THROATED ANT-TANAGER *H. fuscicauda*				•						•	•	•	•	•
ROSE-THROATED TANAGER *Piranga roseogularis*														•
HEPATIC TANAGER *P. flava*		•	•	•	•	•	•	•	•		•	•		
SUMMER TANAGER *P. rubra*	▲	•		•	•	•	•	•	•	•	•	•	•	•
SCARLET TANAGER *P. olivacea*	▲									•			•	•
WESTERN TANAGER *P. ludoviciana*	•	•		•	•	•	•	•	•	•	•	•	•	
FLAME-COLORED TANAGER *P. bidentata*			•	•	•	•	•	•	•		•	•		
WHITE-WINGED TANAGER *Spermagra leucoptera*				•					•	•	•	•		
RED-HEADED TANAGER *S. erythrocephala*					•	•	•	•	•		•			

	1: Baja	2: NW Mexico	3: Monterrey	4: NE Mexico	5: Mazatlán	6: San Blas	7: Colima	8: Cen. Mexico	9: Guerrero	10: Veracruz	11: Oaxaca	12: Cen. Chiapas	13: N. Chiapas	14: Yucatán
CRIMSON-COLLARED TANAGER *Phlogothraupis sanguinolenta*										•	•		•	
SCARLET-RUMPED TANAGER *Ramphocelus passerinii*										•			•	
ROSY THRUSH-TANAGER *Rhodinocichla rosea*						•	•							
COMMON BUSH-TANAGER *Chlorospingus ophthalmicus*								•	•	•	•	•		
GREYISH SALTATOR *Saltator coerulescens*				•	•	•	•		•	•	•	•	•	•
BUFF-THROATED SALTATOR *S. maximus*										•	•		•	
BLACK-HEADED SALTATOR *S. atriceps*				•					•	•	•	•	•	•
BLACK-FACED GROSBEAK *Caryothraustes poliogaster*										•	•		•	
CRIMSON-COLLARED GROSBEAK *Rhodothraupis celaeno*			•	•										
NORTHERN CARDINAL *Cardinalis cardinalis*	•	•	•	•	•					•	•			•
PYRRHULOXIA *C. sinuatus*	•	•	•	•	•	•								
YELLOW GROSBEAK *Pheucticus chrysopeplus*		•			•	•	•	•	•			•		
ROSE-BREASTED GROSBEAK *P. ludovicianus*				•		•	•	•	•	•	•	•	•	•
BLACK-HEADED GROSBEAK *P. melanocephalus*	•	•	•	•	•	•	•	•	•		•			
BLUE-BLACK GROSBEAK *Cyanocompsa cyanoides*										•	•		•	
BLUE BUNTING *C. parellina*				•	•	•	•		•	•	•	•		•
BLUE GROSBEAK *Passerina caerulea*	•	•		•	•	•	•	•	•	•	•	•	•	•
ROSITA'S BUNTING *P. rositae*											•			
LAZULI BUNTING *P. amoena*	•	•				•	•	•			▲			
INDIGO BUNTING *P. cyanea*				•		•	•	•	•	•	•	•	•	•
VARIED BUNTING *P. versicolor*	•	•	•	•	•	•	•	•	•	•	•	•		
ORANGE-BREASTED BUNTING *P. leclancherii*						•	•	•	•		•	•		
PAINTED BUNTING *P. ciris*					•	•	•		•	•	•	•	•	•
DICKCISSEL *Spiza americana*	▲			•	•	•	•			•		•	•	
WHITE-NAPED BRUSHFINCH *Atlapetes albinucha*								•		•	•	•		
YELLOW-THROATED BRUSHFINCH *A. gutteralis*												•		

	1: Baja	2: NW Mexico	3: Monterrey	4: NE Mexico	5: Mazatlán	6: San Blas	7: Colima	8: Cen. Mexico	9: Guerrero	10: Veracruz	11: Oaxaca	12: Cen. Chiapas	13: N. Chiapas	14: Yucatán
RUFOUS-CAPPED BRUSHFINCH *A. pileatus*			●	●	●		●	●	●		●			
CHESTNUT-CAPPED BRUSHFINCH *A. brunneinucha*								●	●	●	●	●		
Plain-breasted Brushfinch *A. [b.] apertus*										●				
GREEN-STRIPED BRUSHFINCH *A. virenticeps*					●		●	●						
ORANGE-BILLED SPARROW *Arremon aurantiirostris*										●			●	
OLIVE SPARROW *Arremonops rufivirgatus*			●	●			●		●	●	●	●		●
GREEN-BACKED SPARROW *A. chloronotus*													●	●
RUSTY-CROWNED GROUND-SPARROW *Melozone kieneri*					●	●	●	●	●					
PREVOST'S GROUND-SPARROW *M. biarcuatum*												●		
WHITE-EARED GROUND-SPARROW *M. leucotis*												●		
GREEN-TAILED TOWHEE *Pipilo chlorurus*	●	●	●		●		●	●			▲			
COLLARED TOWHEE *P. ocai*							●		●		●			
SPOTTED TOWHEE *P. maculatus*	●		●		●			●			●	●		
CALIFORNIA TOWHEE *P. crissalis*	●													
CANYON TOWHEE *P. fuscus*		●	●	●			●	●						
ABERT'S TOWHEE *P. aberti*	●													
WHITE-THROATED TOWHEE *P. albicollis*											●			
BLUE-BLACK-GRASSQUIT *Volatinia jacarina*				●		●	●	●	●	●	●	●	●	●
VARIABLE SEEDEATER *Sporophila aurita*										●			●	
WHITE-COLLARED SEEDEATER *S. [torqueola] morelleti*				●						●	●	●	●	●
Cinnamon-rumped Seedeater *S. [t.] torqueola*					●	●	●	●	●		●			
RUDDY-BREASTED SEEDEATER *S. minuta*						●	●		●			●		
THICK-BILLED SEEDFINCH *Oryzoborus funereus*											●		●	
BLUE SEEDEATER *Amaurospiza concolor*												●		
SLATE-BLUE SEEDEATER *A. relicta*							●	●	●					
YELLOW-FACED GRASSQUIT *Tiaris olivacea*			●	●						●	●		●	●
SLATY FINCH *Haplospiza rustica*												▲		

	1: Baja	2: NW Mexico	3: Monterrey	4: NE Mexico	5: Mazatlán	6: San Blas	7: Colima	8: Cen. Mexico	9: Guerrero	10: Veracruz	11: Oaxaca	12: Cen. Chiapas	13: N. Chiapas	14: Yucatán
CINNAMON-BELLIED FLOWERPIERCER *Diglossa baritula*							●	●	●	●	●	●		
GRASSLAND YELLOW-FINCH *Sicalis luteola*								▲		●			●	
BRIDLED SPARROW *Aimophila mystacalis*								●			●			
FIVE-STRIPED SPARROW *A. quinquestriata*		●			●									
BLACK-CHESTED SPARROW *A. humeralis*							●	●	●					
STRIPE-HEADED SPARROW *A. ruficauda*							●	●			●	●		
SUMICHRAST'S SPARROW *A. sumichrasti*											●			
BOTTERI'S SPARROW *A. botterii*							●			●		●	●	●
CASSIN'S SPARROW *A. cassinii*		●	●	●		●								
RUFOUS-WINGED SPARROW *A. carpalis*		●												
RUFOUS-CROWNED SPARROW *A. ruficeps*	●		●				●	●	●		●			
OAXACA SPARROW *A. notosticta*											●			
RUSTY SPARROW *A. rufescens*				●	●	●	●	●	●	●	●	●	●	
STRIPED SPARROW *Oriturus superciliosus*								●						
BLACK-THROATED SPARROW *Amphispiza bilineata*	●	●	●		▲									
SAGE SPARROW *A. belli*	●													
CHIPPING SPARROW *Spizella passerina*	●	●	●	●	●	●	●	●			●	●		
CLAY-COLORED SPARROW *S. pallida*	●	●	●	●	●	●	●	●		●	●			
BREWER'S SPARROW *S. breweri*	●	●	●											
Timberline Sparrow *S. [b.] taverneri*		▲												
WORTHEN'S SPARROW *S. wortheni*			●											
BLACK-CHINNED SPARROW *S. atrogularis*	●	●	●					●			▲			
VESPER SPARROW *Pooecetes gramineus*	●	●	●	●						▲				
LARK SPARROW *Chondestes grammacus*	●	●	●	●	●	●	●	●		●	●	▲		▲
LARK BUNTING *Calamospiza melanocorys*	●	●	●		▲									
BAIRD'S SPARROW *Ammodramus bairdii*		●												
GRASSHOPPER SPARROW *A. savannarum*	●	●	●	●	●	●	●	●		●	●		●	
NELSON'S SHARP-TAILED SPARROW *A. [caudacutus] nelsoni*	▲			▲										
SEASIDE SPARROW *A. maritimus*				●										
SAVANNAH SPARROW *A. sandwichensis*	●	●	●	●		●	●	●		●			●	●
Large-billed Sparrow *A. [s.] rostratus*	●	●												

	1: Baja	2: NW Mexico	3: Monterrey	4: NE Mexico	5: Mazatlán	6: San Blas	7: Colima	8: Cen. Mexico	9: Guerrero	10: Veracruz	11: Oaxaca	12: Cen. Chiapas	13: N. Chiapas	14: Yucatán
SIERRA MADRE SPARROW *A. baileyi*								•						
SONG SPARROW *Melospiza melodia*	•	•						•						
LINCOLN'S SPARROW *M. lincolnii*	•	•	•	•	•	•	•	•	•	•	•	•	•	
SWAMP SPARROW *M. georgiana*			•	•	•			•		•				
Slaty Fox [FOX] Sparrow *Passerella [iliaca] schistacea*	•													
Sierran Fox [FOX] Sparrow *P. [i.] megarhyncha*	•													
Sooty Fox [FOX] Sparrow *P. [i.] unaleschensis*	▲													
RUFOUS-COLLARED SPARROW *Zonotrichia capensis*												•		
WHITE-THROATED SPARROW *Z. albicollis*	▲													
GOLDEN-CROWNED SPARROW *Z. atricapilla*	•													
WHITE-CROWNED SPARROW *Z. leucophrys*	•	•	•		•		•							
Oregon [DARK-EYED] Junco *Junco [hyemalis] oreganus*	•													
Pink-sided [DARK-EYED] Junco *J. [h.] mearnsi*	•	•												
Grey-headed [DARK-EYED] Junco *J. [h.] caniceps*		•												
BAIRD'S JUNCO *J. bairdi*	•													
Mexican [YELLOW-EYED] Junco *J. [p.] phaeonotus*		•	•		•		•	•	•		•			
Chiapas [YELLOW-EYED] **Junco *J. [p.] fulvescens***												•		
Guatemalan [YELLOW-EYED] **Junco *J. [p.] alticola***												•		
MCCOWN'S LONGSPUR *Calcarius mccownii*		•												
LAPLAND LONGSPUR *C. lapponicus*	▲									▲				
CHESTNUT-COLLARED LONGSPUR *C. ornatus*	▲	•	•											
RED-WINGED BLACKBIRD *Agelaius phoeniceus*	•	•	•	•		•		•		•	•	•	•	•
Bicolored Blackbird *A. [p.] gubernator*							•	•						
TRICOLORED BLACKBIRD *A. tricolor*	•													

	1: Baja	2: NW Mexico	3: Monterrey	4: NE Mexico	5: Mazatlán	6: San Blas	7: Colima	8: Cen. Mexico	9: Guerrero	10: Veracruz	11: Oaxaca	12: Cen. Chiapas	13: N. Chiapas	14: Yucatán
YELLOW-HEADED BLACKBIRD *Xanthocephalus xanthocephalus*	•	•				•	•	•		•				
EASTERN MEADOWLARK *Sturnella magna*				•		•	•	•		•	•	•	•	•
Lilian's Meadowlark *Sturnella [m.] lilianae*		•												
WESTERN MEADOWLARK *S. neglecta*	•	•	•											
MELODIOUS BLACKBIRD *Dives dives*				•						•	•	•	•	•
BREWER'S BLACKBIRD *Euphagus cyanocephalus*	•	•	•	•			•	•						
GREAT-TAILED GRACKLE *Quiscalus mexicanus*	•	•	•	•	•	•	•	•		•	•	•	•	•
BRONZED COWBIRD *Molothrus aeneus*		•		•	•	•	•	•	•	•	•	•	•	•
BROWN-HEADED COWBIRD *M. ater*	•	•		•	•	•	•	•			•			
GIANT COWBIRD *Scaphidura oryzivora*										•			•	
BLACK-COWLED ORIOLE *Icterus dominicensis*										•	•		•	•
BAR-WINGED ORIOLE *I. maculialatus*												•		
ORCHARD ORIOLE *I. spurius*				•	•	•	•	▲	•	•	•	•	•	•
Ochre Oriole *I. [s.] fuertesi*										•		▲		
HOODED ORIOLE *I. cucullatus*	•	•		•	•	•	•	•		•	•		•	•
BLACK-VENTED ORIOLE *I. wagleri*			•		•	•	•	•	•		•	•		
YELLOW-BACKED ORIOLE *I. chrysater*												•		•
AUDUBON'S ORIOLE *I. graduacauda*			•	•				•		•				
Dickey's Oriole *I. [g.] dickeyae*						•	•		•		•			
YELLOW-TAILED ORIOLE *I. mesomelas*										•	•		•	•
STREAK-BACKED ORIOLE *I. pustulatus*		•			•	•	•	•	•		•	•		
ORANGE ORIOLE *I. auratus*														•
SPOT-BREASTED ORIOLE *I. pectoralis*							•		•		•	•		
ALTAMIRA ORIOLE *I. gularis*				•					•	•	•	•	•	•
BALTIMORE ORIOLE *I. galbula*				•			•	•	•	•	•	•	•	•
BULLOCK'S ORIOLE *I. bullockii*	•			•	•	•	•	•	•	•	•	•		
ABEILLE'S ORIOLE *I. abeillei*							•	•			•			
SCOTT'S ORIOLE *I. parisorum*	•		•		•		•	•	•		•			
YELLOW-BILLED CACIQUE *Amblycercus holosericeus*				•						•	•	•	•	•

	1: Baja	2: NW Mexico	3: Monterrey	4: NE Mexico	5: Mazatlán	6: San Blas	7: Colima	8: Cen. Mexico	9: Guerrero	10: Veracruz	11: Oaxaca	12: Cen. Chiapas	13: N. Chiapas	14: Yucatán
YELLOW-WINGED CACIQUE ***Cacicus melanicterus***					•	•	•	•	•		•	•		
CHESTNUT-HEADED OROPENDOLA *Psarocolius wagleri*										•	•		•	
MONTEZUMA OROPENDOLA *P. montezuma*										•	•		•	
PURPLE FINCH *Carpodacus purpureus*	▲													
CASSIN'S FINCH *C. cassinii*	•													
HOUSE FINCH *C. mexicanus*	•	•	•		•	•	•	•	•		•	•		
RED CROSSBILL *Loxia curvirostra*	•		•		•		•	•			•	•		
PINE SISKIN *Carduelis pinus*	•		•			•	•	•						
BLACK-CAPPED SISKIN *C. atriceps*												•		
BLACK-HEADED SISKIN *C. notata*				•	•	•	•	•	•		•	•		
LESSER GOLDFINCH *C. psaltria*	•	•	•	•	•	•	•	•	•	•	•	•		•
LAWRENCE'S GOLDFINCH *C. lawrencei*	•													
AMERICAN GOLDFINCH *C. tristis*			•											
HOODED GROSBEAK ***Coccothraustes abeillei***				•	•			•	•			•		
EVENING GROSBEAK *C. vespertinus*		•						▲						
HOUSE SPARROW *Passer domesticus*	•	•		•		•	•	•	•	•	•	•		

APPENDIX B

SITES FOR ENDEMICS AND SPECIES OF INTEREST

This Appendix cross-references with the birding sites a total of 272 species and well-marked forms endemic to Mexico and northern Central America, regional specialities (e.g., species also occurring in the extreme southwestern U.S.A.), and other sought-after miscellanea.

Slaty-breasted Tinamou. This elusive inhabitant of humid forest in the southeast can be seen at Sites 10.6, 12.7, 13.1, 13.5, 13.6.

Black-vented Shearwater. Endemic breeder on islands off western Baja California; often seen from shore, especially from August through February (Sites 1.1, 1.2).

Townsend's Shearwater. Endemic breeder on Mexico's Revillagigedo Islands (basically inaccessible to birders). Can be seen from boats off southern Baja (Site 1.10) and Oaxaca (Site 11.10).

Black Storm-Petrel. Breeds on islands off Baja and in the Gulf of California. Can be seen from shore (summer) in the Gulf (e.g., Site 2.1) and from boats over inshore waters off western Mexico (Sites 1.2, 6.1e, 7.3, 9.1, 11.10), mainly from August through May.

Least Storm-Petrel. Endemic breeder on islands off Baja and in the Gulf of California. Can be seen from shore in the Gulf (e.g., Sites 1.6, 2.1) or from boats over inshore waters off western Mexico (Sites 1.2, 6.1e, 7.3, 9.1, 11.10), mainly from August through May.

Red-billed Tropicbird. Breeds in the Gulf of California and off western Mexico. Can be seen from shore off Colima (Site 7.2) and Oaxaca (Site 11.10) and from boats off San Blas (Site 6.1e), Colima (Site 7.3), and Acapulco (Site 9.1).

Blue-footed Booby. Breeds commonly in the Gulf of California, wanders to southwestern Mexico. Can be seen from shore in the Gulf (e.g., Sites 1.6, 1.9, 2.1, 2.2) and off Mazatlán (Site 5.2) and San Blas (Site 6.1), less often off Colima (Site 7.3).

Bare-throated Tiger-Heron. Widespread in tropical lowlands. Good sites for this spectacular heron include 6.1d, 6.1h, 13.4, 14.4, 14.8.

American Flamingo. Locally common along the northern coast of the Yucatan (Sites 14.4, 14.6, 14.8).

Muscovy Duck. Widespread in tropical lowlands, but hunted and wary. Good sites include El Naranjo (4.5) and San Blas (Site 6.1d, 6.1h).

Mexican Duck. This dark female-plumaged Mallard is locally common on the Mexican Plateau, e.g., near Janos (Site 2.6) and Ciudad Guzman (Site 7.7).

White-breasted Hawk. Regional endemic of Chiapas highlands (Sites 12.6, 12.7).

White Hawk. The 'all-white' northern race can be seen readily in southern Veracruz (Sites 10.6, 10.7), northern Oaxaca (Sites 11.7, 11.8), and Chiapas (Sites 13.1, 13.2, 13.5, 13.6).

Aplomado Falcon. Fairly common in the Palenque area (Sites 13.3, 13.4); possible elsewhere (Sites 4.5, 10.5, 10.6).

West Mexican Chachalaca. Endemic to western Mexico, especially Colima and Jalisco (Sites 7.2, 7.4, 7.9) and Guerrero (Site 9.2).

Rufous-bellied Chachalaca. Endemic to northwestern Mexico, especially Mazatlán (Sites 5.3, 5.4, 5.5) and San Blas (Sites 6.1h, 6.2).

White-bellied Chachalaca. Regional endemic of southern Chiapas (Sites 12.3, 12.4).

Highland Guan. Regional endemic of central and southern Chiapas (Sites 12.7, 12.9).

Horned Guan. This rare and bizarre-looking cracid is a regional endemic best found at El Triunfo, Chiapas (Site 12.9).

Ocellated Turkey. Regional endemic of the Yucatan Peninsula (e.g., Site 14.3), but protected and far easier to see in Guatemala (Tikal).

Long-tailed Wood-Partridge. This Mexican endemic is elusive but often heard (like all wood-partridges). Can be seen on the Volcán de Fuego (Site 7.8), and in Guerrero (Site 7.2) and Oaxaca (Site 11.5).

Bearded Wood-Partridge. This rare Mexican endemic is best found near Coatepec (Site 10.2) and also occurs near Tlanchinol (Site 8.11).

Buffy-crowned Wood-Partridge. Little-known in Mexico; could be found on Volcán Tacaná in southern Chiapas (Site 12.5).

Singing Quail. This regional endemic is widespread and heard far more often than seen. Good locations to see it are El Naranjo (Site 4.5) and Felipe Carrillo Puerto (Site 14.3).

Ocellated Quail. A regional endemic that requires considerable luck to see; has been found near San Cristóbal (Site 12.8).

Yucatan Bobwhite. Regional endemic of the Yucatan Peninsula (e.g., Sites 14.4, 14.6, 14.7, 14.8).

Banded Quail. Mexican endemic, found in Colima and Jalisco (Sites 7.5, 7.6, 7.8, 7.9, 7.10), central Mexico (Sites 8.6, 8.7), and Guerrero (Site 9.2).

Elegant Quail. Endemic to northwestern Mexico (Sites 2.3, 2.4, 5.2, 5.3, 6.1a, 6.1b.)

Ruddy Crake. Regional endemic, commonly heard. Good places to see it include Palenque (Site 13.1), Coba (Site 14.2), and Cozumel (Site 14.10).

Heermann's Gull. Breeds on islands off Baja and common along northwestern coasts, e.g., Baja California (Sites 1.1, 1.2, 1.6, 1.9) and Sonora (Sites 2.1, 2.2), less so farther south to Mazatlán (Site 5.2), San Blas (Sites 6.1b, 6.1g), and Colima (Site 7.3).

Yellow-footed Gull. Endemic breeder in the Gulf of California, common on coasts of Baja (Sites 1.6, 1.9) and Sonora (Sites 2.1, 2.2).

Elegant Tern. Breeds on islands off Baja and common along northwestern coasts between March and October (e.g., Sites 1.1, 1.2, 1.3, 1.6, 1.9, 2.1, 2.2); transient at San Blas (Sites 6.1b, 6.1g), Colima (Site 7.3), etc.

Xantus' Murrelet. Breeds on islands off the Pacific Coast of Baja; can be seen from boats out of Ensenada (Site 1.2), mainly August to October.

Craveri's Murrelet. Endemic breeder on islands in the Gulf of California, where it can be seen from shore and on the trip to Isla Raza (Site 1.9); also (August to October) from boats out of Ensenada (Site 1.2).

Zenaida Dove. Fairly common along the northern coast of the Yucatan, especially Sites 14.4, 14.6.

Caribbean Dove. Another Yucatan speciality, found readily on Isla Cozumel (Site 14.10); also on the mainland (Site 14.3).

White-faced Quail-Dove. Regional endemic of eastern and southern cloud forests (Sites 8.11, 9.2, 10.2, 11.7, 12.7, 12.9).

Tuxtla [Purplish-backed] Quail-Dove. An endemic taxon (species?) occurs in Los Tuxtlas (Site 10.6).

Green Parakeet. Mexican endemic, found in the east and south (Sites 4.5, 8.11, 10.1, 10.8, 12.1, 12.6).

Pacific Parakeet. Regional endemic found in southern Chiapas (Site 12.3 and probably Sites 12.2, 12.5).

Military Macaw. Local in the northwest (Sites 2.3, 5.3, 6.3) south to near Puerto Vallarta (see Chapter 7) and in northeastern Mexico (Sites 4.4, 4.5).

Scarlet Macaw. Widely extirpated (by the cage-bird trade). Can be found in Chiapas at Yaxchilan (Site 13.6) and perhaps also Bonampak (Site 13.5).

Thick-billed Parrot. Nomadic endemic of northwestern Mexico. Best found from January through April in Jalisco (Site 7.8).

Maroon-fronted Parrot. Nomadic endemic of the Monterrey/Saltillo area (Sites 3.1, 3.4, and, irregularly, 3.5).

Mexican Parrotlet. Endemic to northwestern Mexico (Sites 2.4, 5.4, 6.1a, 6.1h, 7.1, 7.4, 7.5, 7.9).

Yucatan Parrot. Regional endemic, best found on Cozumel (Site 14.10) and at Felipe Carrillo Puerto (Site 14.3).

Red-crowned Parrot. Endemic to northeastern Mexico (Sites 4.4, 4.5).

Lilac-crowned Parrot. Endemic to western Mexico (especially Sites 5.3, 5.5, 7.2, 7.9, 9.2).

Yellow-headed Parrot. Widely extirpated by the cage-bird trade. Can be seen in the northeast (Sites 4.2, 4.5) and Chiapas (Site 13.4).

Yellow-naped Parrot. Another widely persecuted parrot, can be found in southern Chiapas (Site 12.4).

Pheasant Cuckoo. This elusive species can be seen in Guerrero (Site 9.2) and Chiapas (12.1, 12.9), but luck is needed.

Lesser Ground-Cuckoo. Another cuckoo often heard but rarely seen. Good sites are 5.3, 7.2, 7.10, and especially 11.13.

Lesser Roadrunner. Widespread regional endemic of the Pacific Slope and N Yucatan, best seen by luck (especially Sites 7.8, 7.9, 12.1, 14.4).

Balsas Screech-Owl. Mexican endemic found most easily in Colima (Site 7.10) and Guerrero (Site 7.2).

Pacific Screech-Owl. Can be found in Oaxaca (Site 11.12) and around Puerto Arista, Chiapas (Site 12.2).

Bearded Screech-Owl. Regional endemic, found locally around San Cristóbal (Site 12.8).

Mountain Pygmy-Owl. Widespread in montane forests (e.g., Sites 3.1, 3.5, 5.6, 7.8, 8.9, 9.2, 11.5); birds in Chiapas (Sites 12.5, 12.8) may be a different species (voice recordings are needed).

Cape Pygmy-Owl. Endemic to southern Baja (Sites 1.11, 1.12).

Colima Pygmy-Owl. Endemic to western Mexico (e.g., Sites 2.5, 5.5, 6.2, 7.10, 8.6).

Tamaulipas Pygmy-Owl. Endemic to northeastern Mexico, found readily above El Naranjo (Site 4.5).

Fulvous Owl. Regional endemic, found at El Triunfo (Site 12.9).

Stygian Owl. Good areas are the Durango Highway (Site 5.6) and Volcán de Fuego (Site 7.8).

Unspotted Saw-whet Owl. Can be found in Chiapas (Sites 12.6, 12.8).

Eared Poorwill. Endemic to western Mexico (Sites 7.5, 7.8, 9.2).

Yucatan Poorwill. Regional endemic of the Yucatan (Sites 14.2, 14.3, 14.9).

Tawny-collared Nightjar. Endemic breeder in northeastern Mexico (Site 4.2, and see Site 3.5).

Yucatan Nightjar. Regional endemic of the Yucatan (Sites 14.2, 14.3, 14.9, 14.10).

Buff-collared Nightjar. Widespread. Common in Sonora in summer (Sites 2.3, 2.4); other good sites are 5.3, 7.10, 10.4, 12.1.

Mexican Whip-poor-will. Widespread in montane forests (e.g., Sites 1.12, 2.7, 3.5, 5.6, 7.8, 11.5, 12.5, 12.6, 12.7, 12.8, 12.9).

Northern Potoo. Widespread in tropical lowlands, found readily at San Blas (Site 6.1d) and in the Yucatan (Site 14.2).

White-fronted [White-chinned] Swift. This little-known endemic form (species?) has been found at Tacámbaro (Site 8.10).

White-naped Swift. Endemic to western Mexico (Sites 5.3, 5.5, 8.9, 8.10, 9.2).

Great Swallow-tailed Swift. Regional endemic of western and southern Mexico. Good sites are 7.5, 7.9, 8.9, 8.10, 11.9, 12.1.

Mexican Hermit. Endemic to western Mexico (Sites 6.1h, 6.2, 7.4, 9.2, 11.9).

Wedge-tailed Sabrewing. Regional endemic found readily in eastern Mexico (e.g., Sites 4.5, 10.2, 10.3, 11.7, 13.1, 13.5, 14.2, 14.3).

Long-tailed Sabrewing. Mexican endemic, found in Veracruz (Sites 10.6, 10.7).

Rufous Sabrewing. Regional endemic found in Chiapas (Site 12.9).

Emerald-chinned Hummingbird. Regional endemic of cloud forest in Oaxaca (Site 11.7) and Chiapas (Sites 12.7, 12.9).

Short-crested Coquette. Endemic to the Sierra Madre del Sur of Guerrero (Site 9.2).

Black-crested Coquette. The best area for this local and easily overlooked species is Montebello (Site 12.7).

Canivet's Emerald. Regional endemic found readily in eastern Mexico (e.g., Sites 10.4, 10.6, 11.8, 11.15, 12.1, and especially in the Yucatan: Sites 14.2, 14.3, 14.4, 14.5, 14.8, 14.9).

Salvin's Emerald. Occurs in southeastern Chiapas (can be expected at Sites 12.5, 12.6).

Golden-crowned Emerald. Endemic to

western Mexico (especially Sites 5.5, 6.1h, 6.2, 7.4, 7.5, 7.9, 8.6, 8.10, 9.2, 11.9).

Cozumel Emerald. Endemic to Isla Cozumel (Site 14.10).

Dusky Hummingbird. Endemic to southwestern Mexico (especially Sites 8.6, 8.10, 9.2, 11.1, 11.2, 11.3, 11.4).

Doubleday's [Broad-billed] Hummingbird. Endemic to southwestern Mexico (Sites 7.1, 7.2, 11.10, 11.11, 11.13).

Mexican Woodnymph. Endemic to western Mexico, good places are around San Blas (Sites 6.2, 6.4) and near Autlán (site 7.5).

White-eared Hummingbird. Regional endemic common in highlands almost throughout (e.g., Sites 2.7, 5.6, 6.4, 7.8, 8.1, 8.2, 8.5, 8.9, 9.2, 11.5, 12.7, 12.8).

Xantus' Hummingbird. Endemic to southern Baja (Sites 1.10, 1.11, 1.12).

White-bellied Emerald. Regional endemic found readily in southeastern Mexico (e.g., Sites 10.6, 10.7, 11.8, 12.1, 13.1, 14.2, 14.3).

Azure-crowned Hummingbird. Regional endemic found readily in southern Mexico (e.g., Sites 10.6, 11.7, 12.1, 12.5, 12.7).

Berylline Hummingbird. Regional endemic found readily in western and southern Mexico (e.g., Sites 5.5, 6.2, 6.4, 7.5, 7.9, 8.6, 8.9, 8.10, 9.2, 11.1, 11.4, 12.1).

Blue-tailed Hummingbird. Regional endemic of southeastern Chiapas (Site 12.5).

Buff-bellied Hummingbird. Widespread in eastern Mexico (e.g., Sites 4.5, 10.1, 10.4, 10.5, 10.6, 12.1, 13.4), especially Yucatan (all sites except Isla Cozumel).

Violet-crowned Hummingbird. Widespread in western Mexico (e.g., Sites 2.3, 2.5, 5.3, 6.2, 6.4, 7.5, 7.8, 8.6, 8.7, 8.10, 9.2, 11.6).

Green-fronted Hummingbird. Endemic to southwestern Mexico (Sites 9.2, 11.2, 11.12, 11.15, 12.1, 12.2).

Cinnamon-sided [Green-fronted] Hummingbird. Endemic to the Sierra Madre del Sur of Oaxaca (Sites 11.9, 11.11).

White-tailed Hummingbird. Endemic to the Sierra Madre del Sur, readily seen in Guerrero (Site 9.2).

Blue-capped Hummingbird. Endemic to the Sierra Madre del Sur of Oaxaca (Site 11.9).

Green-throated Mountain-gem. Regional endemic of cloud forest in Chiapas (Sites 12.5, 12.6, 12.9).

Amethyst-throated Hummingbird. Regional endemic of humid montane forests of central and southern Mexico (e.g., Sites 7.5, 7.8, 8.9, 8.11, 11.5, 11.7, 12.5, 12.7, 12.8, 12.9; the Mexican endemic taxon (species?) Violet-throated Hummingbird occurs in Guerrero (Site 9.2).

Garnet-throated Hummingbird. Regional endemic in humid montane forests of southern Mexico (Sites 9.2, 11.5, 11.6, 11.7, 12.5, 12.7, 12.8).

Sparkling-tailed Woodstar. Regional endemic of western and southern Mexico (Sites 5.5, 6.2, 7.5, 9.2, 11.9, 12.5, 12.7).

Mexican Sheartail. Mexican endemic, local in Veracruz (Site 10.4) and northern Yucatán (Sites 14.4, 14.6, 14.8).

Slender Sheartail. Regional endemic of Chiapas (Sites 12.1, 12.5, 12.7).

Lucifer Hummingbird. Best found in western and central Mexico (Sites 7.5, 7.8, 8.1, 8.2, 8.6, 8.10).

Beautiful Hummingbird. Endemic to southwestern Mexico, found readily in Oaxaca (Sites 11.1, 11.2, 11.3, 11.4, 11.13).

Bumblebee Hummingbird. Endemic to the highlands of Mexico (especially Sites 9.2, 11.5, 11.7; also Sites 5.4, 6.4, 7.8, 8.11, 10.2, 11.9).

Wine-throated Hummingbird. Regional endemic of central Chiapas (Sites 12.7, 12.9).

Citreoline Trogon. Endemic to western Mexico (e.g., Sites 5.5, 6.1a, 6.1h, 6.2, 7.2, 7.4, 8.1, 11.10, 11.11, 11.13, 11.15).

Mountain Trogon. Regional endemic found widely in highlands (e.g., Sites 2.7, 4.5, 5.6, 7.8, 8.9, 8.11, 11.5, 11.9, 12.7, 12.8, 12.9).

Eared Quetzal. Endemic to northwestern Mexico, best seen in the Barranca del Cobre (Site 2.7); has been found at Sites 5.6, 7.8.

Resplendent Quetzal. This spectacular bird is seen readily at El Triunfo (Site 12.9) and can also be found at Montebello (Site 12.7).

Tody Motmot. This forest denizen can be found in northern Chiapas (Sites 13.1, 13.5, 13.6) and below El Triunfo (Site 12.9).

Blue-throated Motmot. Regional endemic of Chiapas highlands (Sites 12.8, 12.9).

Russet-crowned Motmot. Regional endemic of the Pacific Slope (e.g., Sites 5.5, 6.1a, 6.1h, 7.4, 7.8, 9.2, 11.10, 11.15, 12.1).

Golden-cheeked Woodpecker. Mexican endemic common on the Pacific Slope (e.g., Sites 5.3, 6.1, 7.2, 7.4, 7.9, 8.6, 9.1, 9.2, 11.10).

Grey-breasted Woodpecker. Mexican endemic of the southwestern interior, common in Oaxaca (Sites 11.1, 11.2, 11.3, 11.4, 11.12); also Site 8.7.

Yucatan Woodpecker. Regional endemic, commonest on Isla Cozumel (Site 14.10), but found readily on mainland (Sites 14.1, 14.2, 14.4, 14.7, 14.8).

Arizona Woodpecker. Mountains of northwestern Mexico (Sites 5.6, 6.4, 7.5, 7.8).

Strickland's Woodpecker. Endemic to high elevations of central Mexico (Sites 8.3, 8.4, en route to 8.9).

Bronze-winged Woodpecker. Endemic to northeastern Mexico (Sites 4.4, 4.5, 8.11, 10.2).

Grey-crowned Woodpecker. Endemic to western Mexico (Sites 5.6, 6.2, 6.4, 7.5, 7.9, 9.2, 11.9).

Guatemalan [Red-shafted] Flicker. Regional endemic of Chiapas highlands (Sites 12.5, 12.6, 12.7, 12.8).

Rufous-breasted Spinetail. This skulking regional endemic can be found readily in southern Mexico (Sites 10.5, 10.6, 10.7, 11.8, 12.4, 13.1, 13.4, 13.6, 14.3).

Ruddy Foliage-gleaner. Locally common in eastern and southern Mexico (especially Sites 8.11, 9.2, 11.7, 12.7, 12.9).

Ruddy Woodcreeper. Commonest in the Yucatan Peninsula, especially Felipe Carrillo Puerto (Site 14.3); also Site 13.5.

White-striped Woodcreeper. Mexican endemic of western and central highlands (Sites 5.6, 6.4, 7.8, 8.5, 8.9).

Mexican Antthrush. Regional endemic common in the southeast (Sites 10.6, 10.7, 12.7, 13.1, 13.5, 13.6, 14.3).

Yellow-bellied Tyrannulet. This easily overlooked canopy mite can be found in southern Veracruz (Sites 10.6, 10.7) and northern Chiapas (Sites 13.1, 13.5, 13.6).

Caribbean Elaenia. Most easily found on Isla Cozumel (Site 14.10); also on mainland opposite (Site 14.3, and can be expected at 14.1).

Belted Flycatcher. This localized regional endemic is best found at El Sumidero (Site 12.1).

Pileated Flycatcher. Mexican endemic of the interior southwest, found readily when singing (Sites 8.6, 11.1, 11.3, 11.4, 11.12).

White-throated Flycatcher. Breeds in highlands (e.g., Sites 8.9, and en route to 8.10), winters in lowlands (especially Sites 6.1h, 7.1).

Pine Flycatcher. Regional endemic, somewhat local in highlands. Good sites are 3.1, 3.4, 3.5, 5.6, 7.6, 7.8, 8.5; also Sites 11.5, 12.8.

Yucatan Flycatcher. Regional endemic of the Yucatan (e.g., Sites 14.2, 14.3, 14.7, 14.9, 14.10).

Flammulated Flycatcher. Endemic to western Mexico, found readily when singing (Sites 5.4, 7.2, 7.4, 12.1).

Grey-collared Becard. This widespread regional endemic is often hard to find. The best Sites are 4.5, 7.5, 7.8; also at Sites 5.6, 6.4, 7.9, 8.11, 10.6, 12.1, 13.1, 14.3.

Lovely Cotinga. Good areas are Los Tuxtlas (Site 10.6) and around Palenque (Sites 13.1, 13.2); also Sites 10.7, 11.7, 12.7, 13.5, 13.6.

Sinaloa Martin. Mexican breeding endemic, found in summer on the Volcán de Fuego (Site 7.8); could be found at Site 5.4.

Black-capped Swallow. Regional endemic of Chiapas (Sites 12.5, 12.6, 12.7, 12.8, 12.9).

Ridgway's Rough-winged Swallow. Regional endemic of southeastern Mexico, especially Yucatán (e.g., Sites 14.2, 14.3, 14.5, 14.7, 14.8, 14.9) but not on Isla Cozumel; also Sites 10.7, 11.6, 11.8, 12.1.

Black-throated Magpie-Jay. Mexican endemic of the northwest (Sites 2.4, 5.3, 5.5, 6.1, 6.2, 6.3, 6.4).

Tufted Jay. This spectacular, highly localized Mexican endemic is found readily along the Durango Highway (Sites 5.4, 5.6).

Purplish-backed Jay. Mexican endemic of the northwest (Sites 2.4, 5.3, 5.4, 6.1).

San Blas Jay. Endemic to western Mexico, found easily in Colima and Jalisco (Sites 7.1, 7.2, 7.4), less so around San Blas (Sites 6.1, 6.2).

Yucatan Jay. Regional endemic common in the Yucatan (e.g., Sites 14.2, 14.3, 14.5,

14.7, 14.9), absent from Isla Cozumel; also Site 13.4.

Black-throated Jay. Regional endemic of Chiapas cloud forest (Sites 12.7, 12.8, 12.9).

Dwarf Jay. This highly localized Mexican endemic is found readily in Oaxaca (Sites 11.5, 11.6).

White-throated Jay. Mexican endemic of the Sierra Madre del Sur (Sites 9.2, 11.9).

Grey-breasted Jay. Widespread in northern Mexico (e.g., Sites 3.1, 3.4, 3.5, 7.8, 8.5, 8.9).

Unicolored Jay. Regional endemic of humid montane forest (Sites 8.11, 9.2, 11.6, 11.7, 12.7, 12.8, 12.9).

Tamaulipas Crow. Common endemic of northeastern Mexico (Sites 4.2, 4.5, 10.1).

Sinaloa Crow. Common endemic of northwestern Mexico (e.g., Sites 2.4, 5.1, 5.2, 5.3, 6.1).

Mexican Chickadee. Widespread in mountains (e.g., Sites 2.7, 3.1, 3.4, 3.5, 5.6, 7.8, 8.3, 8.4, 8.5, 9.2, 11.5).

Bridled Titmouse. Widespread in highlands (e.g., Sites 3.1, 3.5, 4.5, 5.6, 7.5, 7.8, 8.9, 9.2, 11.2).

Grey-barred Wren. Mexican endemic of highland forests (Sites 7.6, 7.8, 8.5, 8.9, 11.5).

Giant Wren. Endemic to southern Chiapas (Sites 12.3, 12.4).

Spotted Wren. Endemic to north-central Mexico (Sites 4.5, 5.6, 6.4, 7.5, 7.8, 7.9, 8.9, 8.10).

Boucard's Wren. Endemic to interior southwestern Mexico (Sites 8.7, 9.2, 11.1, 11.2, 11.3, 11.4, 11.6).

Yucatan Wren. Endemic to northern coastal Yucatán (Sites 14.4, 14.6, 14.8).

Sumichrast's Wren. Endemic to southeastern Mexico (Sites 10.3, 11.8).

Nava's Wren. Endemic to southeastern Mexico (Site 10.7).

Happy Wren. Endemic to western Mexico (e.g., Sites 2.4, 5.3, 5.5, 6.1, 7.2, 7.4, 7.9, 8.6, 9.2, 11.6, 11.9, 11.10).

Sinaloa Wren. Endemic to western Mexico (e.g., Sites 2.3, 2.4, 5.3, 5.5, 6.1, 7.2, 7.4, 7.9, 9.2).

White-browed [Carolina] Wren. Regional endemic found readily in the Yucatan (e.g., Sites 14.1, 14.2, 14.3, 14.7, 14.9).

White-bellied Wren. Regional endemic, found readily in Colima and Jalisco (Sites 7.2, 7.4) and the Yucatan (e.g., Sites 14.1, 14.3, 14.5, 14.7, 14.9); also Sites 10.4, 11.8, 13.1, 13.5.

Cozumel Wren. Endemic to Isla Cozumel (Site 14.10).

Rufous-browed Wren. Regional endemic of Chiapas highlands (Sites 12.8, 12.9).

California Gnatcatcher. Found readily in most of Baja (e.g., Sites 1.3, 1.6, 1.7, 1.8, 1.9, 1.10).

Black-capped Gnatcatcher. Endemic to northwestern Mexico, found readily in Sonora (Sites 2.3, 2.4) and near Mazatlán (Site 5.3), harder in Colima (Sites 7.2, 7.5, 7.6).

Brown-backed Solitaire. Regional endemic generally common in highlands throughout Mexico (e.g., Sites 3.1, 3.5, 4.4, 4.5, 5.6, 6.4, 7.5, 7.8, 7.9, 8.5, 8.9, 8.10, 8.11, 9.2, 11.5, 11.7, 11.9, 12.5, 12.6, 12.7, 12.8, 12.9).

Slate-colored Solitaire. Regional endemic of southeastern cloud forest (Sites 8.11, 10.6, 11.7, 12.7).

Russet Nightingale-Thrush. Mexican endemic of highland forests (Sites 5.6, 7.6, 7.8, 8.3, 8.5, 8.9, 9.2, 11.5).

Black-headed Nightingale-Thrush. Can be found in eastern Mexico (Sites 4.5, 8.11, 10.6, 11.7, 12.7).

Spotted Nightingale-Thrush. Can be found in Chiapas (Sites 12.7, 12.9).

Black Thrush. Regional endemic of central and southern highlands (Sites 8.11, 9.2, 11.5, 11.7, 12.7, 12.9).

Rufous-backed Thrush. Endemic to western Mexico (e.g., Sites 2.3, 2.4, 5.3, 5.5, 6.1, 6.2, 7.2, 7.4, 7.9, 8.1, 8.2, 8.6, 8.10, 9.2, 11.10, 11.11).

Rufous-collared Thrush. Regional endemic of Chiapas highlands (Sites 12.5, 12.6, 12.7, 12.8).

San Lucas [American] Robin. Endemic to southern Baja (Site 1.12).

Aztec Thrush. Mexican endemic of highland forests, most easily found from December to April on Volcán de Fuego (Site 7.8); also Sites 3.5, 5.6, 9.2, 11.5.

Black Catbird. Regional endemic common on Isla Cozumel (Site 14.10), also found readily on the mainland (Site 14.3).

Blue Mockingbird. Widespread Mexican endemic (e.g., Sites 4.4, 4.5, 5.3, 5.4, 6.1, 6.4, 7.5, 7.8, 7.9, 8.1, 8.2, 8.5, 8.6, 8.9, 8.10, 9.2, 10.2, 11.1, 11.2, 11.3, 11.4, 11.5, 11.9).

Blue-and-white Mockingbird. Regional endemic of Chiapas highlands (Sites 12.1, 12.5, 12.6, 12.7, 12.8, 12.9).

Long-billed Thrasher. Found readily in northeastern Mexico (Sites 3.1, 4.5).

Cozumel Thrasher. Endemic to Isla Cozumel (Site 14.10).

Grey Thrasher. Endemic to Baja California (e.g,. Sites 1.7, 1.9, 1.10, 1.11, 1.12).

Ocellated Thrasher. Endemic to south-central Mexico (Sites 8.1, 8.2, 11.1, 11.2, 11.4, 11.12).

Grey Silky. Regional endemic widespread in highlands (e.g., Sites 3.1, 3.4, 3.5, 5.6, 6.4, 7.5, 7.6, 7.8, 8.5, 8.9, 8.10, 9.2, 11.2, 11.4, 11.5, 11.9, 12.5, 12.7, 12.8, 12.9).

Slaty Vireo. Mexican endemic best found in Colima (Site 7.9) and Oaxaca (Sites 11.1, 11.4); also Site 8.9.

Mangrove Vireo. The mangrove-inhabiting Pacific Coast form can be found at San Blas (Site 6.1). The scrub-inhabiting form is common in the Yucatan (e.g., Sites 14.3, 14.4, 14.5, 14.6, 14.7, 14.8, 14.9) and replaced by the following species on Cozumel.

Cozumel Vireo. Endemic to Isla Cozumel (Site 14.10).

Black-capped Vireo. Winters along the Pacific Slope, especially Sites 5.5, 6.1h, 6.2, 6.4, 7.9.

Dwarf Vireo. Mexican endemic found in Jalisco (Sites 7.5, 7.8), at least in winter; resident in the Valley of Oaxaca (Sites 11.1, 11.2, 11.4).

Grey Vireo. Common breeder in northern Baja California (Sites 1.4, 1.7), wintering in southern Baja (Site 1.9).

Golden Vireo. Mexican endemic of the west and southwest (e.g., Sites 5.5, 6.2, 6.4, 7.4, 7.5, 7.9, 8.6, 9.2, 11.4).

Yucatan Vireo. Found readily on Isla Cozumel (Site 14.10) and the adjacent mainland (Site 14.1).

Chestnut-sided Shrike-Vireo. Regional endemic of central and southern highlands (Sites 7.5, 7.6, 7.8, 8.5, 8.9, 8.11, 9.2, 11.5, 11.9, 12.9).

Green Shrike-Vireo. Good areas to see this often-heard species in southeastern Mexico are Sites 10.6, 13.1, 13.2.

Colima Warbler. Breeds in the northeast (Sites 3.4, 3.5), winters in the west-central highlands (Sites 7.8, 8.5, 9.2).

Crescent-chested Warbler. Regional endemic, widespread in the highlands (e.g., Sites 3.1, 4.5, 5.6, 6.4, 7.5, 7.8, 8.5, 8.9, 8.11, 9.2, 11.5, 12.5, 12.7, 12.8).

Golden-cheeked Warbler. Found readily (September to February) in Chiapas highlands, especially around San Cristóbal (Site 12.8).

Swainson's Warbler. Winters commonly on Isla Cozumel (Site 14.10), also found in mainland Yucatán (e.g., Site 14.3).

Altamira Yellowthroat. Endemic to northeastern Mexico (Sites 4.3, 4.5, 10.1).

Belding's Yellowthroat. Endemic to southern Baja California (Sites 1.9, 1.10).

Hooded Yellowthroat. Mexican endemic of eastern and southern highlands (Sites 3.4, 3.5, 8.1, 8.2, 11.6).

Black-polled Yellowthroat. Endemic to central Mexico (Site 8.8, and en route to 8.10).

Red-faced Warbler. Breeds in the northwest (Sites 2.7, 5.4), winters in highlands (e.g., Sites 5.6, 6.4, 7.5, 7.8, 8.5, 9.2, 11.5, 12.8).

Red Warbler. Mexican endemic of highland forests (Sites 5.4, 5.6, 7.8, 8.3, 8.4, 8.5, 9.2, 11.5).

Pink-headed Warbler. Regional endemic of Chiapas highlands (Sites 12.5, 12.8).

Fan-tailed Warbler. Widespread but local regional endemic (especially Sites 4.5, 5.4, 6.1h, 7.5, 7.9, 9.2, 12.1).

Rufous-capped Warbler. Widespread and generally common regional endemic (e.g., Sites 2.3, 3.1, 4.5, 5.5, 5.6, 6.4, 7.5, 7.8, 7.9, 8.2, 8.9, 9.2, 10.3, 10.6, 11.1, 11.2, 11.3, 11.4, 11.5, 12.1, 12.7, 12.8).

Golden-browed Warbler. Widespread regional endemic of humid montane forests (e.g., Sites 5.6, 7.8, 8.11, 9.2, 10.2, 11.5, 11.7, 11.9, 12.5, 12.7, 12.8).

Red-breasted Chat. Mexican endemic of the Pacific Slope, best found in spring and summer when singing (e.g., Sites 5.4, 6.1h, 7.2, 7.10, 11.10, 12.1).

Grey-throated Chat. Regional endemic found readily near Felipe Carrillo Puerto (Site 14.3); also Sites 13.5, 14.9.

Bananaquit. Endemic taxon (species?) on Isla Cozumel (Site 14.10).

Cabanis' Tanager. Regional endemic found near El Triunfo (Site 12.9).

Blue-crowned Chlorophonia. Regional endemic of eastern and southern cloud forest (Sites 10.6, 11.7, 12.7, 12.9).

Godman's [Scrub] Euphonia. Endemic to western Mexico (e.g., Sites 5.3, 5.5, 6.1h, 6.2, 7.2, 7.4, 7.9, 9.2).

Yellow-winged Tanager. Regional endemic common in the east and southeast (e.g., Sites 4.5, 10.6, 10.7, 11.7, 11.8, 13.1, 13.5, 13.6, 14.2, 14.3, 14.5).

Stripe-headed Tanager. Endemic subspecies on Isla Cozumel (Site 14.10).

Black-throated Shrike-Tanager. Regional endemic of southeastern rain forest (Sites 10.6, 10.7, 13.1, 13.5, 13.6).

Rose-throated Tanager. Regional endemic of the Yucatan (Sites 14.1, 14.3, 14.10).

Red-headed Tanager. Endemic to western Mexico (Sites 5.4, 5.6, 6.4, 7.5, 7.8, 7.9, 8.9, 9.2, 11.9).

Crimson-collared Grosbeak. Endemic to northeastern Mexico (Sites 3.1, 4.4, 4.5).

Yellow Grosbeak. Regional endemic (assuming South American birds are split) of the west and southwest: Sites 2.4 (summer), 5.3, 5.5, 6.2, 6.4, 7.4, 7.5, 7.8, 7.9, 8.6, 9.2, and (the distinctive 'Golden Grosbeak') in Chiapas (Sites 12.1, 12.9).

Blue Bunting. Widespread regional endemic, mainly western Mexico (Sites 5.3, 5.5, 6.1h, 7.2, 7.4, 9.1, 11.10, 12.1), and the Yucatan (Sites 14.1, 14.3, 14.5, 14.7, 14.9).

Rosita's Bunting. Endemic to the Isthmus of Tehuantepec (Sites 11.15, 12.2).

Orange-breasted Bunting. Endemic to southwestern Mexico (Sites 7.2, 7.4, 7.5, 7.10, 8.10, 9.1, 9.2, 11.10, 11.12, 11.13, 11.15, 12.2).

White-naped Brushfinch. Endemic to eastern Mexico (e.g., Sites 8.11, 10.2, 11.7, 12.7, 12.8).

Rufous-capped Brushfinch. Endemic to Mexican highlands (e.g., Sites 3.1, 4.5, 5.6, 7.6, 7.8, 8.2, 8.4, 8.5, 8.9, 9.2, 11.5).

Plain-breasted [Chestnut-capped] Brushfinch. Endemic subspecies of Los Tuxtlas (Site 10.6).

Green-striped Brushfinch. Endemic to humid montane forests of west-central Mexico (e.g., Sites 5.6, 7.5, 7.6, 7.8, 8.5, 8.9).

Green-backed Sparrow. Regional endemic of southeastern Mexico (Sites 13.1, 13.5, 13.6, 14.3).

Rusty-crowned Ground-Sparrow. Endemic to western Mexico (e.g., Sites 5.5, 6.1h, 6.2, 6.4, 7.5, 7.8, 7.9, 8.6, 8.9, 9.2).

Prevost's Ground-Sparrow. Regional endemic of Chiapas (Sites 12.4, 12.5, 12.7).

Collared Towhee. Endemic to humid montane forests of south-central Mexico (Sites 7.8, 9.2, 11.5).

White-throated Towhee. Endemic to interior southwestern Mexico, mainly Oaxaca (Sites 11.1, 11.2, 11.3, 11.4, 11.6).

Cinnamon-rumped [White-collared] Seedeater. Common endemic taxon (species?) of western Mexico (e.g., Sites 5.1, 5.2, 6.1, 7.1, 7.2, 7.9, 8.6, 8.7, 8.9, 8.10, 9.1, 9.2, 11.2).

Blue Seedeater. This bamboo specialist has often been found at El Sumidero (Site 12.1).

Slate-blue Seedeater. Endemic to southwestern Mexico (Sites 8.6, 9.2).

Cinnamon-bellied Flowerpiercer. Regional endemic of central and southern highlands (Sites 7.6, 7.8, 8.5, 8.9, 9.2, 11.5, 11.7, 12.5, 12.7, 12.8, 12.9).

Bridled Sparrow. Endemic to interior southwestern Mexico, mainly Oaxaca (Sites 11.2, 11.3, 11.4, 11.12); also en route to Laguna San Felipe (Site 8.7).

Five-striped Sparrow. Common and conspicuous in Sonora in summer (Sites 2.3, 2.4), harder to find in winter; also in winter at Site 5.3.

Black-chested Sparrow. Endemic to interior southwestern Mexico (Sites 7.5, 7.10, 8.6, 8.10, 9.2; also en route to Site 8.7).

Sumichrast's Sparrow. Endemic to the Isthmus of Tehuantepec (Sites 11.12, 11.13).

Botteri's Sparrow. Widespread but local (Sites 7.9, 10.4, 10.8, 12.1, 12.7, 13.4, 14.7).

Rufous-winged Sparrow. Found readily near Alamos, Sonora (Site 2.4).

Oaxaca Sparrow. Endemic to interior Oaxaca (Sites 11.12, 11.4, 11.12).

Rusty Sparrow. Widespread (e.g., Sites 4.5, 5.6, 6.4, 7.8, 7.9, 8.6, 8.9, 8.10, 8.11, 9.2, 10.2, 10.6, 11.7, 11.9, 12.1, 12.7, 12.8, 13.1).

Striped Sparrow. Endemic to highlands of central Mexico (Sites 5.4, 8.3, 8.4, 8.8).

Worthen's Sparrow. Endemic to northern Mexico (Sites 3.3; en route to 3.5).

Large-billed [Savannah] Sparrow. Endemic breeder to Gulf of California (Site 2.1); more widespread in winter (Site 1.3).

Sierra Madre Sparrow. Endemic to highlands of central Mexico (Site 8.4).

Baird's Junco. Endemic to southern Baja (Site 1.12).

Chiapas [Yellow-eyed] Junco. Endemic form in northern Chiapas highlands (Sites 12.8).

Guatemalan [Yellow-eyed] Junco. Regional endemic form of southeastern Chiapas (Site 12.5).

Bicolored [Red-winged] Blackbird. Endemic form of central Mexico (Sites 7.7, 8.7, 8.8).

Melodious Blackbird. Regional endemic common in eastern and southern Mexico (e.g., Sites 4.4, 4.5, 11.8, 12.3, 12.4, 13.1, 13.4, and virtually all sites in Veracruz, northern Chiapas, and the Yucatan; not on Cozumel).

Bar-winged Oriole. Regional endemic of interior Chiapas (Site 12.1).

Ochre [Orchard] Oriole. Mexican endemic breeds in the northeast (Sites 10.1, 10.5), winters in the southwest (e.g., site 12.3).

Black-vented Oriole. Widespread regional endemic (e.g., Sites 3.3, 5.3, 6.4, 7.1, 7.4, 7.8, 7.9, 8.7, 8.10, 9.2, 11.1, 11.2, 11.3, 11.4, 12.1, 12.6).

Audubon's Oriole. This (nominate) form occurs in northeastern Mexico (Sites 3.1, 4.4., 4.5, 8.11, 10.1).

Dickey's [Audubon's] Oriole. This distinctive taxon (species?) is endemic to southwestern Mexico (Sites 6.4, 7.8, 7.9, 9.2, 11.2, 11.9); replaced in Chiapas by the similar Yellow-backed Oriole.

Orange Oriole. Regional endemic of the Yucatan Peninsula (Sites 14.1, 14.3, 14.4, 14.5, 14.7, 14.9); not on Cozumel.

Spot-breasted Oriole. Can be found along the Pacific Slope, mainly from Guerrero south (Sites 9.1, 11.10, 12.3, 12.4); also Site 7.1.

Altamira Oriole. Common in eastern and southern Mexico (e.g., Sites 4.4, 4.5, 10.1, 10.4, 10.5, 10.6, 11.8, 11.10, 11.13, 12.1, 12.3, 12.4, 13.1, 13.3, and virtually all sites in Yucatan, except Cozumel).

Abeille's Oriole. Mexican endemic of central highlands (Sites 8.1, 8.2, 8.4, 8.5, 8.9, 8.10; en route to Site 8.11), in winter to Sites 7.8, 11.5.

Yellow-winged Cacique. Regional endemic, common on the Pacific Slope (e.g., Sites 5.3, 5.5, 6.1, 6.2, 7.1, 7.2, 7.4, 8.10, 9.1, 9.2, 11.10, 11.11, 11.12, 11.15, 12.3).

Black-capped Siskin. Regional endemic, local in Chiapas highlands (Site 12.5).

Black-headed Siskin. Regional endemic widespread in highlands (Sites 4.5, 5.6, 6.4, 7.5, 7.6, 7.8, 8.5, 8.9, 8.11, 9.2, 11.5, 11.9, 12.5, 12.7, 12.8).

Hooded Grosbeak. Regional endemic local in montane forests, common and best found in the northeast (Sites 4.4, 4.5); also Sites 5.6, 8.9, 8.11, 9.2, 12.7, 12.9.

INDEX

Note that the index refers only to sites and major cities (cross-referenced to nearby sites and/or relevant chapters). However, Appendix A cross-references all species with chapters, and Appendix B provides specific reference to sites for endemics and other species of interest.

Standard state abbreviations are given with each site to reduce potential confusion between sites with the same or similar names.